Gas Cleaning at High Temperatures

GAS CLEANING AT HIGH TEMPERATURES

Edited by

R. CLIFT
Centre for Environmental Strategy
University of Surrey, Guildford, UK

and

J. P. K. SEVILLE
Department of Chemical and Process Engineering
University of Surrey, Guildford, UK

SPRINGER-SCIENCE+BUSINESS MEDIA, B.V

First edition 1993

© 1993 Springer Science+Business Media Dordrecht

Originally published by Chapman & Hall in 1993

ISBN 978-0-7514-0178-3   ISBN 978-94-011-2172-9 (eBook)
DOI 10.1007/978-94-011-2172-9

A catalogue record for this book is available from the British Library

Library of Congress Cataloging-in-Publication data available

Introduction

This volume comprises the papers presented at the Second International Symposium on Gas Cleaning at High Temperatures, held at the University of Surrey, Guildford, UK on 27–29 September, 1993. The Symposium was organised by the Department of Chemical and Process Engineering and the Centre for Environmental Strategy of the University of Surrey, and co-sponsored by The Institution of Chemical Engineers, The Filtration Society and the Royal Society of Chemistry.

Co-chairmen

R. Clift   Centre for Environmental Strategy, University of Surrey
J. P. K. Seville Department of Chemical and Process Engineering, University of Surrey

Secretary

J. Libaert   Centre for Environmental Strategy, University of Surrey

Scientific Committee

P. Cahill   British Coal Research Establishment, Cheltenham, UK
W. A. Dries   Shell International, The Hague, Netherlands
P. Gäng   University of Karlsruhe, Germany
R. R. Greenfield   Van Tongeren International, Godalming, UK
C. P. Kerton   Blue Circle Technical Centre, Greenhithe, UK
W. S. Kyte   Powergen, Solihull, UK
K. Morris   Separation Processes Service, Harwell, UK
N. Moss   Pall Process Filtration, Portsmouth, UK
G. A. Rimmer   Pilkington, Ormskirk, UK
A. Russell-Jones   Lodge Sturtevant, Birmingham, UK
J. Strickland   Davy McKee, Stockton-on-Tees, UK
C. J. Withers   Glosfume Environmental Controls, Ashleworth, UK

Financial sponsors

Shell International Inc.
BOC Process Plants
Air Products
Lodge Sturtevant
Van Tongeren
European Gas Turbines
NEI International Combustion
Pall Corporation

THE FILTRATION SOCIETY

ROYAL SOCIETY OF CHEMISTRY

INSTITUTION OF CHEMICAL ENGINEERS

Preface

This Second International Symposium on Gas Cleaning at High Temperatures follows the First, held in 1986, also at the University of Surrey, and published by the Institution of Chemical Engineers as no. 99 in their Symposium Series. In the interval since the First Symposium, interest in the technology has grown, driven in part by environmental legislation but also by demands for increases in process efficiency and intensity, notably for power generation and waste incineration. Some techniques for high temperature gas cleaning have now reached practical exploitation, and some of the contributions in this volume describe industrial applications. Other contributors, from both academic and industrial groups, describe research and development for new techniques and applications. The Second Symposium also contains papers in an area which had not emerged at the time of the First Symposium: combined processes for collection of particulates and gaseous components.

We are honoured to have an opening address from Professor Dr.-Ing. Friedrich Löffler of the Institüt für Mechanische Verfahrenstechnik und Mechanik, Universität Karlsruhe. Professor Löffler has pioneered the scientific investigation of all aspects of gas cleaning technology over many years and his group is well-represented at this Symposium. We are delighted to have him with us in the year of his 60th birthday!

As in 1986, the Symposium Organisers are grateful to the authors represented here for the high standard of their contributions. The Co-Chairmen would like to express their gratitude to the Scientific Committee for their constructive ideas and support, and to Mrs Jean Libaert for her usual conscientiousness and reliability. The professional support of The Institution of Chemical Engineers, The Filtration Society, and The Royal Society of Chemistry is acknowledged. The Organisers are also grateful to Shell International and to the other organisations listed on an earlier page for their financial contributions, which have been used to support workers from Eastern Europe and students who would otherwise have been unable to attend.

Roland Clift
Jonathan Seville

Contents

SECTION 3   CHEMICAL SEPARATIONS

Section 1
Inertial Separators and Electrostatic Precipitators

NOVEL CENTRIFUGES FOR HIGH TEMPERATURE GAS CLEANING

A HITCHINGS, T O'DOHERTY and N SYRED
School of Engineering, UWCC
P O Box 925, Newport Road, Cardiff   CF2 1YF, UK

ABSTRACT

A design for a novel technique for separating submicron particles from an exhaust stream of a diesel engine has been developed at Cardiff. It consists of three main sections, an inlet where the exhaust gases are introduced, a rotating cylinder with a speed of 60,000 rpm or more and a two part outlet system, one to remove the particulate laden gas centrifuged out in the cylinder and one to remove the clean gas. The clean gas constitutes up to 90% of the flowrate.

The development has involved the use of FLUENT (a well known finite difference modelling package, solving the approximations to the governing equations) for aerodynamic and particle dynamic predictions. This paper will show a series of predictions including particle tracking of particles 0.2 μm in size and contours of predicted velocities. Much of the development programme has been involved with the bearing system and minimisation of main shaft orbit to prevent bearing damage. Further development of the bearing system to allow higher speeds to be attained will yield greater separative performance. The existing prototype will separate particles down to less than 0.5μm.

NOMENCLATURE

d   particle diameter
M   Mach number
n   vortex exponent
r   radius
u   axial velocity
W   tangential velocity
ϕ density

3

INTRODUCTION

This paper describes initial work at evolving a high speed gas centrifuge for cleaning sub micron particles from gases being primarily directed at the problem of particulate clean up from diesel engine exhaust, although more recent work has indicated a far wider range of applicability. The work originated from the marrying of a number of technologies and the following observations.

a)   A conventional cyclone dust separator is a well known technology for particulate removal with hot or cold gases (Gupta, Lilley et al, 1984). Dirty gas enters a cylindrical section via a tangential inlet and centrifugal forces are used to concentrate fine particulate matter in the wall boundary layer region whilst gas (and very fine matter) passes radially through the device into a central exhaust finder which extends down into the body of the cyclone to a depth greater than that of the tangential inlet. Solid material is removed via a conical base section into a hopper. There are fundamental limitations to the separative performance of cyclone dust separators caused by wall turbulence effects between the stationary wall and the high velocity flow. Typical maximum inlet velocities for best performance are of order 30 m/s (increases above this level usually lead to substantial degradation of separative performance, hence the centrifuge concept). Results obtained from General Motors tests of a cyclone fitted to a coal water slurry fired single cylinder research engine are interesting and show that only particles greater than 2 to 3 µm are separated (Slaughter et al, 1990). As coal ash particles are likely to be denser than that from ordinary diesel engines, being carbon/mineral matter in origin (i.e. $\phi \approx 2000$ kg/m³) engine clean up operation would require 6-10 such cyclones fitted to a large truck to maintain this performance. A large proportion of input particles to the cyclone are about 3 µm in size. The cyclone removes all particles to the cyclone of about 5 µm, but only 50% of 3 µm and possibly 20% at 1 µm. However, there is little likelihood of such a device being able to take the very fine particulate emissions (i.e. 0.1 µm $< d < 0.5$ µm) being produced by the next generation of low emission diesel engines evolving to meet the new emission regulations in now what is effectively a global market, hence the proposal of a gas centrifuge as opposed to a gas cyclone.

Nevertheless there are many lessons that can be learnt from cyclone dust separator cyclone chamber, swirling flow literature for flows where the Mach Number, $M < 1$. In particular they concern the passage of cyclonic or vortex flow down a tube and the influence boundary and walls can have on the shape of the resulting tangential velocity profile, which then, as discussed later considerably influences separative power.

In general terms a number of different tangential velocity profiles can occur (Gupta et al, 1984)

i)   The so-called free or potential vortex where tangential velocity increases with decreasing radius according to the relationship:-
Wr = const (W, tangential velocity; r, radius).
i.e. velocity increases from the wall towards the axis of symmetry.

ii)   Rigid body or forced vortex flow where
Wr^{-1} = const

These relationships can be represented by the following equation
Wr^n = const, where $-1 \leq n \leq 1$ (n is called the vortex exponent).

A typical intermediate case is where n=0 and constant tangential velocity exists for some distance from the wall towards the axis of symmetry. In practice, n can normally vary between -1 and +1 depending on the system. General experience shows that the higher the value of n the better, as owing to the high levels of tangential velocity in the central regions of a device the better the separative performance (i.e. higher centrifugal force fields). The literature on existing nuclear industry cyclones indicates the existence of a rigid body of forced vortex flow owing to the high Mach numbers generated (i.e. $M \geq 5$) (Makihara & Ito, 1974 and Ribando, 1984). Similarly the presence of rotating struts or members internal to a centrifuge will tend to induce rigid body or forced vortex flow (as they themselves rotate as rigid bodies) remembering that the centrifuge walls primarily drive the vortex structure. One final point should be made: the tangential velocity, by its nature and definition of co-ordinate system disappears to zero on the axis.

In systems where the interference of structural bodies, etc with internal flow is minimised, experience shows it is possible to produce a so-called Rankine Vortex which consists of an outer section of flow where the vortex exponent n can reach values of 0.5 to 0.8, whilst the central (very small section of the flow) contains a forced vortex where $n - > -1$; this is a desirable type of distribution to achieve in a gas centrifuge. In simple terms the Rankine type of vortex is the best that can be expected to be achieved with a tangential velocity profile that does at least show some increase towards the central axis before an inevitable decline to zero on the central axis of symmetry.

b) Published work in the literature indicated that over a 20-30 year period gas centrifuge technology for the nuclear industry had evolved to a state where, for instance, not only U_{235} could be separated from U_{238} but for instance a gas such as Xenon 136 could be enriched from 26 to 63% in a single gas pass, despite the atomic weight ratio being only 136 to 124 (Makihara & Ito, 1974 and Ribando, 1984). Typically such gas centrifuges are extremely long, i.e. typically around 4m and rotate with peripheral speeds of up to 600 m/s. As the Mach 1 for U_{238} is only of order 90 m/s, this means high supersonic flow conditions pertain and an expensive form of construction is needed such that the centrifuge can rotate at speeds of up to 100-150,000 rpm with its outer surface under high vacuum condition. For air or combustion products Mach 1 is directly proportional to absolute temperature, being of order 340 m/s at 293 K and 600 m/s at 900 K, thus a gas centrifuge for these gases need not operate under supersonic flow conditions with obvious benefit to parasitic losses, drag, etc as well as constructional costs.

c)   Turbocharger systems consist of a turbine driven by engine exhaust gases at speeds of up to 150,000 rpm and peripheral velocities of up to 600 m/s for an inlet gas temperature of 1000-1100 K which in turn is directly coupled to a shaft, driving a centrifugal compressor which compresses the charge passed into the intake of an engine, either diesel or petrol fuelled (Allard, 1986). This results in quite often dramatic increases in power and torque as well as improved specific fuel consumption. Only 30% of the enthalpy of the gases leaving the engine are typically used to drive the turbine/compressor combination.

5

These observations lead on to the concept of an integrated turbine gas centrifuge to separate sub-micron particulates from gases. The system could be driven in the engine situation by the excess gas enthalpy from the exhaust from which the turbocharger is already driven at speeds of 150,000 rpm. Two concepts have evolved as follows:-

a)   One in which the centrifuge is effectively a continuation of the turbine exhaust being driven by the turbine and maintaining high velocity swirling flow over an appropriate length so as to obtain sufficient residence time for separation.

b)   A system separately driven in its centre by a turbine with its own high enthalpy gas supply. Dirty gas is cleaned by passing it up through the hollow main shaft.

PRELIMINARY STUDIES

These concentrated on examining the performance of a number of alternative configurations in terms of their separation capability and their mechanical stability. The initial concept is shown in Fig 1.

The turbocharger turbine is effectively the entrance of the centrifuge, and provides the high initial tangential velocities to the gas to feed into the centrifuge tube. The long centrifuge tube then provides the strongly rotating flow field with sufficient residence time to concentrate the fine particulates into the wall boundary region. The exhaust section consists of two exits. One is positioned so that it can collect particulate matter from the centrifuge wall and the other centrally located to allow the clean gas, typically up to 90% of the total, to escape from the middle of the centrifuge. In present prototype form the dirty and clean gas exhaust streams are maintained at the same pressure. The ratio between the dirty and clean gas flows can be readily altered by changes to the size of the dirty gas exhaust volute (i.e. exhaust area). Many options exist to clean the dirty gas streams, and are being examined at present. These include the following options:-

a)   As the gas volume is quite small, it can be readily cooled and passed through a disposable cartridge filter. The residual gas could then be passed back to the engine for exhaust gas recirculation.

b)   As a) but only part of the gas stream is cleaned.

c)   The use of ceramic filter elements with electrical heater to periodically burn off the accumulated carbon material.

Experience with conventional cyclones convinced the authors that sub micron particles could probably only be separated from gas flows when a small quantity of gas was used as carrier (i.e. the underflow in conventional cyclone separators).

The separation capability of the system was examined in two main ways:-

i)   Via the use of an available analytical model based on the work of Gupta, Lilley & Syred (1984) with the assumption of high levels of backmixing. Some preliminary results used to demonstrate the feasibility of the system are shown in Figs 2 and 3, including the effects of vortex exponent, length of centrifuge, particulate and mass flow rate for a system based on a 150 kW truck engine. Fig 2 shows results for a centrifuge of length 1.1 m with a tangential velocity of 300 ms^{-1}. Efficiencies have

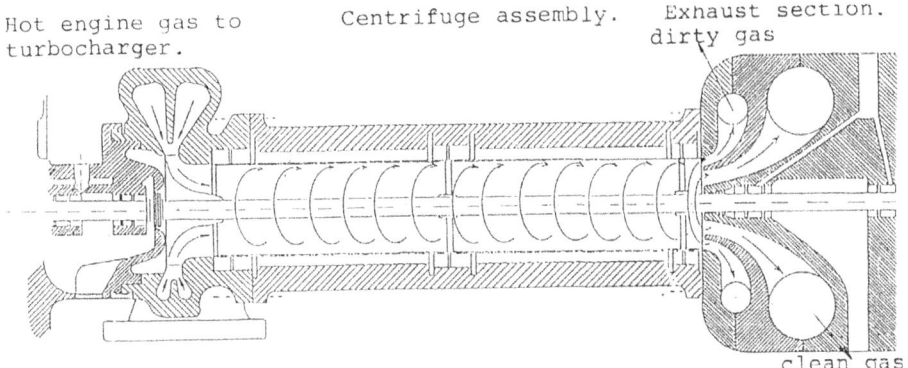

Hot engine gas to
turbocharger.

Centrifuge assembly.

Exhaust section.
dirty gas

clean gas

Fig.1.   Schematic of Turbocharger/Centrifuge.

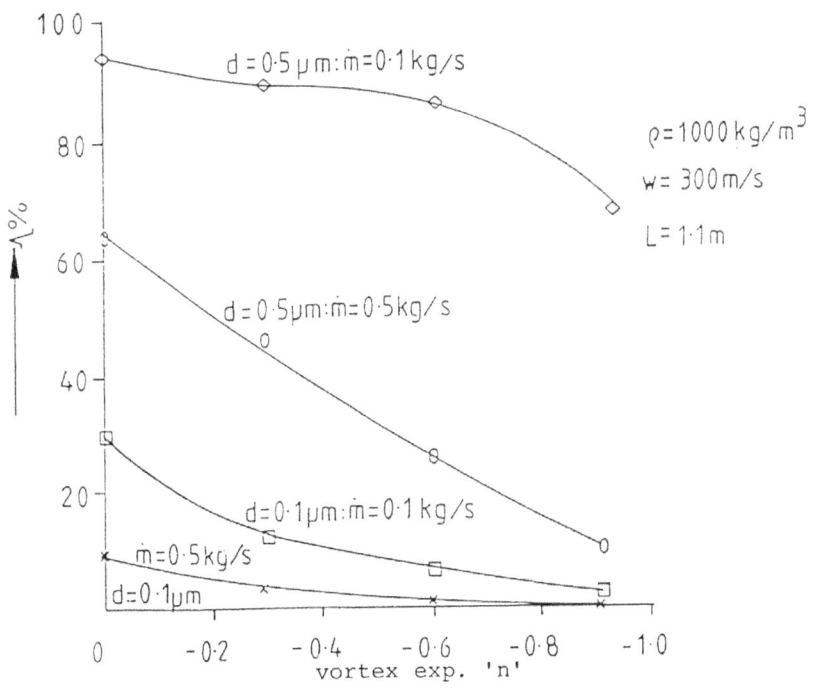

$d = 0.5\,\mu m : \dot{m} = 0.1\,kg/s$

$\rho = 1000\,kg/m^3$

$w = 300\,m/s$

$L = 1.1\,m$

$\Lambda\%$

$d = 0.5\,\mu m : \dot{m} = 0.5\,kg/s$

$d = 0.1\,\mu m : \dot{m} = 0.1\,kg/s$

$\dot{m} = 0.5\,kg/s$

$d = 0.1\,\mu m$

vortex exp. 'n'

Fig. 2.   Effect of vortex exponent, mass flow,
particle size on separation

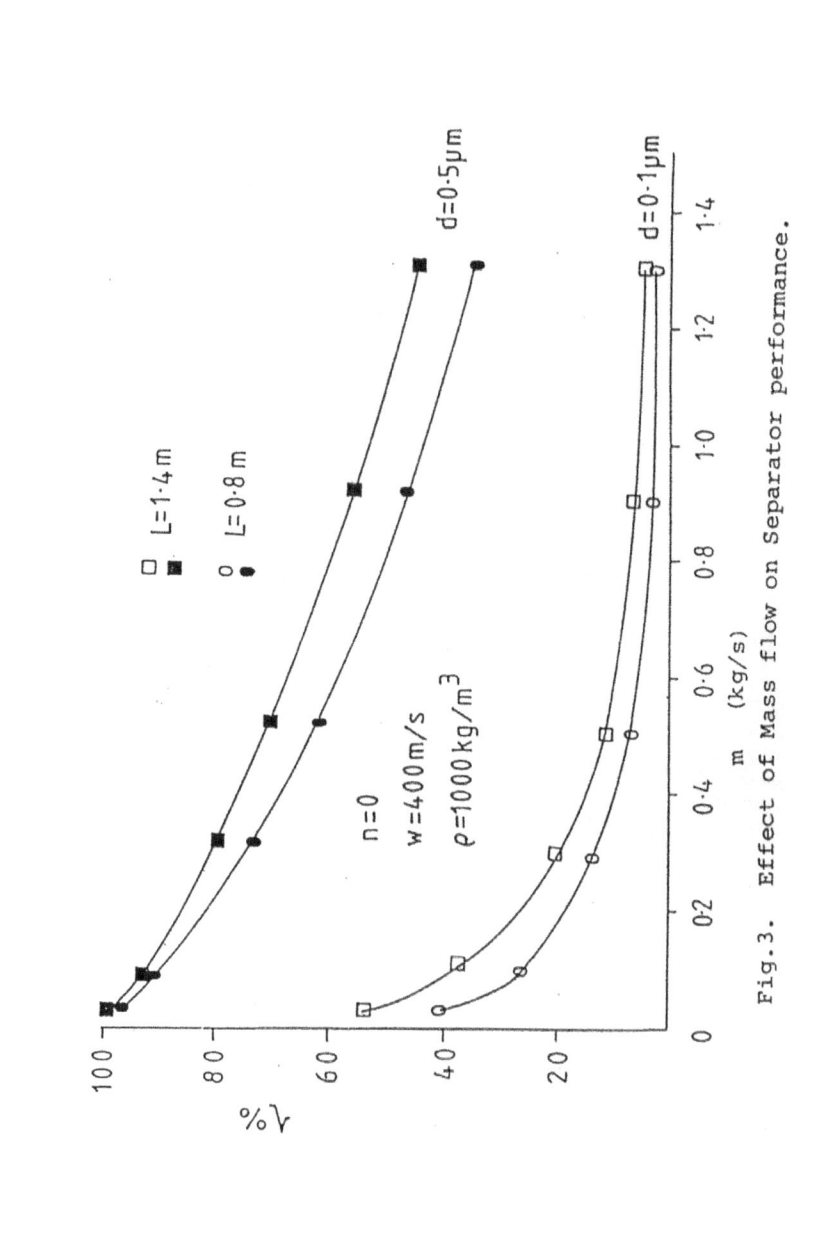

Fig.3. Effect of Mass flow on Separator performance.

been calculated for two particle gas mass flow rates, 0.1 kgs^{-1}, and 0.5 kgs^{-1}, two particle sizes 0.1 μm and 0.5 μm and with varying vortex exponent between 0 and -1. Fig 3 indicates the efficiencies that can be expected with two centrifuge lengths of 1.4 m and 0.8 m, with particle sizes of 0.5 μm and 0.1 μm against a varying mass flow rate. The tangential velocity is assumed to be 400 ms^{-1}. It is clear that high levels of separative performance can be achieved particularly for vortex exponents, n around 0, low mass flow rates (i.e. low axial velocities) with particle sizes in the region of 0.2 to 0.3 μm and with long residence times. Solid to gas loadings are low, as encountered in diesel engines 5-20 mg/kg of gas.

ii) Accepting the limitation of the method described in i) above a finite difference solution of the appropriate Navier Stokes equations for the aero and particle dynamics, was undertaken using the FLUENT package. Full details of the FLUENT package are given in the appropriate manuals. A 2-dimensional axisymmetric representation of a gas centrifuge was made using the turbulent Algebraic stress model. The outer wall of the centrifuge was rotated at 300 ms^{-1} whilst axial velocity of 27 ms^{-1} was used being representative of mass flows used in the size of turbocharger considered, approximately 0.1 kgs^{-1}. Axi-symmetry was achieved with a line of symmetry through the centre line of the centrifuge tube. The dimensions of the centrifuge modelled were 890 mm in length and 54 mm internal diameter. A thin dividing section was inserted near the exit end to represent the partition between the clean and dirty outlet. The exit itself was modelled with one continuous line of outlet cells. This section finished just before the end of the centrifuge to enable one complete row of outlet cells to exist to give the same pressure condition for both clean and dirty outlets. A cruciform was installed at the entrance to more closely match the turbine exhaust tangential velocity profile with the desired vortex distribution (i.e. n = 0) in the main centrifuge tube. Predictions of the typical distribution of axial and tangential velocities are shown in Figs 4 and 5 (particle/gas intersections occurring). It is clear that over a large part of the flow section tangential velocities of order 170 to 180 m/s are occurring giving high separation capability. The axial velocity profiles show that secondary flow, beneficially act close to the inlet forcing particles close to the walls and into the region of the highest centrifugal force field. Particles of size distributions representative of diesel engine particulate matter were randomly fired individually into the inlet and allowed to interact with the flow. After convergence was satisfied a series of particle track plots were obtained and a typical example is shown in Fig 6 for 0.2 μm particles . It is clear that the majority of the particles (even though injected close to the central axis) are rapidly centrifuged/convected by secondary flows to the wall region where they are retained. All particles above 0.3 μm were predicted to be removed by the system, with some separation potential down to 0.1 μm

These results, being at present extended by further work, gave great confidence in the potential separation capability of the system, even at relatively low peripheral velocities of 300 m/s.

In parallel with this programme, work was directed at producing prototype systems which could be economically mass-produced (a typical turbocharger factory price is just over £100). Great problems were found in evolving bearing systems which would allow the

9

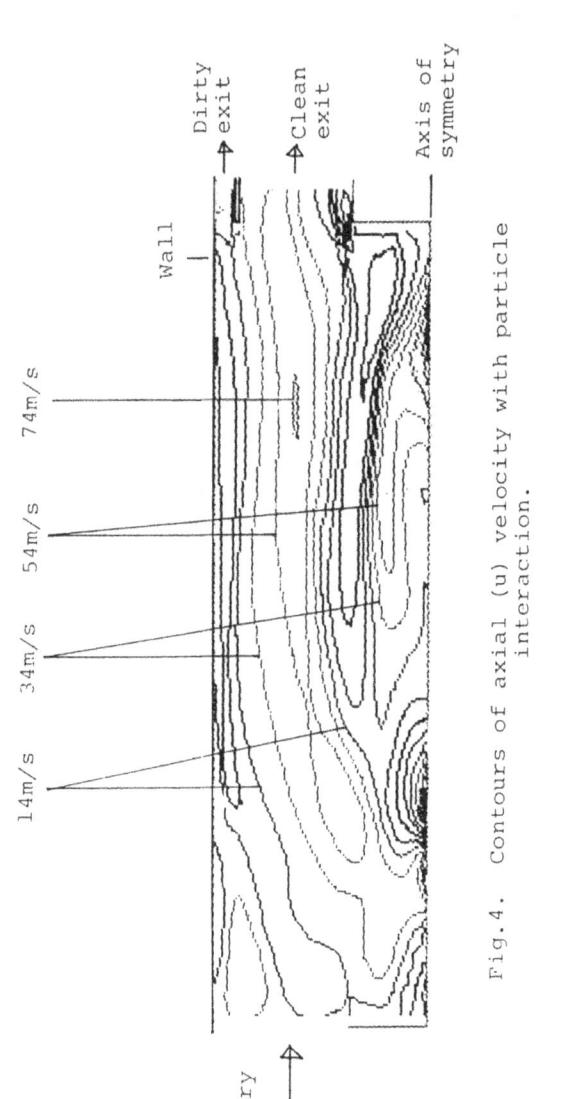

Fig.4. Contours of axial (u) velocity with particle
interaction.

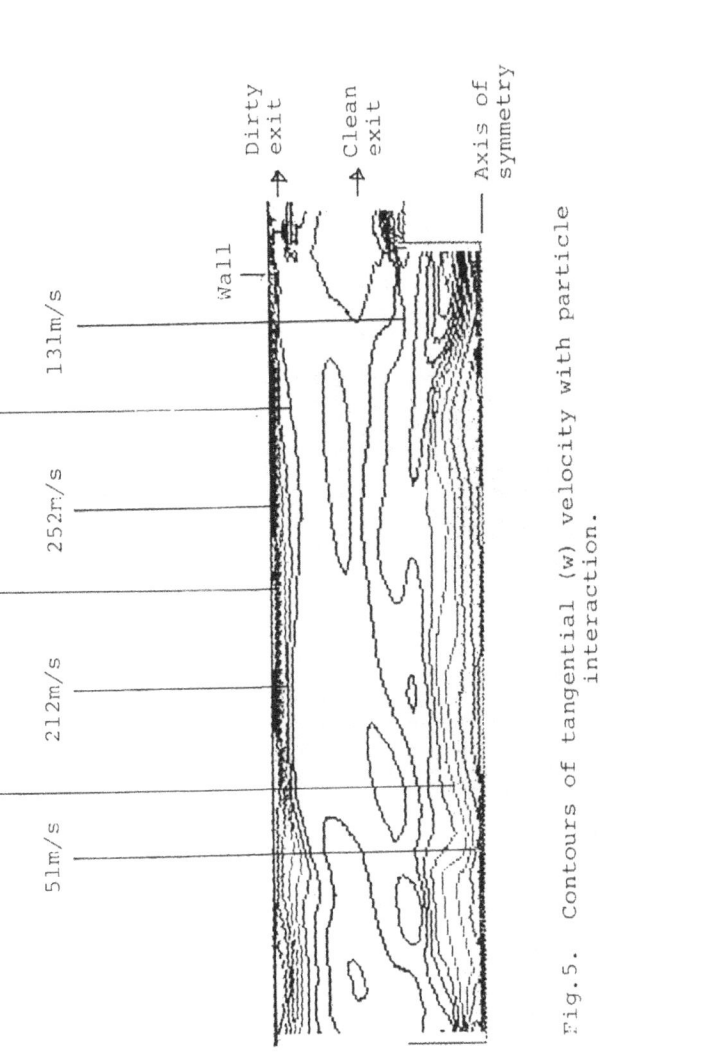

Fig.5. Contours of tangential (w) velocity with particle interaction.

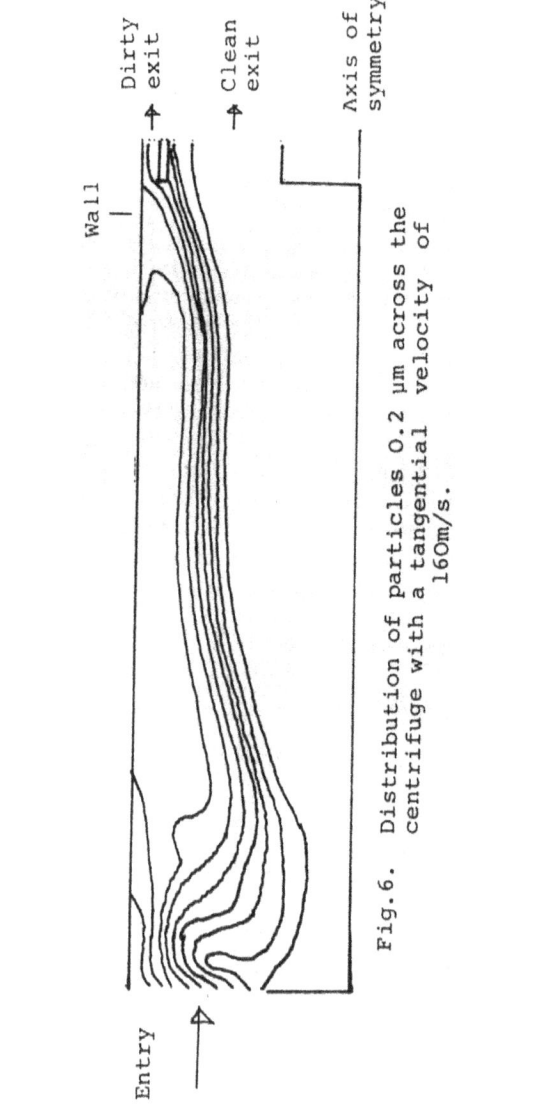

Fig.6. Distribution of particles 0.2 µm across the centrifuge with a tangential velocity of 160m/s.

system to attain the desired speed and have low friction whilst enabling the system to pass through the various critical speeds corresponding to the various critical vibration levels of the system, such as the first whirl speed. Air bearings both journal and thrust of special designs were eventually used both for low friction and to allow sufficient radial displacement for the system to pass through its first critical speed (i.e. operate as a supercritical shaft). ROMAX software was used as an aid to identify critical vibration areas.

DERIVATION OF PROTOTYPE AND INITIAL TESTS

It proved to be impossible to drive the 890 mm long prototype shown in Fig 1 through its first critical speed of 12,000 rpm, despite many changes in bearing configuration. After several redesigns a compromise shortened, 300 mm long prototype was produced as shown in Fig 7. This time the turbine was mounted centrally to even out bearing loads. The bearing configuration comprised two thrust bearing and two sets of air supported foil bearings. Considerable work, has, and still is being directed at sealing the inlet and exhaust areas to prevent particulate matter entering the foil bearing and causing damage (foil clearance is of order 30 to 50 μm). Dirty gas is fed into an inlet torroid vortex chamber which spins it up to a level close to that of the rotating tube. On entering the centrifuge a cruciform more closely matches its rotational speed to that of the centrifuge. After passage down this tube the gas is separated into the two streams, a dirty stream consisting of 10-15% of the total flow into which the majority of the fine particulate matter is concentrated. The clean outlet contains virtually no particulate matter. Rotational speeds of up to 60,000 rpm have been achieved, only limited by compressed air supplies. For the tests this corresponds to an inner peripheral speed of 166 m/s and turbine Mach number of 0.75. High temperature exhaust gases would obviously enable higher speeds to be achieved owing to the increased velocity at Mach 1. Despite the fairly low speed of operation and short length very encouraging preliminary results have been achieved including:

a)   The separation and recovery of sub micron oil based fogs produced by standard fog generators
b)   Separation of fine oxide powder used as a material in the system.

Initial results indicate that a fairly sharp cut is being obtained at around 0.4 to 0.5 μm at low solid/material to gas loadings. The cut closely matches FLUENT predictions made with peripheral wall speeds of 150 m/s or 54,000 rpm.

A problem which has arisen in tests is the formation of a large central recirculation zone which recycles seeding material back through the cruciform region into the inlet. The FLUENT package has been used to successfully predict this phenomena via a three dimensional study of the flow through the cruciform and the centrifuge tube. The effects of centre bodies added to the cruciform to eliminate undesirable recirculation zone have been studied both experimentally and with FLUENT. It was concluded that about 70% of the diameter of the system at inlet needed to be occupied by a shaped centre body to eliminate the recirculation zone.

Further increases in speed are predicted to bring considerable benefits in terms of particle separation. At present we have passed through the first harmonic or critical speed of the

13

Fig.7. Photograph of prototype centrifuge.

14

bearing tube system. Indications are that the system is now approaching a second and possibly a third harmonic. The system is fairly insensitive to initial balance - fairly crude balancing enables high speeds to be obtained.

So far it has not proved possible to start tests at high temperature as the bearing system has not been properly proven. It is anticipated that such trials will commence in the next six to eight months.

It is felt that this programme has produced a prototype with a very high potential for gas particle separation at high temperatures in systems where excess enthalpy (or a high operating pressure system where throttle valves are used for pressure regulation) is available which can be used to drive such systems, the prototype absorbs about 7 kW at 60,000 rpm, and at the moment turbochargers only absorb some 30% of the available enthalpy from diesel engine exhaust and such energy levels can probably be obtained with virtually no reduction in engine power.

CONCLUSIONS

This paper has described a programme of work to evolve a subsonic gas centrifuge for gas and sub micron particles. A first prototype has been successfully constructed and run, and despite many mechanical problems is now giving very successful separation in the sub micron particle range. There seems no doubt that further development of the bearing system to allow higher speeds and a longer length system to be evolved will allow better separative performance to be achieved giving potential for applications in a number of areas. In particular, owing to the relationship between Mach number and temperature, great potential exists for separation at high temperature.

ACKNOWLEDGEMENTS

The authors gratefully acknowledge the permission of the Kudos Corporation to publish this paper. Andrew Hitchings acknowledges the support of the Science and Engineering Research Council via a scholarship.

REFERENCES

Allard, A (1986) Turbocharging and Supercharging. Peter Stephens Ltd, Wellingborough ISBN 0-85059-744-7

Creare Inc. Fluent Manuals, Version 4, 1993.

Gupta, A K, Lilley, D G and Syred N. (1984) Swirl Flows, Abacus Press, Tunbridge Wells, Kent.

Makihara, H and Ito, T. (1974) Separation Characteristics of Gas Centrifuges in the Presence of Light Gas Components, Atomkernergie, Vol 24, No 3, pp 161-170.

15

Ribando, R J. (1984) A Finite Difference Solution of Onsager's Model for Flow in a Gas Centrifuge, Computers and Fluids, Vol 12, No 3, pp 235-252.

Slaughter, P, Cohen, M, Samuel, E, Mengel, M and Gel, E. (1990) Control of Emissions in the Coal-fuelled Diesel Locomotive, Proc Coal-Fuelled Diesel Engines Symposia, pp 11-16, ASME and presented at the 13th Annual Energy Sources Technology and Exhibition, New Orleans, Louisiana, ISBN 07918-0450X.

Wood H, G, Jordan, J A and Ganzburger, M D. (1984) The Effects of Curvature on the Flow Fields in Rapidly Rotating Gas Centrifuges, J Fluid Mechanics, Vol 140, pp 375-395.

Proceedings IMechE. (1991) Worldwide Engine Emission Standards and How to Meet them. Parts 1 and 2, 12-13th February.

Separation efficiency and pressure drop of cyclones at high temperatures

Prof. Dr.-Ing. Matthias Bohnet and Dipl.-Ing. Thomas Lorenz

Institute of Process Technology
Technical University of Braunschweig
Langer Kamp 7, 3300 Braunschweig, Germany

Abstract

The design of aerocyclones operating at high temperatures is still a difficult problem, because practical experiences show that the separation efficiency really obtained is mostly smaller than calculated. For this reason, the pressure drop and grade efficiency have been measured in the temperature range from 293 to 1073 K. Both quantities show a remarkable influence of temperature. With regard to the calculation of the pressure drop, the proposed calculating procedure leads to very good results. The model for calculating the grade efficiency curve still shows differences between calculated and measured results and needs further improvement with respect to the boundary layer flow and the particle movement due to turbulence.

Introduction

The main design parameters of aerocyclones - pressure drop and separation efficiency - can be calculated with a number of models, which were developed on the basis of experimental data received at environmental conditions, that means at ambient pressures and temperatures. But the extrapolation of calculated data with these models over a temperature range of several hundred degrees shows that the results obtained don't agree with experimental data. Fig. 1 shows grade efficiency curves for three cyclone models calculated for a gas temperature of 293 K and ambient pressure. The calculations corresponding to the proposals of Muschelknautz [1] and Mothes and Löffler [2] lead to grade efficiency curves, which show the same tendency but different values for the cut-size diameter d_{p50}. The grade efficiency curve calculated with the model of Leith and Licht [4] shows a considerably smaller cut-size diameter but has a completely different shape. With the model of Leith and Licht a higher separation efficiency is calculated for very small particles with diameters less than $1\mu m$, but the model predicts a lower efficiency for particles with larger particle size. The models were developed and fitted to experimental data measured with different cyclones. The dimensions of the cyclones uesd are for some cases not published in detail, so that it is very difficult to compare the results obtained with these cyclone models even at environmental conditions. Fig. 2 shows the change of the calculated cut-size diameter for the data given in Fig. 1 in the temperature range from 293 to 1173 K.

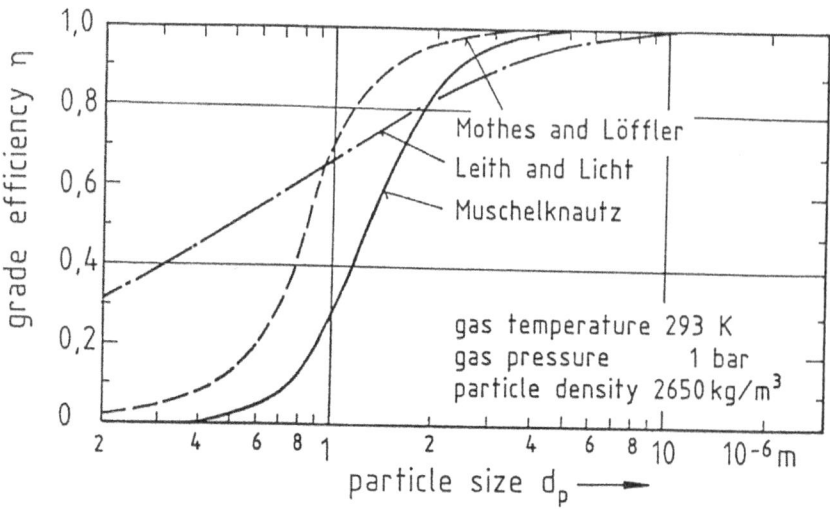

Figure 1: Calculated grade efficiency curves with different models. (Dimensions of the cyclone as shown in Fig.4, outlet velocity $w_i = 20m/s$, dust loading: $100mg/m^3$.

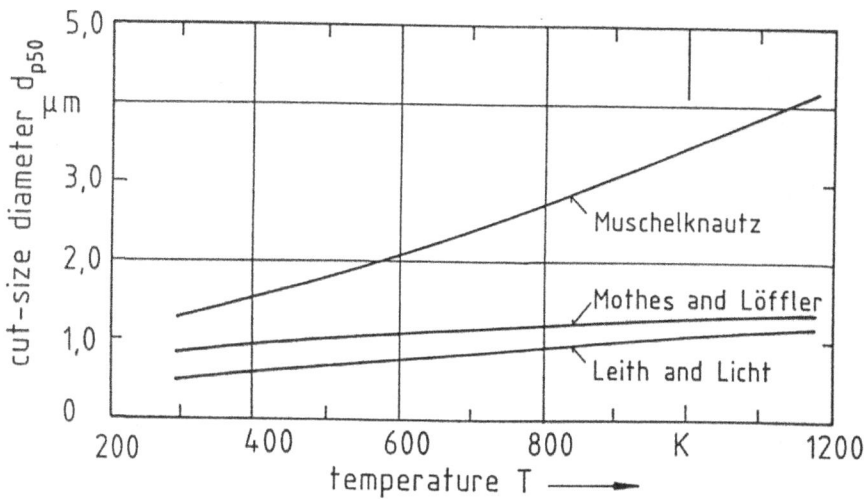

Figure 2: Calculated increase of cut-size diameter with gas temperature

18

The calculation of the cut-size diameter due to the model of Muschelknautz shows the greatest increase with temperature.

Experiences with cyclone separators operating at high temperatures show that the separation efficiency actually obtained is mostly smaller than calculated. Besides the fact that numerous publications over the last ten years have been published [5, 6, 7, 8, 9, 10, 11] the design of hot gas aerocyclones is still an unsolved problem.

Experimental Investigations

At the Institute of Process Technology at the University of Braunschweig an experimental set-up has been installed which allows the measurement of pressure drop and separation efficiency in the temperature range between 293 and 1173 K (Fig. 3). The particle size distributions and the particle concentrations in the inlet and outlet gas streams are measured with light scattering aerosol counters. The measured quantities are scanned and transmitted to a PC. The dimensions of the cyclone used are shown in Fig. 4. The pres-

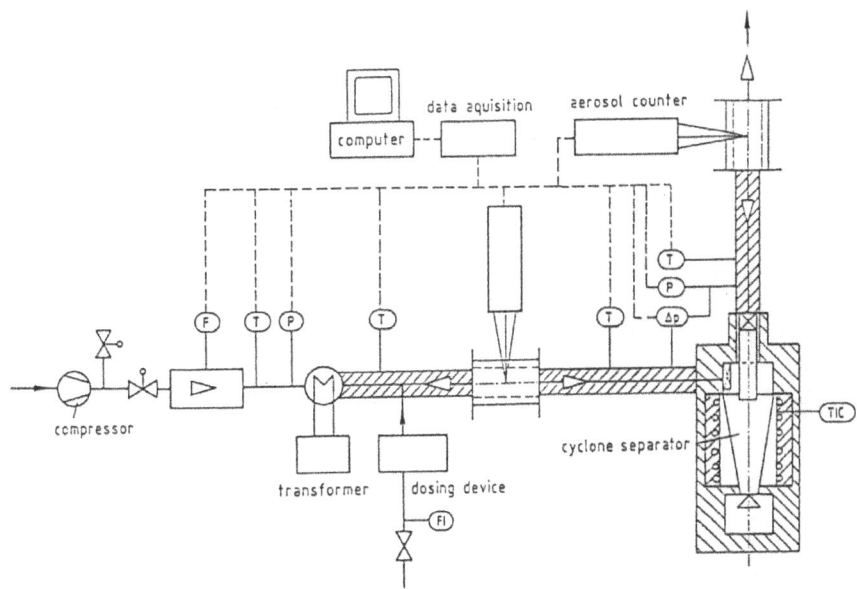

Figure 3: Experimental Set-up

sure drop of the cyclone was measured as difference of the static pressures in the inlet and the outlet duct behind a straightener for reducing the torque of the flow. The straightener was located 300 mm downstream of the gas outlet.

A comparison of the measured pressure drop with calculated data following a proposal of Mothes and Löffler and Meissner [13] is shown in Fig. 5. For a temperature of 293 K

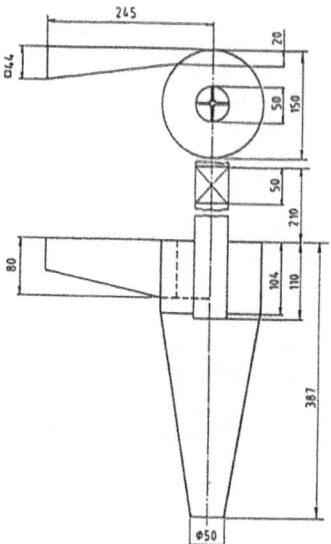

Figure 4: Cyclone dimensions

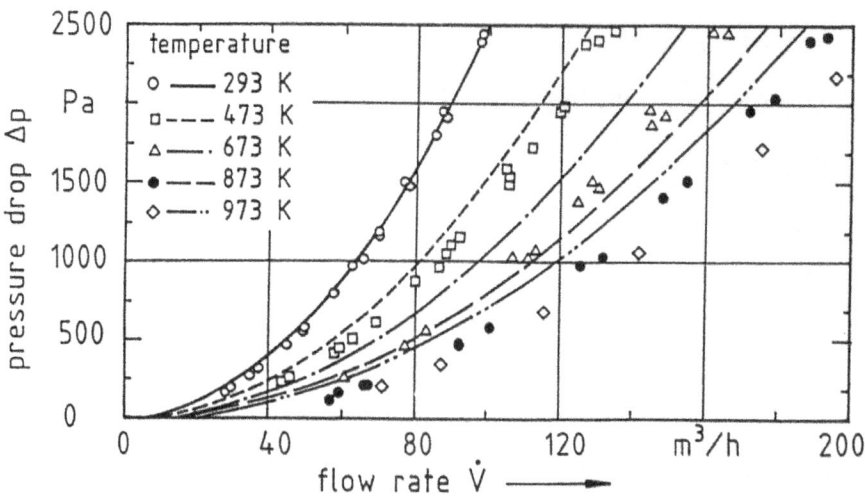

Figure 5: Measured and calculated (Meissner, Mothes and Löffler) pressure drops depending on gas flow rate and temperature

and gas flow rates above 60 m^3/h the measurements agree very well with the calculated data, but at higher temperatures the differences between experiment and calculation are significant. In the calculation a wall fricion coefficient of 0.0075 has been used. The pressure drop coefficient for the cyclone outlet was multiplied by the factor 0.8 as proposed by Meissner for pressure drop measurements behind a straightener.

Fig. 6 shows the inlet particle size distributions of the quartz dust we used. The density was determined to 2650 kg/m^3.

The measured grade efficiency curves at $293, 473, 753, 1003$ K show a strong influence of temperature (Fig.7). The curves are shifted with increasing temperature to larger particle sizes. At the same time the shape of the curves is changing at higher temperatures: the curves become steeper. A comparison with the calculation model according to Mothes shows that the measurements at 293 K are described quite well, whereas the influence of temperature cannot be predicted by the model.

Theoretical Investigations

The model of Mothes shows the greatest potential to describe cyclone performance. For this reason we are working on a modification of this model that allows a calculation of grade efficiency curves in the temperature range from ambient conditions to highest temperatures. As the model of Mothes may not be general known a brief summary is provided in the following.

Mothes and Löffler based their model on considerations of Dietz [20]. They divided the cyclone into four separation regions (Fig. 8) : The inlet region 1, the region of downstreaming gas 2, the region of upstreaming gas 4 and a region 3 which takes into account the turbulent backmixing of particles that have already been deposited in region 2.

From the volume of the cyclone $V_{cyclone}$ and the overall height of the cyclone h_g they calculate a fictitious barrel diameter of a cylindrical cyclone:

$$r_a^* = \sqrt{\frac{V_{cyclone}}{\pi h_g}} \qquad (1)$$

In each region the radial particle concentration is regarded as constant (perfect radial backmixing). Mass balances render differencial equations for the particle concentrations in dependency of the axial coordinate z:

Region of entry (1):

$$\frac{d}{dz}[\dot{V}_e c_1(z)] = -2\pi r_a^* j_1(r_a^*)$$

The particle flux towards the wall $j_1(r_a^*)$ can be calculated as product of the settling velocity of particles in the vicinity of the wall $w_s(r_a^*)$ and the particle concentration $c_1(z)$:

$$j_1(r_a^*) = w_s(r_a^*)c_1(z) \qquad (2)$$

$$w_s(r_a^*) = \frac{\Delta\rho d_p^2 u^2(r_a^*)}{18\mu r_a^*} \qquad (3)$$

21

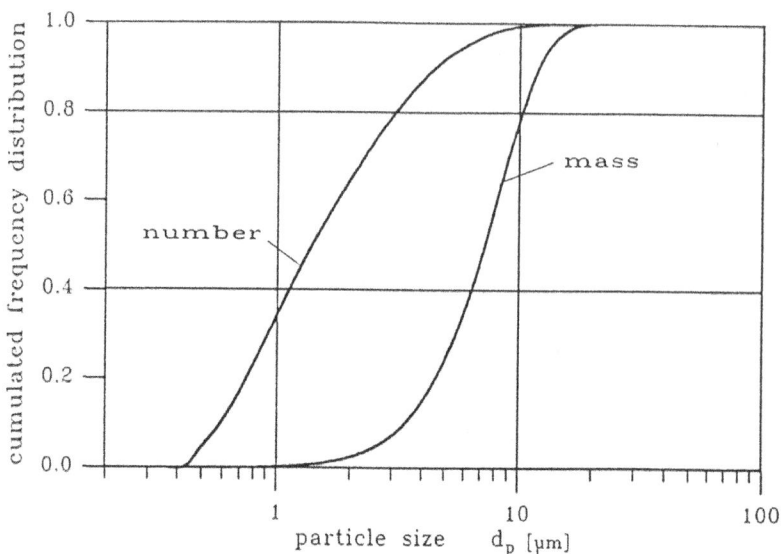

Figure 6: Measured inlet particle size distributions

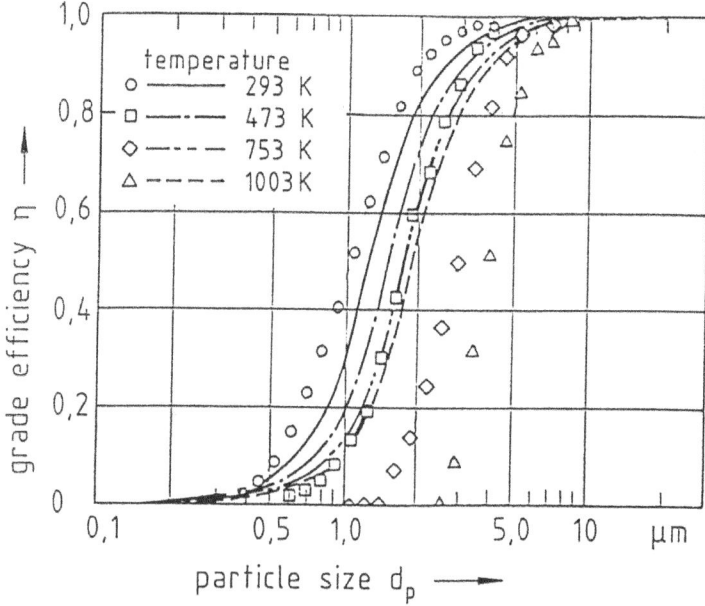

Figure 7: Measured and calculated(Mothes) grade efficiency curves, $V = 80\ m^3/h$, dust loading $\approx 100mg/m^3$

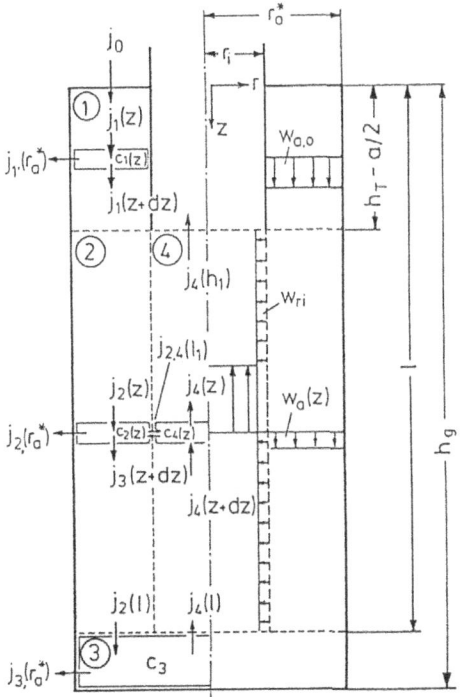

Figure 8: Cyclone geometry according to Mothes and Löffler

Region of downward-directed flow (2):

The concentration of particles in this region is determined by the deposition of particles on the wall $j_2(r_a^*)$ and by the convective and diffusive exchange with region 4 $j_{2,4}(r_i)$:

$$\frac{d}{dz}[\dot{V}(z)c_2(z)] = -2\pi r_a^* j_2(r_a^*) + 2\pi r_i j_{2,4}(r_i) \tag{4}$$

$$j_2(r_a^*) = w_s(r_a^*)c_2(z) \tag{5}$$

$$w_s(r_i) \geq w_r(r_i) \qquad j_{2,4}(r_i) = -D_p\frac{c_2(z) - c_4(z)}{r_a^* - r_i} + (w_s(r_i) - w_r(r_i))c_4(z) \tag{6}$$

$$w_s(r_i) \leq w_r(r_i) \qquad j_{2,4}(r_i) = -D_p\frac{c_2(z) - c_4(z)}{r_a^* - r_i} + (w_s(r_i) - w_r(r_i))c_2(z) \tag{7}$$

$$w_s(r_i^*) = \frac{\Delta\rho d_p^2 u^2(r_i)}{18\mu r_i} \tag{8}$$

In equation 6 and 7 the parameter D_p which Mothes names "Diffusion-parameter" is introduced. Mothes determined the value of this parameter to 0.0125 by fitting his model to his measurements. The difference of the two radii $r_a^*-r_i$ has been chosen as characteristic length.

23

Region of backmixing (3):

The mass balance renders:

$$j_2(l)\pi(r_a^{*2} - r_i^2) - j_4(l)\pi r_i^2 - j_3(r_a^*)2\pi r_a^*(h-l) = 0 \tag{9}$$

with

$$j_3(r_a^*) = w_s(r_a^*)c_3 - \dot{m}_w/2\pi r_a^*(h-l) \tag{10}$$

in which the height of zone 3 ,l, and the mass flow of reentrained particles $\dot{m}_w$ are not clearly defined by Mothes and Löffler. Under the assumption that 10 % of the gas pass through region 3, l can be calculated to:

$$l = h - \frac{h - h_t}{10} \tag{11}$$

Mothes and Löffler calculate for the special case of zero reentrainment with $l = h$ and $\dot{m}_w = 0$.

Region of upward-directed flow (4):

In region 4 only the particle transport to region 2 has to be considered:

$$-\frac{d}{dz}[\dot{V}_z c_4(z)] = -2\pi r_i j_{2,4}(r_i)$$

The particle concentrations in region 1 and 3 can be calculated directly from the mass balances, Region 2 and 4 define a system of differential equations, which can be solved analytically. The solutions are not described here, they can be found by Mothes and Löffler [2, 3]. The separation efficiency can be calculated from the concentration in region 4 that is entering the outlet pipe $c_4(d_p, h_t)$ and the inlet concentration $c_e(d_p)$:

$$\eta(x) = 1 - \frac{c_4(d_p, h_t)}{c_e(d_p)} \tag{12}$$

To describe the tangential velocities Mothes and Löffler use an equation which was derived by Meissner [13] from a differential angular momentum balance:

$$u = \frac{u_a}{\frac{r}{r_a}\left[1 + D\left(1 - \frac{r}{r_a}\right)\right]} \tag{13}$$

The angular momentum parameter D, characterizing the exchange of angular momentum between wall and gas, is obtained from

$$D = \frac{u_a}{w_d}\left(\lambda_D + \frac{\lambda_K}{sin\varepsilon}\right) \tag{14}$$

According to Mothes and Löffler for cyclones with smooth walls the wall-friction coefficients should be chosen between 0.0065 and 0.0075.

The model of Mothes and Löffler has been proved a good tool to design cyclones at ambient temperature. But the influence of temperature on the calculation of collection efficiency and pressure drop does obviously not quite match experimental data. We suppose that there are three reasons for the deviations between measurements and calculation according to Mothes and Löffler:

- the wall-friction coefficient must be related to the gas properties

- the secondary flow over the top of the cyclone is not taken into account

- the backmixing of particles is designed but not realized.

Therefore we replaced the constant values for the wall friction factor that have been proposed by Mothes and Löffler by an approach similar to that derived by Spilger and Brauer [19] on the basis of measurements of Muschelknautz and Krambrock [18]. A two-parameter approach for the wall friction factor was fitted to our pressure drop measurements which gave the following result:

$$\lambda = 0.0049 + \frac{0.87}{Re_z} \tag{15}$$

with

$$Re_z = \frac{w_i \cdot r_i \cdot \rho}{\eta} \left(\frac{r_i/h}{r_a/r_i - 1} \right) \tag{16}$$

and

$$h = \frac{F_r}{2\pi \sqrt{r_a/r_i}} \tag{17}$$

F_r is the total cyclone surface, at which friction losses take place (cylinder and cone, cover plate and outlet duct surface). The Reynolds number takes into account the changes of gas properties with increasing temperature. As the kinematic viscosity increases by an order of magnitude from 293 to 1173 K the Reynoldsnumber decreases. The increasing wall-friction factor diminishes the calculated tangential velocities and results in lower pressure drop and lower collection efficiency of the cyclone. With the modified wall friction factor the measurements of pressure drop can be represented within the measurement accuracy (Fig. 9). Fig. 10 shows the complete experimental data obtained in the temperature range between 293 and 973 K. The pressure drop coefficient

$$\xi_i = \frac{\Delta p}{\frac{\rho}{2} \cdot w_i^2} \tag{18}$$

plotted against the Reynolds number

$$Re = \frac{w_i \cdot d_i \cdot \rho}{\eta} \tag{19}$$

show a strong influence of the Reynolds number. Fig. 11 shows that the representation of the measured grade efficiency data became better though not quite correct. Our attempts to fit the other two mechanisms of particle transport — secondary flow and backmixing — into the model are not yet fully developed. For this reason only the principal ideas shall be described.

Experimental and theoretical investigations on the boundary layer flow which leaves the cyclone over the top and the outer outlet pipe through the outlet without entering the main flow field have been performed by Trefz [14]. According to Trefz approx $10 - 15\%$ of the total gas flow rate take this way out. Due to the lower centrifugal forces they carry even coarse particles in remarkable concentration to the gas outlet. Following the proposals of Ebert [15] the flow rate of the secondary gas flow can be calculated. Dividing

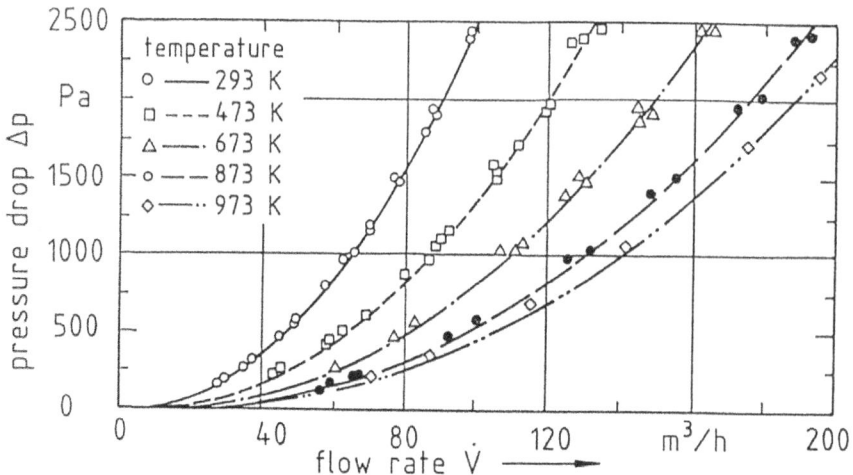

Figure 9: Measured and calculated(Meissner, λ modified by Lorenz[12]) pressure drops depending on gas flow rate and temperature

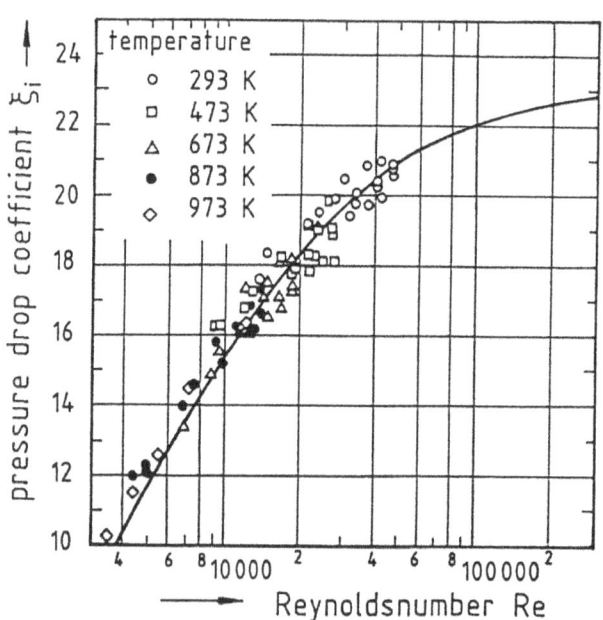

Figure 10: Pressure drop coefficient depending on Reynolds number

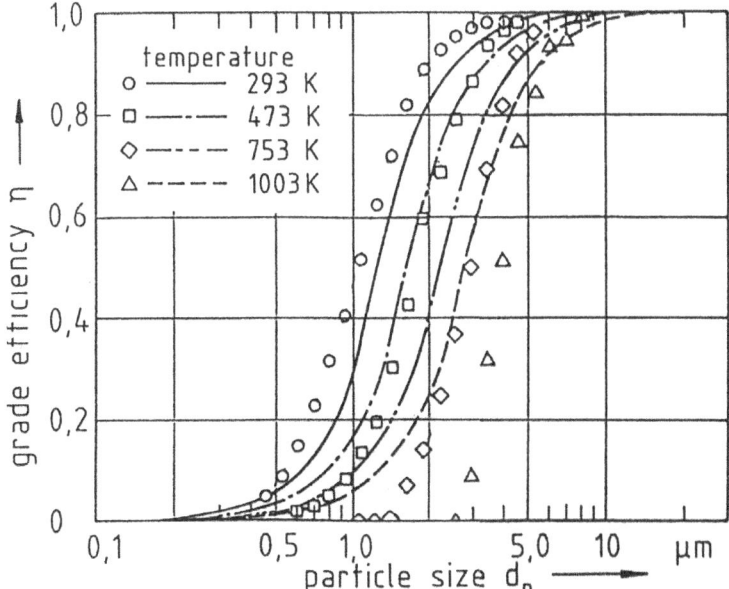

Figure 11: Measured and calculated(Mothes, λ modified by Lorenz[12]) grade efficiency curves, $\dot{V} = 80\ m^3/h$, dust loading $\approx 100mg/m^3$

the entry region defined by Mothes and Löffler it is possible to calculate the solid loading of the secondary flow (proceeding as Mothes and Löffler did) without further assumptions.

The boundary layer flow above the particle outlet carries most of the solid loading (region 3 in Fig. 8). Data on the flow field in this region is not available for this reason a physical separation mechanism cannot be applied. Hence, we try to find a semi-empirical relationship which fits our grade efficiency data.

The efforts to consider these two effects in the calculation have not yet been successful. For a first calculation of the cut-size diameter the following equation may be used:

$$d_{p50} = \sqrt{\frac{18\eta \cdot w_{ri} \cdot r_i}{u_i^2 \cdot (\rho_p - \rho)}} \tag{20}$$

with

$$w_{ri} = \frac{\dot{V}}{2\pi \cdot r_i \cdot h} = \frac{w_i}{2} \cdot \frac{r_i}{h_i} \tag{21}$$

The tangential velocity u_i on the outlet duct radius r_i can be calculated following a proposal of Barth [16]:

$$\frac{u_i}{w_i} = \frac{1}{\frac{F_e}{F_i} \cdot \frac{\alpha}{r_a/r_i} + \lambda \cdot \frac{h}{r_i}} \tag{22}$$

α and r_α can be taken from Bohnet [17].

$$\alpha = 1 - \left(0.54 - \frac{0.153}{F_e/F_i}\right)\left(\frac{b}{r_a}\right)^{1/3} \tag{23}$$

27

temperature	$°C$	20	200	480	730
d_{p50}(clac.)	μm	1.49	2.12	3.28	4.54
d_{p50}(meas.)	μm	1.1	1.8	3.0	3.9

Table 1: Calculated and measured cut-size diameter for different gas temperatures

and

$$r_\alpha = r_a - \frac{b}{2} \tag{24}$$

Tab. 1 compares calculated and measured cut-size diameters. Even though the calculated values have a considerable deviation from the measurements, the magnitude of deterioration of the cyclone performance is predicted right. The pressure drop coefficient can be calculated using the following equations:

$$\xi_i = \xi_{ie} + \xi_{ii} \tag{25}$$

ξ_{ie} describes the pressure drop in the entrance region of the aerocyclone and ξ_{ii} the pressure drop in the outlet duct. It is:

$$\xi_{ie} = \frac{r_i}{r_a} \left(\frac{1}{\left(1 - \frac{u_i}{w_i} \cdot \frac{A}{r_i} \cdot \lambda \right)^2} - 1 \right) \left(\frac{u_i}{w_i} \right)^2 \tag{26}$$

$$\xi_{ii} = 0.8 \left(2 + 3 \cdot \left(\frac{u_i}{w_i} \right)^{4/3} + \left(\frac{u_i}{w_i} \right)^2 \right) \tag{27}$$

The calculation of pressure drop with eq. 4 and 11 to 13 gives results with an accuracy comparable to the calculation according to Meissner shown in Fig. 9 and 10.

Conclusion

Measurements of pressure drop and grade efficiency for hot gas cyclones show a strong influence of temperature. The decrease of pressure drop with increasing temperature can be described by the model of Meissner (with a modified wall friction factor, eq. 1). The measured grade efficiencies indicate that the curves become steeper with increasing temperature. The deviation between calculated and measured data can propably explained with the influence of boundary layer flow and turbulent backmixing in the cyclone. Both these effects are not yet integrated into the calculation model for the grade efficiency. A calculation procedure based on a proposal of Barth allows the calculation of the cut-size diameter with sufficient accuracy.

Nomenclature

a   [m]   entrance height

b   [m]   entrance width

c $[1/m^3]$ particle concentration

d_p [m]   particle diameter

d_{p50} [m]   cut-size diameter

D_p $[m^3/s]$ diffusivity of particles

F_r [m]   inner cyclone surface

$F_{e,i}$ $[m^2]$ inlet, outlet area $F_e = ab, \quad F_i = \pi r_i^2$

h   [m]   cyclone friction height: $h = F_r/(2\pi r_m)$

h_g [m]   overall height

j $[1/m^2 s]$ flux of particles

Δp [Pa]   pressure drop

r_i [m]   outlet radius

r_a [m]   cyclone radius

r_a^* [m]   characteristic cyclone radius

r_m [m]   mean radius $r_m = \sqrt{r_a r_i}$

Re   [-]   Reynolds number, $\mathrm{Re} = w_i d_i/\nu$

Re_z [-]   Reynolds number

$u_{i,a}$ [m/s]   tangential velocity

w_d [m/s]   gas velocity $w_d = \dot{V}/(\pi r_a^2)$

w_i	[m/s]	axial outlet velocity
w_{ri}	[m/s]	radial velocity on outlet radius
w_s	[m/s]	settling velocity of particle
T	[K]	temperature
$\dot{V}$	$[m^3/s]$	gas flow rate
$V_{cyclone}$	$[m^3]$	Volume of cyclone

Greek symbols

ν	$[m^2/s]$	kinematic viscosity
λ	[-]	wall friction factor
η	$[kg/(m \cdot s)]$	dynamic viscosity
η	[-]	grade efficiency
ρ	$[kg/m^3]$	gas density
ρ_p	$[kg/m^3]$	particle density
ξ	[-]	pressure drop coefficient

References

[1] Muschelknautz, E., Theorie der Fliehkraftabscheider mit besondererBerücksichtigung hoher Drücke und Temperaturen, VDI-Berichte Nr. 363, 1980, p.49 - 60.

[2] Mothes, H., Löffler, F., Zur Berechnung der Partikelabscheidung in Zyklonen, Chem.Eng.Process.,18, 1984, p. 323 - 331.

[3] Mothes, H., Löffler, F., Prediction of particle removal in cyclone separators, Int. Chem. Eng. (28) 2, 1988, P. 231 - 240.

[4] Leith, D., Licht, W., The collection efficiency of cyclone type particle collectors - A new theoretical approach, AIChE Symp. Ser., 68(126), 1972, p. 196 - 206.

[5] Wakeman, R.J., Progress in filtration and separation, Elsevier, Amsterdam, 1981.

[6] Patterson,P.A., Munz, R.J., Cyclone collection efficiencies at very high temperatures, Canad.J.Chem.Eng. 67, 1989, p. 321 - 328.

[7] Ernst, M., Hoke,R.C., Siminski,V.J., Parker, R., Drehmel,D.C., Evaluation of a cyclone dust collector for high-temperature high-pressure particle control, Ind. Eng. Chem. Proc. Des. Dev. 21(1), 1982, p. 158 - 161.

[8] Bernard, J.G., Andries, J., Scarlett, B., Pitchumani, B., Cyclone performance at high temperatures and pressures, Proceedings 5th World Filtration Congress, Nizza, 1990.

[9] Wheeldon, J.M., Burnard, G.K., Snow, G.C., Svarovsky L., The performance of cyclones in the off-gas of a pressurised fluidised bed combuster, Inst.Chem. Eng. Symp. Ser. 99, 1986, p. 45-65.

[10] Gulyurtlu,I., Cabrita, I., Bordalo, C., Separation of particles in a cyclone at o temperatures above 400 C, 1. Europ. Symp. Partikelabscheidung aus Gasen, Nürnberg, 1989.

[11] Parker, R., Jain, R., Calvert, S., Drehmel, D.,Abbot, Particle collection in cyclones at high temperatures and high pressures, Env. Sci. Tech. 15(4), 1981, p. 451 - 458.

[12] Lorenz, T., PhD-Thesis, TU Braunschweig, 1993.

[13] Meißner, P., Zur turbulenten Drehsenkenströmung in Zyklonabscheidern, Diss. TH-Karlsruhe, 1978.

[14] Trefz, M., Die verschiedenen Abscheidevorgänge im höher und hoch beladenen Gaszyklon unter besonderer Berücksichtigung der Sekundärströmung, VDI-Fortschrittsber., Reihe 3, Nr. 295, Düsseldorf, 1992.

[15] Ebert, F., Berechnung rotationssymmetrischer, turbulenter Grenzschichten mit Sekundärströmung, Deutsche Versuchsanstalt für Luft- und Raumfahrt, Forschungsbericht 67-56, Freiburg, 1967.

[16] Barth, W., Berechnung und Auslegung von Zyklonabscheidern auf Grund neuer Untersuchungen, Brennstoff, Wärme, Kraft, 8, 1956, p. 1-9.

[17] Bohnet, M., Optimalauslegung von Zyklonen, Chemie-Ing.-Techn. 56 (1984) 5, p. 416-417.

[18] Muschelknautz, E., Krambrock, W., Aerodynamische Beiwerte des Zyklonabscheiders aufgrund neuer verbesserter Messungen, Chemie-Ing.-Techn. 42, 1970, Nr. 5, 247 - 255.

[19] Spilger, H., Brauer, H, VDI-Forschungsheft Nr. 602, VDI- Verlag, Düsseldorf, 1980.

[20] Dietz, P. W., Collection Efficiency of Cyclone Separators, AIChE Journal, 27, 1981, Nr. 6, 888 - 892.

THE ANALYSIS OF THE CYCLONE PERFORMANCES UNDER HIGH TEMPERATURE FOR PFBC UNIT

Shi Mingxian   Liu Juanren   Liu Guorong
University of Petroleum, P.O.Box 902, Beijing,China

Yao Zhibiao   Liu Qianxin
Thermoenergy Engineering Research Institute,
Southeast University,Nanjing,China

ABSTRACT

A high temperature gas cleanup system which consisted of three cyclone separators in series, has operated satisfactorily through 500 hours in a 1 Mwt PFBC pilot test unit. The fly ash concentration in cleaned gas stream reduced to values below 2×10^{-4}kg / m^{3}and there are no particles larger than 10μm. The second stage cyclone is the newly developed Φ240mm PV type cyclone separator with dimensions optimized. Its measured particle cut size is 1.2μm and the resistance coefficient considering pressure drop about 12. The third stage multicyclone separator consists of three Φ100 EPVC−I type cyclone tubes and its collection efficiency of 8μm particle is about 99.5%.

INTRODUCTION

High temperature flue gas cleaning is one of the key problems to operate the advanced PFBC technique. The programme of multistage cyclone separators in series was adopted by many countries. Typical operating conditions are a temperature above 800℃ and a pressure above 0.7MPa. It is required that no particles larger than 10μm present in the cleaned flue gas. The dust concentration not allowed to exceed $2 \times 10^{-4} kg / m^3(N)$. All of the high temperature flue gas cleanup systems of the pilot PFBC units in Leatherhead of British CURL, in Curtiss—Wright, in Exxon CO. and in New York University contained three cyclone separators in series with total collection efficiency between 96.8% and 99% [1-4]. The 1 Mwt PFBC test unit in Southeast university in China also consists of three cyclone separators in series as shown in table 1 and Fig.1.

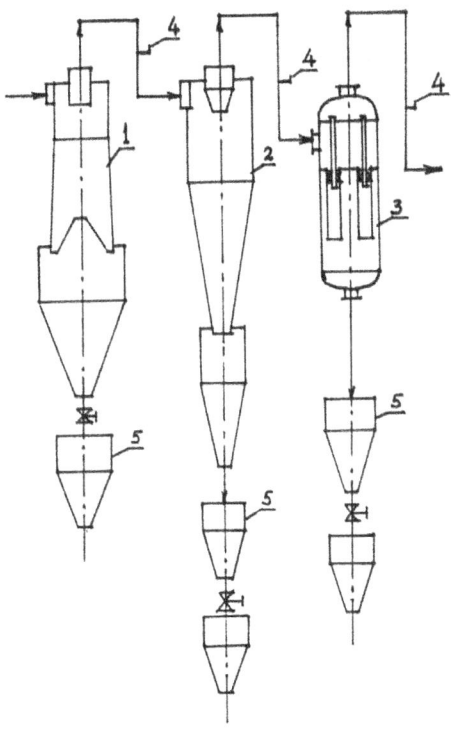

Figure 1. The test rig

1.—1st cyclone   2.—2nd cyclone   3.—3rd cyclone
4.—sampling port   5.—dust hopper

TABLE 1
Three Stage Cyclone Separators

stage	type	main dimensions (mm)
1st.	general	$\Phi 200 \times 1000$, Inlet 200×52
2nd.	PV	$\Phi 240 \times 1370$, Inlet 132×57
3rd.	multicyclone	Three $\Phi 100$EPVC–I type cyclonetubes

The sampling ports with properly designed adjustable samplers are set on the straight sections of the $\Phi 100$mm outlet pipelines of each stage cyclone separator. The fly ash samples were captured by glassfibrous filters and weighed to calculate the dust concentration. The sampled fly ash dispersions, captured by a wet scrubber, were analyzed by Coulter Counter TA–II model, centrifugal particle size analyzer SA–CP3 model, Malvern droplet size analyzer 2600C model and microscope in order to obtain the particle size distribution as accurate as possible.

EXPERIMENTAL RESULTS

The test coals are a bituminous coal of heat value 20.73MJ / kg with an ash content of 30.37% and a high sulphur content coal of heat value 28.56MJ / kg with an ash content of 13.38%. The mean diameters of the coal particles are 1.5–3.7mm and its true density is 1.32–1.5g / cm^3. The mean diameter of the dolomite particles is 1.2–1.75mm. The temperature of inlet flue gas is 730–800°C and the pressure is 0.5MPa (gage). The inlet gas flowrate is 0.26–0.35m^3(N) / s and its dust concentration, calculated from the ash balance, is about 0.0137–0.0253kg / m^3(N). Thus the dust concentration in the outlet cleaned gas of the third stage cyclone separator fluctuated between 1.4×10^{-4} and 2×10^{-4}kg / m^3(N) corresponding to low and high ash content coals respectively. Four tests of each 100 hours duration were carried out at 1990. The total collection efficiency of the three stage cyclone separators was 99–99.3%. These experimental results were summarized in table 2.

The median diameter of the fly ash particles in the outlet cleaned gas was only about 0.92μm and there were no particles larger than 10μm as shown in table 3.

TABLE 2
The Test Results of High Temperature Gas Cleanup System

Stage	Inlet gas Velocity (m / s)	Dust Concentration (kg / m³(N))		Efficiency (%)	Pressure drop (KPa)
		inlet	outlet		
1st	19.8~27	0.016~0.025	$(1.15 \sim 2.9) \times 10^{-3}$	88~94	5.4
2nd	25.8~35.8		$(2.2 \sim 4) \times 10^{-4}$	85~87	5~11
3rd	18.8~25.8		$(1.1 \sim 2.4) \times 10^{-4}$	20~32	11~12

TABLE 3
The Particle Size in Cleaned Gas of 3rd Stage Cyclone

particle diameter (μm)	0.4	0.6	0.8	1.0	1.5 ·	2.0	4.0	5.0	6.0	8.0	10
accumul−ative (%)	100 / 86.9	79 / 74.5	61.6 / 55.4	42.5 / 36	12.2 / 9.1	6.2 / 4.2	6.2 / 4.2	5.5 / 2.9	4.7 / 2	1 / 0.5	0 / 0

THE PERFORMANCE ANALYSIS OF SECOND STAGE PV TYPE CYCLONE SEPARATOR

The characteristics of PV type high efficiency cyclone separator are the simple configuraton with plane top and volute inlet (Fig.2) and higher efficiency than others, obtained by optimum design of their dimensions. The optimum design method is that a dozen dimensions of a cyclone can be divided into three categories which can be optimized separately by different methods. The first category of dimensions have significant influences on efficiency, but almost no influence on pressure drop, such as the diameter of dust discharge dc, the height of the separation space H_s and the insert length of the outlet tube in cyclone, etc. Their optimum values can be specified from flow field studies and performance tests. The third category of dimensions have no distinct influence on both efficiency and pressure drop, such as the dimensions of the dust hopper, etc. Their best values should be established by comprehensively considering the factors of abrasion, operation duration, operation range, and so on. The key is the second category of dimensions. They have significant influences on both efficiency and pressure drop, such as the diameter ratio of the outlet tube entrance to the cyclone $\bar{d}r$ (dr / D) and the area ratio of cyclone cross-section to inlet section $K_A(K_A = \frac{\pi D^2}{4ab})$. If we select the large K_A value, the efficiency will increase and pressure drop decrease under certain inlet gas flowrate, but the cyclone diameter is also increasing. If we select the small $\bar{d}r$ value, both efficiency and pressure drop will increase. It is seen that many kinds of combinations of cyclone diameter D, inlet gas velocity V_i and K_A, $\bar{d}r$

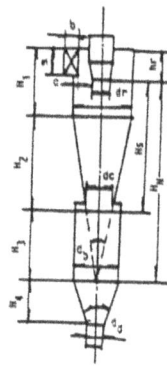

Figure 2. PV type cyclone

values exist at a specified pressure drop. However one of them has the highest efficiency. Based on this idea and our newly developed formulae to calculate the cyclone performances, a computer optimization programme, called "PV type cyclone separator's optimization design programme—PVOD" has been developed. To run the program, firstly input given values as follows: inlet gas flowrate $Qi(m^3 / S)$, gas density $\rho_g(kg / m^3)$, gas viscosity $\mu(pa \cdot s)$, particle density $\rho_p(kg / m^3)$, particle size distribution and concentration in cyclone inlet $Ci(g / m^3)$. Three limitations are then set—permissibe maximum diameter of cyclone, permissible maximum inlet velocity and permissible maximum pressure drop. Then an optimum result of D, K_A, $\bar{d}r$, V_i as well as the efficiency and pressure drop at this matching case are obtained [5]. The PV type cyclone separator was used satisfactorily in our PFBC pilot test unit. Its Collection efficiency was stable at 85–87%. The measured particle size distribution in the inlet and outlet gas streams are approximately logarithmic probablity distributions as shown in Fig.3. The curve at the upper right corner of this figure is its grade efficiency curve, calculated by the following formula:

$$\eta_i(\delta) = 1 - (1 - \eta)\frac{f_o}{f_i} \tag{1}$$

where η is the measured collection efficiency, f_i and f_o are the measured particle size weight frequencies in inlet and outlet gas streams respectively. This curve is represented by the following equation:

$$\eta_i(\delta) = 1 - exp\ (-25St^{0.546})$$

(2)

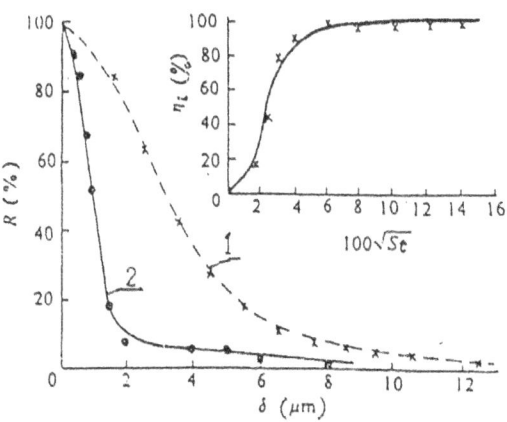

Figure 3.   The grade efficiency of second stage cyclone
1.—Inlet,   2.—outlet

where $St = \dfrac{\rho_p \delta^2 V_i}{18\mu D}$ and δ is the particle diameter in m. The particle cut size of this $\Phi240mm$ PV type cyclone was about $1.2\mu m$ and the grade efficiency of $10\mu m$ particles was already 99%, so that there were a little particles larger than $10\mu m$ in the outlet gas of the 2nd stage PV type cyclone. It is shown obviously that the collection performance of the PV type cyclone for the fine fly ash is excellent.

The pressure drop $\triangle P$ (pa) can be calculated by following:

$$\triangle P = 14.5\ (K_A \tilde{dr}^2)^{-0.83} \tilde{dr}^{-0.08} D^{0.2} (\frac{\rho_g V_i^2}{2})$$

(3)

THE PERFORMANCE ANALYSIS OF THIRD STAGE MULTICYCLONE SEPARATOR

The 3rd stage multicyclone separator consists of three Φ100mm EPVC– I type cyclone tubes which are characterized by the orthogonal guide vanes and the patented split–flow core tube [6]. The calculating formula of the grade efficiency for this cyclone tube is following:

$$\eta_i(\delta) = 1 - exp\ (-30St^{0.55})\qquad(4)$$

and the pressure drop can be calculated by:

$$\triangle P = 90\rho_g V_o^2\ (Pa)\qquad(5)$$

where V_o is the superficial velocity of the cyclone tube in m / s. The measured particle size distributions in the inlet and outlet gas streams of the 3rd stage multicyclone separator are shown in Fig.4. Its collection efficiency fluctuated

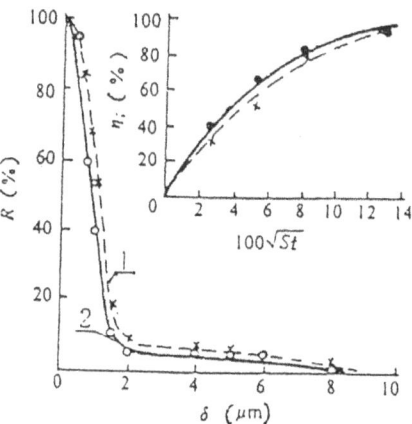

Figure.4   The grade efficiency of the 3rd multicyclone separator
1.—Inlet,   2.—Outlet

between 20% and 32% which was unstable and lower because the particle size in its inlet gas stream was already very fine. Its grade efficiency curve is shown at the upper right corner of Fig.4. The dotted line is the experimental result and the solid line is the calculated result of formula (4). These two curves coincided well to show that this multicyclone separator operated at its estimated performance. The particle cut size of the 3rd stage multicyclone separator was $1.6\mu m$ and the grade efficiency of $8\mu m$ particle was 99.5%, so again no particles larger than $10\mu m$ were found in the outlet cleaned flue gas.

CONCLUSIONS

The high temperature gas cleanup system which consisted of three cyclone separators in series as shown in Fig.1 was operated successfully in our 1 Mwt PFBC test unit. The newly developed PV type high efficiency cyclone separator is suitable to use at PFBC conditions due to its excellent collection performance for fine fly ash. It is noticed that the performance of the PV type cyclones also drops off with scale up, so we should design the PV type cyclone separators with large K_A and small $\tilde{d}r$ values in parallel as 1st and 2nd stage cyclone separators for a future commercial PFBC facility. Then a high efficiency multicyclone separator must be used as the 3rd stage to obtain the cleaned flue gas free of particles larger than $10\mu m$. But the performance of the present EPVC– I type cyclone tubes for very fine fly ash particles smaller than $10\mu m$ is not very satisfactory as shown in Fig.4, we are devoting ourself to develop the new high efficiency cyclone tubes used in the multicyclone separators to suit the needs to built the commercial PFBC–CCG power plant in the future.

REFERENCES

1. Henry, R.F., Podolski, W.F. & Saxena, S.C. (1982). Gas cleanup systems for fluidised–bed coal combustors. Filtration & separation, 6, 502.
2. Wheeldon, J. M. & Burnard, G.K. (1987). Performance of Cyclones in the off–gas path of a pressurized fluidized–bed

combustors. Filtration & Separation, 3, 178–187.
3. Ernst, M., et al. (1982) Evaluation of a cyclone dust collector for high–temperature, high–pressure particulate control. Ind. End. Chem. Process Des. Dev., 21, 158–161.
4. Parker, R., et al. (1981). Particle collection in cyclones at high temperature and high pressure. Environmental Science and Tech., 15(4), 451.
5. Shi Mingxian, et al. (1991). Optimum design of cyclone separators for FCC units. In Proc. of Int. Conf. on Pet. Ref. and Petrochem. Processing, ed, Hou Xiang Lin. Int. Academic Publishers, Beijing, PP.331–337.
6. Wei Yaodong, et al. (1991). Research and development of the 3rd stage multicyclone separator in FCC power recovery system. In Proc. of Int. Conf. on Pet. Ref. and Petrochem. Processing, ed, Hou Xianglin. Int. Academic Publishers, Beijing, PP.1086–1091.

UTILIZATION OF TURBULENCE FOR SEPARATION OF FINE PARTICLES – PRELIMINARY TESTS WITH THE "TURBULENT PRECIPITATOR"*

F.A.L. DULLIEN, WILL KWAN and ANN COLLINS
Department of Chemical Engineering, University of Waterloo,
Waterloo, Ontario, Canada   N2L 3G1

ABSTRACT

A device suitable for the removal and collection of fine particulates from gases also at high temperatures, called "turbulent precipitator", has been built and tested.   The principle of operation consists of placing a plurality of vertical collector plates, occupying the entire width of the housing, with their planes perpendicular to the gas flow, underneath the open and unobstructed passage in which the gas flows in a highly turbulent mode.   Turbulent eddies carry the dust into the gaps between the vertical collector plates where the dust deposits on the plates.   Using a dust of about 2.5 μm median diameter about 90% collection efficiencies have been reached in a unit of about 3 m length at gas velocities of about 25 m/s with a pressure loss of about 2.5 kPa.

1. INTRODUCTION

The "turbulent precipitator" is an efficient and inexpensive device that can be used over a wide range of conditions, including very hot gases, for the removal of fine solid or liquid particles from gas streams. It may be described as an "electrostatic" precipitator without using electricity that, on the basis of the preliminary tests, appears to be more effective than an electrostatic precipitator at a fraction of the cost.

The idea underlying the construction of the turbulent precipitator owes a great deal to the experiments of Friedlander & Johnstone (1957) who showed that from a gas

---

*Invention is protected by UK Informal Patent "Method and Apparatus for Removing Suspended Fine Particle from Gases and Liquids" 9203437.

in turbulent flow dust deposits on the wall of a duct coated with a sticky substance. Consideration of the conditions existing in an electrostatic precipitator have led to the realization that there is a resemblance between the collection mechanisms in this device and in the experiments of Friedlander & Johnstone. The reason for this similarity is that the turbulence in the core of electrostatic precipitators overwhelms the effect of the electric field, and electrostatic forces become effective only after the particles have penetrated the so-called viscous sublayer (e.g., Flagan & Seinfeld, 1988). Hence, because the effect of electrostatic forces is limited to capturing particles after these have approached the collector electrode to within a very short distance, their net effect is that of a very special kind of "sticky substance" whose effectiveness depends also on the electrical properties of the particle captured.

The "turbulent precipitator" captures dust or mist particles, without using either a sticky substance or electricity, by purely mechanical means. The basic idea consists of dividing the gas into two parallel zones: an upper zone containing flowing gas and a lower zone with stagnant gas. The boundary between the two zones is merely a mathematical plane so that there is free interchange of gas between the two zones. The stagnant, lower zone is in effect a very deep sublayer into which turbulent eddies penetrate and then die out.

2. THEORETICAL BACKGROUND

2.1 The Deutsch-Anderson Equation

Some important aspects of the mechanism of dust collection by the turbulent precipitator have counterparts in the functioning of electrostatic precipitators. Hence first the authors' interpretation of the well-known Deutsch-Anderson equation (Anderson, 1919; Deutsch, 1927) is presented.

As a preamble, a brief and very sketchy review of the state of the art concerning this equation is given. The Deutsch-Anderson equation was derived on the basis of theoretical considerations, using several simplifying assumptions, including uniformly distributed dust throughout the flow cross section and constant electrical drift, or migration velocity of the dust all the way from the discharge electrode to the surface of the collector electrode. The theoretical value of the drift velocity was based on Stokes'

Law. At a later stage, some authors have used the assumption that there is a constant electrical migration velocity, based on Stokes' Law, only near the surface of the collector electrode in the region sometimes referred to as viscous sublayer. At the same time, the assumption of uniform distribution of the dust throughout the flow cross section was maintained. In the derivations of the Deutsch-Anderson equation uniform gas velocity all the way to the collector electrode surface, i.e. plug velocity profile, was assumed (Batel, 1972; Flagan & Seinfeld, 1988). As the gas velocity goes to zero at the wall, i.e. at the point where the dust is assumed to be deposited at the theoretical migration velocity, the assumption of uniform gas velocity is not consistent with physical reality. The theoretical migration velocities are well-known to be different from the effective values, back calculated from measured collection efficiencies and, perhaps even more importantly, their theoretically predicted trend in dependence of particle diameter is quite different from the measured trends. In a very recent study (Riehle & Löffler, 1992) it has been demonstrated visually that the dust distribution is not uniform over the flow cross section.

In view of these inconsistencies pertaining to the derivation of the Deutsch-Anderson equation, it seems best to discard the concept of theoretical electrical migration velocity altogether, and regard the effective migration velocity as an empirical parameter in the Deutsch-Anderson equation, until such time that a better theoretical interpretation, including a consideration of turbulent transport, can be found. The derivation which is consistent with this conception is based on a simple differential steady state mass balance on the dust, as sketched in Figure 1. The mass of dust removed from

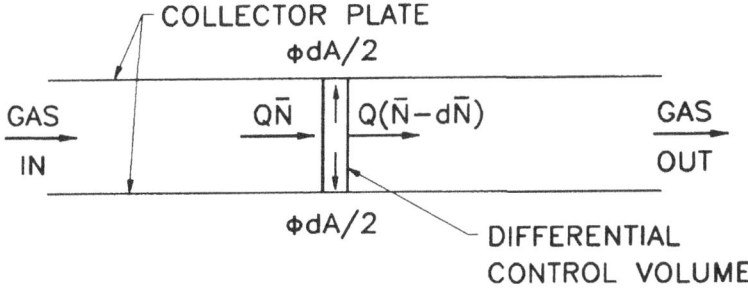

Figure 1. Differential mass balance on the dust in a passage of an electrostatic precipitator.

the differential control volume per unit time is equal to the mass of dust deposited on the differential electrode surface per unit time:

$$-Q \, d\bar{N} - \phi \, dA \tag{1}$$

where Q is volumetric gas flow rate (m^3/s), $\bar{N}$ is the average dust load (kg/m^3) over the flow cross section, A is collector surface area (m^2), and ϕ is mass flux of dust (kg/m^2s) depositing on the electrode surface. The value of ϕ is determined by a variety of factors of aerodynamics, aerosol mechanics, electricity, etc., in addition to the value of the dust loading $\bar{N}$. As the dust loading at the wall is not equal to the average dust loading $\bar{N}$ over the flow cross section, the relationship

$$\phi - \bar{N} \, w \tag{2}$$

is purely formal and it may be regarded as the definition of w, the effective particle drift velocity, which is thus not equal to the actual particle migration velocity at electrode surface.

The mechanism of particle migration, qualitatively speaking, appears to be as follows. The particles migrate from the mainstream to near the collector surface by turbulent diffusion, with an eddy diffusion coefficient practically equal to that of the gas. Eddies approach the wall to within a short distance while they gradually decay. As a result, dust particles are catapulted into the relatively quiescent gas near the wall, sometimes referred to as viscous sublayer, to varying depths that tend to increase with particle size as does the stopping distance of particles of the same shape and the same density (Friedlander, 1977; Dullien, 1989). As soon as a particle is no more subject to eddy diffusion, the effect of the electrical field becomes important and it helps to transport the particle all the way to the collector surface.

Combination of Equations (1) and (2):

$$-Q d\bar{N} - \bar{N} w dA \, , \tag{3}$$

whence, on integration between the inlet and the outlet of the precipitator

$$\int_{in}^{out} d\bar{N}/\bar{N} - -(\bar{w}/Q) \int_{in}^{out} dA \tag{4}$$

or

$$\ell n \ (\overline{N}_{out}/\overline{N}_{in}) - -(\overline{w}A/Q) \tag{5}$$

In Equations (4) and (5) $\overline{w}$ is the value of w averaged between the inlet and the outlet. There is no a priori justification to assume that w is independent of $\overline{N}$ and the value of $\overline{N}$ varies significantly between the inlet and the outlet.

Finally, using the definition of collection efficiency and Equation (5), the Deutsch-Anderson equation is obtained:

$$\eta_{ESP} - 1 - \frac{\overline{N}_{out}}{\overline{N}_{in}} - 1 - \exp\left\{-\frac{\overline{w}_{ESP} \ A_{ESP}}{Q_{ESP}}\right\} \tag{6}$$

In Equation (6) the subscript ESP refers to electrostatic precipitator.

2.2 Similarities Between Turbulent Precipitator (TP) and Electrostatic Precipitator (ESP)

The basic design features of the turbulent precipitator (TP) are shown in Figure 2. The gas, after entering the device, flows in the passage above the vertical plates occupying the lower portion of the housing. The turbulent eddies penetrate to a certain depth into the gaps between the collector plates, as sketched in Figure 3, carrying the dust along, while they gradually decay. A large portion of the dust carried by the eddies into the gaps is trapped there and it deposits on the walls by inertial impaction and Brownian

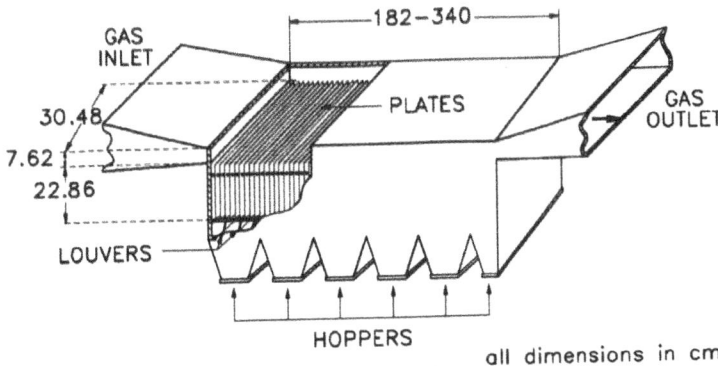

Figure 2. Schematic view of turbulent precipitator.

45

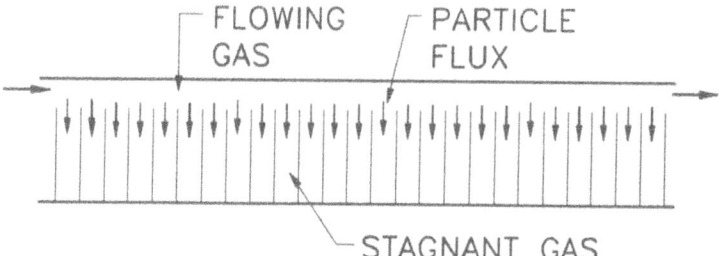

Figure 3. Schematic diagram of principle leading to dust deposition in turbulent precipitator.

Figure 4. Photograph showing the pattern of dust deposited on collector plates of turbulent precipitator.

diffusion in the forms shown in the photograph in Figure 4. The result of these phenomena is a net flux ϕ of particles crossing the plane separating the gas flowing in the passage from the stagnant gas contained between the vertical plates. As a result, the Deutsch-Anderson equation may be expected to apply also to the TP:

$$\eta_{TP} = 1 - \exp\left\{-\frac{\overline{w}_{TP} \, A_{TP}}{Q_{TP}}\right\} \tag{7}$$

46

where A_{TP} is the area of the bottom face of the passage and $\bar{w}_{TP}$ is the effective migration velocity of the dust in the TP, the value of which is determined by a variety of parameters.

2.3 Comparison Between the Geometries of Electrostatic Precipitator (ESP) and Turbulent Precipitator (TP)

A single passage of a plate-type ESP is compared with a passage of the TP in Figure 5. An inspection of the diagrams shows that the quantities in Equation (6) are: $A_{ESP} = 2 H_{ESP} L_{ESP}$ [m^2], $Q_{ESP} = 2 S_{ESP} H_{ESP} v_{ESP}$ [m^3/s], whereas the quantities in Equation (7) are: $A_{TP} = H_{TP} L_{TP}$ [m^2], $Q_{TP} = S_{TP} H_{TP} v_{TP}$ [m^3/s].

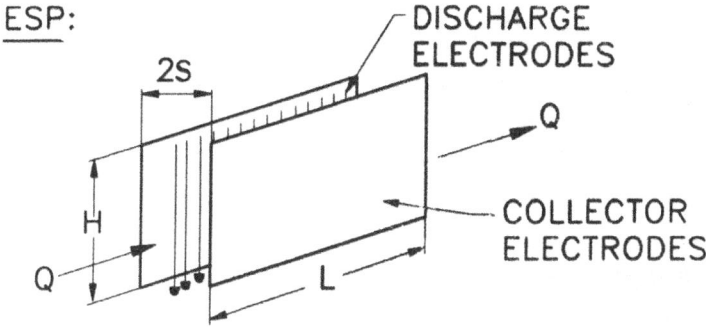

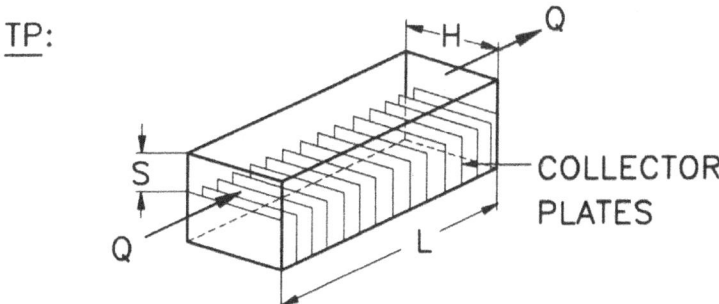

Figure 5. Comparison of schematic geometries of plate-type electrostatic precipitator and turbulent precipitator.

A reasonable basis for a comparison of equipment size is $\eta_{ESP} = \eta_{TP}$ for $Q_{ESP} = Q_{TP}$ of the same dusty gas. Thus

$$\frac{2\,H_{ESP}\,L_{ESP}}{H_{TP}\,L_{TP}} - \frac{A_{ESP}}{A_{TP}} - \frac{(-Q_{ESP}/\overline{w}_{ESP})\,\ell n\,(1 - \eta_{ESP})}{(-Q_{TP}/\overline{w}_{TP})\,\ell n\,(1 - \eta_{TP})} - \frac{\overline{w}_{TP}}{\overline{w}_{ESP}} \tag{8}$$

and the ratio equipment volumes:

$$\frac{V_{ESP}}{V_{TP}} - \frac{2\,H_{ESP}\,L_{ESP}\,S_{ESP}}{H_{TP}\,L_{TP}\,(S_{TP} + \text{plate height} + \text{dust discharge space height})}$$

$$- \frac{\overline{w}_{TP}}{\overline{w}_{ESP}}\,\frac{S_{ESP}}{S_{TP} + \text{plate height} + \text{d.d.s.h.})} \tag{9}$$

The experimental results presented in the next section show that for a dust of a median particle size of about 2 μm, $\overline{w}_{TP}$ has been found to be in the range of 40 to 90 m/s, whereas according to literature values, for the same particle diameter $\overline{w}_{ESP}$ is on the order of a few cm/s. Thus the value of the ratio $\overline{w}_{TP}/\overline{w}_{ESP}$ is in the range of about 20 to 40, leaving a comfortable margin for space required for discharging and transporting away the dust collected on the bottom of the housing of TP. The high value of the ratio $\overline{w}_{TP}/\overline{w}_{ESP}$ is probably due to several operating and constructional differences, including the following: $v_{TP} \gg v_{ESP}$, $S_{TP} \ll S_{ESP}$. In addition, the stagnant gas between the collector plates of the TP is a much deeper "sublayer" than the viscous sublayer on the collector electrodes' surface and, hence, it may constitute a more effective "trap" for the dust than the surface of the collector electrode plates.

3. EXPERIMENTAL PROCEDURE AND RESULTS

In all the tests the same ASP-100 aluminium silicate dust was used as in the previous work of the first author (Dullien & Munro, 1973; Douglas et al., 1976; Dullien, 1977; Dullien & Spink, 1980; Douglas et al., 1982; Dullien & Collins, 1992). The median particle size, determined by Anderson 2000 cascade impactor, was about 2.5 μm. As in previous work, a standard sampling train was used to obtain collection efficiencies. A schematic diagram of the experimental set-up is shown in Figure 6. The inlet and the

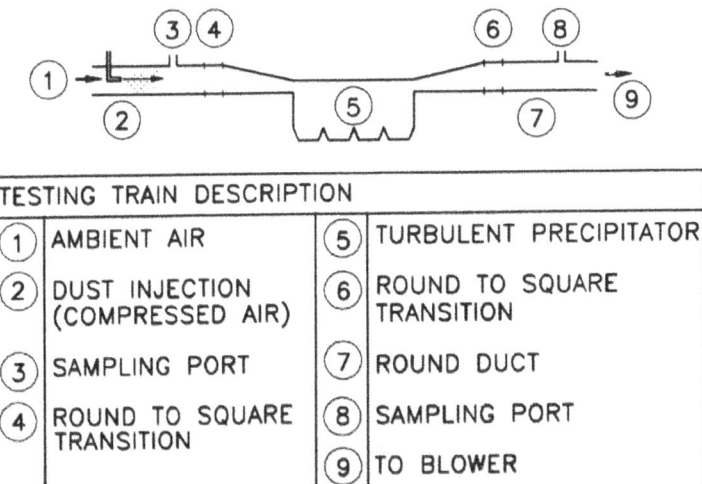

Figure 6. Schematic diagram of experimental set-up.

outlet ducting consisted of 30.5 cm (1 ft) i.d. pipes, whereas the TP housing was a 30.5 cm by 30.5 cm (1 ft x 1 ft) rectangular box. The other relevant dimensions of the TP are indicated in the table and the figures containing the data. Details of the sampling procedure and the calculations are given in the Environment Canada publication, Standard Reference Methods for Source Testing. Measurement of Emissions of Particulates from Stationary Sources, EPS-1-AP-74-1. The dust feeder was a screw conveyor type made by AccuRate. A suction fan was used to draw the air through the system.

The time elapsed from the first test to the last one reported in this paper was only 12 months and, therefore, the results reported here are of a preliminary character: it is not likely that the optimal design and the optimal operating conditions have been established at the time of writing this paper.

Some of the results obtained so far have been compiled in Table 1.

On the basis of the results shown in Table 1 it can be concluded that

- smaller passage height S, accompanied by a higher gas velocity in the passage, has resulted in a higher collection efficiency,
- greater passage length L, at a comparable passage height, has resulted in a higher collection efficiency.

TABLE 1
Samples of experimental results and conditions of tests with TP

Collection Efficiency (%)	System configuration	Passage length L (m)	Passage height S (cm)	Velocities (ms^{-1})		ΔP (kPa)	Plate spacing (cm)	Inlet loading (gm^{-3})
				duct	passage			
50.8	constant S	1.8	7.6	4.2	~13	0.3	2	4.2
51.0	constant S	1.8	7.6	4.2	~13	0.3	2	5.3
69.1	T-4	1.8	2.9	~3	25	1.5	2	5.4
67.3	T-4	1.8	2.9	~3	25	1.5	2	5.6
65.8	T-4	1.8	2.9	4.2	~36	2.3	2	6.5
73.9	T-4	1.8	2.5	3.0	~38	2.8	2	5.1
73.3	T-4	1.8	2.5	3.0	~38	2.7	2	5.4
82.2	T-4	1.8	1.3	~3	~56	2.6	2	6.5
82.0	T-4	1.8	1.3	~3	~57	2.6	2	6.1
86.2	TL-9	3.0	2.0	~3	~36	2.7	2	5.5
86.5	TL-9	3.0	2.0	~3	~36	2.7	2	5.6
85.9	TL-9	3.0	2.0	~3	~36	2.7	2	6.2
85.4	TL-9	3.0	2.0	~2	~36	2.7	2	6.1
85.4	TL-9	3.0	2.0	~3	~36	2.7	2	6.3
85.6	TL-9	3.0	2.0	~3	~36	2.7	2	6.3

The tests at the bottom of Table 1 were carried out without cleaning the plates between tests. The dust fell off the plates spontaneously, in lumps, to the louvred bottom of the unit, whence it was discharged into the collection hoppers, after opening the louvres. With the plate spacings used, there has never been any clogging or any troublesome dust build up. The results indicate that the system can be operated continuously without a change in efficiency.

Some more recent results have been plotted in Figures 7 and 8. In Figure 7 the collection efficiency is plotted vs. passage length L for two different gas velocities in the passage. The passage height was S = 3 cm and the plate spacing was 2 cm. It can be concluded from these results that

- the effect of L is approximately consistent with the Deutsch-Anderson equation
- for a fixed passage height S, lower gas velocity in the passage results in higher collection efficiency
- the values of $\bar{w}$, calculated from the results at the higher and the lower gas velocity, were in the ranges of 55 to 61 cm/s and 43 to 50 cm/s, respectively.

In Figure 8 the collection efficiency is plotted vs. the gas velocity in the passage

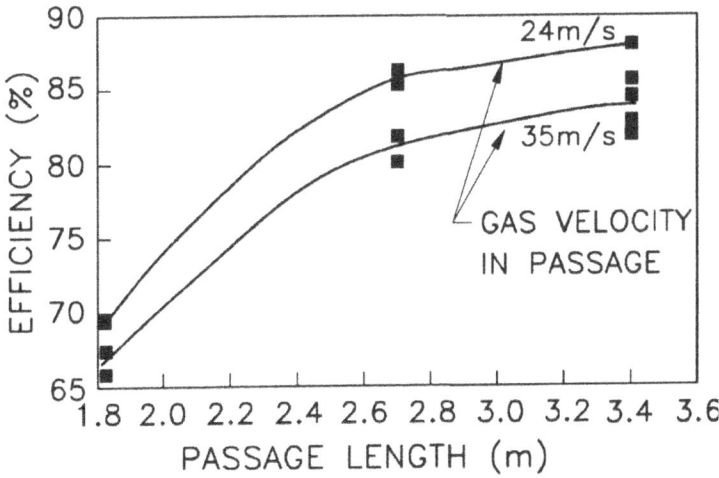

Figure 7. Collection efficiency of a configuration of turbulent precipitator as a function of passage length.

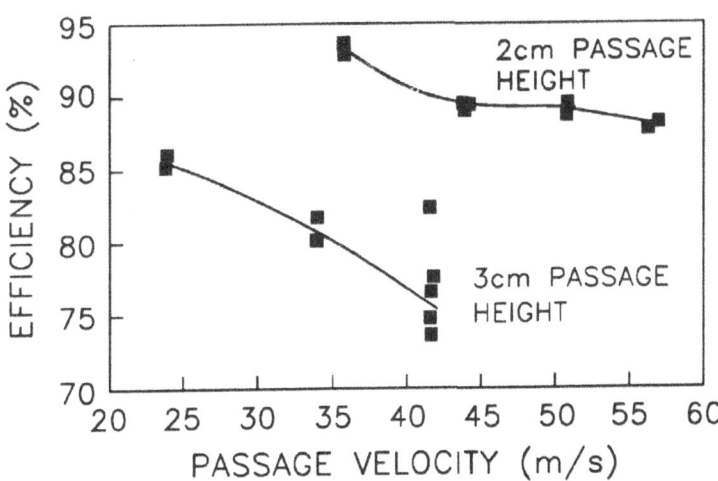

Figure 8. Collection efficiency of a configuration of turbulent precipitator as a function of gas velocity for two different passage heights.

for two different passage heights S, a passage length L of 2.7 m and with a plate spacing of 2 cm. From the results presented it can be concluded that

- the collection efficiency has been much greater for smaller passage height S at the same gas velocity.

- the collection efficiency has increased with decreasing gas velocity at a fixed passage height S.
- the value of the effective migration velocity $\bar{w}$ for the greater and the smaller passage height was in the ranges of 52 to 67 cm/s and 71 to 88 cm/s, respectively.
- $\bar{w}$ decreases with decreasing gas velocity at a slower than linear rate.

Figures 9 and 10 contain histograms of particle size distribution of the inlet and outlet dust and of the fractional collection efficiency of the TP, determined with the help of Anderson 2000 cascade impactor, for passage length L = 2.7 m, passage height S = 3 cm and passage velocity v = 42 m/s. The fractional efficiency levels off, on the average, for particle sizes greater than about 5 μm, at about 90%. This trend is interpreted by hypothesizing that all particles greater than about 5 μm that are carried in the gaps by the eddies are deposited by inertial impaction on the collector plates, whereas an increasing proportion of smaller particles is not deposited but is carried back into the passage by the backflow of air. An important consequence of the observed trend of efficiency with particle size is a similar trend of the effective migration velocity $\bar{w}$. Hence, the rate of increase of the overall collection efficiency η with passage length L is expected to become slower and slower.

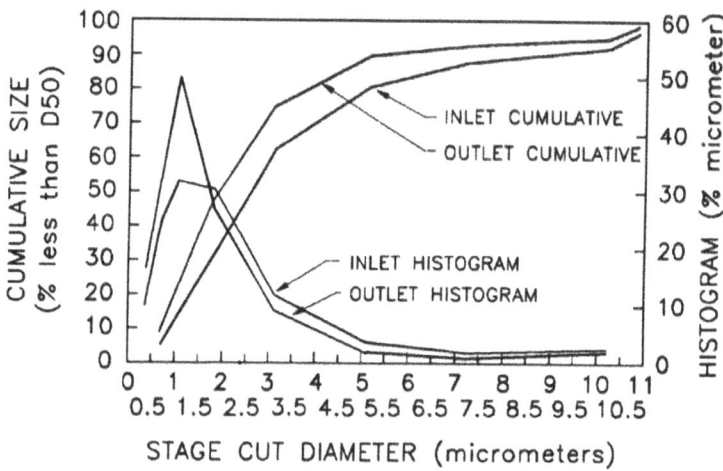

Figure 9. Inlet and outlet particle size distribution of ASP-100 aluminium silicate dust measured by Anderson 2000 cascade impactor.

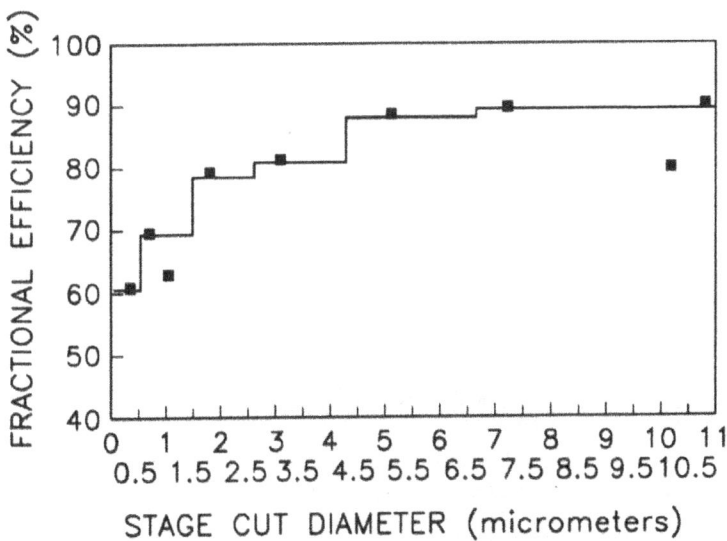

Figure 10. Fractional collection efficiency measured in a configuration of turbulent precipitator by Anderson 2000 cascade impactor.

4. DISCUSSION

For the purpose of qualitative analysis of the trends observed it is convenient to recast the Deutsch-Anderson equation (Equation (7)) as follows:

$$\eta = 1 - \exp\left(-\frac{A\overline{w}}{Q}\right) = 1 - \exp\left(-\frac{HL\overline{w}}{HSv}\right) = 1 - \exp\left(-\frac{L\overline{w}}{Sv}\right) \qquad (10)$$

At first, in a first approximation, it is assumed that each parameter L, v, S and $\overline{w}$ can be varied independently. Subsequently, the variation of $\overline{w}$ with L, v and S is considered.

i) It has been observed, as it is also predicted by Equation (10) that η increases with L because, with everything else kept equal, more particles are collected from the gas over a greater passage length.

ii) It has been observed, as it is also predicted by Equation (10), that η decreases with increasing v because, with everything else (including $\overline{w}$) kept equal, fewer particles are collected over a shorter residence time, implied by a greater v.

iii) It has been observed, as it is also predicted by Equation (10), that η decreases

53

with increasing S because, with everything else (including the dust loading) kept equal, the total number of particles present in the passage at any moment is directly proportional to S, whereas the rate of removal of particles is independent of S. Hence, for larger values of S, the same number of particles are removed from a larger total number of particles.

With everything else kept unchanged, $\bar{w}$ decreases with L because smaller particles are collected with a lower efficiency and as L increases the proportion of smaller particles in the dust, remaining in the gas, increases.

The dependence of $\bar{w}$ on v and S can be explained by writing

$$w\,\bar{N} - \phi - -\mathcal{D}_{eddy}\,\frac{\partial N}{\partial y} \tag{11}$$

where $\mathcal{D}_{eddy}$ is the eddy diffusion coefficient of the gas and y is the position coordinate perpendicular to the direction of flow.

1. As the turbulence intensity increases with v, there follows that

a) $\mathcal{D}_{eddy}$ increases with v, and

b) $|\partial N/\partial y|$ decreases with increasing v everywhere.

As a result, ϕ and also $\bar{w}$ vary with v. It has been found that, at a fixed S, $\bar{w}$ increases with v at a slower than linear rate.

2. At a fixed v, a larger value of S implies a smaller $|\partial N/\partial y|$ everywhere, resulting in a decrease of ϕ and $\bar{w}$ with increasing S, in accordance with Equation (11).

A quantitative analysis of these relationships would require accurate efficiency data with monodisperse dust.

5. CONCLUSIONS

1. Fine particles carried by a turbulent gas stream are transported by eddy diffusion into adjoining spaces, containing stagnant gas, where they are captured by inertial impaction and Brownian diffusion mechanisms.

2. The principle outlined in (1) has been brought to fruition by placing a large number of sheet metal plates, spaced a few centimetres apart, standing on their edges in upright position on the bottom of a rectangular conduit, occupying its entire width but leaving an unobstructed passage for the gas flow in the upper part of the conduit.

3. Tests with a 2.5 μm mean particle diameter dust have shown that the dust collection efficiency in this device, called Turbulent Precipitator (TP), can be represented by the Deutsch-Anderson equation.

4. At a fixed gas velocity v in the passage, the collection efficiency increases with decreasing passage gap height S. The effective migration velocity $\bar{w}_{TP}$ is a very weak decreasing function of S.

5. At a fixed value of passage gap height S, the collection efficiency has been found to increase with decreasing gas velocity v. The effective migration velocity $\bar{w}_{TP}$ varies in the same sense as v, but at a slower rate. As a result, $\bar{w}_{TP}/v$ increases with decreasing v.

6. With everything else kept unchanged, the collection efficiency increases with the active length L of the unit. For a polydisperse dust $\bar{w}_{TP}$ decreases with L, because the fractional collectional efficiency of smaller particles is less.

7. The effective migration velocity $\bar{w}_{TP}$ has been found to be greater by an order of magnitude than the published values of the effective migration velocity on an ESP, $\bar{w}_{ESP}$, for the same particle size.

8. The collection efficiency of the TP can, in principle, be increased indefinitely by increasing the active length of the unit L, decreasing the passage gap height S and the passage gas velocity v. In the test unit the limits imposed by practical considerations on the commercially permissible values of L and v have not even been approached.

9. The TP can be operated continuously, without cleaning, at a constant rate, a constant pressure drop and a steady collection efficiency.

10. The TP has a very simple construction and contains no moving parts.

11. The TP is suited for use at high temperatures because it can be constructed entirely of metal.

12. The TP can also be used as a wet, nozzle-type scrubber for the removal of both particulates and noxious gases.

ACKNOWLEDGEMENT

The authors wish to acknowledge support of an operating grant from NSERC of Canada.

REFERENCES

Anderson, E. (1919). Report, Western Precipitator Co., Los Angeles, California. See Trans. Amer. Int. Chem. Eng. 16, 69.

Batel, W. (1972). Entstaubungstechnik. Springer, Berlin – Heidelberg – New York.

Deutsch, W. (1922). Bewegung und Ladung der Elektrizitätsträger im Zylinder kondensatos. Ann. Phys. (Leipzig) [4] 68, 335-344.

Douglas, P.L., Dullien, F.A.L. & Spink, D.R. (1982). On the deposition of aerosols – A review and recent theoretical and experimental results. In Waste Treatment and Utilization, ed. M. Moo-Young & G.J. Farquhar. Pergamon Press, Oxford and New York. pp. 391-408.

Douglas, P.L., Dullien, F.A.L. & Spink, D.R. (1976). An investigation of the operating parameters of a low energy wet scrubber for fine particulates. Can. J. Chem. Eng. 54, 173-176.

Dullien, F.A.L. (1977). The role played by eddy diffusion in the aerodynamic capture of particles. In Proceedings of Particle Technology Nuremberg European Congress, "Transfer Processes in Particle Systems", ed. H. Brauer and O. Molerus. pp. 30-55.

Dullien, F.A.L. (1989). Introduction to Industrial Gas Cleaning. Academic Press, Inc.

Dullien, F.A.L. & Collins, A. (1992). Utilization of turbulence for separation of fine particles from gas streams. In Preprints of Particle Technology "2. European Symposium: Separation of Particles from Gases", Nuremberg, Germany, ed. F. Löffler, pp. 85-96.

Dullien, F.A.. & Munro, T.T. (1973). Fractional mass efficiency measurements on a wet dust scrubber. Powder Technology, 8, 57-68.

Dullien, F.A.L. & Spink, D.R. (1980). The Waterloo scrubber part I: the aerodynamic capture of particles. In Waste Treatment and Utilization, ed. M. Moo-Young & G.J. Farquhar. Pergamon Press, Oxford and New York. pp. 469-486.

Flagan, R.C. & Seinfeld, J.H. (1988). Fundamentals of Air Pollution Engineering, p. 414. Prentice Hall, New Jersey.

Friedlander, S.K. (1977). Smoke, Dust and Haze. John Wiley & Sons.

Friedlander, S.K. & Johnstone, H.F. (1957). Deposition of suspended particles from turbulent gas streams. Ind. Eng. Chem. 49, 1151-1155.

Riehle, C. & Löffler, F. (1992). A revision of the Deutsch-model in electrical precipitators based on grade efficiency measurements, similarity laws and cinematographic studies of particle transport. In Preprints of Particle Technology "2. European Symposium: Separation of Particle from Gases", Nuremberg, Germany, ed. F. Löffler, pp. 5-19.

ELECTROSTATIC SEPARATION OF FINE PARTICLES AT HIGH TEMPERATURE

HIDEO YAMAMOTO

Dept. of Bioengineering, Soka University
Tangi-cho, Hachioji-shi, Tokyo 192, JAPAN

ABSTRACT

The experiments of electrostatic precipitation at high temperature above 500°C were demonstrated for the separation of the ultra-fine particles synthesized by the thermally activated CVD (chemical vapor deposition) method directly from the reaction gas stream. Silicon-nitride particles were prepared from a $(SiCl_4 + NH_3)/N_2$ gas system at 1200°C. This process produces NH_4Cl as by-product which is solid particles at normal temperature. Using the high temperature electrostatic precipitation technique, it si expected to collect only the Si_3N_4 particles directly from the reaction gas stream because the sublimation temperature of NH_4Cl is about 340°C. The new electrode system for corona discharge was developed, which can generate a stable corona discharge even in the reaction gas atmosphere at high temperature. The X-ray diffraction analysis shows that the particles collected at 500°C contain no NH_4Cl particles.

The new electrostatic precipitation system was developed. The device has the particles discharging zone and the collection zone, separately. The boxer charger was applied for charging particles. It stably worked at high temperature above 700°C. Silicon-nitride particles prepared by the CVD method were collected on the collection plate with 100%. This technique is expected to be useful for hot gas cleaning.

INTRODUCTION

Ultra fine (under submicron size) particles prepared by the PVD (physical vapor deposition) or CVD (chemical vapor deposition) methods have generated considerable interest recently as advanced industrial materials because of their superior chemical and physical properties. For the present, however, although they are extensively studied in the laboratories as materials for research, there are not so many applications to industry probably because of their high cost and handling difficulties in such as separation, dispersion, classification, mixing etc. In this research, we aimed to develop a new handling technique to separate ultra fine particles prepared by the CVD method (hereafter referred to as CVD ultrafine particles) from reaction gas stream under high temperature conditions.

For example, the reaction system in gaseous process of metal chloride and ammonia is one of useful methods to produce ultra fine particles of metal nitrides such as Si_3N_4, BN, AlN, ZrN, TiN, VN etc. This reaction being carried under an ammonia gas excess condition, ammonium chloride is generated as the by-product, for example, following the reaction:

$$3SiCl_4 + 16NH_3 \longrightarrow Si_3N_4 + 12\underline{NH_4Cl}$$

Ammonium chloride being solid particles at room temperature, separation of the product particles under room temperature conditions by using such as the filtration collects a mixture of Si_3N_4 and NH_4Cl particles. Although the by-product of NH_4Cl can be removed from the product by heating in a non-oxidizing atmosphere above 340°C, the sublimation temperature of NH_4Cl, it may not achieve to purify the product of high purity. Separating the product directly from the reaction gas stream under high temperature condition above 340°C, it is possible to collect only the product of high purity. Although some kinds of fibrous filter can be used at high temperature, there are troubles such as gas pressure drop rising due to clogging, and after-handling difficulties such as separation of the particles collected in fibrous layers. If particles can be charged effectively at high temperature, electrostatic separation technique is available to collect the product from such a reaction system because of easily controlling electrostatic field not depend upon temperature.

In this report, high temperature electrostatic separation technique is determined to separate the Si_3N_4 product directly from a reaction gas stream, in case of being prepared from the $SiCl_4/NH_3$ gas reaction system by the thermally activated CVD method.

EXPERIMENT (1)

Experimental procedure

The experimental device is constituted of a reactor and an electrostatic separator heated at 1200° and 500°C, respectively, which are connected by a quartz tube kept above 500°C by a heat insulator. The reactor comprises three coaxial quartz tubes. $SiCl_4$ carried with nitrogen gas, and NH_3 gas were introduced into the inner and the middle tubes, separately, and they are flew together into the reaction zone of the middle tube to produce Si_3N_4 particles. Flowing out from the middle tube, the prepared particles sheathed with nitrogen gas from the outer tube flew toward the separation zone, like a beam flow with diameter of a few millimeters. This aimed to avoid the particles deposition on the tube wall and to transport them closely to the discharging electrode.

The most important factor of electrostatic separation is to generate stable corona discharge. The characteristic of DC-corona discharge extremely depends upon temperature, gas pressure and gas components. When a DC-corona discharging system using a wire-cylinder electrode was used under the reaction gas atmosphere at 500°C, spark discharge directly occurred and there was no stable corona discharge range.

Then a new electrode system using a high frequency surface corona discharge that can stably occur even at high temperature and in complex gas compositions was developed as shown in **Figure 1**. It consists of three electrodes, a cylindrical exciting electrode(A), a spiral discharging electrode(B) and a cylindrical collection electrode(C). The electrodes (A) and (B) are insulated each other with a thin quartz tube. When a high frequency excit-

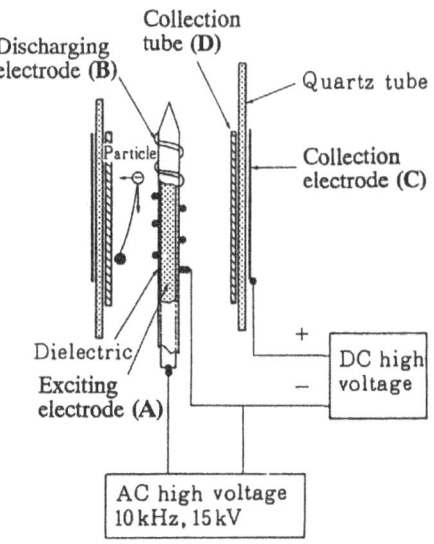

Figure 1. The new discharge electrode system using a high frequency surface corona discharge

ing voltage is applied between electrodes (A) and (B), a surface corona discharge occurs to form a plasma ion source along the electrode assembly. A DC–high voltage is applied between the electrodes (B) and (C) so that a DC charging field is produced between them. As a result, positive or negative uni–polar ions are emitted from the plasma source along (B), and travel across the charging zone to be collected finally by the electrode (C). There- fore, particles passing through the charging zone are charged uni–polar and travel across the charging zone to be collected on the surface of collection quartz tube(D). A high fre- quency voltage of 10–20 kVpp and frequency of 10 kHz was applied between the elec- trodes (A) and (B), and a DC–voltage of 0.2–1.0 kV, was applied between the electrodes (B) and (C).

Experimental Results and Discussion
Figures 2 and 3 are views of particles collected on the surface of the collection tubes directly from reaction gas stream at 500°C using the electrode system shown in Figure 1. The high frequency voltage applied between the exciting and the discharging electrodes was 10 kV, and the DC–voltage applied to the charging zone was 1kV. The particles were prepared under the reaction conditions of $SiCl_4$ concentration =1.4%, ratio of NH_3 to $SiCl_4$ =6, and reaction temperature =1200°C. The quantity of the particles deposited were uni- formly in the circumferential direction of the tubes and were distributed in flow direction. As no particle deposition on the fifth tube wall from the left side being the upper stream in Figure 3, it is satisfactory to consider that particles could be collected completely on this

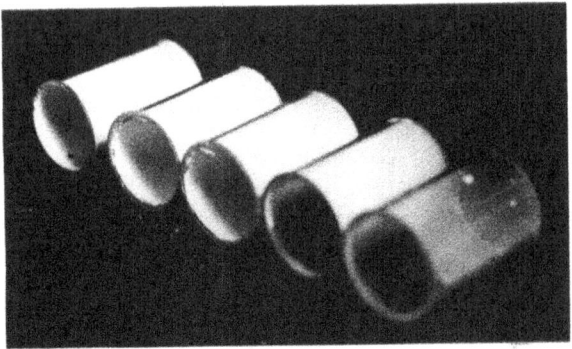

Figure 2. View of the particles collected on the inner surface of tubes

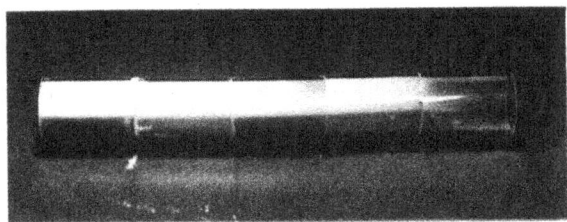

Figure 3. View of the particles collected on the inner surface of tubes
(left side is the upper stream)

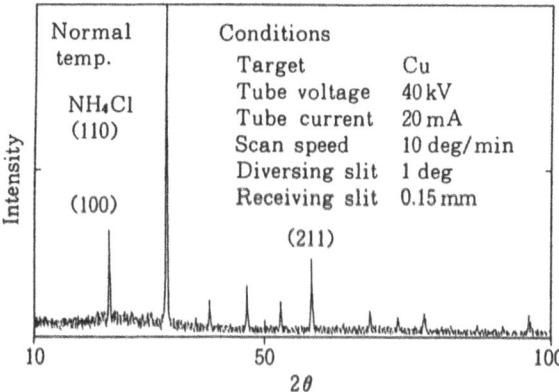

Figure 4. X–ray diffraction patterns of the particles collected
at normal temperature

side. As amount of the particles collected was approximately similar to theoretical yield of Si_3N_4 particles calculated from the inlet gas concentrations, it is supposed that only Si_3N_4 particles as the product were collected, without NH_4Cl particles.

Figures 4 and 5 show the X-ray diffraction pattern of the particles collected at room temperature and that of the particles collected at 500°C, respectively. In Figure 4, the diffraction peaks of NH_4Cl appear. On the other hand, the X-ray diffraction pattern in Figure 5 indicates that the particles collected at 500°C contains no NH_4Cl particle and are amorphous silicon nitride. Dropping $0.1N-AgNO_3$ into the water suspension of the particles collected at 500°C, no precipitate of AgCl was observed. These results indicate that the high temperature electrostatic separation technique using the new electrode system is

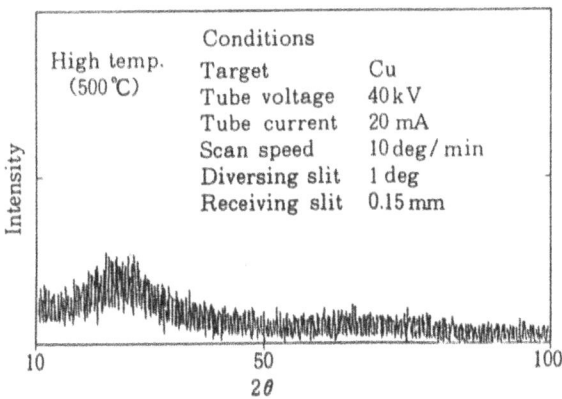

Figure 5. X-ray diffraction patterns of the particles collected at high temperature of 500°C

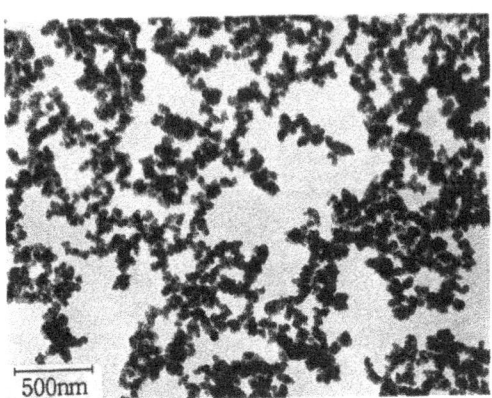

Figure 6. TEM photograph of the particles of Si3N4 prepared

61

available to collect the product particles without the by–product, directly from the reaction gas system.

Figure 6 shows a TEM photograph of the particles collected on the tube wall, and **Figure 7** shows the size distribution of them measured by the photo extinction centrifugal sedimentation method. Observed in Figure 6, the particles are uniform diameter of several tens nm, and the weight median diameter is about 70 nm according to Figure 7. Using these particles, the relationship between the position of particles deposition on the tubes and the intensity of the DC–field applied between the electrodes (A) and (B) was investigated. **Table 1** represents the results. When the applied DC–voltage is higher, the particles were collected at nearer position from the inlet of the collection zone because the mobility of particles was larger. This result indicates that an electrostatic behavior of fine particles can be well controlled by a DC–voltage even at high temperature. In this experiment, as shown in Figure 1, the zone of electrophoresis combines with the particle charging zone. The charge quantity of particles is larger when the DC–field intensity is high, because the ion current density is high as shown in Table 1. Therefore, in this system, it is difficult to estimate a position of particle deposition. To quantitatively investigate the characteristics of particle collection, it is necessary to improve the device to have a particle charging zone and an electrophoresis zone.

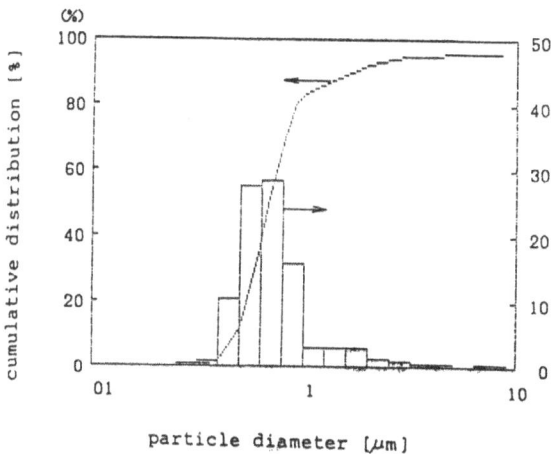

Figure 7. Size distribution of the particles of Figure 6 measured with the centrifugal sedimentation method

Table 1

Weight distribution of particles collected on the collection tubes

		Experiments			
		No.1	No.2	No.3	No.4
DC voltage supplied to electrodes [V]		1000	600	380	200
Current density of ion [µA]		12.4	8.1	4.7	3.2
Tube number	Distance from inlet [cm]	Weight of particles collected on the tube [mg]			
(1)	0 – 4	31.6	25.6	9.1	3.1
(2)	4 – 8	29.1	21.8	22.3	12.3
(3)	8 – 12	6.2	13.7	19.5	11.0
(4)	12 – 16	0	2.8	10.2	7.3
(5)	16 – 20	0	0.2	3.9	6.0
(6)	20 – 24	0	0	0.1	5.5
(7)	24 – 28	0	0	0	8.3
(8)	28 – 32	0	0	0	9.6
Total [mg] (Collection efficiency)		66.9 (98.1%)	64.1 (94.0%)	65.1 (95.5%)	63.1 (92.5%)

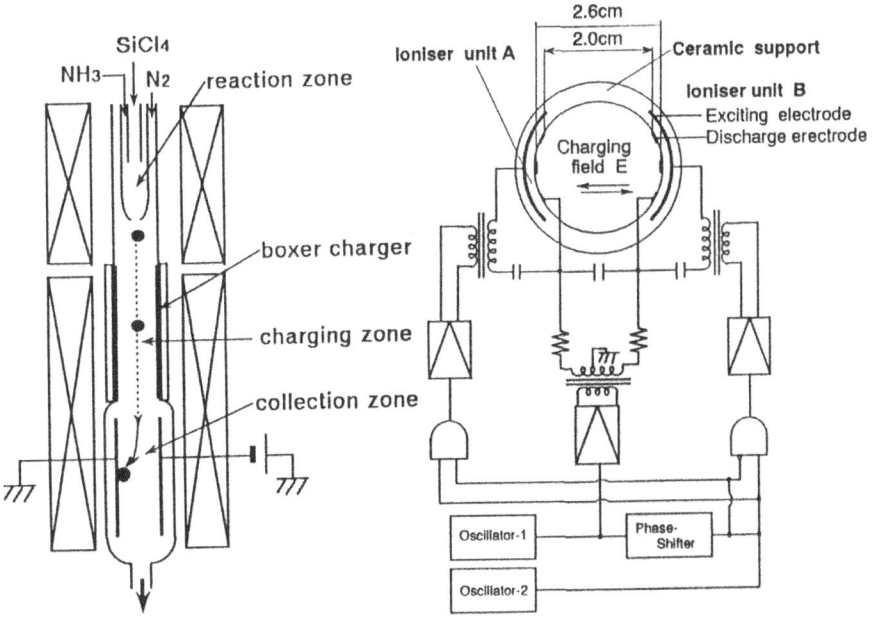

Figure 8. Experimental apparatus (2)   **Figure 9.** Circuit of the boxer charger

EXPERIMENT (2)

Experimental procedure

To separate a charging zone and an electrophoresis zone, the experimental apparatus as shown in **Figure 8** was constructed. It consists of a reactor, a precharger and a collector. A boxer charger developed by S. Masuda which is a high speed charger (Masuda *et al.*, 1978, 1980, 1984) was applied for a precharger. **Figure 9** represents schematically the construction and circuit of a boxer charger used in the experiment(2). Each ionizer unit, A and B, consists of a discharge and exciting electrodes. When a high frequency, high voltage (10 kHz, 10 kVpp) is applied between the discharge and exciting electrodes, high frequency surface corona discharge occurs along the discharge electrode so that a plane plasma appears over the surface of the discharge electrode to form an ion source. This is referred as the EXCITING voltage. A low frequency square wave voltage (500 Hz, 5–10 kVpp) is applied between two discharge electrodes to form a charging field. This voltage is referred as the MAIN voltage. Synchronous to the MAIN voltage, the EXCITING voltage is applied alternately to two ionizer units. As a result, a plane plasma is produced alternately on the surface of the ionizer units, switching between the opposite ionizer unit with the MAIN frequency. Depending on the polarity of the MAIN voltage on the ionizer, either positive or negative ions are forced to travel across the charging zone. As a result of this procedure, particles passing through the charging zone are bombarded alternately from two sides by uni–polar ions and charged to uni–polar, positive or negative.

The boxer charger is considered as a potential option for pre–charging even at high temperature condition, because the electrode assembly is made from ceramics of alumina and a high frequency surface corona discharge stably occurs even at high temperature and in complex gas compositions.

The collector in Figure 8 consists of two plate electrodes to form a DC–field. Particles charged by the boxer charger are collected on the positive or negative electrode plate due to their polarity.

Results and Discussion

Ultra–fine particles of silicon nitride were prepared in the CVD reactor. They were similar to the particles shown in Figure 6. The boxer charger and collector were kept at 700°C by the furnace. They worked well stably. Particles were almost completely collected on the electrode in the collector. When the boxer charger produced negative ion source, particles were deposited on the positive electrode.

Further investigations are being performed to measure a charge quantity of particles passing through the boxer charger, to determine a position of particle deposition on the collection electrode according to the DC–field intensity in the collector, and to apply the electrostatic separation system for submicron dust collection at high temperature.

CONCLUSION

High temperature electrostatic separation technique was investigated for separating ultra fine particles of silicon nitride prepared by the thermally activated CVD method using the $NH_3/SiCl_4$ gas reaction system. The results obtained lead to the following conclusions:

1) DC-corona discharge can not be used to charge particles at high temperature and in complex gas components of the reaction system, because its characteristic is extremely influenced by conditions of temperature and gas compositions.

2) A new discharging electrode system using a high frequency surface corona discharge was developed. Using the electrode system, the ultra fine particles of silicon nitride could be collected directly from the reaction gas stream at a high temperature of 500°C. The particles collected were under submicron size and contained no particle of NH_4Cl by-product.

3) A new electrostatic precipitation system was developed. The system has a particle charging zone and a collection zone, separately. The boxer charger was applied for pre-charging particles at high temperature. It well worked at 700°C. CVD ultrafine particles of silicon nitride were collected on the collection electrode with an efficiency of approximately 100% according to their polarity. The system is expected to be applied for hot gas cleaning.

REFERENCES

Masuda, S., Washizu, M., Mizuno, A. & Akutsu, K.(1978). Boxer
Charger – a novel charging device for high resistivity powders.
Conf. Rec. IEEE/IAS Ann. Meeting Vol. 1B pp.16–22

Masuda, S.(1980). Device for electrically charging particles.
US Patent 4,210,949

Masuda, S.(1984). State of the art in precharging. Proc. 2nd. Int.
Conf. on Electrostatic Precipitation Kyoto (JAPAN) pp.177–193

HOT GAS CLEANUP FOR ADVANCED POWER GENERATION SYSTEM

JAY M. QUIMBY
Manager-Eastern R&D Labs
and
K. SAM KUMAR
Manager-Applications Development
Research-Cottrell Companies, P.O. Box 1500,
Somerville, NJ   08876

BACKGROUND

About 800 million tons per year of coal is consumed each year in the United States.   Electric generating utilities account for approximately 680 million tons (85%) of coal use leaving 15%, or 120 million tons, available for other applications.   Considering that average operating thermal efficiency of the coal fired powerplants in the U.S. is approximately 33%, an increase of this efficiency towards 50% by the use of modern high efficiency systems can reduce coal consumption by 250 million tons per year while maintaining the same electric power output. Emissions of CO_2, SO_2, organics and trace metals would be reduced in proportion, alleviating the reduction control requirements in downstream pollution control systems.

Pressurized fluidized bed combustion (PFBC) with an estimated efficiency of 42%, advanced PFBC including a topping cycle with estimated efficiencies of 49%, and direct coal fired turbines (48 to 52% efficiency) are among three of the advanced power generation technologies under development in the United States, primarily funded by the U.S. Department of Energy.   Costs of these systems are expected to be 15 - 20% lower than conventional P-C fired units.

In all these processes particulate laden hot gases at high pressure and high temperature must be cleaned, commensurate with turbine tolerance requirements. For PFBC system hot gas cleanup must be accomplished ahead of the turbine at 700 to 850 deg C at pressures between 7 to 15 atmospheres, depending on the specific type of PFBC design. Gas cleanup in PFBC exhaust takes place after combustion in oxygen rich conditions. Gas cleanup in the topping cycle scheme takes place at lower temperatures, typically in the 500 to 700 deg C range; but the gases are now in fuel rich condition and may contain some organic condensibles. In

DCFT systems operating temperature of gas cleanup systems have to be in the 1000 deg C to 1400 deg C range.

Simultaneous high temperature and high pressure conditions present serious challenges to the commercialization of gas cleanup technologies because of a) the limited choices available in the selection of materials that can provide protection from high temperature erosion and corrosion, b) the lack of adequate design data to allow for thermal embrittlement of internals and resistance to thermal shock, and c) the potential long term loss of strength due to high temperature and cyclic operation.

Several hot gas cleanup devices have been under development to perform this particulate collection under very hostile conditions. Future commercial success of high efficiency power generating systems will ultimately depend on the performance and reliability of these devices, let alone the issue of competitive economics.

Particulate Collection At High Temperature And High Pressure

Many different types of particulate collection devices are being developed for applications in advanced power generation systems. Some of the more prominent ones are:
- high performance mechanical collectors
- high efficiency barrier filtration filters
  - bag filters
  - ceramic candle filters
  - ceramic cross flow filters
- electrostatic precipitators
- sonic agglomerators ahead of mechanical collectors
- electrostatic agglomerators ahead of mechanical collectors

Mechanical Collectors

Mechanical collectors such as cyclones utilize the inertial separating force on the particles resulting from high velocity axial or tangential entry of flue gas into the collector vessel. The inertial force drives the particles to the collector walls, and its magnitude is dependent on the particle size. The commercial practice of high temperature high pressure cyclones has been known since the early 60's, especially in the collection of catalyst dust ahead of single stage gas expanders in Fluid Catalyst Cracker applications. The cyclones are considered to be the most reliable and least expensive of all devices considered in this paper, although they are the lowest performing of all for the collection of fine particulates.

The collection efficiency of cyclones suffers for size ranges below 5 to 10 microns because of the rapid decrease of inertial forces. Experience with PFBC applications suggests that it is difficult to limit the overall particulate loadings below 100 to 200 ppm ahead of the gas turbine at practical tangential velocities around 70 to 90 feet/second. Both the operating pressure drop and materials erosion become objectionable beyond these velocities. The tolerance requirements for DCFT's may likely turn out to be more stringent. It is seen that while the turbine tolerance goals can be met with cyclones it cannot meet the environmental protection requirement levels of 20 to 30 ppm. Additional particulate control is therefore necessary to meet the environmental requirements. The PFBC system installed on American Electric Power System's 77 MWe Tidd plant in Ohio utilizes a downstream electrostatic precipitator for meeting the particulate control requirements, while relying on HPHT cyclones only for turbine protection.

The development of high efficiency HPHT particulate control technologies is predicated on the potential gains in turbine life that accompanies the order of magnitude reduction of turbine inlet loadings. The environmental protection goals can now be achieved without a dedicated downstream control system. In the following sections the performance of several high efficiency particulate control devices are discussed.

Barrier Filtration

Barrier filtration devices are different from mechanical collectors in that unlike the cyclones, there is a mechanical medium between the dirty and clean gas streams. Fabric filters, candle filters and cross flow filters are among the most widely studied of the barrier filtration devices.

Particulate collection in the case of fabric filtration occurs initially by interception and impaction on the bag fibers which act as the barrier. Subsequent filtration occurs on the dustcake itself which typically has smaller interstitial pores than the bag weave.

Candle filters and cross flow filters are built rigid with very narrow pore sizes (5 to 100 microns), and these rely less on the formation of dustcake and more on the barrier itself for collection. Pressure drop across rigid candle or cross flow filters is generally higher than that of fabric filters: 0.8 to 8 kPa for candle filters versus 1.6 to 2.5 kPa at ambient conditions. Operating pressure drop will be in direct proportion to operating gas density.

Both the pressure drop and filter element life depend on the cleaning technique employed. Adequate cleaning of filters is essential to restore pressure drop balance across the system. As the cake layer builds up on the filters, the pressure drop across the bag will increase and at the point of peak acceptable pressure drop some type of cleaning is necessary. Conventional techniques such as reverse gas or shake deflate cleaning are not suitable because of the need for isolation of filters in compartments at high temperature and high pressure. Reverse gas flow cleaning requires complex piping and is not practical for HPHT applications. On-line pulsed jet cleaning is the only practical technique available. Pulse jet cleaning has been the only cleaning technique employed in all the test programs on barrier filtration designs.

In pulse jet filtration, the dirty flue gas enters from outside of the bags. High pressure compressed air is used in short burst of pulses from the clean, inside of bags or candle tubes to dislodge the cake for subsequent gravity transport down into the hopper. Dusts with strong cohesive properties will be more difficult to remove, and tend to require higher pulse jet pressure and increased frequency of cleaning. The wear and tear on the bags, or tubes in the case of candle filters is likely to be influenced by the cleaning cycle and intensity.

In the United States Ciliberti and Lippert of Westinghouse have conducted some of the most comprehensive testing on fabrics, crossflow filters and candle filters (Ciliberti and Lippert). PFBC ash has shown ideal cohesive properties for optimal collection in FF's and candle filters. At air to cloth ratios of about 6:1 ft/min for fabrics and up to 10:1 for candles and cross flow filters, all of these were able to show near absolute collection, leading to particulate outlet loadings in the 5 to 10 ppm. Operating temperatures typically varied between 700 and 850 deg C (1292 and 1562 deg F) for the PFBC application.

Highly carboneous ash from coal gasification processes behaved very differently. Ash from coal gasifiers were characteristically uncohesive, and fabric filter efficiencies at air to cloth ratio of 3.6 ft/min were limited to a range between 92 to 98% both due to unstable dustcake and particle bleedthrough (Dellefield and Bedick). Candle filters with finer pore sizes were able to function at the 99% levels, although at the cost of higher pressure drop than the fabric filter as noted earlier.

Conventional fabrics such as fiberglass, ryton, etc. utilized in commercial baghouse installations are limited to a temperature of 260 deg C (500 deg F). Fabric materials utilized for PFBC applications are made of ceramic woven fabrics to resist flexural stresses at high temperatures. 3 M's Nextel AB 312 and Carborundum's Fibresil fabrics were two of the fabrics tested by Ciliberti and Lippert.

In contrast to the limited number of fabrics that were tested, candle filters were made with a wide variety of designs as shown in Table I. Some of the key design features of candle filters were a) candles with pore sizes ranging between 5 and 100 microns, and b) cleaning flexibility through intertwining of ceramic fibers in the rigid candle elements.

A more recent development in filters is the ceramic tube design advanced by Asahi Glass Company of Japan. A schematic of the ceramic tube filtration concept is shown in Figure 1. The material of construction of these tubes is β Cordierite, a crystalline mixture of magnesium, aluminum and silicon oxides. While individual elements come in approximately 6 foot lengths, these can be joined together through complicated mating arrangements to provide lengths up to 18 feet long, approaching commercial bag lengths in operation today. In the operation of Asahi ceramic tubes, the flue gas enters from the inside of hollow tubes, in contrast to fabric filter and ceramic candle filter designs. The performance of ceramic tube filters is expected to be similar to that of candle filters. Additional data on materials durability issues are expected from the EPRI sponsored evaluation program at Ahlstrom's HTHP facility located in Finland (Sellakumar, Isaksson, and Provol).

The development of ceramic cross flow filters as shown in Figure 2 has largely been pioneered by Westinghouse Electric. The construction of these elements is significantly different from candle or ceramic tube filter elements. The materials of construction are typically alumina and sometimes in combination with silica.

Because of the arrangement of the porous parallel plates, gas is forced upwards through the porous layer where filtration takes place. Clean gas moves upward and out in a direction perpendicular to the entering dirty flow. Pulsed jet cleaning is used to expel the dust catch. Several cross flow filter elements can be mounted together on a single headersheet inside a pressure vessel.

Ceramic cross filters have consistently shown high collection efficiency, limiting outlet loading below 10 ppm.

Materials Problems With Barrier Filters

While the barrier filtration devices have shown high collection efficiencies during short term tests, they have also exhibited a consistent pattern of component failures that is disturbing. Some of the reported failures are described below.

a)   Fabric Filtration

Failure of Néxtel and Fibresil bags due to pulsed jet cleaning at 800°C (1472 deg F) has been reported by Ciliberti and Lippert. Mechanical degradation of ceramic threads that were used to stitch the bag seams resulted in failure. Fabric degradation when exposed to high temperatures has also been reported by Chang (Chang, Sawyer, Lips, Bedick, and Dellefield). Fabric had exhibited mechanical degradation that was thought to be time and temperature dependent. Frictional abrasion and

70

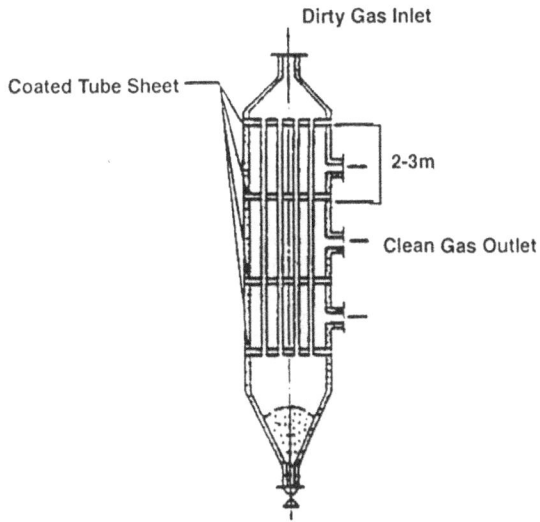

Figure 1: Schematic of the Asahi advanced ceramic tube filters

Dirty Gas Inlet

Coated Tube Sheet

2-3m

Clean Gas Outlet

Figure 2: Schematic of the ceramic cross-flow filter

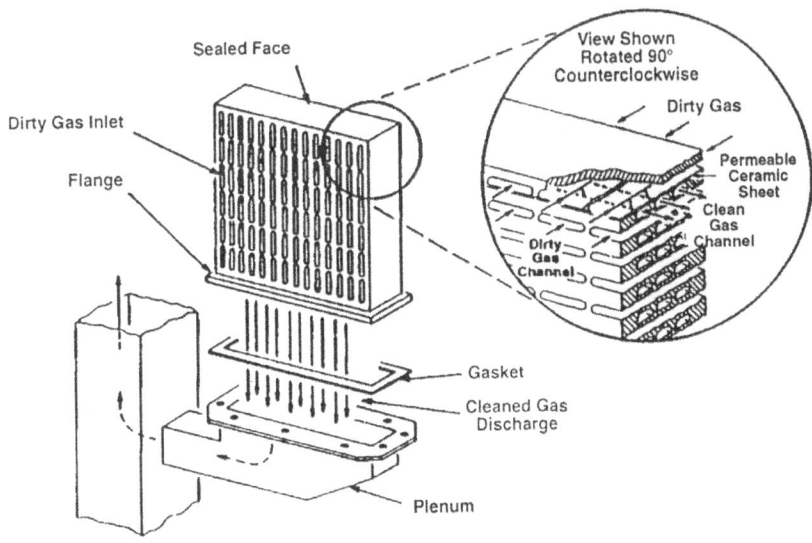

View Shown Rotated 90° Counterclockwise

Sealed Face

Dirty Gas Inlet

Flange

Dirty Gas

Permeable Ceramic Sheet

Clean Gas Channel

Dirty Gas Channel

Gasket

Cleaned Gas Discharge

Plenum

reaction with constituents of flue gas have also been suggested as reasons for the loss of strength.

To date there have been no reports of long term testing at the 800-900 deg C (1472-1651 deg F) levels concluding the viability of fabric filtration at PFBC and DCFT operating environments.

Additional concerns are the ability to clean the fabric over the long term, and the impact of cleaning pulses of compressed air on bag life and pressure drop.

b)   Candle Filters

While the high collection efficiency of candle filters has been consistently reported during short term tests, there have not been many long term test programs with candle filters on PFBC or DCFT flue gas derived from coal firing.

The testing at British Coal's Grimethorpe (Stringer and Clark) facility on candle filters, partially funded by EPRI, represents the longest cumulative trial reported in literature.   A cumulative operation of 800 hours was undertaken.

The operating data reveals excellent collection efficiency initially, with dust penetrations of less than 10 ppmw at a face velocity of 0.07 m/s. Operating pressure drop stabilized around 23.3 kPa.

While the collection efficiency of candle filters was very promising, internal inspection of pressure vessel revealed many failures. It was found that the headersheet supporting 130 candles may not hold up in much longer service because of high temperature effects of thermal expansion and material creep.

Distortion of headersheet caused the misalignment of candles.   Pulsed jet cleaning physically displaced and dropped the candles causing breakage at the candle's supporting collars.   Mechanical degradation of candles occurred throughout the filter elements.   Microcracks identified at the binder medium and silicon carbide granular surfaces suggest thermal shock effects resulting from cold air impingement during pulsed jet cleaning cycles.

c)   Ceramic Cross Flow Filters

The short term results of ceramic cross flow filters are similar to that of candle filters. Dust penetration during short term tests never exceeded 10 ppm. However, when testing of 15 filter elements was conducted at New York University's PFBC facility, failure of filters resulted in outlet loadings exceeding 100 ppm after only 11 hours of operation.

Failures of dust seals installed between filter body and mounting flange were numerous. Five of the 15 elements showed delamination of the filter body - a consistent occurrence in many short term test programs.

Results of other test programs on gasifiers revealed crushed corners and longitudinal cracks on cross flow filter bodies. Because the design of cross flow filter elements involve many sharp corners, stress concentration is considered to be especially severe at these places, making them vulnerable to localized failures.

Additional programs are underway to develop improved ceramic materials, improved mounting design and more dedicated quality control procedures.

Electrostatic Precipitators

Electrostatic precipitators have been used for particulate collection in coal and oil fired power plant applications for more than 50 years. Current applications have been limited at atmospheric pressure, at temperatures between 121 to 427 deg C (250 to 800 deg F). During World War II ESP applications at pressures up to 2 atmospheres were developed for the recovery of catalyst dust in oil refineries at temperatures up to 427 deg C (800 deg F). These were large gas volume applications requiring large pressure vessels built to code standards. In most of the large volume applications (those exceeding 50,000 acfm) collecting electrodes are of the parallel plate type with round wire or rigid discharge electrodes hanging vertically from the center. Most of the smaller gas volume applications utilize circular or hexagonal tubes as collecting electrodes.

The development of ESP applications for higher pressures and higher temperatures began in the early 60's, with the purpose of protecting turbine from erosion created by particulates in coal fired exhaust. Since then several pilot plant projects have been completed. Table II summarizes the ESP experience in the United States at high temperature and high pressure conditions covering projects up to the late 80's (Taksicker). Most of the reported experience comes from work conducted by Research-Cottrell through government funded programs. Work was also conducted at Denver Research Institute in cooperation with Research-Cottrell under a DOE funded program as shown in Table II. HPHT ESP research has also .been conducted in Europe by the University of Essen under sponsorship by German Government in cooperation with the Lurgi Corporation (Riepe and Wiggens).

All the HPHT ESP testing to date in the U.S. have been short term test programs accompanied by frequent starts and stops. Testing was conducted at temperatures up to 1650 deg F. at pressures up to 10 atmospheres. Collection efficiency tests have shown a performance range between 90 and 99.6%. Tests in the lower collection efficiency range were generally associated with lower residence times in the ESP and higher gas velocities. Most ESP operating data has been with PFBC ash after at least two stages of cyclones. Mean particulate size has been typically 5 microns, with about 20% below 2 microns.

At operating temperatures in the 704 to 900 deg C (1300 to 1652 deg F) range the particulate resistivities have been shown to be in the 10^5 to 10^6 ohm.cm. There have been some concerns as to whether this low resistivity range presents serious limitations to ESP performance. Precipitator literature has many case histories of low operating resistivities in conventional ESP's that have shown two extremes of results: either excellent ESP performance or collection efficiency limitation due to excessive particulate reentrainment. Both tendencies are possible. The results would largely be determined by the cohesive nature of particulates. The associated ESP parameters such as operating current density and gas velocity influence the ESP performance if ash is of low resistivity and not cohesive.

Figure 3 shows the relationship between operating power and collection efficiency of ESP operated at DOE funded Curtiss-Wright PFBC facility in the early 1980's. Operating temperatures were in the 815 deg C (1500 deg F) range, at an operating pressure of 84 psia. Operating resistivity was 10^5 ohm.cm. Collection efficiency of ESP is seen to be higher at the higher operating power. Frequent rapping of collector tubes reduced the collection efficiency suggesting particulate reentrainment from rapping. Tests without rapping resulted in highest ESP collection efficiency. Collection efficiencies as high as 99.6% were obtained at operating gas velocities of about 3 feet/sec.

The ESP operating power densities in the Curtiss-Wright test program were about ten times higher than the conventional ESP levels. However, this does not reflect an order of magnitude increase in power consumption. For a given MWe output, higher gas density results in proportionally lower gas volume compared to ESP's in pulverized coal-fired applications. Consequently, there will be lower overall plate area requirements at HPHT conditions for a given precipitator size. The overall ESP power levels based on Curtiss-Wright data are thus estimated to be 2 to 3 times that of conventional applications per MWe output.

Operating data indicates that HPHT ESP operating mechanism is not significantly different from conventional applications. Electrical forces influencing particulate charging and collection will continue to be important and this brings the focus on the influence of temperature and pressure on operating field strengths. Figure 4 shows the relationship between relative gas density (relative to 21 deg C (70 deg F) and 1 atmosphere) of operation and operating field strength at three different operating temperatures of 704, 815, 900 deg C (1300, 1500 and 1652 deg F). The operating voltage or field strength increases

Figure 3: Collection efficiency as a function of corona for power various rapping levels

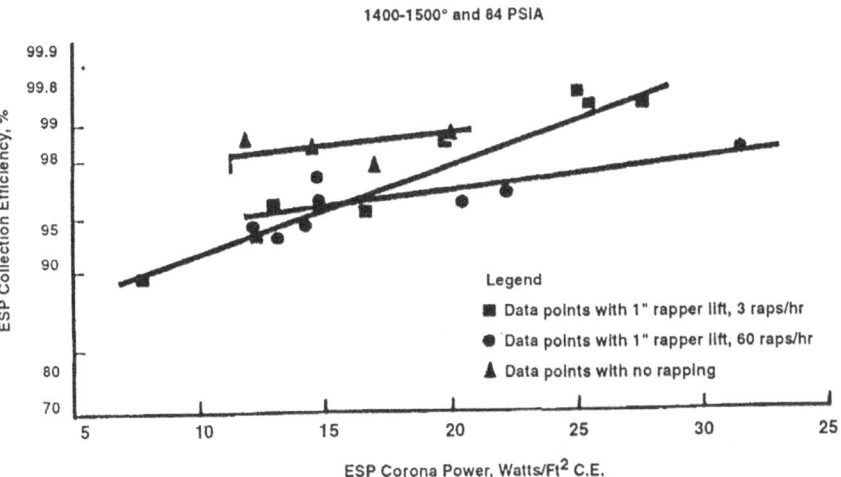

1400-1500° and 84 PSIA

Legend

■ Data points with 1" rapper lift, 3 raps/hr

● Data points with 1" rapper lift, 60 raps/hr

▲ Data points with no rapping

ESP Corona Power, Watts/Ft^2 C.E.

Figure 4: HTHP ESP operating field strength as a function of gas density

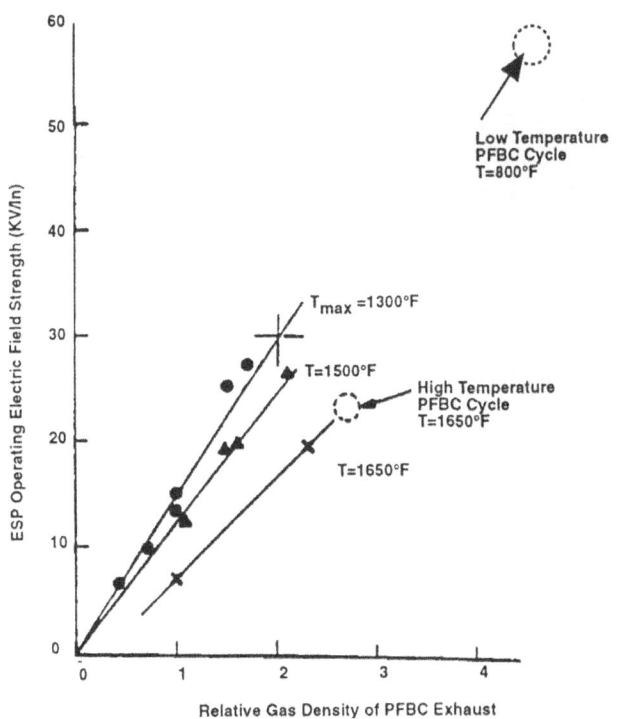

proportionally with gas density at all these temperatures. However, there is an independent temperature effect noted at temperatures above 704 deg C (1300 deg F). For a given gas density, the operating voltage level is seen to be lower at 900 deg C (1652 deg F) than at 704 deg C (1300 deg F).

During the 1970's tests were conducted at 900 to 982 deg C (1652 to 1800 deg F) at 10 atmospheres to evaluate this independent effect of temperature on lowering the applied voltage. It was found that sodium and potassium concentrations in flue gas influenced the electrical behavior of ESP, driving it towards higher current and lower voltage depending on the vapor phase concentrations. Tests at Denver Research Institute conducted in the 80's also confirmed this trend.

Will the corona discharge become unstable past a given temperature, whatever the operating gas density? Answer to this question will become important in determining the upper bounds of temperature at comparable gas densities with respect to corona stability and operability of the ESP. Unfortunately, ESP performance data on coal based applications at temperatures of 900 deg C (1652 deg F) and above are not yet available and no conclusions can be made about ESP's viability at these higher temperatures.

Materials Problems with HPHT Electrostatic Precipitator

The experience gained from pilot plants pinpointed several aspects of design that are critical for the success of HPHT precipitator operation.

1) Because ESP internals for PFBC application are made of steel and not ceramic, designs must consider high strength steels that will withstand the mechanical stresses at operating conditions. Poor choice of materials in early pilot programs have resulted in material warping and poor interelectrode clearances.

Relatively recent developments in nickel/chromium alloys have extended the range of applicability to higher temperatures: up to 982 deg C (1800 deg F) in oxidizing environments. Operating experience during the 80's have confirmed the viability of these materials during the short term tests. In some test programs such as at Curtiss-Wright materials were exposed for more than 700 hours at 815 to 871 deg C (1500 to 1600 deg F). The materials were also able to withstand serious impact stresses imposed by the collecting and discharge electrode rappers.

2) Allowance has to be made in the design for thermal expansion of materials exposed to high temperature. Earlier designs in the 60's did not include this effect. Thermal expansion of steel resulted in serious buckling of headersheet, resulting in deflection of collector tubes and attendant precipitator performance problems.

All recent projects have included this design concept, and ESP operating problems from thermal expansion have been practically eliminated.

3) Proper design of high voltage insulator assembly is critical for the success of ESP operation. Alumina insulators with at least 90% alumina content are recommended. Because these insulators can function adequately at exposure temperatures only up to 426 deg C (800 deg F), care must be taken to locate these insulators away from the gas stream and cooled to assure that temperatures are limited to safe operating levels.

While the insulators utilizing the above design principles generally operated well in the Curtiss-Wright project in 1981, repeated failure of insulators in the NYU project in 1984 prevented any meaningful testing of ESP for PFBC application (Kumar, Marx, and Feldman). Curtiss-Wright design allowed for proper control of insulator temperature through start-up and shut down. In the NYU design however, the insulators were located directly above the gas flow. In spite of the inclusion of cooling coils around insulators, the free convection flow of hot gas overheated the insulators resulting in failure. In addition to overheated bushings, there was some evidence of thermal shock on the insulators due to the frequent rapid start-up and shut down of NYU PFBC facility during the test program.

Upon reverting to the Curtiss-Wright insulator design in a more recent HPHT project on electrostatic agglomeration we were able to operate the ESP without any high voltage insulator failures.

Future research needs in ESP technology development

ESP material component issues, at least covering the short term, have been successfully addressed for the PFBC application. It is now important to evaluate the performance capability of ESP at PFBC conditions. The DCFT application is even more severe because of its much higher temperatures in relation to PFBC's. Several questions still need to be addressed as follows:

- can the ESP's operate with high collection efficiency, say 99.5 to 99.9%, without the high temperature related corona stability problems for PFBC applications?
- is there a potential for long term electrical stability problems resulting from molten flyash, due to difficulties in cleaning the discharge and collecting electrodes?
- can the material strength of ESP internals be maintained adequately over the long term?
- what should be the ESP size to achieve the high particulate collection efficiencies for PFBC applications?
- what is the relationship between operating power and collection efficiency for various commercial HPHT conditions ahead of the gas turbine?

- are new concepts of electrical energization necessary for the control of corona power?
- what should be the material of construction of ESP internals for DCFT applications?
- will ceramic electrodes work for ESP application?

In comparison to the number of research programs on barrier filtration, the ongoing research on HPHT ESP is relatively sparse. Because of the continuing uncertainties in the long term performance viability of barrier filters for the 1650 deg F and higher temperatures it would be prudent to continue research on alternative particulate control concepts such as the HPHT ESP filtration.

Agglomeration Concepts

Agglomeration concepts are geared to improve the performance of high efficiency cyclones. By the application of external forces such as sonic or electric fields inter-particulate contact is established in the agglomerator (Koopmann and Reethof), (Kumar, Quimby, and Helfritch). Subsequent growth of fine particulates through inter-particulate contact is essential for the success of agglomeration concepts. For successful agglomeration, the particulates have to be sticky or molten enough for cohesive growth to take place. Further, the larger aggregates must be robust enough to withstand the mechanical forces experienced in a cyclone operating with tangential velocities in the 70 to 90 feet per second. Such velocities would be necessary for achieving high particulate control efficiencies to meet the environmental compliance limits discussed earlier.

Sonic Agglomeration

Tests conducted at Pennsylvania State University on redispersed PFBC flyash confirmed the feasibility of the sonic agglomeration concept. An air siren was used to generate high frequency sound waves at 800 to 900 cycles per second, and at amplitudes up to 157 dB. Particulate growth due to sonic agglomeration was significant at .1 to 10 micron sizes. However, no collection tests with cyclones were conducted to demonstrate enhanced cyclone performance. Additional DOE sponsored projects are underway for gathering performance data on this concept for PFBC and DCFT applications.

Electrostatic Agglomeration

In the concept of electrostatic agglomeration both particle collision and particle adhesion are utilized to the fullest extent possible. The electrostatic agglomerator (ESA) performs the following functions necessary for effective removal of particulate matter.

1. promotes inter-particle contact by particle charging and particle migration to a collection surface,

2. provides a collection surface for adhesion by electrical, capillary and molecular forces,

3. utilizes a smaller volume, as compared to an ESP, and a high velocity gas to scour the surface of the collection electrode so that agglomerates can be captured downstream by a cyclone.

A schematic of the experimental ESA is illustrated in Figure 5. The ESA consists of a particle concentration section followed downstream by an agglomeration section, each comprised of 4 grounded electrode tubes as illustrated in Figure 6. High voltage sections are located concentrically within the collection tubes. In the particle concentration section corona discharge is used to charge the particles. Particle charging by corona discharge provides maximum particle migration velocity so that particles reach the collection (ground) electrode surface. The design philosophy for particle capture is similar in principle to ESP design but the larger velocity through the gas passage enables the size of the ESA to be much smaller as compared to an ESP. The high gas velocity in the particle concentration section will scour the surface of the collection electrodes to cause particle reentrainment. Reentrainment of agglomerated particles reduces the number concentration of particles as compared to the inlet number concentration. Sufficient residence time (0.6 sec) is allowed for small particles to migrate to a collection surface. Particles of sub-micrometer diameter have the lowest migration velocities. Therefore, the length of the concentrator section must be sized to provide sufficient residence time for sub-micrometer particles to reach the collection electrode surface.

In the agglomerator section, downstream of the particle concentrator section, very high electric field strengths are maintained for charged particles. High field strengths are achieved without corona discharge. The purpose of the agglomerating section is.to promote inter-particle contact by changing the particle charging mechanism. Particle charging is carried out by induction. Particles of negative polarity will reach the positive electrode, reverse charge, and be accelerated towards the negative electrode. Particle charge reversal occurs with greater frequency in the agglomerator section as compared to the concentrator section. It is expected that particles in the 0.1 to 5 um size range will be preferentially retained on inner surfaces through adhesion by contact charging, capillary attraction, and Van der Waal's forces

79

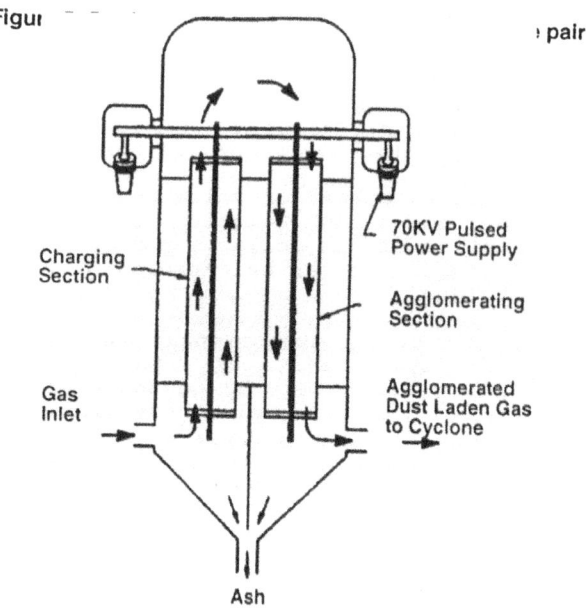

Figure [...] pair

Charging Section

70KV Pulsed Power Supply

Agglomerating Section

Gas Inlet

Agglomerated Dust Laden Gas to Cyclone

Ash

Figure 6: Arrangement of corona producing charging section and non-corona producing agglomeration section of an electrostatic

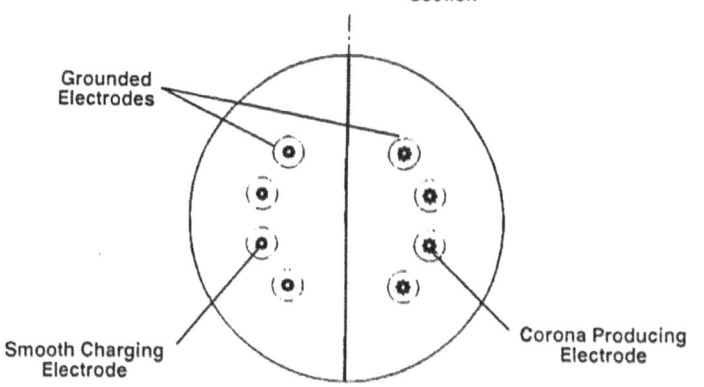

Agglomeration Section

Corona Charging Section

Grounded Electrodes

Smooth Charging Electrode

Corona Producing Electrode

until the drag and gravitational forces become large enough to reentrain agglomerates.

Performance of the ESA will be dependent on: 1. the adhesive forces holding particles together, 2. the charging electric field strength in the concentrator section, 3. the residence time of particles in the ESA, and 4. the physico-chemical characteristics of the particulate matter. A problem will develop if particles become too sticky to be reentrained by the gas. If the charging field strength is too low little particle charging will result. Also, excessive currents caused by materials of low ionization potential at 1371 deg C (2500 deg F) may cause thermal ionization. However, thermal ionization may be used to an advantage by employing pulsed ionization, where short voltage pulses (up to 700 pulses per second) of less than 10 nanoseconds duration can be applied without gas breakdown.

The experimental investigation of this concept at DCFT conditions was attempted in early 1992 by Research-Cottrell under a US DOE funded program. The program was to evaluate the concept of electrostatic agglomeration of fine particulates at temperatures exceeding 1093 deg C (2000 deg F) by using a tubular agglomerating surface made of silicon carbide. Silocon carbide was chosen to withstand the high operating temperature of DCFT system. In addition, the electrical performance of pulsed method of energization was to be evaluated in contrast to the direct d.c. method of conventional energization. The purpose was to determine if cornoa power could be limited to practical operating levels by this method and still effect good agglomeration.

Unfortunately testing was not completed because of problems with the operation of coal water slurry nozzles and plugging of lines from the relatively lower melting point ash for the coal used at the DCFT operating temperatures. Further, high nickel/chromium alloys had to replace the ceramic agglomerating surface because of problems in the manufacture of 4 inch diameter silicon carbide tubes.

Stable HPHT operation with gas firing was achieved in conjunction with the injection of alumina dust. Successful electrical operation was possible with alumina dust at 4 to 8 ft/s ESP gas velocities, resembling the ESP operation at 982 deg C (1800 deg F) and 10 atmospheres. Operation beyond this temperature was not attempted in order to protect the alloy steels. Collection efficiencies of 92 to 98% at grain loading of 1 to 6 gr/dscf are consistent with the data gathered in the 70's on a ESP of similar size.

Future research needs in ESP technology development

The electrostatic agglomerator concept deserves further evaluation because of its promise in effecting the growth of find particulates at PFBC and higher temperatures and thus allowing the use of simpler and more reliable cyclones. ESP material component issues, at least covering the short term, have been successfully addressed for the PFBC application. It is now important to evaluate the performance capability of ESP at PFBC conditions. The DCFT application is even more severe because of its much higher temperatures in relation to PFBC's. Several questions still need to be addressed as follows:

- can the ESP's operate with high collection efficiency, say 99.5 to 99.9%, without the high temperature related corona stability problems for PFBC applications?

- is there a potential for long term electrical stability problems resulting from molten flyash, due to difficulties in cleaning the discharge and collecting electrodes?

- can the material strength of ESP internals be maintained adequately over the long term?

- what should be the ESP size to achieve the high particulate collection efficiencies for PFBC applications?

- what is the relationship between operating power and collection efficiency for various commercial HPHT conditions ahead of the gas turbine?

- are new concepts of electrical energization necessary for the control of corona power?

- what should be the material of construction of ESP internals for DCFT applications?

- will ceramic electrodes work for ESP application?

In comparison to the number of research programs on barrier filtration, the ongoing research on HPHT ESP is relatively sparse. Because of the continuing uncertainties in the long term performance viability of barrier filters for the 900 deg C (1652 deg F) and higher temperatures it would be prudent to continue research on alternative particulate control concepts such as the HPHT ESP filtration.

CONCLUSIONS

It is seen that high temperature high pressure particulate collection is beset with many challenges. While the barrier filters have shown superior collection at temperatures up to 871 deg C (1600 deg F), they have been practically untested at the higher temperatures of interest. Molten ash could pose problems of cleanability and pressure drop control.

Besides the problem of cleanability of barrier filters at higher temperatures, materials survivability of ceramic based barrier filters have repeatedly come up. Failures of candles, cross flow filters, filter bags routinely get mentioned in most of the reports on these devices. Loss of long term strength of ceramics is also a cause for concern. Thermal shock of ceramic barrier filter materials from compressed air jets during cleaning cycles may present a serious threat to long term materials survivability.

Other devices such as electrostatic precipitators and agglomerators have been relatively untested at the 900 deg C (1652 deg F) and higher temperatures. In addition, serious questions regarding corona stability, operating power consumption and materials choice at DCFT temperatures remain to be addressed for electrostatic precipitators and agglomerators.

In spite of these seemingly daunting challenges, the potential for high pay-off resulting from the development of hot gas cleanup devices is real, and a high performance HGCU may be essential for the commercialization of high efficiency power generating systems. It is hoped that research continues to bridge the current gaps in the know how of hot gas cleanup.

REFERENCES

Chang, R., Sawyer, J., Lips, H., Bedick, R., and Dellefield, R., "The Testing and Evaluation of Ceramic Filter Fabrics", I Chem E Symposium, Univ. of Surrey, Guildford, U.K., 1986.

Ciliberti, D.F., and Lippert, T.E., "Gas Cleaning Technology for High Temperature, High Pressure Gas Streams", EPRI CS-4859, EPRI, October 1986.

Dellefield, R.J., and Bedick, R.C., "Evaluation of Three High-Temperature Particle Control Devices for Coal Gasification", I Chem E Symposium Series, Univ. of Surrey, Guildford, U.K., 1986.

Koopmann, G.H. and Reethof, G., "Evaluation of Acoustic Agglomerator for HTHP Particulate Control", DOE/MC/22012-2916, US DOE Report, March 1989.

Kumar, K.S., Marx, J.A., and Feldman, P.L., "Testing and Evaluation of Electrostatic Precipitator at New York University, DOE Final Report, August 1988.

Kumar, K.S., and Quimby, J.M., "Integrated Low Emissions Control of Direct Coal-Fueled Turbines - Electrostatic Agglomeration", US DOE Proceedings of Seventh Annual Gasification and Gas Stream Cleanup Contractors Meeting, August 1987.

Riepe, T. and Wiggens, H., "Pilot Tests of Electrostatic Precipitators at High Pressures and High Temperatures", I Chem E Symposium Series, Univ. of Surrey, Guildford, U.K. 1986.

Sellakumar, K.M., Isaksson, J., and Provol, S.J., "High Pressure High Temperature Gas Cleaning Using an Advanced Ceramic Tube Filter", Eleventh International Conference on Fluidized Bed Combustion, Montreal, Canada, April 1991.

Stringer, J., Leitch, A.J., and Clark, R.K., "EPRI Hot Gas Filter Plant at Grimethorpe: What Worked, What Broke and Where Do We Go Now?", Eleventh International Conference on Fluidized Bed Combustion, Montreal, Canada, April 1991.

Taksicker, O.J., "High Temperature High Pressure Electrostatic Precipitator for Electric Power Generation", I Chem E Symposium, Univ. of Surrey, Guildford, U.K. 1986.

Table 1   Summary of the HTHP ESP tests conducted in the USA

	CES UNION CARBIDE	BU. OF MINES CES	CES COMBUSTION POWER	CES EPA	CES CURTISS-WRIGHT	DENVER RESEARCH INSTITUTE
DATES TESTED	1962-1964	1963-1968	1966-1968	1976-1977	1982-1983	1982-1985
TEST LOCATION	INSTITUTE, W.V.	MORGANTOWN, W.V.	MONTEBELLO, CA	BOUND BROOK, NJ	WOODBIRDGE, NJ	DENVER, CO
COMBUSTION	PFBC, COAL	NATURAL GAS	METHANOL	AIR, FUEL GAS	PFBC, COAL	METHANOL
DUST TYPE	FLY-ASH	FLY-ASH	ALUMINA & ASH	NO DUST	FLY-ASH	CW FLY-ASH
TEMPERATURE, C	500-700	800	900	540-1090	790-850	700-900
PRESSURE, bar	3-8.1	4.5-6.5	4.6-11.2	1-35.5	5.4-6.4	6.4-10
GAS FLOW, m^3/s	0.1	0.16	0.11	--	0.53-.59	.054-0.97
GAS VELOCITY, m/s	0.2-1.2	0.3-1.1	2.9	--	0.83-0.93	.76-1.37
NO. COLLECTOR TUBES	19	16	1	1	9	1
LENGTH OF TUBES, m	1.83	1.83	4.6	0.76	4.6	2.1
DIA. OF TUBES, m	0.15	0.15	0.20	0.073	0.30	0.30
COLLECTION AREA, M^2	16.4	13.8	2.9	0.17	39	2
DISCHARGE ELECTRODES	WIRE	WIRE	TWISTED WIRE	WIRE	MASI, VANES	CYL, VANES
DIA DISCH ELECT. mm	2.1,3.4	2.1	3x2.8	1.58,2.34,3.18	25.4,76	16,70,83
SCA, $m^2/m^3/S$	166	45	26.6	--	66.1-73.6	36
DUST LOAD, ppm	--	550-1700	--	--	2300-3700	1500-2000
DUST DIA. MICRON	2.6	30	6-10	--	5	5
TYPICAL COLLECTION EFFICIENCY, %	98.8	91.96	80.82	--	95-99.5	95-99.5

Section 2
Filters

OPERATION AND PERFORMANCE OF THE EPRI HOT GAS FILTER AT GRIMETHORPE PFBC ESTABLISHMENT : 1987-1992

G. K. Burnard[1], A. J. Leitch[1], J. Stringer[2], R. K. Clark[3]
and P. Holbrow[3]

1   Hoy Associates Ltd., Office Unit 12, Woodend Business Centre, Cowdenbeath, Fife, KY4 8HG, UK.
2   Electric Power Research Institute, 3412 Hillview Ave., Palo Alto, California 94304, USA.
3   Grimethorpe PFBC Establishment, Grimethorpe, Barnsley, South Yorkshire, S72 7AB, UK.

ABSTRACT

This paper summarises and compares the operation and performance of the EPRI filter during the 1987 and 1991/92 periods of operation at the Grimethorpe PFBC Establishment. During the latter period the filter was operated as an integral component of the Grimethorpe Topping Cycle Project, which formed part of British Coal Corporation's overall strategy in the development of their Topping Cycle.

The periods described cover a total of 2369 hours operation at 7-10.5 bar and 780-860 °C. In particular, the modifications and changes carried out prior to the 1991/92 operation are assessed in the light of the experience gained during 1987. The modifications carried out during 1991/92 are reviewed and appraised.

Specific items which are addressed include :
* effect of feedstocks on filter performance
* performance of the ceramic filter elements
* performance of new pulse valves
* effectiveness of preheating the pulse air
* improvements to the element hold-down system.

In addition, a summary of the outstanding issues relating to the commercial development of rigid ceramic filter units, highlighting the main issues and putting forward suggestions as to how best these can be addressed, is presented. Of particular importance is the long-term durability of the ceramic elements and the cleanability of the dust cakes generated by advanced fossil fuel power plants.

INTRODUCTION

Fundamental studies on pressurised fluidised bed combustion and associated systems have been carried out in the UK since the late 1960's. In 1975, the decision was officially taken to build an experimental PFBC plant at Grimethorpe, South Yorkshire, which was jointly funded by the governments of the Federal Republic of Germany, the UK and the USA, under the auspices of the International Energy Agency (IEA). The plant was commissioned in 1980, and between then and 1984 over 3600 h of operation was achieved.

The potential of PFBC was recognised, and a further programme, funded by the British Coal Corporation and the UK Central Electricity Generating Board (CEGB), ran from 1985 until 1988. Additional funding was contributed by the US Department of Energy (US DoE) and the US Electric Power Research Institute (EPRI), the latter funding the design, construction and testing of a hot gas filter (HGF) to clean up the exhaust gases under high temperature, high pressure (HTHP) conditions. Around 2000 h of PFBC operation was accomplished during this phase.

Finally, from 1989 to March 1993, the Grimethorpe PFBC was in use again, though this time, primarily, to appraise the feasibility of operating a commercial gas turbine on coal-derived gas at elevated temperature [Clark et al., 1991]. The programme, known as the Grimethorpe Topping Cycle Project (GTCP), was jointly funded by British Coal, the UK Department of Trade and Industry, GEC-Alsthom and PowerGen. Additional EPRI funding was targeted towards hot gas filtration studies, with further testing of the HGF unit. Well in excess of 1500 h of PFBC operation during 1991 and 1992 brought the total operation on coal at Grimethorpe to somewhere in the region of 7000 hours.

During the latter two programmes, operation of the EPRI-owned HGF unit resulted in over 2300 hours of definitive process data being gathered. This paper is a summary of the experiences gained and the outstanding issues which require addressing before this technology can be considered commercially viable.

HOT GAS FILTRATION

Critical to the success of PFBC, and other so-called advanced cycles, is an effective hot gas clean-up (HGCU) system to protect the gas turbine. In commercial PFBC plants today, cyclones are used as the only particulate removal equipment which can operate reliably under HTHP conditions [Pillai, 1988]. Configured of several trains in parallel, the efficiency of these cyclones is such that the gas turbines must be 'ruggedised' to make them sufficiently robust to withstand the erosivity of the particulates emitted. Much effort throughout the world in recent years, not least during the latter two programmes at Grimethorpe, has been devoted to the development of a more efficient clean-up system, using rigid ceramic elements as the filtration media [Stringer et al., 1991; von Wedel et al., 1992; Clark et al., 1993; Mudd & Hoffman, 1993; Brown & Leitch, 1993]. In March, 1992, EPRI sponsored a workshop on dust filtration from coal-derived gases at high temperature, where the current status of many of these programmes was reviewed. A filter employing ceramic elements is a barrier filter and, hence, particulate emissions should be low, both with respect to concentration and to size.

Apart from protecting the gas turbine, an effective HGF confers the advantage that no further clean up of the exhaust gas is required before it is released to the atmosphere. This is significant as regulations governing dust emissions to the environment are often much more stringent than those suggested for the protection of gas turbines. Furthermore, the development of more advanced cycles becomes feasible with the possibility of increasing the gas turbine inlet temperature.

The dust emission limits required by environmental legislation vary from country to country, but a figure of less than 20 ppmw is a reasonable guideline. In the case of the erosion tolerance of the gas turbine, the desired limits will also depend on the detailed design of the turbine and, in particular, the degree to which design changes are made to make it more tolerant - this is generally termed 'ruggedizing'. Furthermore, the limits refer both to concentration and size. A typical specification might be: less than 0.1 ppmw particles over 20 microns diameter, less than 1 ppmw over 10 microns, and less than 10 ppmw of sizes between 4 and 10 microns. Particles smaller than 4 microns are not generally regarded as damaging. However, deposition is also an issue and this may require restrictions on smaller particles. Cyclone trains are capable of matching these requirements, but only just. Previous experience has shown that a positive filter operating properly transmits less than 1 ppmw; but, in an overall system, the specification of the acceptable limit may have to take account of the possibility of defects such as leaking seals, and broken or cracked filter elements.

For some of the proposed advanced cycles, the filtered combustion gas from the PFBC is further heated by firing it with a supplemental fuel prior to its entering the gas turbine expander. This will require even tighter limits on the particulate loading of the gas exiting the clean-up system, since the dust may be melted within the auxiliary combustor and deposit on the downstream components. In addition, alkali may be released from the dust, promoting hot corrosion of the turbine.

In order to reach commercial viability, an HGF must be proven to be efficient, reliable and cost-effective. The objectives of the test plans during the last two programmes at Grimethorpe have been focused primarily on investigating these attributes.

Experimental objectives of the HGF programmes

As the HGF experimental programmes were parts of the overall experimental programmes, they assumed a different priority during each of the two phases. During the British Coal/CEGB programme, there were a number of investigations performed in parallel. Apart from the HGF programme, there were investigations regarding the PFBC, for example, into tubebank wastage and mode of coal/sorbent feed. Each investigation co-existed and, in effect, assumed equal priority. This was not the case, however, during the GTCP where investigation of the performance of materials exposed in the gas turbine was of primary concern. The efficient performance of the filter, in this case, was required to protect the turbine, and an experimental programme on the filter was of secondary priority. The filter was, of course, investigated to the fullest extent possible within these limitations.

The major experimental objectives of the HGF programmes during each phase were essentially the same, and may be summarised as:

For a representative range of operating conditions,

1. To establish the durability of the filter elements.
2. To investigate the pressure drop and permeance over the filter.
3. To investigate dust emissions from the filter.
4. To determine the characteristics of the reverse cleaning system.

PLANT LAYOUT

As the major objectives of the two programmes differed, the plant was arranged slightly differently for each case. Figures 1 and 2 show schematically the plant layouts used. The design and duty of the EPRI HGF, however, remained principally the same during both programmes.

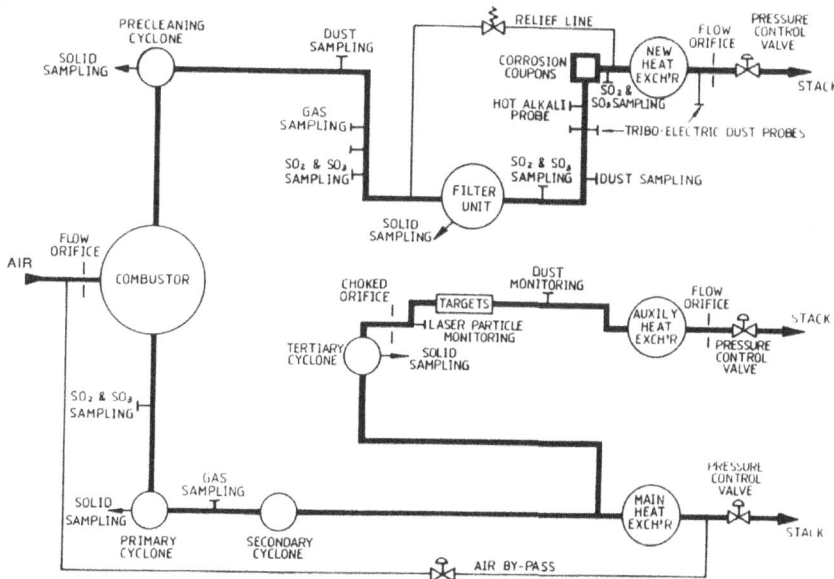

Figure 1. Plant layout during the British Coal/CEGB programme.

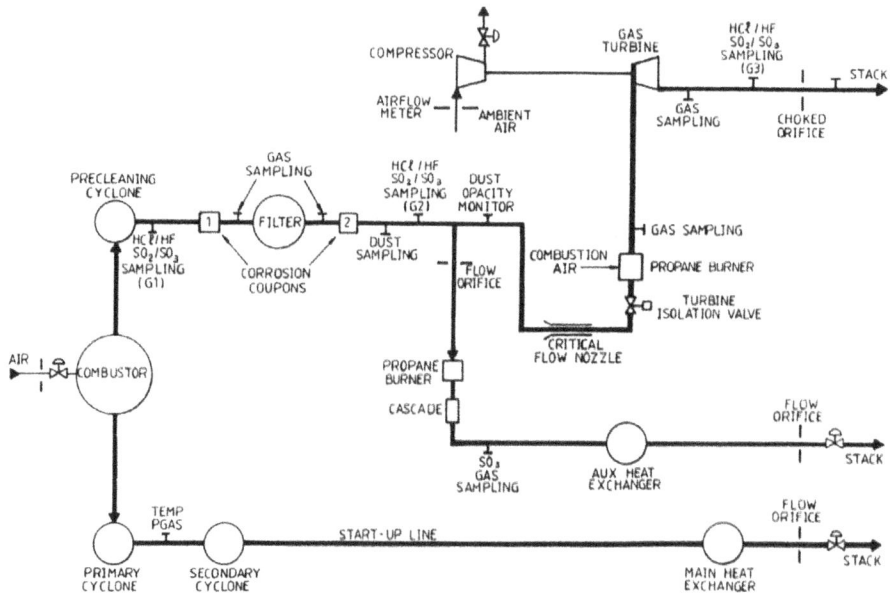

Figure 2. Plant layout during the GTCP.

THE EPRI HOT GAS FILTER

The EPRI HGF has been well described in recent literature [Stringer *et al.*, 1991; Clark *et al.*, 1993]. The 1987 programme is definitively described in four EPRI reports, published September 1992, which are detailed in the references. Designed to hold up to 130 ceramic elements, it was until late 1992, when the 10 MW-equivalent side-stream filter [Mudd & Hoffman, 1993] on the Tidd PFBC plant began in service, the largest of its type in the world.

Filter vessel

Figure 3 shows the main components and major dimensions of the filter. The 'clean' and 'dirty' sides of the filter were separated by a 50.8 mm thick RA 333 alloy tubesheet, from which the ceramic elements were suspended. 'Dirty' gas entered the filter below the level of the elements and, just inside the entrance, encountered an impingement plate which protected components in the vessel from damage. The inner walls of the vessel were predominantly lined with 3 mm thick Type-310 stainless steel, which covered and protected the insulation. A debris grid, covering the full cross-section of the vessel and, located below the gas inlet, protected the ash screw-conveyor from damage, e.g. from filter elements which might inadvertently be broken during operation. As the debris grid would periodically provide a site for ash to collect and build-up, an 'air cannon' was installed, during the GTCP, to pulse it intermittently with a blast of air. The filter vessel remained essentially unchanged throughout the two programmes, though some features, notably the element hold-down and the reverse cleaning systems, did go through a significant evolutionary process.

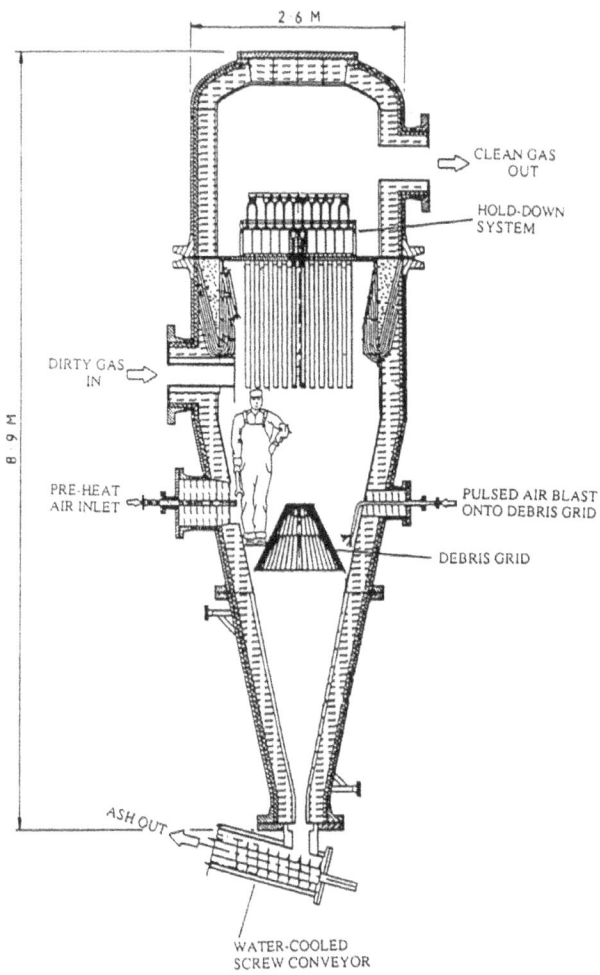

Figure 3. EPRI hot gas filter vessel.

Filter elements

Diaschumalith F40 rigid ceramic filter elements, made and supplied by Schumacher GmbH, Germany, were used in both programmes. The cylindrical elements were 1.5 m long, with an external diameter of 60 mm and a wall thickness of 15 mm. Silicon carbide granules, bonded with clay, supported a thin outer layer, of 50 to 100 microns thickness, composed of fine alumina fibres and silicon carbide grains, also bonded with clay.

The method by which the elements were sealed into the tubesheet was of critical importance throughout. During the first programme, a relatively simple arrangement of counter weights and ceramic seals was employed, as shown in Figure 4. It was felt, however, that this system was unsuitable because of the periodic tendency during operation for counter weights to become displaced, allowing seals to be blown. Once seals were gone, elements were free to move up and down, resulting in so-called 'chatter', and the heads of some elements would suffer damage. It was further felt that higher pulse pressures may need to be applied in the future, and the existing system had already been tested to its limits in this respect.

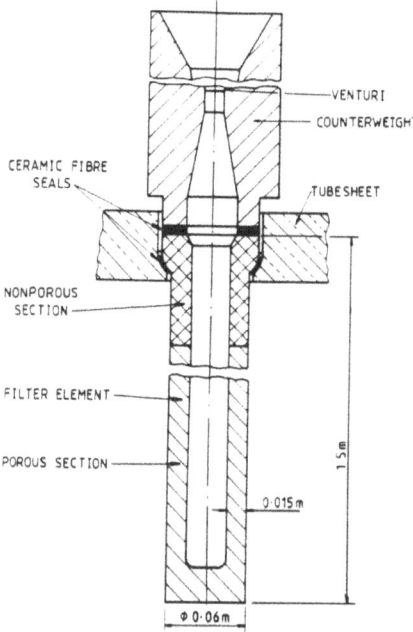

Figure 4. Element hold-down system employed during the British Coal/CEGB programme

For the GTCP, the system was re-designed to provide the elements with a more positive location in the tubesheet, and even the heads of the elements were re-modelled. The design, unfortunately, suffered a number of drawbacks and had to be radically altered. After a further 3 re-designs, and around 250 hours operation later, an effective system was finally available - using the standard Diaschumalith F40 elements. The final design proved capable of allowing much higher pressures to be developed in the cavities of the elements, and thereafter no seals were damaged during the normal course of operation. British Coal are currently seeking a patent to cover use of this device in such applications.

Reverse cleaning system

The appropriate design of the reverse cleaning system is a critical ingredient for a successful filter. It has ramifications on the operability and performance of the filter, as well as having a significant impact on the overall economics of the plant. Consequently, investigation of this system was ongoing throughout each programme. Investigation considered both the performance of the system, with regard to effective cleaning of the filter elements, and also to the mechanical integrity of the materials used in its construction.

Effective cleaning is thought to be dependent on the development of a high rate of pressure rise within the cavity of an element during reverse cleaning. On this premise, the original reverse cleaning system, as designed by Westinghouse, was perceived to be inefficient and largely ineffective. It contained numerous bends, elbows and pipe size transitions, as well as an excessive length of narrow bore piping between reservoir and nozzle tip, which were considered to attenuate the pulse severely. Cleaning of the elements was poor, which seemed to support this perception and, as a result, the system was partially re-built after only 300 hours operation. Apart from some rationalisation of the piping, two secondary air reservoirs were added close to the top of the filter vessel and the pulse valves were installed downstream of these reservoirs, as close to the filter vessel as possible. Subsequent element cleaning appeared markedly improved, though other important factors had also been altered before resuming operation, e.g. the feedstocks were different and the temperature of the fluidised bed had been increased.

Prior to embarking upon the GTCP, further modifications were made. It had become clear during the earlier programme that the particular Atkomatic globe-type pulse valves being used were very sluggish, and not consistent with an effective system according to the definition given above. After trials with a number of valves, a Mecafrance 2" ball valve with a Kinetrol actuator was selected as providing the most suitable characteristics based on the assessment criteria. Comparison of the test characteristics of the two valves is shown in Figure 5. Furthermore, extra valves were added to allow the Mecafrance valves to be isolated for maintenance during operation, if required.

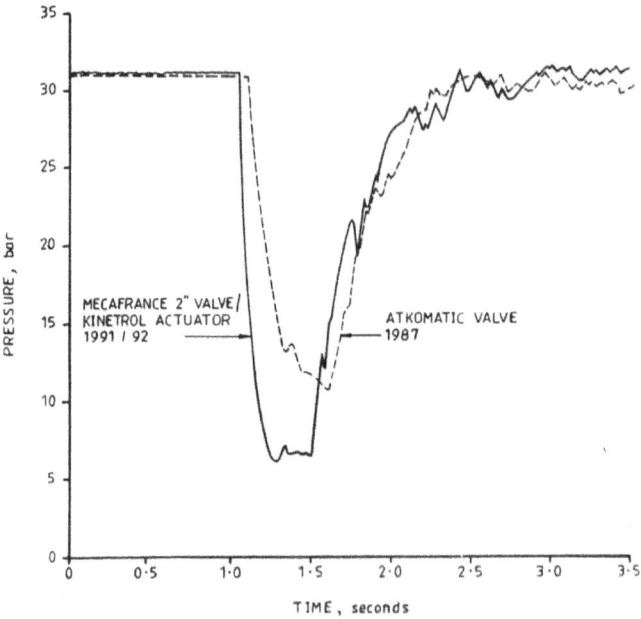

Figure 5. Valve differential pressure profiles for a 0.4 s pulse

Studies at RWTH, Aachen, were sponsored by EPRI to develop mathematical models of the pulse systems employed during the two programmes [Pitt & Leitch, 1991; Pitt & Grüttner, 1991]. These models were used to investigate their performances, identified inherent losses and inefficiencies due to the designs of the systems, and recommendations were made for design optimisation.

Mechanical failure of materials, particularly those of which the manifold and pulse pipes were constructed, was a constant feature of the programmes. The material chosen in the initial design was Type-310 stainless steel. This is known to be prone to the precipitation of a chromium-rich sigma phase which results in embrittlement of the metal at low temperatures, and can result in enhanced internal oxidation of the chromium-depleted material adjacent to the sigma-phase precipitates at elevated temperatures. The sigma-phase precipitation is most severe at temperatures close to 850°C. The types of failure observed in the early stages were shown to be related to sigma-phase formation [Stringer *et al.*, 1991]. These components were gradually replaced by ones fabricated from Incoloy 800H. This material was selected because it was readily available, but it would have been better to use a low chromium version of Incoloy 800. The 800H components also proved unsuitable after long term exposure. This is not an unusual problem, although the environment is undoubtedly demanding. The next material selected was Inconel 625, but this was introduced fairly late during the GTCP and,

96

consequently, had little time to prove its resilience to the thermal and mechanical stresses imposed on it during pulsing.

Ash removal system

Dry ash removal was accomplished by using a water-cooled screw conveyor to transport ash from the bottom cone of the filter vessel to waste via a lockhopper system. The screw conveyor performed the subsidiary function of cooling the ash to a temperature appropriate for disposal, whilst the lockhoppers essentially allowed a means of reducing the pressure to atmospheric. Prior to the latter programme, the GTCP, a single-lockhopper system was instituted to replace the original double, primarily because of space constraints. A diagram illustrating the main features of the British Coal/CEGB arrangement is shown in Figure 6.

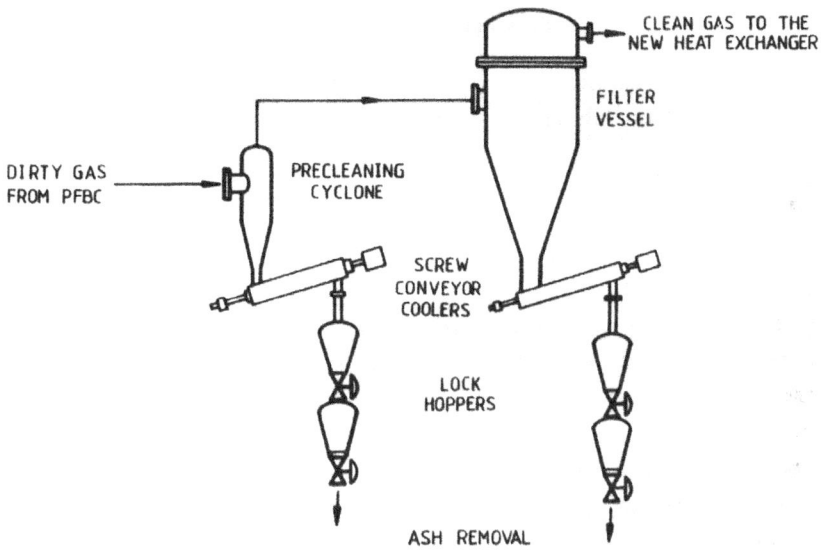

Figure 6. Ash removal system for the British coal/CEGB programme

Apart from some minor commissioning problems, the ash removal systems employed during each programme operated successfully and reliably throughout.

Instrumentation

The filter was comprehensively instrumented for the monitoring of temperatures, pressures and flows. Some of these instruments, and particularly those associated with the reverse cleaning system, were fast response and connected to rapid data collection and storage systems.

97

In addition, dust and gas sampling devices were installed in the ductwork upstream and downstream of the filter. Ash captured by the filter could be collected for weighing, to determine the rate of capture, and analysis.

Selected ceramic elements were withdrawn at stages during each programme and a representative number sent for non-destructive and destructive analysis following completion of the programmes. However, some non-destructive testing was performed on-site to complement the off-site activities. One such test was to measure the ultra-sonic time-of-flight (ToF) of the elements whenever the opportunity arose, e.g. during any inspection when elements would be removed from the vessel. ToF measurements have been shown [Morrell *et al.*, 1990] to correlate with certain aspects of the mechanical strength of the elements, although whether these relate to eventual failure modes of candles in practice has still to be demonstrated.

OPERATING CONDITIONS

Temperatures and pressures in the filter system were predominantly dictated by the prevailing PFBC conditions, although modified to some degree by the effects of the reverse cleaning operation. Dust concentrations to the filter were dependent predominantly on the efficiency of the precleaning cyclone.

During the British coal/CEGB programme, feedstocks, feed mode and operating conditions were varied, basically according to the 3 sets of conditions which constituted the 3 test series, shown in Table 1. In contrast, the GTCP was operated at substantially constant conditions, dictated by the gas turbine experimental requirements. These conditions are also shown in Table 1.

TABLE 1
Filter operating conditions during the two experimental programmes

Programme	Gas flows kg/s	Inlet gas pressures bara	Inlet gas temperatures °C	Filtration velocities cm/s	Inlet dust concns. ppmw
British Coal / CEGB:					
-Test Series A2.1	2.3 - 4.0	8.0 - 10.5	793 - 804	3.0 - 4.2	2500 - 3500
-Test Series A2.2	2.3 - 4.5	7.2 - 10.6	780 - 872	3.0 - 7.0	1500 - 3500
-Test Series A2.3	0.9 - 4.3	10.2 - 10.5	796 - 851	1.0 - 5.8	900 - 2100
GTCP	3.2 - 7.0	10.5	780 - 850	3.0 - 6.5	1100 - 2900

FILTER PERFORMANCE

Performance of the filter elements

The filter elements are obviously the key components of the technology, and as such their performance was keenly monitored throughout the programmes. Though 4 elements failed on-line during the British Coal/CEGB programme and a further 13 during the latter programme, this was not unduly worrying because each failure could be attributed to some form of plant excursion, see Tables 2 and 3.

TABLE 2
Causes of ceramic element failures during the British Coal/CEGB programme

Run	Number of failed elements	Reasons for failure
129	1	A pulse valve failed open during the run.
137	4	High pulse pressures allowed elements to 'chatter'.

TABLE 3
Causes of ceramic element failures during the GTCP

Run	Number of failed elements	Reasons for failure
148	9	Trip during turbine start-up disturbed the hold-down assembly.
151	1	Thermocouple on tubesheet suspected of dislodging a counterweight.
154	1	High pulse pressures allowed elements to 'chatter'.
165	2	A pulse valve failed open during the previous run.

Of significantly greater concern was the progressive loss in strength with exposure. It was speculated, after the 1987 programme, that this loss of strength was due to the thermal shock of the relatively cold pulse air [Morrell *et al.*, 1990]. In Figure 7, the variation of the average ToF's of the elements is plotted as a function of the number of pulses they experienced. In both cases, the ToF values initially increase with number of pulses, indicating strength loss, to a maximum, after which the rate of strength loss appears to cease. The numbers of pulses, after which the ToF values no longer increase, correspond fairly closely in each case. It is interesting to note that these points correspond to around 612 hours and 1022 hours exposure for the 1987 and the 1991/92 programmes, respectively. This was further investigated during the GTCP by allowing the pulse air in one of the 10 manifolds, Manifold-A, to be pre-heated. Figure 8 shows both Manifold-A and the standard manifold design. Comparison of the strength of the Manifold-A elements with those from the other manifolds demonstrated, without doubt, that reducing the temperature drops experienced during a pulse, by pre-heating, was beneficial. Figure 9 compares the in-cavity temperatures experienced by the pre-heated manifold with a non-heated manifold. Measurements showed that after almost 700 h of exposure, the ToF's of the standard manifolds increased by about 60 microseconds whilst that of Manifold-A increased by about 45 microseconds, indicating the reduced strength loss incurred by the Manifold-A elements.

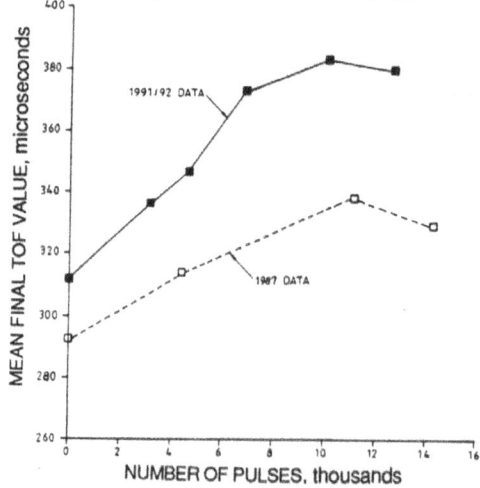

Figure 7. Ultra-sonic time of flight data as a function of number of pulses experienced.

Ash accumulation around the elements and bridging of ash between the elements was a problem experienced during both programmes, though predominantly during the GTCP. The worst example of this occurred during the GTCP, Run 175, when an estimated 50 to 60% of the cross-sectional area of the elements was covered with ash. This problem was never adequately resolved. It was most probably due to a combination of effects involving ineffective cleaning, ineffective dust shedding from the body of the elements and, perhaps, insufficient element separation. Furthermore, preliminary investigations by British Coal suggest a propensity for the ash cakes partially to sinter at these temperatures.

Also of significant concern was the tendency of elements to bend during operation, some of them to an extraordinary degree. For example, if the neck of one element was held on a flat surface, its foot would be 150 mm above the surface. A number were bent so badly that the surface of the element in tension was cracked. This phenomena is believed to have been due to a combination of the previous problem, ash accumulation around the elements, and pulsing. It is hoped that, once ash accumulation and bridging have been successfully addressed, this issue will also be resolved.

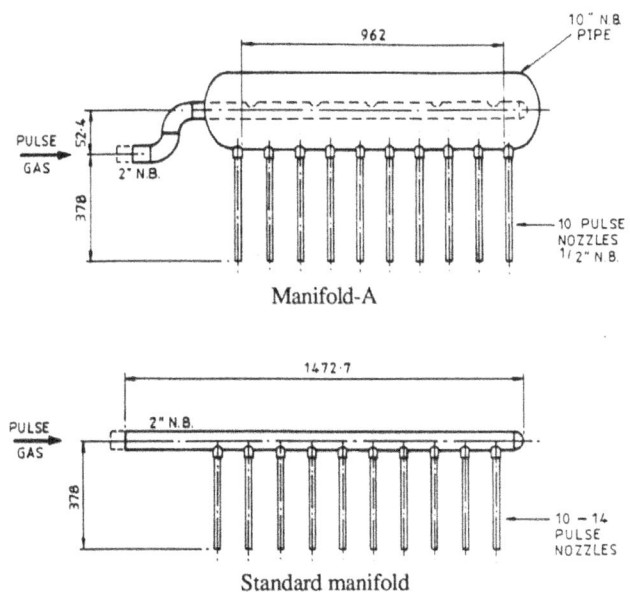

Figure 8. Comparison between Manifold-A and a standard manifold.

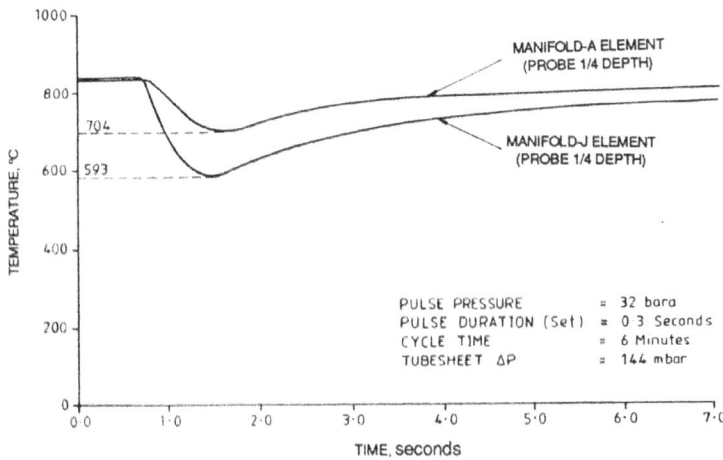

Figure 9. Comparison of element in-cavity differential temperatures measured

Effect of feedstocks

The critical effect of feedstocks on the performance of the filter only became apparent during the GTCP. It is now clear that the nature of the ash from certain feedstock combinations results in a dustcake which is highly tenacious and extremely difficult to dislodge.

During the GTCP, the initial feedstocks were Kiveton Park run-of-mine coal and Middleton limestone. Problems were encountered very early in the test programme, manifesting themselves as continuously increasing pressure drops regardless of the reverse cleaning conditions. For example, at the nominal filtration velocity of 5.5 cm/s, the rate of increase of pressure drop could accelerate from 1 or 2 mbar/h, at the start, to 15 mbar/h over the course of a continuous run of only 40 or 50 hours duration. As the maximum pressure drop possible was around 380 mbar, above which the elements and counter weights would theoretically be able to lift, sustaining operation with a rising pressure drop was not viable for long periods, and run lengths were obviously restricted. The rate of increase could not be abated, even when pulsing to the maximum pressure possible within the limits of the existing design of the reverse cleaning system, i.e. at 20 bar over plant pressure (producing in-cavity differential pressures of around 900 mbar). Unfortunately, as it was not until later in the programme that the reverse cleaning system was modified to be capable of pulsing at substantially higher pressures (enabling in-cavity differential pressures of up to 1800 mbar to be generated), it was not possible to discover whether higher pressures would have been capable of controlling the pressure drop.

As part of the investigation to determine how this restriction to operation might be overcome, tests were performed feeding excess limestone to the PFBC. The limestone feedrate was increased by factors of between 5 and 15, and some degree of success was achieved. At the higher value, feeding limestone at around 1000 kg/h, the pressure drop was almost stabilised. It is not yet known whether the effect of the additional limestone was simply due to dilution with unreacted limestone or whether the physical/chemical properties of the ash were modified, making it less tenacious.

Substitution of dolomite, a more reactive sorbent, instead of limestone finally overcame the problem. With the Cadeby dolomite, it became possible to clean the elements, albeit still requiring rigorous pulsing conditions. Some of the data collected during the GTCP are shown in Table 4, and illustrate clearly the dependency of pressure drop stability on the sorbent employed.

The feedstocks used initially during the GTCP were nominally the same as those used during a substantial part of the British Coal/CEGB programme, which probably illustrates just how sensitive the nature of the dustcake formed may be to small differences in the minerology of the feedstocks. During the earlier programme, both dolomite and limestone were also used, and it was clear that cleaning of the elements was less effective with limestone, as illustrated in Figure 10. However, as there were a number of other variables altered during the same period, making a correlation between the feedstocks and pressure drop was not at all obvious. In fact, at the time, it was thought that the differences in pressure drop were more likely to be associated with changes made to the design of the reverse cleaning system.

TABLE 4
Effect of sorbent on pressure drop stability during the GTCP

Run	Sorbent	Typical sorbent feedrate, kg/h	Pulse pressure, bara	Filtration velocity, cm/s	Typical pressure drop rate of rise, mbar/h
159	Middleton limestone	1000 360 68	40 30 30	4.6	0 4.0 10.0
160	Cadeby dolomite	320	40	5.0	0.7
161	Cadeby dolomite	360	40	5.0	0.25
178	Cadeby dolomite	355	29	3.2	0

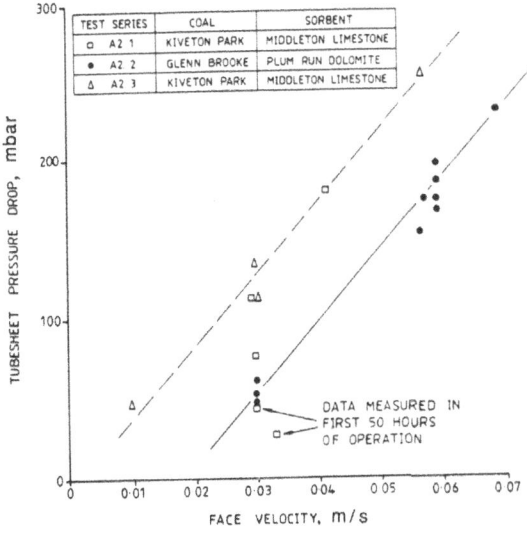

Figure 10. Pressure drop as a function of filtration velocity for data gathered during the British Coal/CEGB programme

103

Pressure drop and permeance

Pressure drops across the tubesheet were recorded on a continuous basis throughout operation. It was important to monitor that the reverse cleaning strategy applied was effective and that the pressure drop did not rise above the critical level, allowing lifting and damaging of the elements.

It was also important to compare operation of the filter under various conditions and to monitor the long-term performance of the elements. For these latter considerations, pressure drop alone is of limited value and it is preferable to use permeance. This is defined as:

$$\text{Permeance} = \frac{\text{Face velocity}}{\text{Pressure drop}}$$

Permeance is, however, dependent on the operating temperature, which varied throughout each of the test programmes. Though the temperatures did not vary significantly in this respect, permeance was usually normalised to 20°C for comparison, using the following expression:

$$\text{Permeance (at 20°C)} = \frac{\text{Face velocity}}{\text{Pressure drop}} \times \frac{\mu_T}{\mu_{20}} \quad \text{, where}$$

μ_T = dynamic viscosity of gas at the operating temperature, T°C
μ_{20} = dynamic viscosity of gas at 20°C

To determine whether permeance continues to decrease with continued exposure or whether a plateau is reached, where the permeance stabilises, is very important for any particular 'brand' of elements. Continuous reduction of permeance with exposure would essentially pronounce the elements under test unsuitable for this particular function. Permeance trends during the two programmes are reproduced in Figures 11 and 12. Starting at high values with new elements, the permeance decreases and appears gradually to level off. Prior to each subsequent run, the permeance was partially regenerated by either pulsing or off-line manual cleaning to remove most of the existing dustcake. Though somewhat subjective because of the relatively short periods of continuous operation, the curves do appear to plateau, at 0.26 m/s/bar for the British Coal/CEGB programme and at around 0.2 m/s/bar for the GTCP.

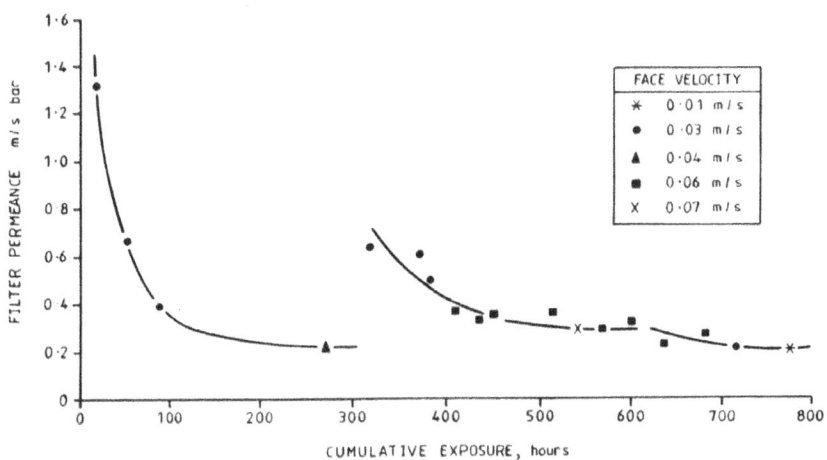

Figure 11. Filter permeance (at temperature) for the British Coal/CEGB programme

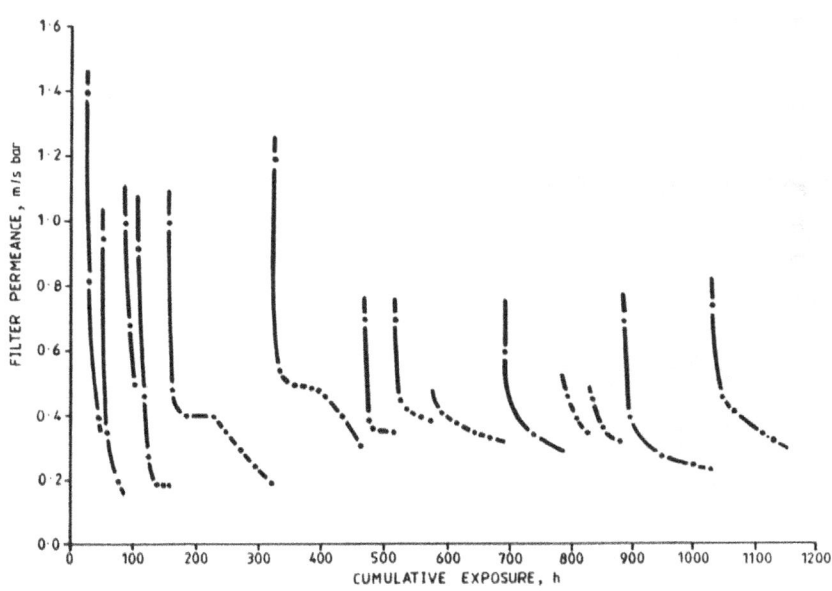

Figure 12. Filter permeance (at temperature) for the GTCP.

Dust emissions

Apart from the properties of the dust, measured emission values depend on the following:

- The integrity of the ceramic elements.
- The integrity of the sealing of the elements in the tubesheet.
- Contamination of the sample by either metallic scale deposits arising from local corrosion of the instrument or dust in the ductwork downstream of the filter as a consequence of previous breakthrough.
- The reliability of the dust measurement device.

During the 1987 programme, measured emissions led to the belief that values of less than 20 ppmw were characteristic of a well-assembled filter, operating properly. It was shown, however, in the later programme, that values of this magnitude had probably been due to inadequate sealing and/or contaminated ductwork. In fact, under normal circumstances, values of less than 1 ppmw should be expected. Any greater value would probably indicate the existence of a flawed seal, downstream contamination or breakthrough caused by a broken element.

Reverse cleaning parameters

The major parameters which could be varied during reverse cleaning were the reservoir pressure, the cycle time and the pulse duration. Broadly speaking, if the cleaning strategy was effective, the pressure drop, after a pulse, would not increase significantly during the course of a run. This coarse monitoring of the system, however, would not identify whether excessive air was being consumed nor would it allow the effects of the individual parameters to be examined. To gather such information it was necessary to measure, using high resolution equipment, in-cavity differential pressures developed during a pulse.

During the GTCP, for the Mecafrance pulse valves, investigation showed that a pulse duration (open time for the valve) existed, below which the maximum in-cavity differential pressure did not have time to develop. In this case, the set time on the programmable logic controller was 300 ms, corresponding to a closed-to-closed time for the valve of 640 to 700 ms. Above this value, there appeared to be no benefit. In fact, it is likely that it could be disadvantageous as it may simply cause further cooling of the elements. The Atkomatic valves, used during the earlier programme, were too sluggish for such data as measured to be meaningful.

It was demonstrated, from data gathered during the British Coal/CEGB programme, that the in-cavity differential pressures gradually reduced for elements situated along the length of a manifold, being higher at the inlet end.

In general, it was necessary for in-cavity differential pressures to be in excess of 1000 mbar for effective cleaning, though on occasion values as high as 1800 mbar were used.

Values in excess of about 900 mbar could only be consistently applied, without blowing seals in the hold-down system, after the final hold-down system had been developed.

The air consumption during reverse cleaning is a penalty insofar as the overall economics of the process are concerned. Minimising the air consumption whilst maintaining acceptable cleaning of the elements, i.e. optimisation of the system, is, therefore, an important issue for investigation. Unfortunately, dealing with problems related to the reverse cleaning system, e.g. achieving a reliable design and attending to the materials problems, precluded opportunity for a systematic investigation of air consumption being performed. Thus, although air consumption was actually measured during both programmes, optimisation was never achieved.

CONCLUSIONS

The EPRI hot gas filter at the Grimethorpe PFBC Establishment, over the 5 year period during which it was operated, was the largest of its kind in the world. Throughout this period all the major issues regarding filtration using rigid ceramic filters were addressed. As expected with any new technology, problems were encountered. If, however, the technology is to become commercial, economic solutions to these problems must be found.

The key components of the technology are, of course, the filter elements themselves. Though 17 elements failed during service, this does not detract from the technology, as the failures can be attributed to plant excursions or filter maloperation. Progressive strength loss is a problem which has not been resolved and must be addressed. If strength loss reduces with continued exposure (or pulse cycles), it is important to establish when the residual strength becomes 'insufficient' to support further operation. To what extent strength loss can be attributed to adverse thermal effects during reverse cleaning is not known. Tests showed that pre-heating the pulse air reduces the rate of strength loss but does not eliminate it. Further investigations regarding the effects of exposure to hot gases, of appropriate composition, and pulse cycles must be made to prove the suitability of these elements for this application. Unsuitability would mean further materials development, i.e. modifying the existing properties of the elements or turning to alternative materials. Many companies are presently engaged in such developments, including Schumacher, Westinghouse, Refractron Technologies, Industrial Filter and Pump, BWF Textil, CeraMem and 3M. [All of these, and more, participated in the Second EPRI Workshop on Filtration of Dust from Coal-Derived Reducing and Combustion Gases at High Temperature, March, 1992.]

A programme to investigate the dynamics of the gas and ash distribution around the elements could lead to measures which would eliminate ash accumulation and bridging. Bending of the elements during operation is also unacceptable: it is expected, however, that this should not occur once the problem with ash accumulation and bridging between the elements has been successfully addressed.

A reliable and effective reverse cleaning system is vital. The duty of this system, to cope with regular thermal and mechanical stresses, is particularly onerous. However, solutions to develop an appropriate design from appropriate materials should be possible from conventional engineering sources, and should not present an obstacle to the future of the technology. The experience from these two programmes suggest that gravity hold-down systems are not reliable, and that some means of positively locating the elements in the tubesheet should be developed in preference. Westinghouse [Lippert, 1992] and Deutsche Babcock [von Wedel *et al.*, 1992], among others, are companies currently investigating novel forms of hold-down system. Optimisation of the air consumption needs to be addressed for a reliable economic assessment of reverse cleaning to be made.

Ash removal, considering the nature of the ash to be handled, was initially predicted as likely to generate significant difficulties. However, the system employed, comprising a water-cooled screw conveyor and lockhopper(s), provided no cause for concern throughout the 2369 hours.

The tenacity of the dust resulting from certain feedstocks came as a complete surprise and was certainly not predicted. The phenomenon, though under current investigation, is not yet understood. It has, however, been recognised as an issue of obvious concern and is acquiring prompt attention.

Dust emissions of < 1 ppmw are a significant advantage of this technology. Whilst well within any proposed specification for a gas turbine, the levels also satisfy any existing or proposed statutory emissions requirements world-wide.

The EPRI filter has been successfully employed to identify the major issues confronting hot gas filtration, and has addressed many of them. Those currently engaged in the development of hot gas filters have taken advantage of the experience gained at Grimethorpe and have put themselves, consequently, in a better position to advance the commercialisation of this technology.

ACKNOWLEDGEMENTS

The authors would like to acknowledge the contributions made by all those, too numerous to mention, who made Grimethorpe PFBC Establishment the 'Mecca' for anyone involved with any aspect of advanced coal-fired power generation over the last 15 years.

REFERENCES

Brown, R.A. & Leitch, A.J. (1992). Filtration of PFBC dusts using a rigid ceramic tube-type filter. In Proceedings of the ASME 12th Fluidised Bed Combustion Conference, La Jolla, CA., USA., to be published.

Clark, R.K., Holbrow, P., Oakey, J.E., Burnard, G.K. & Stringer, J.S. (1993). Some recent experiences with the EPRI hot gas rigid ceramic filter at Grimethorpe PFBC Establishment. In Proceedings of the ASME 12th Fluidised Bed Combustion Conference, La Jolla, CA., USA., to be published.

Clark, R.K., Arnold, M.St J., Fackrell, J.E., Mordecai, M. & Dawes, S.G. (1991). Advanced clean coal technology for power generation. In Proceedings of the Institute of Energy's 5th International Fluidized Combustion Conference, London, pub. Adam Hilger, Bristol, UK, pp. 353-362.

Grimethorpe High-Temperature/High Pressure Experimental Program. Volume 1: Design, Commissioning, and Modification of a Large Hot-Gas Filter on the Grimethorpe PFBC. EPRI Report No. TR-100499, Prepared by Grimethorpe PFBC Establishment under Contract RP 1336-08, September 1992

Grimethorpe High-Temperature/High Pressure Experimental Program. Volume 2: Preliminary Report on the Filter Performance. EPRI Report No. TR-100499, Prepared by Grimethorpe PFBC Establishment under Contract RP 1336-08, September 1992

Grimethorpe High-Temperature/High Pressure Experimental Program. Volume 3: An Investigation of the Behavior of Schumacher Filter Elements in the EPRI Hot-Gas Filter on the Grimethorpe PFBC. EPRI Report No. TR-100499, Prepared by Grimethorpe PFBC Establishment under Contract RP 1336-08, September 1992

Grimethorpe High-Temperature/High Pressure Experimental Program. Volume 4: Performance Evaluation of the EPRI Large-Scale Hot-Gas Filter on the Grimethorpe PFBC. EPRI Report No. TR-100499, Prepared by Grimethorpe PFBC Establishment under Contract RP 1336-08, September 1992

Lippert, T.E. (1992). 10 MW Advanced Particle Filter for installation on the AEP PFBC Tidd plant. In Proceedings of the Second EPRI Workshop on Filtration of Dust from Coal-Derived Reducing and Combustion Gases at High Temperature, San Francisco, CA., USA.

Morrell, R., Butterfield, D.M., Clinton, D.J., Barratt, P.G., Oakey, J.E., Reed, G.P., Durst, M. & Burnard, G.K. (1990). The mechanical performance of ceramic dust filter elements in the tertiary dust capture filter of the Grimethorpe fluidised bed combustor (PFBC). In Proceedings of the 1st International Conference on Ceramics in Energy Applications, Sheffield, U.K.

Mudd, M.J. & Hoffman, J.D. (1993). Operating data from the Tidd hot gas clean up program. In Proceedings of the ASME 12th Fluidised Bed Combustion Conference, La Jolla, CA., USA., to be published.

Pillai, K. K. (1988). Pressurised fluidised bed combustion. In Electricity: Efficient end-use and new generation technologies, and their planning implications, ed. T.B. Johansson, B. Bodlund & R.H. Williams. Lund University Press, Lund, Sweden, pp. 555-593 (ISBN 0-7966-065-7).

Pitt, R.U. & Grüttner, K. (1991). Pulse cleaning system design support calculations for the EPRI hot gas filter pilot plant: 1991 British Coal Topping Cycle Project at Grimethorpe PFBC Establishment. Report prepared for EPRI under Contract RP 3161-04 (Task 1). To be published.

Pitt, R.U. & Leitch, A.J. (1991). A simple method to predict the operation of flue gas filter pulse cleaning systems. In Proceedings of the ASME 11th Fluidised Bed Combustion Conference, Montreal, Canada, pp. 1267-1281.

Stringer, J., Leitch, A.J. & Clark, R.K. (1991). The EPRI hot gas filter pilot plant at Grimethorpe: What worked, what broke and where do we go now ? In Proceedings of the ASME 11th Fluidised Bed Combustion Conference, Montreal, Canada, pp. 971-974.

von Wedel, G.-W., Rehwinkel, H. & Kitchen, W.A. (1992). Deutsche Babcock pressurized circulating fluidized technology test facility experience and outlook for market applications. In <u>Proceedings of the EPRI Conference of Fluidized Combustion for Power Generation</u>, Cambridge, MA., USA.

FILTRATION PROPERTIES OF DUST
FROM FLUIDISED BED GASIFICATION SYSTEMS

P CAHILL, G RASMUSSEN, M TUSTIN AND D ROBERTSON
Coal Research Establishment, British Coal Corporation,
Stoke Orchard, Cheltenham GL52 4RZ

ABSTRACT

Advanced coal-based combined cycle power generation technologies, such as the British Coal Topping Cycle, require hot clean gas to be produced in order to maximise their thermal efficiency. The cleaning is needed to prevent wear of the heavy duty industrial gas turbines which burn the gas and to meet emission requirements.

The British Coal Topping Cycle is based on a pressurised air blown spouted bed gasifier to produce fuel gas with a circulating fluidised bed combustor to burn the char. Hot gas filtration using ceramic filter elements is used to remove particulate material from the fuel gas. Performance of the hot gas filter will be affected by the design of the system, operating conditions and also the nature of the dust from the fluidised bed gasifier.

The work reported in this paper is part of a programme of hot gas filtration studies being carried out at the Coal Research Establishment, Cheltenham. Satisfactory cleaning was achieved for a range of UK and internationally traded coal/sorbent combinations. There was some variation in cleaning behaviour for the dusts but further work is required in order to quantify these effects in terms of the impact upon operation of a multi-element filter unit. The results have highlighted properties of the dust and coal which may have an impact on filtration performance. A programme of additional tests is recommended in order to investigate further the effect of dust properties on filtration performance. Reducing filtration temperature from 600°C to 400°C did not lead to difficult to clean dust cakes and this result suggests that filtration temperature for the demonstration plant will not be a major issue as far as cleaning is concerned.

INTRODUCTION

British Coal is developing an advanced combined cycle power generation system known as the British Coal Topping Cycle. The cycle involves partial gasification of coal in a pressurised air blown fluidised bed gasifier and the residual char is burned in a circulating fluidised bed combustor. A controlled amount of sulphur is removed from the fuel gas by addition of either limestone or dolomite to the fluidised bed. The fuel gas is cooled and cleaned prior to being burned to drive a gas turbine with exhaust heat recovery. A hot gas filter incorporating ceramic filter elements will be used for dust removal at temperatures up to 600°C.

A progr.mme of hot gas filtration research and development is being carried out at the Coal Research Esta ,lishment (CRE) (Cahill et al, 1991.). The work addresses design and operating issues for a hot gas filter for use in coal based gasification combined cycles and in particular for the British Coal Topping Cycle. The main areas of work are: optimisation of filter cleaning, durability of ceramic filters, design of filter hold-down/sealing systems, hot gas pulse cleaning and optimisation of flow of dust laden gas in the filter vessel. The work described in this paper forms part of this programme and addresses how changing coal and sorbent properties will affect the filtration behaviour of the dust.

Potential problems for the filter resulting from the nature of the dust are:

- A dust cake which requires high pressures to clean from the ceramic filters.
- A dust cake which re-disperses on cleaning resulting in dust being carried back to the ceramic filters.

It may be possible to design filter units to cope with difficult dusts; for instance a dust cake requiring high cleaning pressures could be accommodated by designing a cleaning system to deliver a high pressure cleaning force and matching this by components designed to cope with the stresses generated. Alternatively, the design of the filter could be conservative in terms of, for instance, using a low filtration velocity. This will result in a larger vessel and an increased number of ceramic filter elements and therefore will have capital cost implications for plant. For a filter once installed, problems with difficult dusts could result in restrictions on the feedstocks used, outages, or down-rating of plant, which would impose cost penalties on the plant.

Recent tests at the Grimethorpe PFBC Establishment (GPFBCE) showed that performance of the ceramic barrier filter was affected by the nature of the coal/sorbent combination used (Holbrow, 1992.). Using an untreated power station grade fuel and limestone feed to the pressurised fluidised bed combustor (PFBC) led to formation of a dust which was difficult to clean. This resulted in a high rate of increase in pressure drop for the filter and may have contributed to failure of filter components due to excessive forces. Use of dolomite instead of limestone resulted in a lower rate of rise in pressure drop. The effect of dust type has been observed by Westinghouse Corporation, USA on their filter connected to a Texaco gasifier at the Montebello site, USA (Lippert, 1992). The nature of the dust from the oxygen blown gasifier was such that it was found necessary to operate with a low filtration velocity of 1 cm/s. It is believed that problems were caused by poor cohesion of the dust and this was leading to re-entrainment. It is important that the hot gas filter is designed to cover the range of dusts which may be experienced for a given type of plant. Therefore, prior to design of the filter, an investigation needs to be carried out to determine the filtration characteristics of the dust and to identify potentially difficult to clean dusts. The study reported in this paper forms part of such an assessment for the British Coal Topping Cycle.

Dust properties which may affect filtration performance include particle size/shape, permeability of the dust, and potential physical/chemical interactions between dust cake and gas vapours. Temperature, filtration velocity and the filter medium used will also have significant implications for filter performance. A knowledge of the filtration behaviour of the range of dusts likely to be encountered on a given filter will lead to improved design for many aspects of the filter; for instance the number of filter elements, pulse system design pressure, maximum pressure drop across the tubesheet, pressure to be tolerated by the hold-down device and design of dirty gas flow.

In order to investigate the performance of different dusts and filter materials it is necessary to simulate as closely as possible the gaseous atmosphere and particle size of the dust for the given application. Use of a simulator rig using re-entrained dust may not give the performance experienced on the actual filter (Lippert, 1992). There are a number of possible reasons for the poor simulation of real plant performance: for instance the process of dust feeding in the simulator does not break up agglomerates of fine particles and processes such as vapour deposition are not simulated.

The principal aim of the work described in this paper was to investigate, for the British Coal Topping Cycle process, the sensitivity of filtration behaviour to changes in coal/sorbent feed to the gasifier and therefore to the nature of the dust. An existing rig, the Hot Coupon Test Rig, connected to an atmospheric pressure gasifier was used to carry out the investigation. The rig provides a comparatively simple hot gas filtration test.

Filtration of the gasifier-derived dust is carried out in the rig using a flat disc of ceramic filter medium (the coupon). The filtration behaviour of the dust is studied by monitoring pressure drop across the coupon as the dust collects on the surface. In this way a "dust cake" is formed on the coupon. The cleaning behaviour of the dust cake is evaluated in two ways:

i. Regular removal of the dust cake using short pulses of pressurised nitrogen and monitoring the changes in the pressure drop across the coupon immediately before and after cleaning.

ii. Cake detachment tests. After a number of regular cleaning events have been carried out then the dust cake is allowed to build-up for a specified period but is then not pulse cleaned. The detachment test comprises carrying out a sequence of cleaning pulses, each with a higher pressure than the previous. The amount of dust removed after each pulse is weighed and the resistance to flow of the residual cake is studied. The percentage by weight removed is plotted against cleaning pressure (referred to as a cake detachment profile). This "cake detachment test" provides information on the amount of force required to remove a given dust cake.

The nature of the dust removed in the tests is important, ie. dust cake removed as lumps or flakes indicates that the dust is relatively cohesive and therefore less likely to be re-entrained into the gas stream than for a dust cake which is removed as individual particles or very small agglomerates.

The information required from the tests was to identify potentially difficult to clean dusts, to investigate how dust/coal properties affect filtration performance and to assess the effect of filtration temperature. The programme was not designed to provide guidelines to operating or cleaning strategy for a hot gas filter. These issues are being addressed through the main programme of filtration work at the Coal Research Establishment.

FILTRATION TESTS

All filtration tests were carried out on a hot coupon test rig installed on an atmospheric pressure fluidised bed gasifier at the Coal Research Establishment. Cake detachment tests were carried out for a range of coal/sorbent combinations using filtration temperatures of 400°C and 600°C.

Equipment

A schematic of the equipment used is shown in Figure 1. A hot coupon test rig (Cheung et al 1989, Cheung 1989.), connected directly to an air blown fluidised bed gasifier, was used to study the filtration behaviour of dust from a range of different coal/limestone sorbent combinations. The dust laden gas to the coupon rig was taken from the gasifier immediately after the cyclone. In normal operation the cyclone efficiency was around 95 %. The cyclone and all lines leading to the coupon rig were trace heated to ensure that the temperature of the dust laden gas did not fall below the filtration temperature resulting in loss of volatile components.

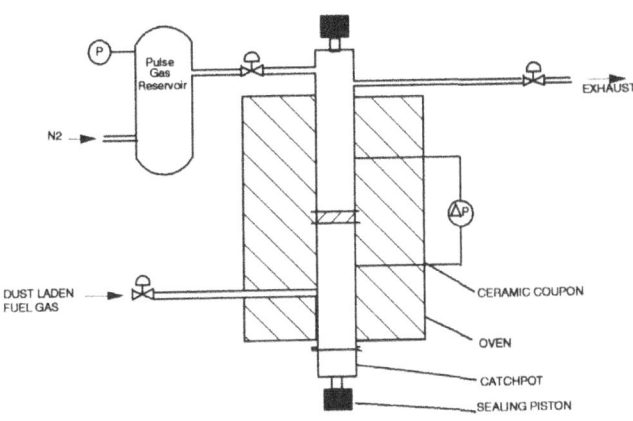

Figure 1   Hot Coupon Test Rig

The coupons, made from clay bonded silicon carbide with a fibrous surface coating, had dimensions of 60 mm outer diameter and 15 mm thickness. The coupons were sealed between two steel discs which were welded together at their outer edges leaving the central part of the ceramic exposed. This assembly was then sealed into a steel vessel so that it separated into two chambers. Dirty gas was admitted from below the coupon, depositing dust on its lower face. Compressed nitrogen at ambient temperature was pulsed

115

periodically from the clean, upper, side of the coupon to remove the accumulated dust cake and the pressure drop across the coupon was logged. When the pressure drop stabilised, the medium was considered to be "conditioned" and the cake removal stress was determined. This was done by a sequence of reverse pulses at increasing pressures. The cake detached was collected and weighed after each of the pulses.

Coal/Sorbent Combinations

A range of coal and sulphur retention sorbents was used to vary the characteristics of the dust from the gasifier. The coals were selected to give a range of properties, typical for UK and internationally traded coals used in power generation. Both deep mined and opencast mined coals were included. The composition of the coals selected for the tests are given in Table 1. Tests were carried out at both 400°C and 600°C.

Ash contents varied between 8.9% (El Cerrejon) and 22.4% (Kiveton Park run of mine,[ROM]). Kiveton Park part treated coal and the Coalfield Farm coals contained highest iron in ash, 16.7% and 16.0%, respectively. The Coalfield Farm coal had much higher calcium in ash than the other coals, 13.5% compared with an average of about 3%. Thoresby and Kiveton Park coals contained the highest sodium in ash contents of 2.1% and 1.2%, respectively. Thoresby and Hunter Valley had the highest swelling numbers. All coals were prepared to 100% less than 3 mm. The limestone used was Grangemill limestone which is similar in composition to the Middleton limestone used at Grimethorpe PFBC Establishment. The Cadeby dolomite was used at Grimethorpe. The limestone and dolomite had similar size distributions and both were low in iron ($< 0.5\%$ ad) and sodium ($< 0.2\%$ ad).

TABLE 1
Composition of Coal (Preparation to 100% <3mm)

	Kiveton Park (Part treated)	Kiveton Park (ROM)	Coalfield Farm	Daw Mill	Thoresby	El Cerrejon	Hunter Valley
Moist % ad	3.0	3.4	6.0	3.8	2.6	3.1	2.5
Ash % ad	10.4	22.4	9.6	12.2	16.4	8.9	9.0
VM % daf	38.5	38.5	41.8	41.7	38.5	39.4	37.7
C % db	70.5	60.6	71.0	70.5	68.3	72.6	75.4
H % db	4.70	3.85	4.39	4.26	4.32	4.69	4.73
N % db	1.60	1.37	1.39	1.15	1.53	1.53	1.74
S % db	1.89	1.69	1.81	1.51	1.91	0.93	0.47
Cl % db	0.28	0.19		0.26	0.51		
Ash Composition % ash							
SiO_2	51.2	58.6	31.7	46.1	51.2	59.0	66.1
Al_2O_3	20.2	21.8	20.8	24.9	26.5	21.6	24.2
Fe_2O_3	16.7	10.7	16.0	8.8	11.3	9.5	5.0
CaO	2.9	2.0	13.5	7.6	2.8	3.1	0.8
MgO	1.6	1.6	2.6	2.0	1.4	2.0	0.6
Na_2O	1.2	0.8	0.2	0.8	2.1	0.6	0.3
K_2O	2.1	3.2	1.3	2.1	3.8	2.4	1.6
Swelling No	1	1	0.5	1	5	1	4

All of the coals were tested with limestone addition of calcium to sulphur ratio (Ca/S) of 2:1. In addition, the Kiveton Park part treated coal was tested with limestone (Ca/S 1:1), with dolomite (Ca/S, 2:1) and without sorbent. The Kiveton Park coal combinations were chosen to give change in ash composition and sorbent content/type for one coal.

Tests were carried out, at 600°C and 400°C which represent possible filtration temperatures for the Topping Cycle.

The coal/sorbent combinations used in the 600°C filtration tests are listed below.

i.   Kiveton Park (no limestone)

ii.   Kiveton Park (Ca/S, 1:1)

iii.   Kiveton Park (Ca/S, 2:1)

iv.   Kiveton Park/Cadeby Dolomite (Ca/S, 2:1)

v.   Run of Mine (ROM) Kiveton Park Coal (Ca/S, 2:1) - high ash content

vi.   Coalfield Farm (Ca/S, 2:1) - high iron, high calcium ash content

vii.   Daw Mill (Ca/S, 2:1) - low swelling properties

viii.   Thoresby (Ca/S, 2:1) - high sodium in ash

ix.   El Cerrejon (Ca/S, 2:1) - Columbian internationally traded coal

x.   Hunter Valley (Ca/S, 2:1) - Australian internationally traded coal

For the 400°C tests the number of coal/sorbent combinations used were: Kiveton Park (Ca/S 2:1), Kiveton Park/Cadeby dolomite (Ca/S, 2:1), Thoresby (Ca/S, 2:1) and Hunter Valley (Ca/S, 2:1). Hunter Valley and Thoresby, respectively, were chosen as the low and high chlorine content coals. The range of chlorine contents were to allow investigation of the effect of chlorine adsorption on cleaning of the dust cakes (chlorine adsorption by the dust is favoured more at 400°C than at 600°C). The Kiveton Park combinations were chosen to investigate the effects of sorbent type at the two temperatures.

Experimental Procedure

The experimental procedure for each of the tests was:

Coal/sorbent mixture was fed to the gasifier and the bed temperature controlled at around 980°C. When the gasifier had achieved stable conditions, the inlet to the coupon test rig was opened and the first filtration cycle commenced. In order to achieve a dust loading on the coupon surface of 200 g/m^2, a filtration velocity of 10 cm/s was used and the coupon was pulsed every fifteen minutes using a pulse pressure of 2.5 bar$_a$. These conditions were used for all of the tests carried out. The nature of the dust cake is dependent upon the dust and the filtration velocity used. Thus, 10 cm/s was used both as an expedient to carry out tests in an acceptable time, to accentuate compaction of the dust cake and to maximise the probability of dust penetrating the coupon material. Pressure drop across the coupon was monitored throughout the conditioning period and the detachment test. Up to ten hours was allowed for "conditioning" of the coupons. Conditioning was judged as a slow rate of rise in residual pressure drop (ie pressure drop immediately after a cleaning event). The conditioning period was much shorter than would be required for conditioning on a filter unit (several hundred hours). The effect of coal type for longer duration tests will be addressed in separate work on a twelve element filter rig at CRE. After conditioning, the coupon was pulsed one final time and the rig was isolated from the gasifier. The accumulated dust was removed. A clean, dry and weighed filter paper was inserted and fuel gas was re-introduced. After 15 minutes the filtration was stopped and in this period no pulse took place. The detachment test was then carried out which comprised pulsing the filter with incremental pulse pressures in the range 0 to 200 kPa, collecting and weighing the accumulated dust after each pulse. In this way it was possible to construct a cake detachment profile for each coal.

The dust collected during the conditioning period for each coal was analyzed for particle size (Coulter Counter), carbon content, hydrogen content, ash content, ash constituents (7) and surface area (BET). This was to identify properties of the dusts which could be related to the filtration performance. Scanning electron microscopy with electron dispersive x-ray analysis (SEM/EDX) was carried out for all dusts in order to provide qualitative information on the morphology and surface composition of the dusts. The residual dust on all coupons from the tests was analyzed using SEM and optical microscopy. A resin impregnated polished section was prepared and examined by optical microscopy for dust ingress.

RESULTS AND DISCUSSION

Duplicate cake detachment tests were carried out for the Kiveton Park coal without limestone and for Kiveton Park coal with 2:1 limestone. In both cases reproducibility was found to be acceptable (less than 10% difference in the amount of material collected per pulse).

Typically the "conditioned" residual pressure drop was around 30 mbar and the peak pressure drop was 100 - 130 mbar. Figure 2 shows typical pressure drop and temperature recordings for the conditioning period. Typical "saw tooth" shaped pressure drop traces were observed for all of the tests. The peak pressure varied according to the coal /sorbent combination used, the state of conditioning and also when operations such as fines removal from the cyclone catchpot were carried out.

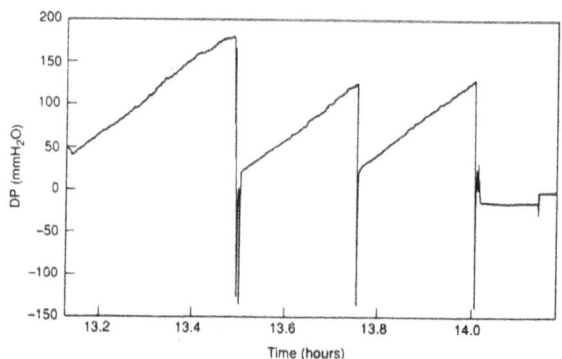

Figure 2   Typical Pressure Drop Trace During Filtration

Effect of Coal Type

Figure 3 shows the effect of coal type on cake detachment (Grangemill limestone was added to all coals in the ratio Ca/S 2:1). A range of cake detachment stresses were observed. The Kiveton Park run of mine and the Thoresby coals gave the most difficult to detach cakes whereas the Hunter Valley and Coalfield Farm coals gave easiest to detach cakes.

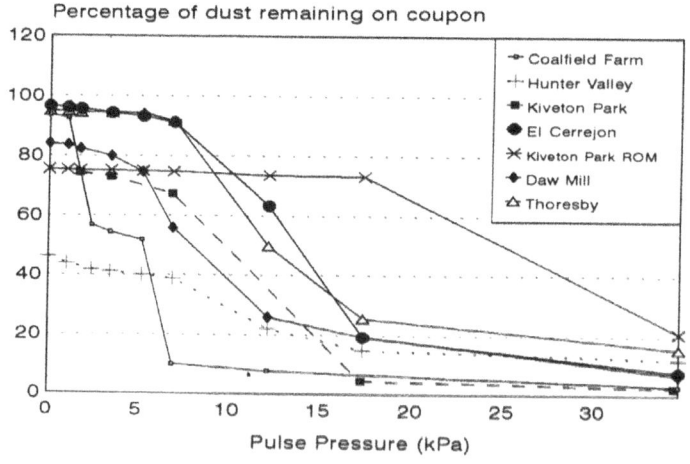

Figure 3   Cake Detachment Results for the Range of Coals

120

For all coal/sorbent combinations, the dust samples collected during the detachment tests comprised individual dust particles and flakes varying in size between about 2-4mm across. The flakes broke-up easily when disturbed. The results suggest that for each of the coal types it should be possible to remove a significant amount of the dust cake as flakes or lumps, thereby reducing the possibility for re-entrainment of dusts. The flakes broke up relatively easily therefore turbulence flows which could break up the flakes must be avoided in a filter vessel. This removal of the dust cake in flakes or patches was confirmed by studying the pressure drop across the coupon for forward flow after each cleaning pulse. Analysis of values for this residual pressure drop against the quantity of cake removed allows investigation of the nature of detachment. The results for the detachment tests shows that for all coals the initial detachment of dust is consistent with removal of patches of dust ("patchy cleaning").

The analysis results for the dusts collected during the conditioning period for each coal are given in Table 2.

The median size for the dust samples varied between 3.8 and 6.2 microns. Surface areas for the dusts varied considerably between 7 m^2/g (Thoresby) and 122.3 m^2/g (Coalfield Farm). The variation in surface area cannot be accounted for by differences in particle size and the SEM analysis indicates that all dust samples comprise irregular shaped char and ash fragments. Coalfield Farm appears to contain a high proportion of angular fragments. However, it is likely that variation in porosity of the char, rather than particle shape, accounts for the major differences in surface area.

A relationship between the dust properties and the detachment behaviour is required in order to obviate the need for an empirical test (possibly based on the cake detachment test) to investigate filtration behaviour. The varying shape of the detachment curves makes it difficult to relate the filtration behaviour to the dust properties. Therefore, the pulse pressure for 80% removal has been interpolated from Figure 5 for each of the coals and used for comparison with dust properties.

The order of ease of removal was:

Coalfield Farm

Hunter Valley

Kiveton Park

Daw Mill

El Cerrejon

Thoresby

Kiveton Park ROM

TABLE 2
Analysis of Dust from Effect of Coal Type Tests

	Kiveton Park 2:1	Daw Mill 2:1	El Cerrejon	Hunter Valley 2:1	Thoresby 2:1	Kiveton Park ROM	Coalfield Farm	Lignite
Median Size (microns)	4.7	5.9	4.8	6.2	4.8	5.4	3.8	4.1
Carbon Content (% ad)	50.9	56.0	52.1	54.6	44.0	37.7	48.4	15.6
H (% ad)	1.3	1.3	1.0	1.1	1.1	0.9	0.9	0.6
Ash Content (% ad)	47.8	42.7	46.9	44.3	54.9	61.4	50.7	83.8
Ash Composition (% dust)								
SiO_2	15.4	16.8	25.4	17.3	25.7	27.5	30.1	36.4
Al_2O_3	9.2	11.0	11.0	9.3	14.0	13.5	11.5	25.3
Fe_2O_3	5.2	2.6	3.1	4.0	4.8	5.7	2.9	4.4
CaO	9.7	5.8	2.0	7.5	3.9	4.5	1.9	8.6
Na_2O	0.6	0.5	0.3	0.2	1.3	0.5	0.2	0.3
K_2O	1.1	1.1	1.3	1.0	2.2	2.0	1.2	2.1
Surface Area m^2/g	16.2	53.6	19.6	41.5	7.0	12.1	122.3	1.9

The dust from the Thoresby and Kiveton Park ROM had the highest carbon contents. There was no relationship between the 80% removal force and silica, iron, calcium, sodium or potassium contents. Generally, the more difficult to clean dusts had lower surface areas than the easiest to clean dusts, particularly for the pulse pressures less than 15kPa. A hypothesis which would explain the surface area effect is condensation of alkali metal vapours (as the gases cool from around 1000°C to 600°C) affecting cake detachment behaviour. Surface concentration of condensed alkali metal vapours would be less for high surface area dusts, ie. the vapours will deposit in the pores as well as on the exposed surfaces where inter-particle contact takes place resulting in less alkali on particle surfaces to promote sticking.

The EDX spectra did not indicate enhanced surface concentration for any of the ash constituents. For all of the samples no significant amounts of sodium were detected by EDX. The SEM analysis showed that all dust samples comprised angular fragments of char and ash with a small proportion of spheres. It is concluded that there were no major observable differences for the dusts investigated.

Micrographs from dust samples collected during a sequence of detachment pulses at different pressures showed that the sample collected at the lowest pulse pressure contained more of the large particles than for the samples collected at higher pressures. The presence of large particles may lead to more easily detached patches of dust cake.

Kiveton Park Coal Combinations

Figure 4 shows cake detachment results for the Kiveton Park (KP) coal combinations. Lowest detachment stresses were obtained for KP/no limestone and the test where Grangemill limestone was added at 1:1. Addition of limestone to the KP coal in the ratio 2:1 resulted in a more difficult to detach dust. Addition of dolomite instead of limestone made the dust easier to remove, the effect seen during the tests for PFBC conditions at Grimethorpe.

The dust samples varied in ash content, from approximately 40 to 60%, but the median particle size for the samples varies little, 4.1 to 5.4 microns. The absence of a significant variation in particle size is a result of using a cyclone upstream of the coupon rig. The dust from the run of mine coal contained more silica and alumina than the other dust samples. There was little variation in iron, and sodium content for the ashes. The dust from the run of mine (ROM) coal, which proved most difficult to detach, had the lowest carbon content for the samples tested. SEM investigation showed that this dust contained many angular particles.

123

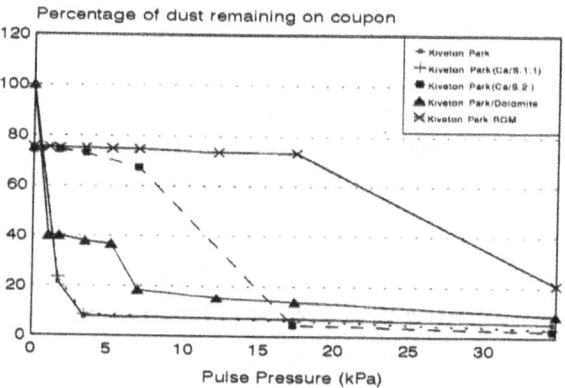

Percentage of dust remaining on coupon

Legend:
- Kiveton Park
- Kiveton Park (Ca/S 1:1)
- Kiveton Park (Ca/S 2:)
- Kiveton Park/Dolomite
- Kiveton Park ROM

Pulse Pressure (kPa)

Figure 4   Cake Detachment Results for Kiveton Park Coal Mixtures

The principal differences between the dust from Kiveton Park/no limestone and Kiveton Park/limestone (2:1) was that the former contained more carbon, median particle size was higher, calcium content was lower, silica content was higher and had twice the surface area. The high surface area for the Kiveton Park coal without limestone is probably a result of the high carbon content for this dust. The surface area effect could be related to condensation of sodium vapours, as discussed previously.

Comparing dusts from the KP/limestone and the KP/dolomite, the principal differences were that the KP/dolomite dust contained more ash (reflecting more dolomite is added than limestone for a given calcium to sulphur ratio) and the dust contained less potassium. SEM analysis of the coupon surfaces after the cake detachment tests did not indicate any significant differences in the composition and size of the residual dust (which would have comprised part of the conditioning layer).

It is difficult to relate the detachment behaviour to the composition or morphology of the dust. Evidence from the tests using Kiveton Park only, KP/limestone and KP ROM/limestone suggests that increased carbon content and surface area of the dust favours ease of cleaning. However, this relationship does not explain differences between the KP/limestone 2:1 and the KP/dolomite 2:1.   This suggests that several factors are significant to cleanability in the test procedure described.

Effect of Filtration Temperature on Cake Detachment

Reducing filtration temperature from 600°C to 400°C for Kiveton Park (no limestone), Hunter Valley (2:1), Thoresby (2:1) and Kiveton park/Cadeby dolomite (2:1) had little effect on pressure drops during the conditioning. Figure 5 shows the cake detachment results for these coals at both 400°C and 600°C. Reducing temperature appears to have made the dust from Thoresby coal easier to detach but the opposite appears to be true for the other coals.

Chemical analysis of the dusts showed that they were similar in composition to those produced at 600°C. One possible effect of reducing filtration temperature is to increase or change the nature of alkali metal vapour deposition. Previous work on this gasifier gave alkali concentrations of a few ppm at 600°C (Fantom 1993), therefore increased condensation at 400°C is unlikely to result in a measurable increase in alkali metal content for the dusts compared with dusts from 600°C tests.

EDX analysis of the dusts from 400°C filtration indicated the presence of significant amounts of chlorine for the Thoresby and Kiveton Park whereas little chlorine was present for the dusts from the 600°C tests. Only a small amount of chlorine was detected for the Hunter Valley dust from the 400°C test. The results suggest that chlorine is being deposited onto the dust in significant quantities at 400°C but not at 600°C. The chlorine is most likely to have been retained in the dust as calcium chloride which is favoured thermodynamically at 400°C rather than at 600°C. The amount of chlorine in the dust samples appears to reflect the amount of chlorine in the coal rather than the amount of calcium in the dust.

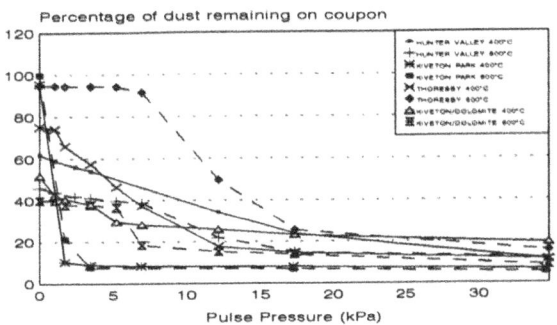

Figure 5   Effect of Filtration Temperature on Cake Detachment

CONCLUSIONS

The work is part of a programme of hot gas filtration studies being carried out at the Coal Research Establishment, Cheltenham. Satisfactory cleaning was achieved for a range of UK and internationally traded coal/sorbent combinations. There was some variation in cleaning behaviour for the dusts. Increased limestone added to the coal feed and high coal ash content both made the dust cake easier to clean. It was not possible to identify a single property of the coals or dusts which could be measured in order to determine filtration behaviour. There was some evidence to suggest that easiest to clean dusts had highest surface area and lowest ash content. A hypothesis which would explain the surface area effect is condensation of alkali metal vapours affecting the properties of the dust cake. A programme of additional tests is recommended in order to investigate further the effect of dust properties on filtration performance. Reducing filtration temperature from 600°C to 400°C did not lead to difficult to clean dust cakes and this result suggests that filtration temperature for the demonstration plant will not be a major issue as far as cleaning is concerned.

ACKNOWLEDGEMENTS

The work was sponsored by the United Kingdom Department of Trade and Industry as part of the appraisal of the British Coal Topping Cycle.

REFERENCES

Cahill P, Hudson D M and Minchener A J, Hot Gas Clean up for the British Coal Topping Cycle. EPRI 10th Annual Conference on Gasification Power Plant. San Francisco, USA, 1991.

Holbrow P., Second EPRI Workshop on Filtration of Dust from Coal-Derived Reducing and Combustion Gases at High Temperature, San Francisco, USA, 1992.

Lippert T, Second EPRI Workshop on Filtration of Dust from Coal-Derived Reducing and Combustion Gases at High Temperature, San Francisco, USA, 1992.

Cheung W, Seville J P K, Clift R, Bower C J and Twigg A N, "Filtration and Cleaning Characteristics of Ceramic Media". Inst. Energy 4th FBC Conference, 1989.

Cheung W, Phd Thesis, "Filtration and Cleaning Characteristics of Ceramic Media, University of Surrey, 1989.

Fantom I R, "Measurement and control of Alkali Metal Vapours in Coal-Derived Fuel Gas", 2nd International Conference on gas Cleaning at High temperatures, University of Surrey, 1993.

FILTRATION OF FLYSLAG FROM THE SHELL COAL GASIFICATION PROCESS USING POROUS CERAMIC CANDLES

J.N. Phillips and H.W.A. Dries
Shell Internationale Petroleum Maatschappij, B.V.
P.O. Box 162, 2501 AN, The Hague, the Netherlands

ABSTRACT

Ceramic candle filters were used to remove flyslag from the syngas produced at the Shell Coal Gasification Process demonstration plant, SCGP-1, near Houston, Texas, USA. Over 5000 hours of operation at 2.4 MPa and temperatures ranging from 493 to 553 K [220 to 280°C] were achieved without a candle failure. The 44 candles were cleaned on-line by periodic blowback pulses of recycled syngas. Most of the operation occurred with a cyclone upstream, but 220 hours of tests were performed with the cyclone bypassed. Both with and without the cyclone, the candles performed as near absolute filters with typically 5-10 ppmw [5-10 mg/Nm^3] solids slip. The permeability of the candles after blowback has been calculated, and the decline is well represented by an exponential decay model.

INTRODUCTION

The Shell Coal Gasification Process (SCGP) is based on a dry feed, entrained-flow, high pressure, high temperature slagging design. A simplified block flow diagram of the SCGP is presented in Fig. 1. The coal is first dried and pulverized. Next it is pressurized to the gasifier operating pressure (typically 2.5-3.0 MPa) using lockhoppers and high pressure nitrogen. The pressurized coal is pneumatically transported to the gasifier with nitrogen. The coal, oxygen, and, if necessary, steam enter the gasifier through pairs of opposed burners. The oxygen and nitrogen are supplied by an air separation plant.

In the gasifier the combustible fraction of the coal is converted into synthesis gas, or syngas. Part of the heat of reaction is recovered as steam in the surrounding membrane wall. Most of the mineral content of the coal is converted into molten slag. The slag drains into a water bath, is frozen, depressurized, dewatered, and may later be used in the production of concrete or other materials (Salter et al., 1991).

The hot syngas leaving the gasification zone is quenched with cooled, recycled syngas to convert any entrained molten slag into solid "flyslag".
The syngas cooler recovers high-level heat from the quenched syngas by generating high-pressure steam. By doing so, up to 95% of the coal's energy is recovered in a useful form (syngas or steam).

127

The flyslag contained in the gas leaving the syngas cooler is removed from the gas using a cyclone and/or a filter. The recovered flyslag can be recycled back to the gasifier to provide almost complete carbon conversion.

The syngas then goes to a water wash system, where water soluble contaminants such as ammonia are removed, and thereafter to an acid gas removal system, where an amine-based solvent, such as Shell's Sulfinol, typically removes more than 99% of the sulphur species. These sulphur species (primary H_2S) are then converted into marketable elemental sulphur using the Claus process. The clean syngas may then be used as fuel in highly efficient gas turbine combined cycle power plants.

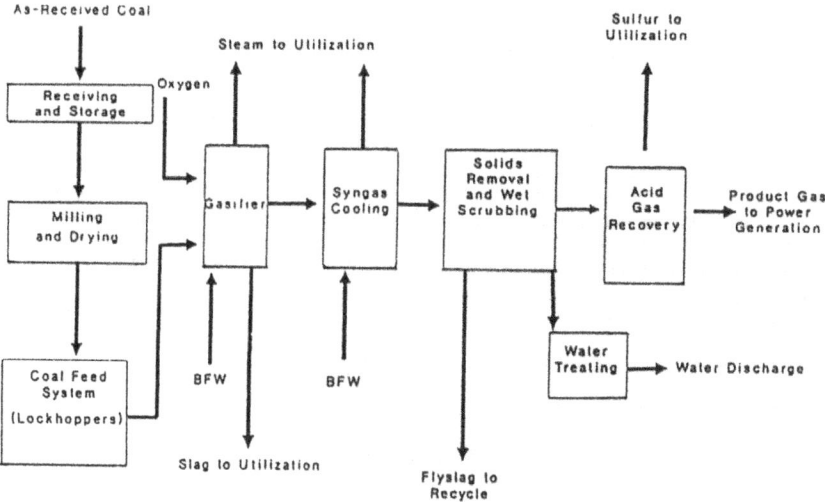

Figure 1. Block Flow Diagram of the Shell Coal Gasification Process.

Importance of Filtration

The use of dry solids filtration to remove flyslag from the syngas is a critical part of the process. Dry flyslag removal simplifies the design of the downstream water wash system and improves the reliability of the wet system by virtually eliminating the possibility of process piping or columns becoming plugged with fines. It makes it economical to recycle the recovered flyash. If a wet scrubbing system was used to remove the flyslag, the energy requirement for evaporating the water in the resulting slurry would tend to counteract any efficiency gains from improved carbon conversion. The use of dry solids filtration also allows the use of warmer quench gas which maximizes the quality of the recovery of the syngas sensible heat.

128

For those reasons, Shell has placed considerable priority on testing medium temperature, high pressure filtration technologies.

History of SCGP and the Use of Ceramic Filters

The development work on SCGP began in 1972 using the experience gained from the Shell Gasification Process for heavy oil (SGP). SGP was introduced in 1956 and has been applied on a wide scale throughout the world (over 150 gasifiers).

A 0.07 kg/s [6 metric ton/day] coal gasifier was built at Shell's Amsterdam Research Laboratories and was commissioned in 1976. This unit has been used to gasify a wide range of coals and to test new concepts for process improvement. Shell's first test with ceramic candle filters were conducted here in 1986, and subsequent tests have been conducted in the years following.

In 1978, a 1.7 kg/s [150 t/d] pilot plant was started up in Deutsche Shell's Harburg Refinery. During its program, which ended in 1983, this unit established the viability of the process as indicated by a 1000 hour continuous run.

In 1987, a more advanced demonstration unit called SCGP-1 was placed in service in the US at Shell's Deer Park Manufacturing Complex near Houston, Texas. SCGP-1 had a capacity which ranged from 2.5 kg/s [220 t/d] on high sulphur bituminous coal (as received basis) to 4.2 kg/s [360 t/d] on high moisture, high ash lignite. SCGP-1 was operated from July 1987 until April 1991, and the results of the program have been extensively reported elsewhere (Phillips et al., 1993, Mahagaokar et al., 1991). Highlights of its accomplishments included: the gasification of a variety of feedstocks including bituminous and sub-bituminous coals, lignite, and petroleum coke; the ability of the process to meet and surpass all current environmental standards for gas, water and solids streams; and the successful test of ceramic candle filters for over 5000 hours (Salter & Monsavoir, 1992).

Demkolec Project

SEP (the Dutch Electricity Generating Board) have announced that coal gasification will play an important role in their overall generating plan (Zon,1990, Demkolec,1990). A 250 MWe integrated coal gasification combined cycle (IGCC) power plant is under construction at the site of an existing coal fired plant on the Maas River at Buggenum, near Roermond. This IGCC is scheduled to start-up in the second half of 1993.

On the basis of a feasibility study which compared three gasification technologies, the Shell Coal Gasification Process was selected in April 1989. The plant will be built and operated by Demkolec, a SEP subsidiary. A single gasification train, designed for approximately 23 kg/s [2000 t/d] of a range of imported coals, will fuel a combined cycle power plant with a 156 MWe gas turbine and a 128 MWe steam turbine, both supplied by Siemens. The design coal-to-busbar efficiency is 43.2% on LHV basis and 41.4% on HHV basis. Based on the successful results from SCGP-1, ceramic candles will be used downstream of the syngas cooler to remove fine solids from the full stream of syngas.

SCGP-1 FILTRATION RESULTS

Equipment Layout

A schematic of the SCGP-1 dry solids removal system is shown in Fig. 2. SCGP-1 had both a cyclone and a filter. The cyclone was designed to remove approximately 90% of the flyslag from the syngas. The filter, located downstream of the cyclone, was originally designed for fabric bag filters, but was later revised to accommodate ceramic candle filter elements. An isokinetic sample station located downstream of the filter allowed the rate of solids slip to be measured. Isokinetic sample points were also located upstream and downstream of the cyclone.

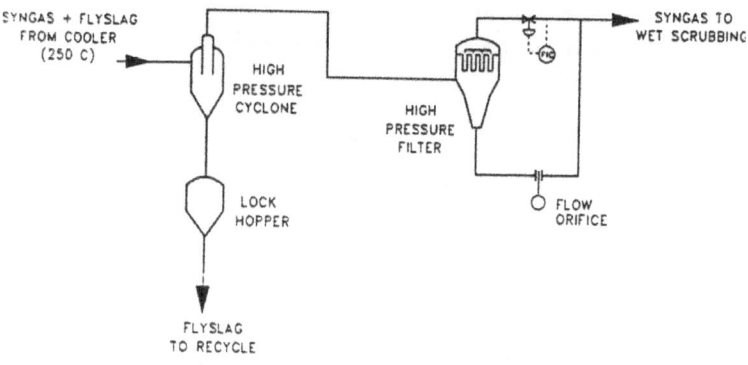

A) NORMAL DRY SOLIDS REMOVAL LINE-UP

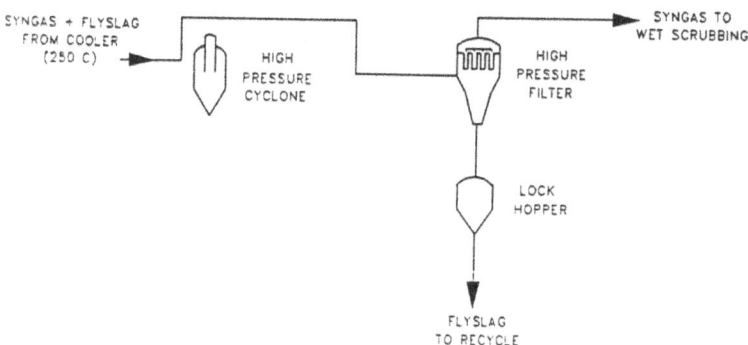

B) DRY SOLIDS REMOVAL LINE-UP WITH CYCLONE BY-PASSED

Figure 2. Simplified Schematic of SCGP-1 Dry Solids Removal System.

130

Flyash Properties

A particle size distribution of flyslag in SCGP-1 cyclone overflow, as determined by Coulter Counter measurement, is shown Fig. 3. The mean particle size was between 2 and 3 μm.

The compositional analyses of the flyslag produced at SCGP-1 has been extensively reported in (Mahagaokar & Krewinghaus, 1990), (Mahagaokar et al., 1990) and (Phillips et al., 1992). The carbon content of the flyslag typically varied from approximately 5% wt to approximately 30% wt depending on the reactivity of the feed to the gasifier and the gasifier operating conditions. A typical carbon content for flyslag derived from bituminous coal of average reactivity was 20%.

The bulk density of the flyslag was influenced by the carbon content and coal type. Bulk density ranged from 400 to 1000 kg/m³ with 600 kg/m³ being typical for a flyslag with 20% carbon.

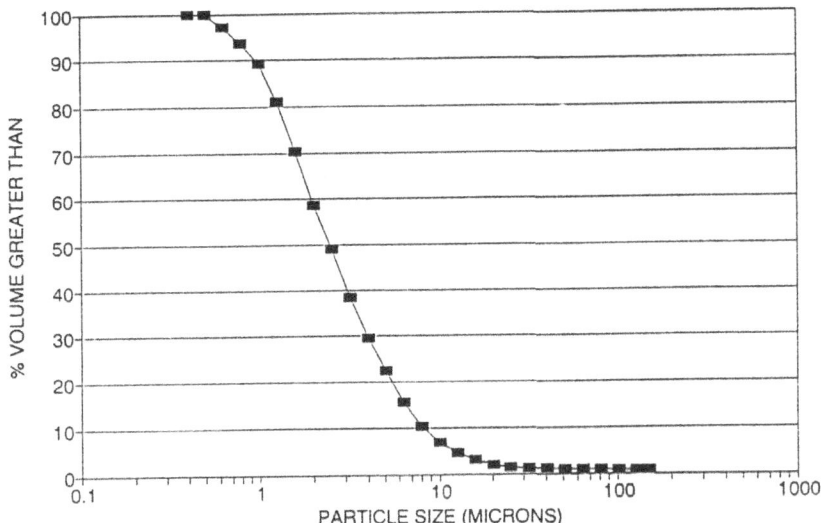

Figure 3. Flyslag particle size distribution of an isokinetic sample taken from the overflow of the SCGP-1 cyclone. Distribution was measured with a Coulter Counter.

Conversion from Fabric to Ceramic Filters

The original SCGP-1 filter was designed to accommodate fabric bag filters which were cleaned by periodic blowback pulses of recycled syngas. Typical operating conditions were 513 K [240°C], 2.4 MPa with an inlet gas dust loading of 2 g/Nm³, a face velocity of 1.0 cm/s and a pressure drop of 2.0 kPa. In total almost 8000 hours of operating experience were accumulated with bag filters on-line, and while their solids removal performance was good (solids slip was normally < 20 mg/Nm³), their reliability was less than desired.

Six different sets of bags were needed during the course of the tests due to leaks which developed. The vendor-supplied pulsing system provided inadequate pulse gas distribution and caused the initial bag failures. A Shell-designed pulsing system featuring higher over pressure (1.7 MPa) solved the back pulsing problems, and the new design performed well for the rest of the project, a period that included over 60,000 blowback cycles. Increased care during installation and emphasis on inspecting bags for defects prior to installation also improved performance. As a result of these changes, the final set of bags was working well when they were removed after more than 1300 hours of service. Nevertheless, there was concern that fabric bags would not be robust enough to withstand typical plant upsets particularly if higher operating temperatures were desired.

Based on this conclusion, and the successful scouting runs with ceramic candle filters at the Amsterdam pilot plant, the decision was made in late 1989 to adapt the filter vessel to accommodate 44 ceramic candle filters. Fig. 4 shows the configuration of the candles in the filter tube sheet. Each of the eight rows of candles had a blowback gas header pipe. Each pipe had nozzles located at the centerline of each candle and a pneumatically-actuated pulse valve between a buffer tank and the filter vessel. The control system for the pulse valves allowed the pulsing to be triggered by a clock or by pressure drop across the filter tube sheet.

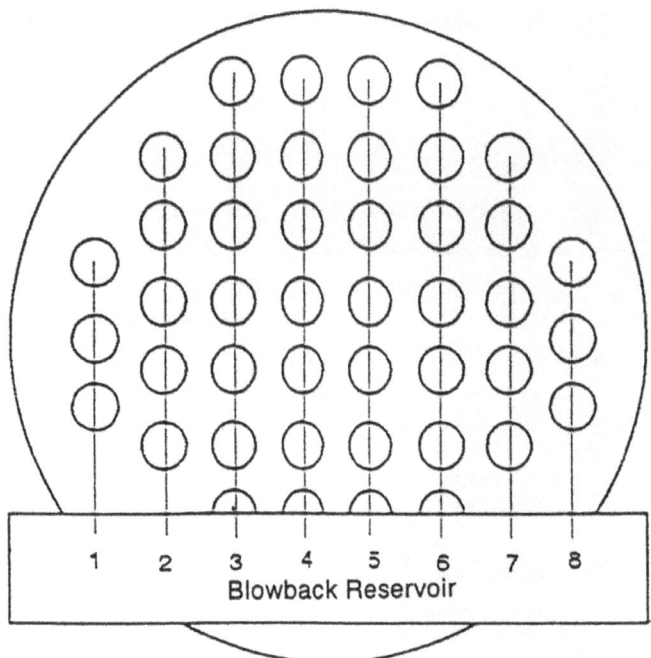

Figure 4.   Layout of 44 element candle filter tubesheet.

132

Two sets of candles were tested at SCGP-1. The first was supplied by Schumacher (Crailsheim, Germany), and the second was supplied by Industrial Filter and Pump or IFP (Cicero, Illinois, USA).

Dia-Schumalith Candles

Each of the 44 Schumacher candles were 1.5 m long with an outer diameter of 60 mm and an inner diameter of 30 mm. The candles consisted of a relatively large pore silicon carbide substrate with a thin (100-200 μm), outside layer of ceramic material with much finer pores. Schumacher uses the tradename of "Dia-Schumalith" for these filters (Durst et al., 1988).

A total of 4400 hours of operation were accumulated on a single set of candles. Included in this are 220 hours in which the upstream cyclone was bypassed. During this period the solids loading of the gas to the filter was significantly greater. Filter performance during this period will be discussed in the following section, but first the performance during the 4173 hours of normal operation will be reviewed.

Performance behind Cyclone

Operating pressure was typically 2.4 MPa. Operating temperature ranged from 503 to 553 K [230° to 280°C]. Face velocity, except during periods of start-up and shutdown, ranged between 2.1 and 3.5 cm/s, and the inlet solids loading ranged between 0.5 and 3 g/Nm^3 with an average value of 2 g/Nm^3 (2000 ppmw).

During the initial operation, the blowback system was controlled by a timer. In this mode the candles were cleaned on a continuous basis with a fixed interval of time between the pulsing of each row of candles. This resulted in an initial pressure drop across the filter of 9.0 kPa.

Solids slip, based on isokinetic samples, during the initial run ranged between 10 and 20 mg/Nm^3 (10-20 ppmw). This was similar to the slip obtained with properly functioning bag filters, and represents a filtration efficiency of 99.0 to 99.5%.

Upon start up of the second run with the ceramic candles, the blowback operation was changed from timer to pressure drop triggering, and from that point on it became the preferred mode of blowback control. As more operating hours were accumulated on the candles, the pressure drop set point for triggering the blowback had to be periodically increased in order to maintain a minimum cycle time dictated by the capacity of the blowback gas compressor.

Particle slip through the candles usually decreased over the course of a run. Typically values around 20 mg/Nm^3 in the first 24 hours after start-up would eventually settle to the range of 5 mg/Nm^3 or less.

Performance without Cyclone

During the period in which the cyclone was bypassed the lockhopper normally utilized by the cyclone was used to depressure the solids captured by the candle filters (see Fig. 2b). This increased both the solids loading in the flow to the candles (to 18-20 g/Nm^3) and the mean particle size of the solids. In addition the face velocity reached up to 5.3 cm/s. The operating temperature varied between 511 and 536 K [238 and 263°C].

133

Filter performance under these conditions was excellent with the filtration efficiency exceeding 99.9%. Isokinetic samples showed that the solids slip was no different than that obtained when the cyclone was in the line-up.

IFP LayCer Candles

While the Schumacher candles were still providing excellent filtration efficiency after 4400 hours of service, there was a desire to qualify a second candle supplier before the SCGP-1 program ended. The second set of candles were supplied by IFP under the tradename LayCer 70/3. Like the Dia-Schumalith candles the LayCer candles consisted of a coarse silicon carbide substrate with a less porous outer layer and were 1.5 m long with an outer diameter of 60 mm. However, the substrate of the IFP elements was only 10 mm thick, and this resulted in a lighter element (Zievers et al., 1990). The permeability of a new LayCer candle was similar to that of the Dia-Schumalith elements.

The IFP candles were in service for 821 hours of operation at temperatures ranging from 493 to 543 K [220 to 270°C]. The face velocity varied from 2.0 to 5.8 cm/sec. The cyclone was always in the line-up and thus the solids loading at the inlet of the filter did not exceed 3 g/Nm³. Fig. 5 shows the pressure drop history of the IFP candles. The face velocity was considerably higher during the period after 550 hours and thus pressure drop was also higher. As with the Dia-Schumalith candles, the initial solids slip averaged 20 mg/Nm³.

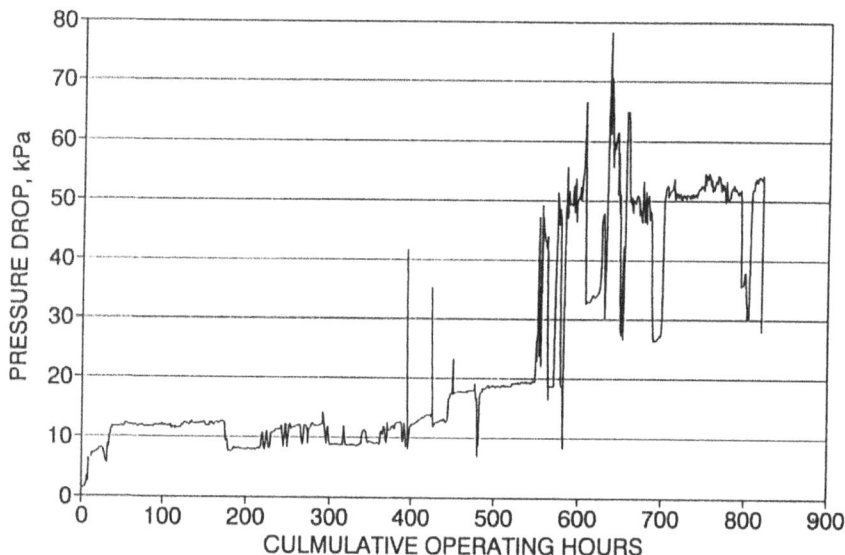

Figure 5. Pressure drop history of LayCer ceramic candles from hourly average data.

PERMEABILITY CALCULATIONS

Permeability, K, is defined as:

$$K = \mu * v/\Delta P \tag{1}$$

where μ is the dynamic viscosity of the gas, v is the filtration face velocity, and ΔP is the pressure drop across the filter.

Ideally ΔP should be the pressure drop immediately after blowback, otherwise one cannot distinguish between a true decline in candle permeability and the influence of a build-up of the temporary dust cake on the candle. Unfortunately the data retrieval system at SCGP-1 was not set up to obtain the cleaned candle ΔP. The ΔP meter was scanned every 2 to 4 seconds, but only the instantaneous value every two minutes was stored for retrieval. Thus, only if the blowback events coincided with the two minute storage interval, would the ΔP data represent the cleaned candle value. Furthermore, during the Dia-Schumalith campaign the blowback control scheme initiated the cleaning of only one row at a time. Consequently there was never a moment when all of the 44 candles were simultaneously clean.

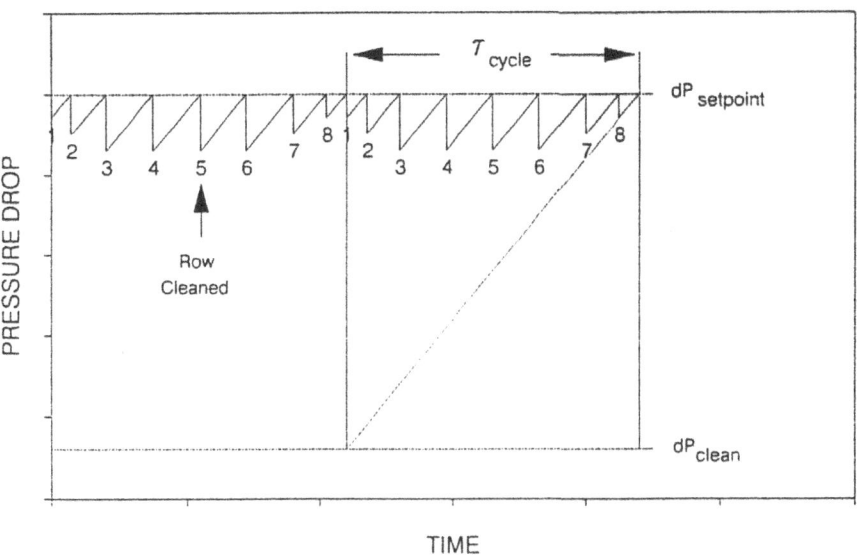

Figure 6   Idealized pressure drop versus time curve for SCGP-1 filter in single row cleaning mode showing the relationship between the pressure drop setpoint for blowback control and the pressure drop if all rows were cleaned simultaneously, dP_{clean}.

135

During the IFP campaign, the blowback control was changed to a "burst" mode. In this mode once the ΔP set point or the fixed time interval was reached, all 8 rows of candles were quickly cleaned one after another in a "burst" of 8 blowback pulses. Nevertheless even in this burst mode it took 60 seconds to blow back all the candles, since some recovery time was needed between pulses to insure that the pressure in the pulse tank remained relatively constant. After the eighth row had been cleaned, the candles in the first row had already been building up a new dust layer for almost a minute.

Nevertheless, the ΔP data together with data on the rate of increase of ΔP between blowback events and the frequency of blowback events can be used to estimate what the cleaned ΔP would have been. This is illustrated in Fig. 6 for the case of single row cleaning and in Fig. 7 for burst cleaning. In both figures it is assumed that the blowback is triggered when a DP set point is reached. For brevity, the situation when blowback is triggered by a timer will not be covered.

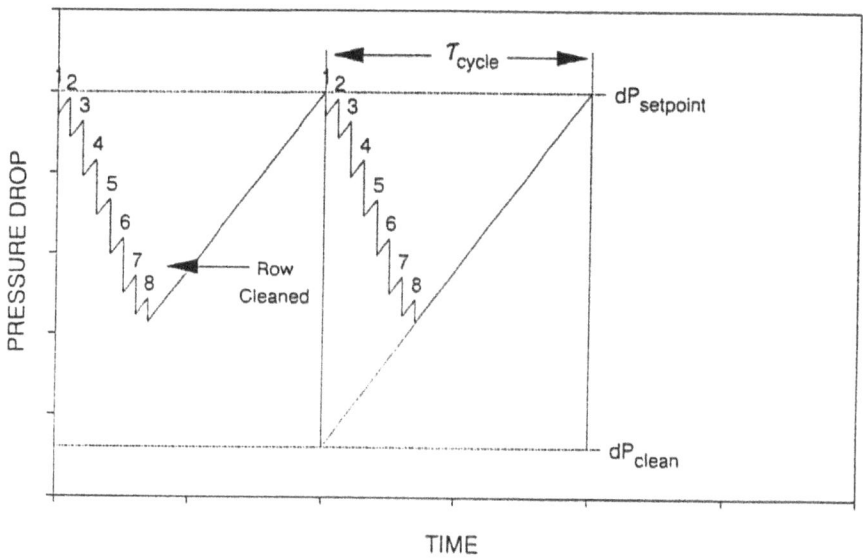

Figure 7. Idealized pressure drop versus time curve for SCGP-1 filter in "burst" cleaning mode showing the relationship between the pressure drop setpoint for blowback control and the pressure drop if all rows were cleaned simultaneously, dP$_{clean}$.

136

The cleaned candle ΔP, ΔP$_{clean}$, can be estimated by:

$$\Delta P_{clean} = \Delta P_{setpt} - \tau_{cycle} * d(\Delta P)/dt \qquad (2)$$

where $d(\Delta P)/dt$ is the time rate of change of the pressure drop across the filter between blowback events, ΔP_{setpt} is the pressure drop set point for blowback, and τ_{cycle} is the time between cleaning of a given row of candles. ΔP_{setpt} is available from the database, and τ_{cycle} can be determined from the output history of a counter which tallied the number of open/close cycles of one of the eight blowback valves.

$d(\Delta P)/dt$ can be estimated by examining ΔP data during periods in which τ_{cycle} was much greater than the time interval of the data points (i.e. 2 minutes). One such period is shown in Fig. 8. During this period the blowback system was intentionally disabled for approximately a half hour. Once the blowback system was again allowed to operate, the pressure drop returned to the normal level after one cycle through the 8 candle rows.

The $d(\Delta P)/dt$ during the period shown in Fig. 8 averaged 12 Pa/s. This value, naturally, is dependent on operating conditions. To be able to estimate how $d(\Delta P)/dt$ would change as operating conditions change, one must examine the parameters which influence $d(\Delta P)/dt$.

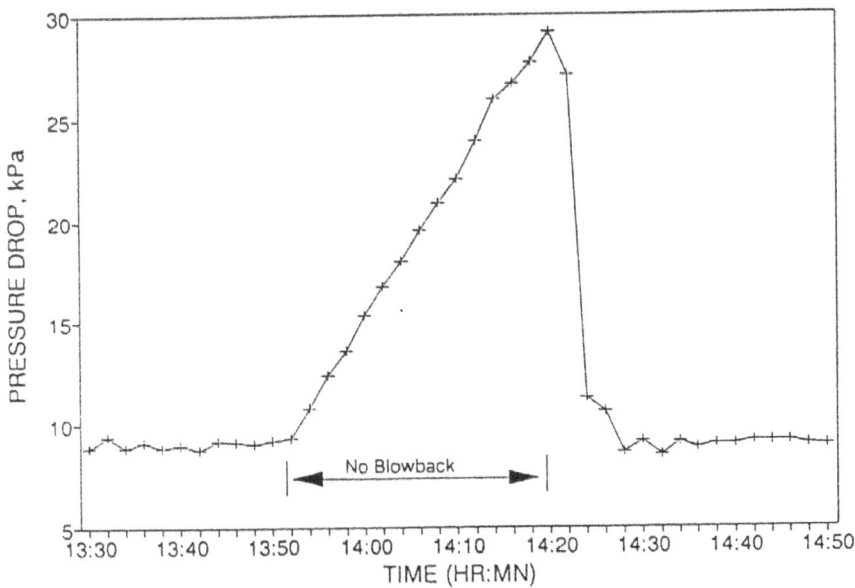

Figure 8. Pressure drop across SCGP-1 ceramic filter during test period in which blowback was temporarily suspended.

From Lippert et al., 1991, the time rate of change in pressure drop across filter dust cake can be written as:

$$d(\Delta P_c)/dt = \frac{(\mu * \rho_g * X * v^2)}{(k_c * \rho_c)} \qquad (3)$$

where: μ = gas viscosity

ρ_g = gas density

ρ_c = dust cake density

X   = solids mass loading (kg/kg)

v   = filtration face velocity

k_c = dust cake permeability per meter = $\mu * v * \delta_c/\Delta P_c$, δ_c is dust cake thickness

If one assumes that k_c, ρ_c and X are all constants, and one ignores the small temperature dependent changes in ρ_g and μ, then the time rate of change in pressure drop can be written as:

$$d(\Delta P)/dt = [d(\Delta P)/dt]_o * (v/v_o)^2 \qquad (4)$$

where the subscript represents reference case values. It is acknowledged that this is a gross simplification, and that the dust-related parameters were certainly not constant during the operating period, however, since face velocity has a quadratic influence and since it varied by almost a factor of three during the test periods, this simplification should account for the strongest influence on d(ΔP)/dt.

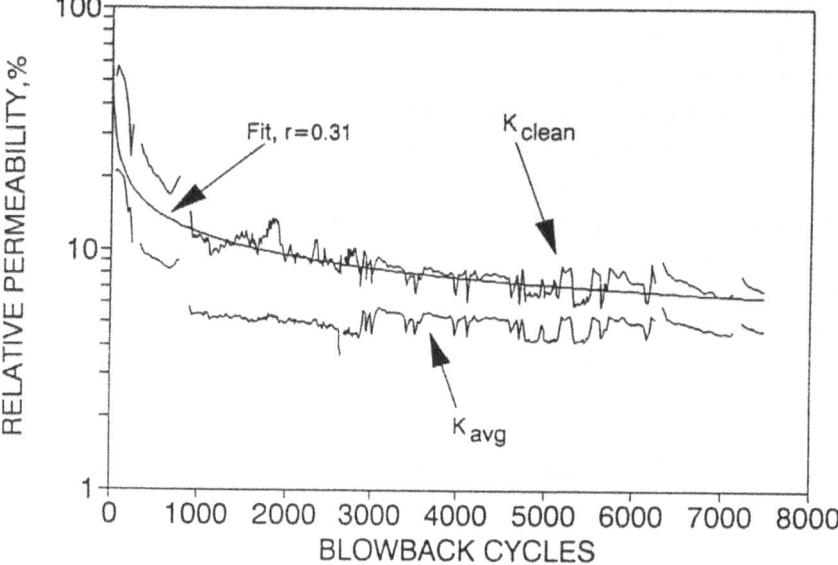

Figure 9. Permeability history of LayCer ceramic candles and calculated fit using model in equation 5.
K-clean uses estimates of DP if all candles had been cleaned simultaneously (equation 2),
and K-avg uses one-hour average data for DP.

138

By using equations 2 and 4 to estimate ΔP_{clean}, an estimate can also be derived for the permeability of the cleaned candles. The results of these calculations are shown in Fig. 9 for the IFP candles. (In this figure the data have been normalized by the initial filter permeability, thus a permeability of 10% means the permeability was one tenth of the original value.) Also shown are the permeabilities calculated from the average filter pressure drop. It is clear that the correction for the temporary dust layer contribution to permeability became less important over time. This is understandable since the pressure drop due to the candle itself increased over time while the pressure drop due to the dust cake remained approximately the same.

Modelling Permeability Decline

In (Durst et al., 1989) it is suggested that the decline in candle permeability can be accurately modelled by:

$$K = K_0 * N^{-r} \tag{5}$$

where K_0 is the original permeability of a new candle, N is the number of cleaning cycles, and the exponent r is an empirical parameter used to best fit experimental data. It can also be used to characterize the clogging behaviour of the filter material. The larger the r, the more severe the clogging.

As shown in Fig. 9, an adequate fit to the LayCer permeability data can be obtained by using the model in equation 5 with $r = 0.31$. It is very apparent from both the SCGP-1 data and the model fit that the bulk of decline in permeability occurs during the initial operating period. After 2000 blowback cycles there was relatively little change in the permeability. The permeability data from the Dia-Schumalith test show a similar trend.

It is believed that this initial permeability decline is caused by the fraction of very fine solids clogging the pores of the outer membrane of the candles. Data on the cumulative filter flow area as a function of pore size for a LayCer candle are presented in Sheckler, 1992. It shows that approximately 90% of the flow area is made up of pores larger than 5 microns. Since a significant fraction of gasification flyash is smaller than 5 microns, it seems logical that over time the filter pores larger than 5 microns would become clogged. This alone would reduce the candle permeability to 10% of its original value.

According to (Durst et al., 1989), an r value greater than 0.3 is typical for a deep bed filter while values between 0.1 and 0.3 reflect genuine surface filtration. Thus the SCGP-1 results fall on the border between deep bed and surface filtration and indicate that gasification flyslag is a challenging solid to filter. Nevertheless, this does not mean that ceramic candles are inappropriate for filtering flyslag. To the contrary, ceramic candles are now the preferred dry solids removal option for any future SCGP plant. It does show, however, that the filtration system must be designed to accommodate the higher pressure drops which result from the permeability decline. The ability of the SCGP-1 filter to handle a wide variety of operating conditions while maintaining a filtration efficiency of 99% or better proves that such systems are feasible.

SUMMARY

Porous ceramic candles were used to filter flyslag from syngas at high pressure and moderate temperature for over 5000 hours at SCGP-1. A summary of the operating parameters is presented in Table 1 for both sets of candles tested. The candles proved to be robust and acted as virtually absolute filters. Unlike the fabric bag filters which had been used earlier for flyslag removal, there were no failures of any filter elements during the ceramic candle tests. Solids slip through the candles was typically 5-10 mg/Nm³ [5-10 ppmw]. The permeability decline of the candles followed an exponential decay model.

TABLE 1
Summary of Operating Parameters for the Ceramic Candle Filters Tested at SCGP-1

Parameter		Schumacher Dia-Schumalith	IF&P LayCer
Number of Candles		44	44
Filter Temp	(K)	503 - 553	493 - 543
	[°C]	[230 - 280]	[220 - 270]
	[°F]	[446 - 536]	[428 - 518]
Filter Pres	(MPa)	2.4	2.4
	[psia]	345	345
Face Velocity	(cm/s)	2 - 5	2 - 6
Inlet Solids Load			
Maximum	(ppmw)	20,000	3,000
Minimum	(ppmw)	500	1,000
Dust Size, d_{50}	(μm)	2 - 5	2 - 3
Total Duration	(h)	4400	821

ACKNOWLEDGEMENTS

The authors would like to acknowledge the efforts of their colleagues who were instrumental in accomplishing the work described in this paper, particularly: J.L. Monsavoir, J.A. Salter, P.E. Unger, and H.A. Dirkse.

The SCGP-1 Demonstration program was jointly sponsored by Shell Internationale Research Maatschappij, Shell Oil Company, and the Electric Power Research Institute.

REFERENCES

Demkolec B.V. (1990). The world's largest coal gasification plant for electricity generation is being realized in the Netherlands. Demkolec News. 1. January 1990, p. 2.

Durst, M., Muller, M., & Vollmer, H. (1988). Hochleistungsentstaubung von Prozess und Abgasen zwichen Raumtemperatur und 1000°C mit Hilfe asymmetrischer poroser Keramik. Staub-Reinhaltung der Luft. 48. 197-202.

Durst, M., Reinhardt, A., & Vollmer, H. (1989). High Efficiency Particle Collection with the Aid of Ceramic Filter Media. Paper presented at the First European Symposium on Separation of Particles from Gases, Nurmberg, Germany, 19-21 April, 1989.

Lippert, T.E., Bachovchin, D.M., Smeltzer, E.E., Alvin, M.A., Meyer, J.H., & Hughes, C.A. (1991). Subpilot Scale Gasifier Evaluation of Cross Flow Filters. In Proceedings of the Eleventh Annual Gasification and Gas Stream Cleanup Systems Contractors Review Meeting, ed. V.K. Venkataraman, L.K. Rath, J.W. Martin, & R.C. Burdick. U.S. Dept. of Energy, Morgantown, WV, DOE/METC/91/6123, Vol. 2, pp. 396-405.

Mahagaokar, U. & Krewinghaus, A.B. (1990). Shell Coal Gasification Project: Gasification of SUFCo Coal at SCGP-1, Electric Power Research Institute, Report GS-6824, Palo Alto, CA.

Mahagaokar, U., Krewinghaus, A.B., & Kiszka, M.B. (1990). Shell Coal Gasification Project: Gasification of Six Diverse Coals, Electric Power Research Institute, Report GS-7051, Palo Alto, CA.

Mahagaokar, U., Phillips, J.N., & Krewinghaus, A.B. (1991). Shell's SCGP-1 Test Program - Final Overall Results. Paper presented at the Tenth EPRI Conference on Coal Gasification Power Plants, San Francisco, CA, 15-18 October, 1991.

Phillips, J.N., Kiszka, M.B., Mahagaokar, U., & Krewinghaus, A.B. (1993). Shell Coal Gasification Project: Final Report on Eighteen Diverse Feeds, Electric Power Research Institute Report TR-100687, Palo Alto, CA.

Phillips, J.N., Mahagaokar, U., & Krewinghaus, A.B. (1992). Shell Coal Gasification Project: Gasification of Eleven Diverse Feeds, Electric Power Research Institute Report GS-7531, Palo Alto, CA.

Salter, J.A. & Monsavoir, J.L. (1992). SCGP-1 High Temperature High Pressure Filtration Experience, In Proceedings: Second EPRI Workshop on Filtration of Dust from Coal-Derived Reducing and Combustion Gases at High Temperature, ed. M. Epstein & R.A. Brown, Electric Power Research Institute, Palo Alto, CA, Chapter 5.

Salter, J.A., Gantz, S.H., Tang, W.T., Tijm, P.J.A., DuBois, J.B., & Perry, R.T. (1991). Shell Coal Gasification Process: By-Product Utilization. Paper presented at the Ninth International Coal Ash Utilization Symposium, Orlando, FL, 22-25 January, 1991.

Sheckler, C., Critical Components of Hot Gas Filter Design, In Proceedings: Second EPRI Workshop on Filtration of Dust from Coal-Derived Reducing and Combustion Gases at High Temperature, ed. M. Epstein & R.A. Brown, Electric Power Research Institute, Palo Alto, CA, Chapter 30.

Zievers, J. F., Eggerstedt, P.M., Aguilar, P. & Zievers, E. C. (1990). Comparative Physiochemical Studies of Porous Ceramics for Filtration of Hot Gases. Filtration & Separation. 27. 353-353.

Zon, G.D. (1990). IGCC future under test at Buggenum. Modern Power Systems. 10 (August 1990). 39-45.

SUBSTITUTION OF LIGHTWEIGHT CERAMICS FOR ALLOY AND SILICON CARBIDE IN A HOT GAS FILTER

E. C. Zievers
Universal Porosics, Inc.
La Grange, IL, U.S.A.

J. F. Zievers, P. Eggerstedt & P. Aguilar
Industrial Filter & Pump Mfg. Co.
Cicero, IL, U.S.A.

ABSTRACT

Vacuum formed aluminosilicate ceramic fiber shapes have often been post-treated to add density and strength. Using relatively inexpensive raw materials, service operation temperatures of up to 1100°C (2012°F) have been realized and, at these temperatures, the material strength rivals that of costly alloys. Castable, high modulus of rupture refractory has many interesting properties that can be used to advantage when designing ceramic structures that, additionally, work well in conjunction with vacuum formed ceramic materials.

INTRODUCTION

The use of porous ceramics to remove particulate in the filtration of hot gases was first reported in literature in 1978[1]. The porous materials involved were alternately silicon carbide (SiC) and mullite. The mean pore of the filter element (candle) was about 20-40 micrometers and pore volume was about 38%. Face velocity was in the range of 4-5 cm/s (8-10 ft/min). Since that time, a good deal has been written about the use of porous SiC, mullite, and cordierite for the filtration of high temperature gases[2-6]. Within the last few years, the use of vacuum formed ceramic fiber filter candles has also been reported[7].

Ceramics, in the form of refractory, have long been used to protect metal structures from temperature and abrasion and there is an extensive body of literature that deals with this subject[8][9]. Castable refractories can be used to produce special shapes that exhibit excellent strength at high temperatures[10]. Recently, some literature has dealt with the use of lightweight ceramics for filter media as well as structural members[11][12].

In Table 1, the authors have attempted to show a rough chronology for various material combinations. Very little change occurred between 1978 and 1988 and it is important to note that the data are those of first known reports in literature. Also note that the first entry, namely SiC candles and alloy tubesheet with an alloy jet pulse delivery system, is still the most common and widely used.

The Ceramic Material "Arsenal"

In Table 2, the authors have set forth an array of ceramic materials that have been used in conjunction with hot gas filtration. The materials are grouped by filter component and some pertinent characteristics are shown. It should be noted that in some cases the ceramic materials can have multiple uses. For example, Vacuum Formed Ceramic Fiber (VFCF) has been used in various forms as filter elements as well as tubesheets which support the filter elements[13]. Castable refractory is also being used for tubesheets and structural tubesheet support members[14]. Ceramic sponge and ceramic cloth have both been employed as "safety filters" in conjunction with a flow control device made from castable refractory[15].

Table 2, read in conjunction with Figure 1, illustrates some of the ways in which ceramics have been used. The letter suffixes following the identity legends in Figure 1 coordinate with the categories shown on Table 2. Figure 1 shows a tubesheet (VFCF or castable refractory) supported by a strongback (castable refractory), and a filter candle (VFCF or SiC) sealed by means of a ceramic gasket and held in place with a hold-down plate (VFCF). Located on the hold-down plate, beneath the jet pulse introduction point, is a pulse diverter cone made of castable refractory. Above the tubesheet/hold-down plate assembly is a plenum, formed from VFCF or castable refractory, in which an individual nozzle type of jet pulse delivery system, also constructed of castable refractory, may be substituted. Located

143

<u>TABLE 1</u>

Chronology of Ceramic Filter Material Combinations

YEAR	CANDLES	TUBESHEET	JET PULSE DELIVERY/ PLENUM
1978 — 1988	SiC	ALLOY	ALLOY
1990	VFCF	VFCF	ALLOY
1991	SiC	VFCF	ALLOY
1991	VFCF	VFCF	VFCF
1992	VFCF	VFCF	CASTABLE REFRACTORY
1992	SiC/VFCF	VFCF	CASTABLE REFRACTORY

<u>TABLE 2</u>

Ceramic Materials and Properties as used in Hot Gas Filtration Applications

USE CLASS	MATERIAL	AS USED DENSITY (g/cm^3)	MODULUS OF RUPTURE (kPa)		
			AMBIENT	1000 °C	1500 °C
(A) CANDLES	SiC	1.8 — 2.0	3,000–5,000	4,000–7,000	3,000–5,000
	ReSIC	2.5 — 2.6	12,000–18,000	14,000–20,000	10,000–12,000
	MULLITE	1.8 — 3.0	15,000–45,000	18,000–25,000	500–1,000
	ALUMINA	1.9 — 2.1	2,000–5,000	1,500–4,000	1,000–3,000
	VFCF I*	0.3	350–380	120–140	100–120
(B) TUBESHEET	VFCF II*	0.55	900–1,600	270–450	190–300
	VFCF III*	0.85	2,000–2,500	900–1,100	700–850
	HIGH M.O.R. CASTABLE	2.7	18,000–24,000	32,000–42,000	7,600
(C) STRUCTURAL	HIGH M.O.R. CASTABLE	2.7	18,000–24,000	32,000–42,000	7,600
	Al$_2$O$_3$/MULLITE	2.3		20,000 +	
(D) SAFETY	HIGH M.O.R. CASTABLE	2.7	18,000–24,000	32,000–42,000	7,600
	SPONGE	0.2			
	TEXTILE	542 gm/m^2			

* VACUUM FORMED CERAMIC FIBER

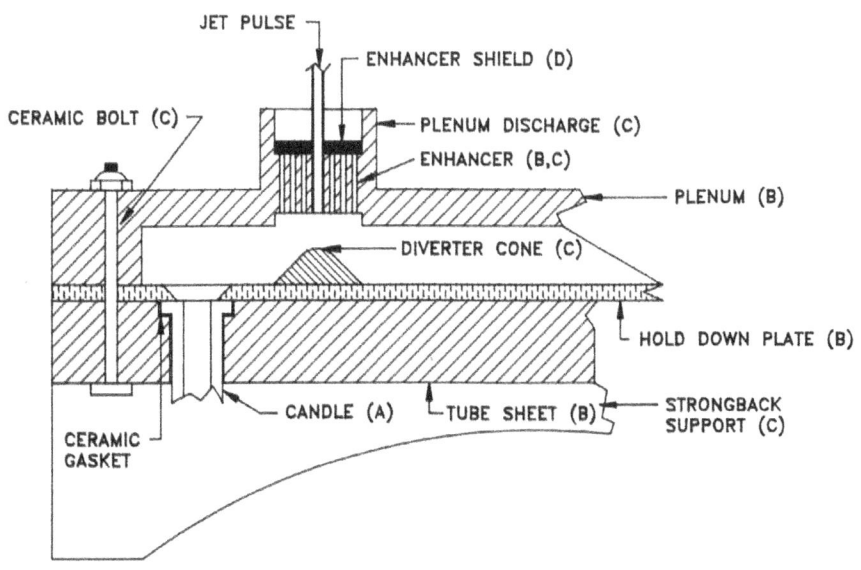

FIGURE 1

Typical Ceramic Filter Cross–Sectional View

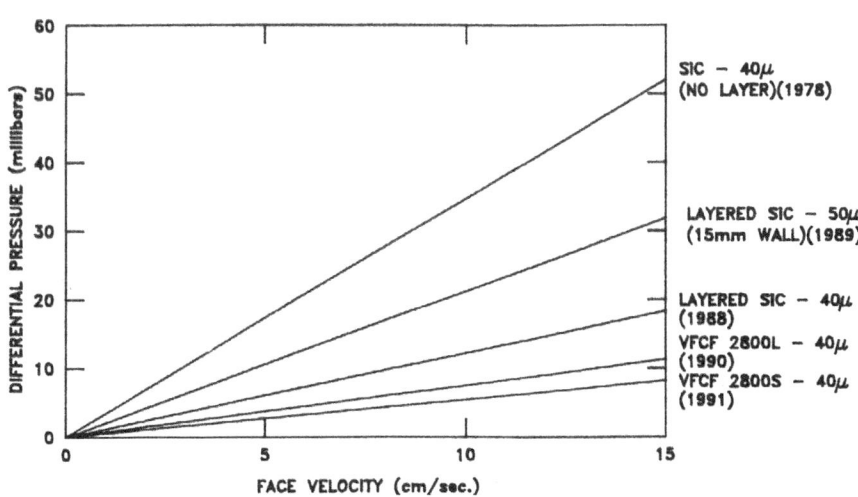

FIGURE 2

Pressure Drop vs. Face Velocity Relationships for
Various Filter Candles

in the top of the plenum is a flow control device (VFCF, castable refractory, or alumina) which in turn is fitted with a ceramic sponge or ceramic cloth "safety filter". For handling purposes the entire "package" is first secured with ceramic (SiC or alumina) bolts and nuts and then, when in use, this "package" is mounted in compression between the head and chamber.

Filter Element Possibilities

As indicated in Table 2, hot gas filter elements made from silicon carbide[16], mullite[17], alumina[18], cordierite[19], and VFCF[7] have been reported. All of the materials can be layered so that a relatively coarse porous structural layer supports a fine porous "skin". The typical wall thickness is 10mm, although candles with a 15mm wall have also been used. VFCF elements are approximately 1/6 the weight of the conventional "hard" ceramic elements and, because of their relatively light density (high pore volume), they exhibit lower pressure drop characteristics and excellent resistance to thermal shock. Figure 2 shows relative pressure drop characteristics of some selected ceramic filter element structures and the approximate date they first appeared in the literature.

Corrosion of ceramic filter elements and thermal shock seem to be among the most frequent causes of premature failure[20][21]. According to this literature, correct selection of the base material and formulation of the bonding system is essential to fit the ceramic filter element to the job it will face.

Tubesheet and Hold-Down Plate

VFCF densified to about $0.88g/cm^3$ (55 lbs./ft.^{3}) was reported in service as tubesheet material more than 2 years ago where an oxidizing atmosphere and temperatures up to 1010°C (1850°F) were encountered[22]. When properly formulated, the VFCF shows a temperature capability to 1535°C (2800°F). Typically, VFCF tubesheets are 75mm (3 inches) thick and experience has shown that the VFCF material machines more than 10 times quicker than alloys.

VFCF hold-down plates are generally 25mm (1 inch) thick and of a lesser density than the tubesheets. The lighter density of the hold-down plate enhances the dust tight integrity of a locked-up system because, when locked tightly in compression, the elevated density of the tubesheet and plenum slightly crush the hold-down plate, ensuring a better seal.

Tubesheets have also been fabricated from castable, high MOR (modulus of rupture) refractory by contractors of Industrial Filter & Pump Mfg. Co.   To date, pieces larger than 1.5m (60 inches) in diameter have not been reported.

Table 3 shows some typical MOR data for VFCF at ambient temperature and at elevated temperature. Note that the MOR will vary with the degree to which the material was densified.   The last line on Table 3 shows data for hold-down plate grade VFCF.

Strongbacks and Supports

VFCF tubesheets have been supported by transverse rods of 80% Al_2O_3 (BTC - ML80-P)[23] (See inset A, Figure 3.). The alkali content of this material is extremely low and it s maximum operating temperature is 1650°C (3000°F).   Alternately, arched strongbacks, with a center plate used as a deflector for incoming dirty gas, have been formed of high MOR castable refractory (Plicast Hy MOR 3100S) (See inset B, Figure 3.).

The bar chart in Figure 3 shows hot and cold MOR for the castable refractory used in making strongbacks and cast tube-sheets. The increased MOR, which coincides with the increase in service temperature, is a very welcome characteristic. The data shown are actual quality control (QC)[24] data and typically, as a part of QC, at least 2 MOR test bars are cast from each pour.

Plenums

Plenums, with or without integral individual jet pulse nozzles, have been made from VFCF and from high MOR castable refractory. Figure 4 illustrates the two basic designs. If the individual jet pulse nozzles and passages are preferred the "spider" (passage/nozzle matrix) is generally preformed from combustible material buried in the "pour" and burned out during cure.   The "pie shape" plenum design, shown in Figure 4, simplifies filter scale-up and stocking of spares.

Ceramic Fasteners

Table 2 and Figure 1 shows ceramic bolts and nuts being used as assembly aids. Both high purity alumina and silicon carbide

TABLE 3
Vacuum Formed Ceramic Fiber Tubesheet Specifications

MATERIAL	TEMPERATURE °C	MOR kPa	DENSITY (g/cm³)
2800 AS FORMED	24	362	0.30
2800 DENSIFIED	24	2445	0.84
2800 DENSIFIED	982	957	0.84
2800 PARTIALLY DENSIFIED	24	1527	0.55

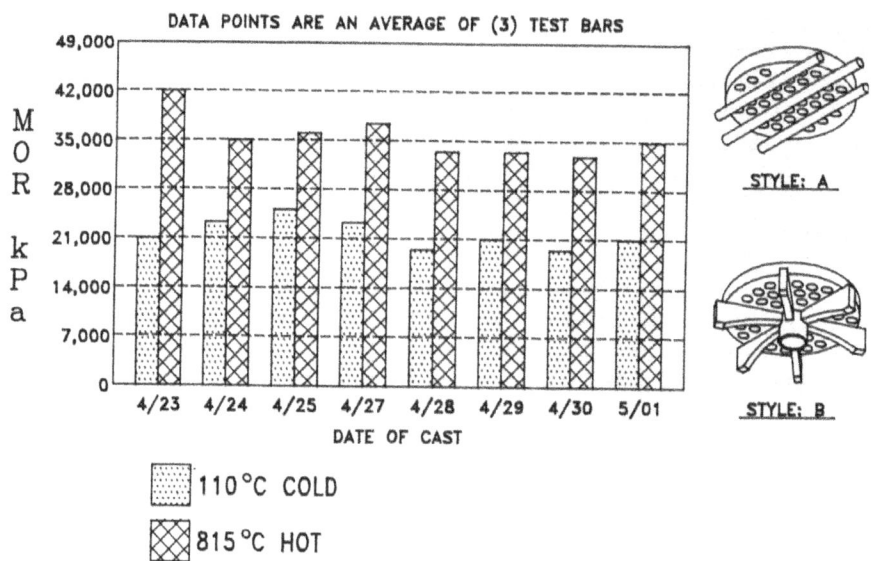

FIGURE 3

Castable Refractory Modulus of Rupture Data

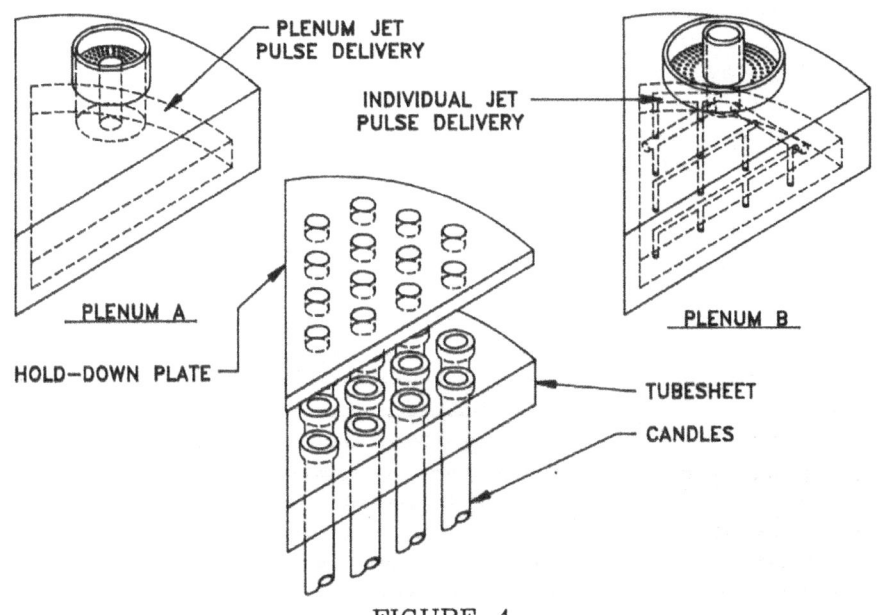

FIGURE 4
Jet Pulse Cleaning Plenum Configurations

TABLE 4
Ceramic Bolt Specifications

	SiC (85%)	Al_2O_3 (80%)
M.O.R. (kPa)	34,500	29,000
APPARENT DENSITY (gm/cc)	2.5	2.3
POROSITY (%)	16	12
COEFFICIENT OF THERMAL EXPANSION (mm x 10^{-6}/mm/°C)	5.2	5.3
RELATIVE COST	1.0	2.5

have been used. While the tensile strengths of these ceramic fasteners are low, they can be used with great success for assembling and handling the parts of a plenum/tubesheet package until the package can be placed in compression. Table 4 compares alumina bolts and nuts with those of silicon carbide. Obviously, there is a difference in cost and it is simply a matter of choosing the proper material for a given application. Please note that the figures given are typical and not necessarily specific.

Jet Pulse Enhancer and Safety Filter

While face velocity during candle filtration might be in the range of 3.5-7 cm/s (7-14 ft/min), the corresponding velocity through the bore of a 1.5 meter long candle will be on the order of 760-3000 cm/s (25-100 ft/sec), depending upon it s inside diameter. This upward velocity creates a momentum which must be overcome during jet pulse cleaning. In baghouse technology this momentum does not present a problem since the bags are often cleaned "off-line" (momentarily shutting off inlet gas to the bags requiring cleaning) and tend to require less pulse gas volume due to the bag's ability to flex and shake off the accumulated dust.

Unlike baghouse technology, hot gas candle filters must be cleaned in an "on-line" fashion since valves capable of repeatedly shutting off the flow of 982°C (1800°F) combustion gases simply do not exist. Additionally, due to their brittle nature, ceramic candles cannot be subjected to shaking, flexure, or even acoustical (such as sonic horns) methods of dust displacement. As a result, the only effective means of candle filter cleaning is by overcoming the upward clean gas velocity and momentum.

The jet pulse enhancer[15] is installed on the clean gas side of the candle filter and it s function is to create a slight back pressure that aids in offsetting the clean gas momentum and discourages jet pulse "bounce". The principle behind the enhancer is to create a small, tolerable pressure drop during service that will in turn aid jet pulse cleaning. This is accomplished by providing a multitude of calibrated orifices through which the clean gas must pass. In candle filtration the "clean-side" gas is virtually particulate free due to the inherent high particulate removal efficiency of the candles which allows for a wide range orifice sizes. The candle filters also tend to operate within a limited range of flow (face velocity) so the fixed orifice size can be calculated and specified as part of the filter design.

Work performed under a U. S. Dept. of Energy Research Contract[26] has shown that while pressure drop as a result of using a jet pulse enhancer is slightly increased, the volume of the jet pulse is about doubled, and the time to initiate the pulse is reduced by more than 60%. The enhancer device has been fabricated from densified VFCF and also cast in castable refractory and alumina. Figure 5 illustrates the device, it s performance, and the typical effect of the enhancer with respect to the jet pulse as cited above. The time of pulse pressure initiation, as indicated by the recording instrumentation, occurs much sooner, is more intense, and has greater duration when an enhancer is used. This indicates that, depending upon the individual hot gas filter application, more energy can be delivered to the filter candle wall with the same input. Alternately, less input is now required to achieve an equal energy delivery to the filter candle wall.

Figure 6 shows pressure within the filter chamber as a function of time, both with the enhancer and without. On each of the first two strip charts pressure differential during normal filtration service is represented by the nearly horizontal line at the bottom of each chart. During jet pulsing pressure differential momentarily surges to dislodge accumulated dust, as evidenced by the regular "spikes" on each chart. While the strip charts themselves are not impressive, close examination of the "spikes" during jet pulse cleaning reveal the higher pressure and jet pulse volume delivered as a result of the enhancer. These increases are illustrated by the diagram in the upper right quadrant of Figure 6.

It is important to note that the enhancer device is fitted with a low pressure drop ceramic "safety filter" which serves as protection from catastrophic failure and consequent particulate introduction to critical downstream areas such as turbines. The safety filter has been made from a zirconia toughened Al_2O_3 sponge and ceramic textile.

Economic Merits of Ceramic Filter Components

It is indeed possible to combine ceramic and alloy components to achieve a well performing hot gas filtration system. Just a few years ago a typical bill of materials included silicon

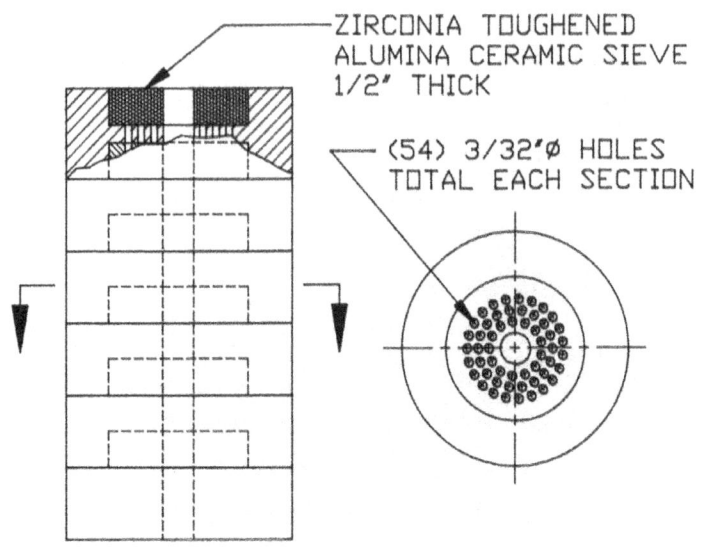

ZIRCONIA TOUGHENED
ALUMINA CERAMIC SIEVE
1/2″ THICK

(54) 3/32′⌀ HOLES
TOTAL EACH SECTION

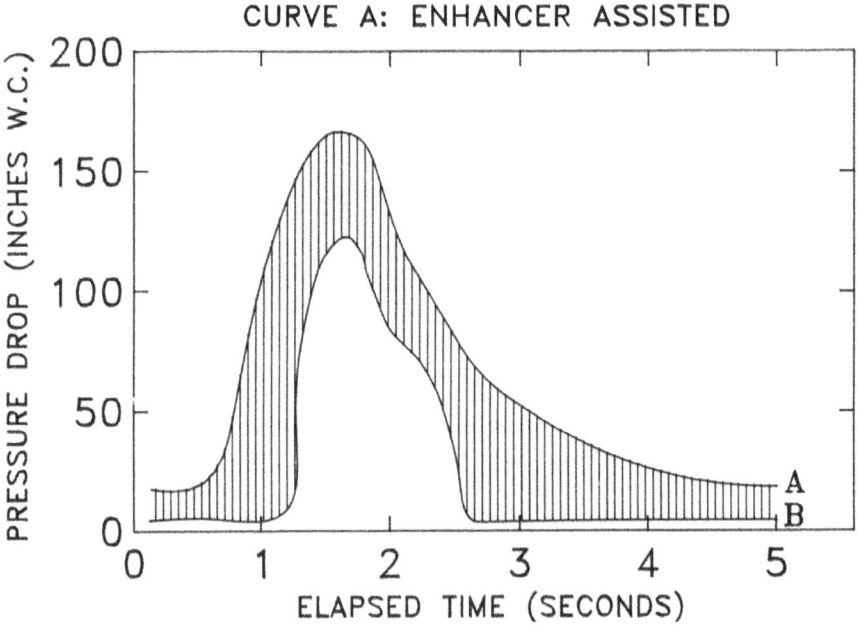

CURVE A: ENHANCER ASSISTED

FIGURE 5

Jet Pulse Enhancer Design & Typical
Performance Characteristics

TEST NUMBER II – A

PARAMETERS:

ENHANCER:	PROTOTYPE II
	(W/ 3/32" HOLES)
CANDLE:	LayCer 70/3, 1.5m
FLOWRATE:	20 SCFM–AIR (6.67 fpm)
TEMP.:	AMBIENT, 75 °F
PULSE GAS:	12 L N_2 @ 90 PSIG
CYCLE:	60 seconds
DURATION:	1 second
DATA LOG:	200msec @ 10 kHz

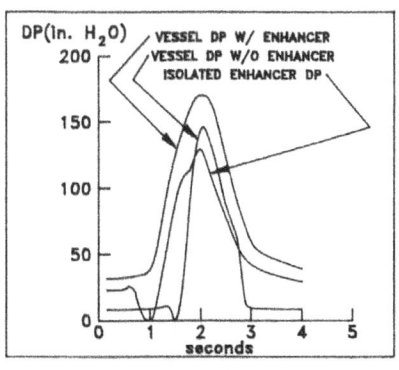

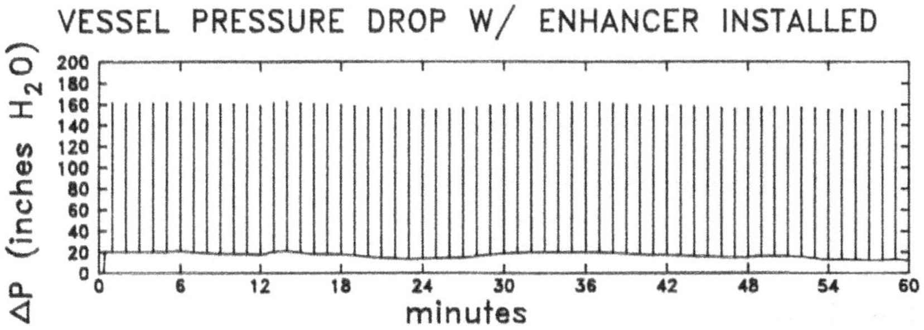

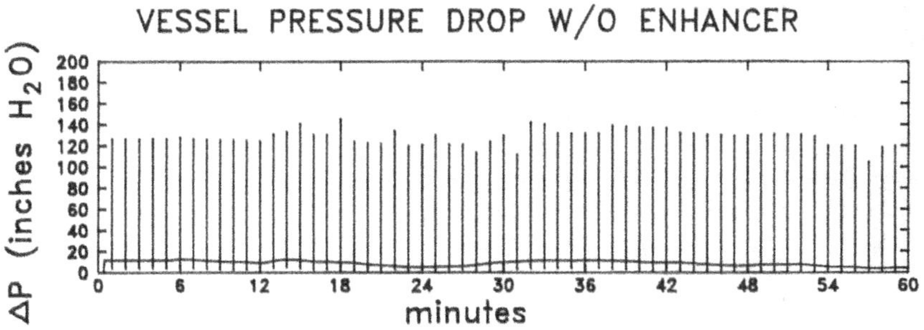

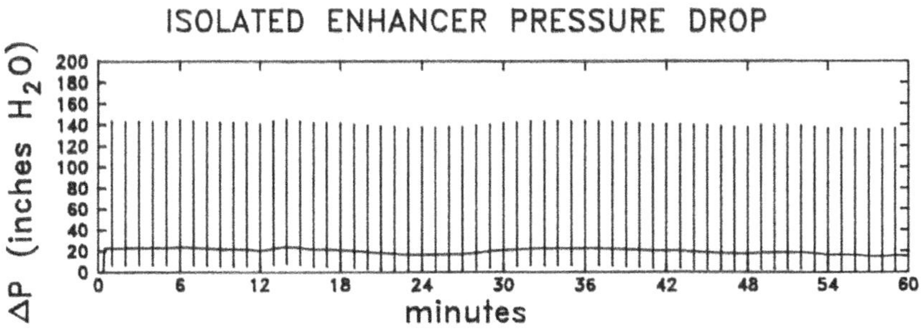

VESSEL PRESSURE DROP W/ ENHANCER INSTALLED

VESSEL PRESSURE DROP W/O ENHANCER

ISOLATED ENHANCER PRESSURE DROP

FIGURE 6
Jet Pulse Enhancer Developmental Test Data

153

carbide (SiC) candles, an alloy tubesheet, (cooled with water or steam if the operating temperature so necessitated), and an alloy jet pulse delivery system. Although relatively effective these components are inherently expensive.

VFCF, first used as material for filter elements, is now subsequently being used for tubesheet and other internal component manufacture. Extensive use of VFCF and castable refractory can have a significant affect on the overall capital cost of a hot gas filter station. A series of recent cost calculations[27][28][29] for stations built to operate at low pressure, and alternately at high pressure, in systems utilizing from 100 candles (60mm O.D. x 40mm I.D. x 1500mm long) to more than 4,000 candles consistently indicated an average savings of approximately 25%. Figure 7 clearly illustrates this in that it considers four different combinations of interior components for hot gas filters.

Conclusion

The use of ceramics for hot gas filter components other than filter elements, until lately, have largely consisted of relatively dense materials like silicon carbide, mullite, alumina and cordierite.

Lightweight ceramics such as VFCF appear to have definite promise as a relatively low cost, easily processed material. Not only in the case of filter elements, but also for certain structural purposes. Recent progress using light-weight ceramics in hot gas filtration applications has been brisk and soon VFCF will undoubtedly gain greater acceptance, especially when combined with high MOR castable materials

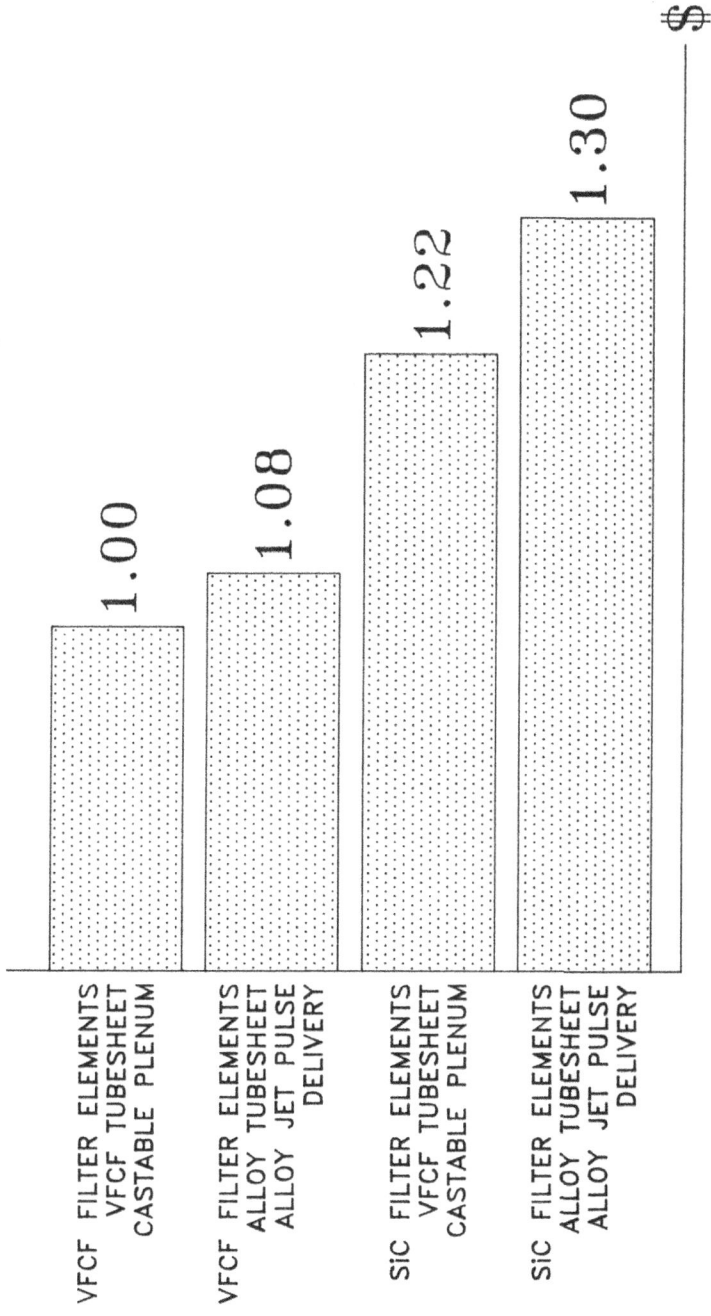

1.00 — VFCF FILTER ELEMENTS VFCF TUBESHEET CASTABLE PLENUM

1.08 — VFCF FILTER ELEMENTS ALLOY TUBESHEET ALLOY JET PULSE DELIVERY

1.22 — SiC FILTER ELEMENTS VFCF TUBESHEET CASTABLE PLENUM

1.30 — SiC FILTER ELEMENTS ALLOY TUBESHEET ALLOY JET PULSE DELIVERY

FIGURE 7

Relative Capital Cost Ratios for Various Hot Gas Filter Configurations

REFERENCES

1. Hempelmann, W. (1978). Incineration of Low Level Radioactive Waste, Report to Kernforschungszentrum, Karlsruhe, Germany.

2. Leibold, H., Dirks, F., Rudinger, V. (1988). Particulate Emissions from a LLW Incinerator and Off-Gas Cleaning with a New Type of Ceramic Candle Filter. Proceedings of International Conference on Incineration of Hazardous, Radioactive and Mixed Wastes, San Francisco, California.

3. Tassicker, O. (1982). Gas Stream Filtration for Pressurized Fluidized Bed Combustion - An EPRI Perspective. Proceedings of Second Annual Contractors Meeting on Contaminant Control in Hot Coal Derived Gas Streams, Morgantown, WV.

4. Ciliberti, D., Lippert, T. (1984). Gas Cleaning Technology for High-Temperature, High Pressure Gas Streams. Procedings of Fourth Symposium on the Transfer and Utilization of Particulate Control Technology: Volume 1.

5. Brown, J., Brown, N., Zievers, J., Eggersdedt, P. (1991). Anticipated Advances in High Temperature Ceramic Barrier Filters for Particulate Control in Power Generator Systems. Proceedings of Particle Control Technology Symposium, Williamsburg, VA.

6. Zievers, J., Eggerstedt, P., Zievers, E., (1991). Porous Ceramics for Gas Filtration. American Ceramic Society Bulletin: Volume 70.

7. Zievers, E., Zievers, J., Eggerstedt, P., Aguilar, P., (1991). Porous Ceramics in Medical Waste Incineration. Proceedings of Incineration Conference, Knoxville, TN.

8. Plibrico Japan Company Limited (1984). Technology of Monolithic Refractories.

9. Green, A. P., Industries, Inc. Refractory Pocket Catalog, Mexico, MO.

10. Private communication with John Lukasik, Plibrico Co., Chicago, IL.

11. Zievers, J., Eggerstedt, P., Zievers, E., Nicolai, D., (1992). What Affects the Cost of Hot Gas Filter Stations?. Proceedings of ASME Turbo Conference (IGTI), Koln, Germany.

12. Zievers, J., Eggerstedt, P., Zievers, E., Nicolai, D., (1991). Lightweight Ceramic Materials Make High Temperature Gas Filtration Simpler. Deutsche Keramische Gesellschaft e.V. (DKG), Erlangen/Nurnberg, Germany.

13. Zievers, E., Zievers, J., Eggerstedt, P., (1992). A Report on Novel Materials of Construction After More Than a Year of Operation in a Hot Gas Filter. _Proceedings of Incineration Conference_, Albuquerque, NM.

14. Industrial Filter & Pump Mfg. Co. Internal Research Report #161, May 1992.

15. U. S. Patent No.: 4,909,813.

16. Reinhardt, E., (1986). Ceramic Filter Elements for High Pressure - High Temperature Gas Filtration. _Proceedings of Workshop on Pressurized Fluidized-Bed Combustion_, Milwaukee, WI.

17. Didier Filtertechnik, D6719 Eisenberg/Pfalz, Bulletin V3.45, 1983.

18. Eggerstedt, P., (1993). Advanced Ceramic Materials for Use in Hot Gas Filtration Applications. _Chemical Engineering Progress Magazine_.

19. Oda, N., (1988). New Ceramic Tube Filter Technology for Hot Gas Cleaning. _Proceedings of Seminar on Fluidized-Bed Combustion Technology for Utility Applications_, Palo Alto, CA.

20. Sawyer, J., Vass, R., Brown, N., Brown, J. (1990). Corrosion and Degradation of Ceramic Particulate Filters in Direct Coal-Fired Turbine Applications. _Proceedings of Gas Turbine and Aeroengine Congress and Exposition_.

21. Alvin, M., Lippert, T., Lane, J., (1990). Assessment of Porous Ceramic Materials for Hot Gas Filtration Applications. _Proceedings of AIChE Annual Meeting_.

22. Zievers, E., Zievers, J., Eggerstedt, P., (1990). A Comparison of Cylindrical Porous Ceramic Elements Used for Hot Gas Filtration. _Proceedings of Incineration Conference_, San Diego, CA.

23. Bolt Technical Ceramics, Conroe, TX - ML80P.

24. Private communication with John Lukasik, Plibrico Co., Chicago, IL.

25. U. S. Patent No. 4,713,174.

26. U. S. Department of Energy SBIR Grant No. 20409-92-I.

27. Industrial Filter & Pump Mfg. Co. work file 92-0723.

28. Industrial Filter & Pump Mfg. Co. work file 92-0891.

29. Industrial Filter & Pump Mfg. Co. work file 92-0973.

EXPERIENCES OF A FIBROUS CERAMIC CANDLE FILTER FOR HOT GAS CLEANING IN PRESSURIZED FLUIDIZED BED COMBUSTION

JUKKA JALOVAARA, ILKKA HIPPINEN & ANTERO JAHKOLA
Helsinki University of Technology
Department of Energy Engineering
Otakaari 4, SF-02150 Espoo, Finland

ABSTRACT

A fibrous, layered Al_2O_3/SiO_2-based ceramic candle filter has been tested for pressurized fluidized bed combustion (PFBC) hot gas cleaning at the Otaniemi PFBC test rig. The fuels used have been bituminous coal and peat. The exhaust gas from a pressurized fluidized bed reactor was passed through two cyclones before entering the ceramic filter unit, which was operated at a temperature range from 853 to 1013 K (580 to 740 °C) and at a pressure range from 0.6 to 1.0 MPa. Total exposure time of the two candle filter elements was about 340 hours.

This paper presents measured data and operating experiences of the ceramic filter. The inlet particle loading was in the range of 1 - 20 g/m^3n depending on operating conditions and the properties of the fuels and sorbents. The mass median diameter of particles entering the filter unit was 6 - 8 μm. The outlet particle concentration was always lower than 5 mg/m^3n. The collection efficiency of the studied filters was very high, typically higher than 99.9 %. These results indicate clearly that they will meet particle removal requirements of most gas turbines and environmental emission limits.

The permeance of the filter behaved as expected: it first dropped quickly and then seemed to stabilize to a constant level or about 15 % of the original permeance of the clean element. The effect of pulse pressure on cleaning of the candles, strength of the filter material and dust penetration into the filter element have also been investigated.

INTRODUCTION

Power production with pressurized fluidized bed combustion (PFBC) makes it possible to construct gas turbine plants based on different solid fuels. These PFBC combined cycle processes improve the overall efficiency of energy production and the power to heat ratio of a cogeneration power plant. At the same time they have a lower environmental impact compared with conventional combustion processes of solid fuels.

In addition to sulphur and nitrogen compounds and alkali metals, the exhaust gas of a pressurized fluidized bed combustor contains particles elutriated from the bed that cause erosion of the gas turbine blades. For this reason flue gases have to be effectively cleaned before entering the gas turbine to ensure a reasonable life time of the gas turbine.

Hot gas cleaning is one of the most difficult problems facing commercialization of PFBC technology today. Up to now several types of filters have been developed for high-temperature, high-pressure (HTHP) particle removal. Due to very specific operating conditions caused by high temperature and pressure various ceramic materials seem to be the most promising. Satisfying the ever stricter limits proposed by the gas turbine manufacturers and environmental legislation is possible by means of ceramic candle filters. These have been quite widely tested and studied recently in many countries, including Finland.

This paper reports on a study of a fibrous ceramic candle filter, which has been made at Helsinki University of Technology during 1991 - 1992.

EXPERIMENTAL

The Otaniemi PFBC test rig has a maximum thermal input of 130 kW. The reactor operates at pressures up to 1.0 MPa and at temperatures up to 1273 K (1000 °C). Figure 1 shows the simplified flow diagram of the rig. A detailed description of the test rig and its operation can be found elsewhere (Hulkkonen et al., 1989). The technical data for the test rig are presented in Table 1.

TABLE 1
The technical data for the test rig

Max. thermal input	130 kW
Max. operating pressure	1.0 MPa (10 bar abs)
Bed temperature	1023-1223 K (750-950 °C)
Fluidizing velocity	0.7-1.3 ms^{-1}
Bed diameter	0.15 m
Bed height	1.2 m
Freeboard diameter	0.25 m
Freeboard height	2.8 m

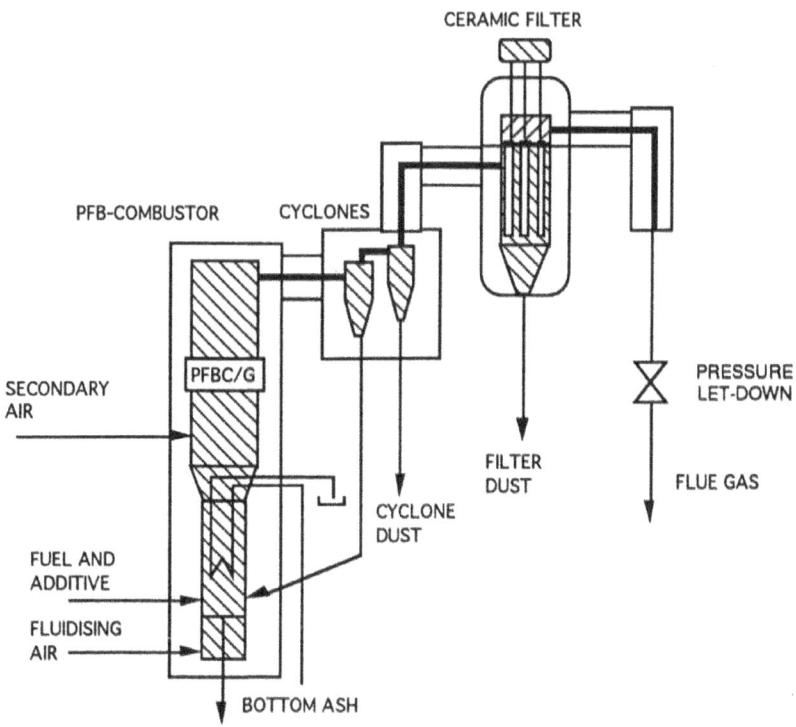

Figure 1. The Otaniemi PFBC test rig

Hot flue gas cleaning equipment includes two cyclones and a ceramic filter unit, which consists of a pressure vessel housing up to five ceramic candle filter elements. Due to the heat losses between the reactor and the filter unit the temperature of the flue gases is at a level of 873 - 1023 K (600 - 750 °C) before filtration. A schematic diagram of the filter unit is shown in Figure 2.

The pressure vessel is 0.7 m in diameter and 2.8 m in height. It is internally insulated by refractory. The tube sheet, on which the filter elements have been mounted, is made of steel. The candles and the tube sheet are bolted to the counter plate to allow a maximum pressure drop of about 1 bar across the tube sheet without any leakage. The filter elements and the tube sheet are surrounded by metal lining, which is electrically heated to compensate for heat losses. The inside diameter of the metal lining is 0.4 m (Jahkola et al., 1991).

A periodical on-line cleanup is provided for each candle element during filtration. The interval between cleaning pulses for two individual elements is approximately 10 s (Kurkela et al., 1991). The candles are cleaned by high pressure nitrogen on-line pulse cleaning system that can be used either automatically or manually. The automatic pulsing can be set based on either a constant cycle or a maximum pressure drop across the wall of the filter element. The pressure pulse for each candle is provided by a separate valve with a pulse duration of 30 - 300 ms (Jahkola et al., 1991). Dust cake on the filter, which is removed by pulse cleaning, drops to the bottom of the filter unit and is removed from the vessel by a lock-hopper system.

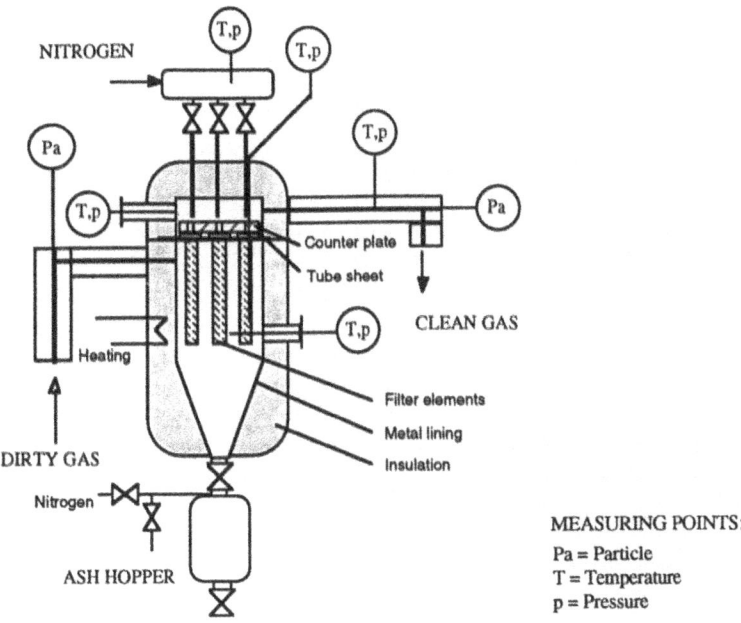

Figure 2. The ceramic filter unit

The filtration tests were carried out using one type of fibrous layered Al_2O_3/SiO_2-based ceramic candle filter (Foseco Cerafil 2000i). The element is made of Al_2O_3 fibres with Al_2O_3/SiO_2-based binder. The surface layer of the element is made of SiO_2 (Withers et al., 1990). Table 2 shows properties of the tested filter elements (Jalovaara, 1992).

TABLE 2
Properties of the filter elements

Type	Layered, fibrous
Material	Al_2O_3/SiO_2
Filtration area, m^2	0.18
Weight, kg	0.95
Fibre diameter, μm	2.5
Dimensions:	
- length, mm	1000
- diameter, mm	60
- material (wall) thickness, mm	10

SAMPLING AND MEASUREMENTS

Particle measurements are based on isokinetic sampling. The sample flow is extracted from the flue gas after the second cyclone and after the filter unit and cooled down to a level of 473 - 573 K (200 - 300 °C). Particle concentrations were determined using a quartz fibre filter (an absolute filter), which was placed in an electrically heated casing. After leaving the quartz filter the gases are cooled and dried in an icebath condenser. Finally, the gas flow is measured using a dry gas flow meter. Typical sampling times are 1 - 5 minutes after the second cyclone and 30 - 60 minutes after the filter unit (Jahkola et al., 1991; Kurkela et al., 1991).

The particle size distribution of dust samples entering the ceramic filter and ash samples from the bottom of the ceramic filter, is determined by Coulter Counter analysis. This is performed in a water based electrolyte (2 % NaCl) after 5 minutes of ultrasonic dispersion (Jahkola et al., 1991). The amount of particles after the ceramic filter unit was so low that it was not possible to determine the particle size distribution.

The pressure drop across the tube sheet is measured continuously by a pressure cell with an electronic pressure transmitter. Due to the specific material and structure of the filter element it was not possible to measure the pressure drop across the wall of the individual elements during pulsing. During these filtration tests average cleaning cycle durations were set automatically based on a constant cleaning cycle duration.

After the filtration experiments the used filter elements were analysed using Scanning Electron Microscope (SEM) and Energy Dispersive x-ray (EDX) techniques to find out any changes in the structure of the filter and to examine dust and particle penetration into the filter material. Corresponding SEM and EDX analyses were also made for the unused filter element for comparison.

Strength of the used and unused filter elements was studied making so called Diametral Compression (DC) test with full rings (often called O-ring test) to find out any changes in strength between the unused element and the element exposed to PFBC conditions. The DC-test was done with 10 parallel ring samples.

OPERATING CONDITIONS

The filtration study is one main objective of a larger PFBC research project that studies combustion and emission properties of different solid fuels which has been carried out at Helsinki University of Technology. A wide variety of operating conditions with different fuels has been used during the test period. However, most of the experiments were carried out at 1.0 MPa pressure and at filtration temperatures around 973 K (700 °C).

The whole test period included four test weeks that gave a total exposure time of the filters of about 340 hours. The total number of cleaning pulses was approximately 1570. In the first three test weeks bituminous coal was used as the fuel for 260 hours with some 1410 cleaning pulses. During the first week very fine dolomite at a Ca/S-molar ratio of 2.0 was used for sulphur capture. During the next two weeks limestone at a Ca/S-molar ratio of 1.5 was used. The fourth test week was carried out without sorbent using peat as a fuel. The number of cleaning pulses in the peat combustion tests was about 160 during 80 hours of exposure. Some properties of the fuels, fuel ashes and sorbents are given in Table 3.

TABLE 3
Properties of fuels, fuel ashes and sorbents

Fuel or sorbent	Bit. coal	Peat	Dolomite	Limestone
Moisture content, w-%	2.0	12.8		
Dry matter composition				
Volatile matter, w-%	33.5	68.2		
Ash, w-%	13.4	4.8		
Ca, w-%	3.4	5.7	19.7	38.4
Mg, w-%	0.4	1.4	10.9	0.4
Na, w-%	0.8	1.0	0.06	0.06
K, w-%	2.1	1.2	0.39	0.13
S, w-%	2.9	0.9		
Mass median diameter, μm			40	100

Before and after each test week there were start up and shut down procedures that are included in the numbers for total exposure time and cleaning pulses. All the test runs were carried out using two similar fibrous ceramic candle filter elements with a total filtration area of 0.36 m^2. The elements were not replaced by new ones during the whole test period.

In the third test week (4/92) the operating pressure was 0.6 MPa all the time. The face velocity of the filter element was some 20 % higher than during normal operation. During the first test run of each test week the filtration temperature varied between 823 and 923 K (550 and 650 °C) because the filter unit was not completely warmed up. These test runs were not included in more detailed analysis of filter behaviour. Typical operating conditions of the filter during each test week are summarized in Table 4.

TABLE 4
Operating conditions of the filter

Test week	46/91	48/91	4/92	8/92
Fuel	Bit. coal	Bit. coal	Bit. coal	Peat
Sorbent	Dolomite	Limestone	Limestone	-
Exposure time (hours)	80	90	90	80
Number of cleaning pulses	≈ 480	≈ 370	≈ 560	≈ 160
Operation pressure, MPa	1.0	0.6	0.6	1.0
Filtration temperature, K	883-1013	873-1003	863-963	853-953
°C	610-740	600-730	590-690	580-680
Gas flow rate, m^{3}n/h	150-170	100-160	110-150	90-160
Face velocity, cm/s	3.7-5.3	3.5-5.4	4.3-6.4	3.6-5.2

Figure 3 shows typical particle size distributions of the dust samples entering the ceramic filter during bituminous coal and peat combustion. The mass median diameter of dust particles was between 6 and 7 μm and between 7 and 8 μm, respectively.

Some characteristics of filter ash during three test weeks has been presented in Table 5. The most probable composition of filter ash was found to be $CaSO_4$ and MgO during test week 46/91 (bit. coal + dolomite), $CaSO_4$ and SiO_2 during test weeks 48/91&4/92 (bit. coal + limestone) and SiO_2 and Fe_2O_3 during test week 8/92 (peat) (Brown, 1992).

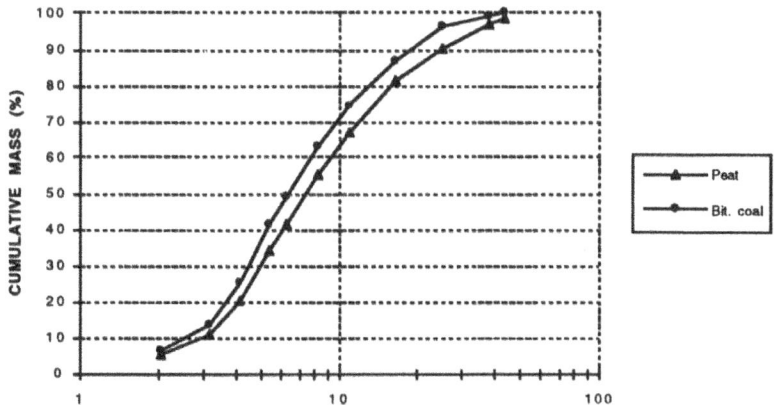

Figure 3. Particle size distribution of dust entering the filter

TABLE 5
Some characteristics of filter ash

Test week	46/91	48/91 & 4/92	8/92
Filter ash			
Ca, w-%	20.0	18.8	7.1
Mg, w-%	10.6	2.1	1.9
Al, w-%	1.6	2.9	4.8
Si, w-%	9.8	13.9	20.5
Fe, w-%	3.7	5.6	11.6
S, w-%	6.8	6.3	0.1
Na, w-%	0.04	0.12	0.34
K, w-%	0.37	0.63	0.53
Mass median diameter, μm	30	35	20

RESULTS AND OPERATING EXPERIENCES

Ranges of the most important operating characteristics of the filter during each test week are summarized in Table 6.

165

TABLE 6
Ranges of operating characteristics of the filter

Test week	46/91	48/91	4/92	8/92
Cycle duration, min	10	10-15	5-20	30-60
Pulse duration, ms	100	100	100	100
Pressure drop, kPa				
- after cleaning (=baseline)	6.0-9.0	7.0-10.0	8.0-11.0	7.0-9.0
- before cleaning	9.0-14.0	10.0-13.0	11.0-15.0	9.0-12.0
Inlet particle loading, g/m^3n	19-21	7-9	13-15	1-3
Outlet particle loading, mg/m^3n	1-2	1-3	1-5	1-2
Collection efficiency, %	> 99.9	> 99.9	> 99.9	> 99.9
Pulse pressure, MPa	1.4-1.6	1.4-1.5	1.0-1.2	1.2-1.4
Permeance, [(m/s)/bar]	1.1-3.6	0.9-1.5	0.8-1.3	1.0-1.3

The inlet particle loading during bituminous coal combustion seemed to be very high, about 20 g/m^3n, especially during the first test week. This was probably due to the very small particle size and high amount of the sorbent that could not be removed by the two cyclones. Despite this the outlet particle loading was very low, always lower than 5 mg/m^3n. With both fuels over 99.9 % collection efficiency of the filter was achieved.

The baseline pressure drop was always rather low, typically 6.0 - 8.0 kPa, just after the start up procedure of each test week. This increased to an almost constant level of 9.0 - 11.0 kPa by the end of each test week. The change of the baseline pressure drop from 9.0 - 11.0 kPa to 6.0 - 8.0 kPa is the result of the change of the thickness of the dust cake layer on the surface of the candles caused by temperature changes between the test weeks. The pressure drop before cleaning was typically 2.0 - 5.0 kPa higher than that of the baseline pressure.

It is difficult to compare the baseline pressure drops to each other because of varying conditions. A better picture of filter operation is given by the permeance of the filter, which is defined by the following formula:

$$\text{Permeance} = (U/\Delta p) * (\mu/\mu_{20})$$

where

U = face velocity (m/s)
Δp = baseline pressure drop (bar); (1 bar = 0.1 MPa)
μ = dynamic viscosity of the gas at filter operating temperature (kg/ms)
μ_{20} = dynamic viscosity of the gas at 293 K (20 °C) (kg/ms)

Figure 4 shows the development of the permeance of the filter over 1570 cleaning pulses during 340 hours of exposure. The shut down and start up procedures can be seen as an increase of the permeance between the test weeks after 80, 170 and 260 hours of exposure. The permeance was higher (1.2 - 1.5 m/s/bar) in the beginning of the second, third and fourth test weeks (and the baseline pressure drop was naturally lower, 6.0 - 8.0 kPa) reaching an almost constant level of 1.0 - 1.2 m/s/bar (which corresponds the baseline pressure drop values of 9.0 - 11.0 kPa) by the end of these test weeks. The permeance after cleaning pulses behaved as expected: first it dropped quickly after about 300 cleaning pulses and 50 hours of exposure and then it seemed to stabilize to a constant level of about 1 (m/s)/bar or about 15 % of the permeance of almost 7 (m/s)/bar of the clean element before the filtration experiments.

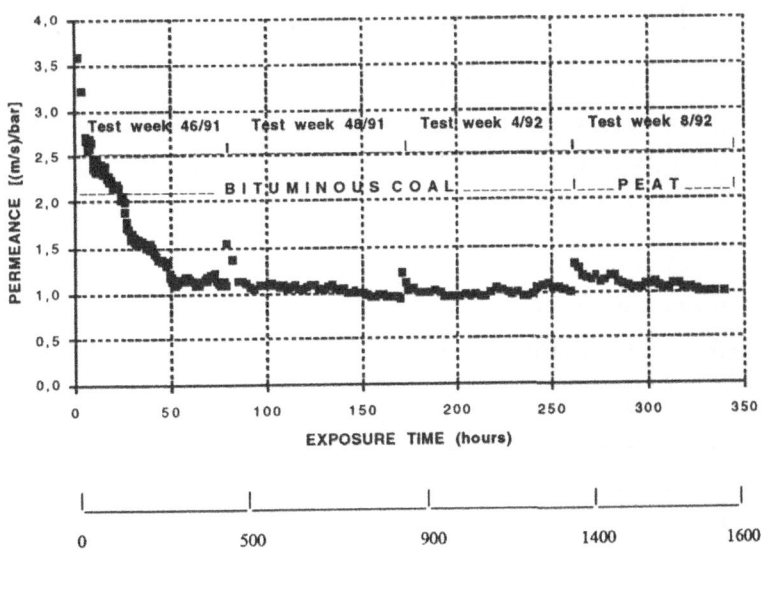

Figure 4. Development of filter permeance

The cleaning of the candle elements cannot be effective unless the pressure inside a candle overcomes the system pressure plus the pressure drop of the wall and allows the pulse gas to flow through the wall. During normal operation with bituminous coal combustion it was noticed that the pulse reservoir pressure had to be at least 0.4 MPa higher than the filtration pressure. This pressure difference between the nitrogen reservoir and the filter unit seemed to

167

be a minimum value for this type of ceramic candle filter and pulsing system in order to accomplish efficient cleaning of the candles (Jalovaara et al., 1993).

It was observed that after 50 - 60 hours of exposure the cleaning of the candle elements became the most effective because the differences between pressure drops before and after cleaning were the highest (4.0 - 5.0 kPa). Probably just at this stage a thin dust cake layer formed on the surface of the candle to ensure effective filtration. This phenomenon, which is considered a necessity for good filtration by ceramic candle filters, is often called absolute or surface filtration.

The pulse duration was set to be 100 ms during the whole operation time of the filters because in earlier experiments in 1989 it was found that no clear effect of pulse duration could be noticed if the duration was at least 100 ms (Jahkola et al., 1991). Also, it is very difficult to optimize the pulse duration with the Otaniemi PFBC test rig because the real pulse duration is longer due to the opening and closing times of the pulsing valve.

It is difficult to determine exactly from the SEM studies whether or not dust particles have penetrated into the filter element. This can be seen from Figure 5, which shows a comparison between the clean and used element taken from near outside surface (about 1 mm from outer surface). There is almost no difference between these two photographs. For this reason it is more convenient to assess this by analysing the EDX result, which shows the presence of only very small amounts of sulphur, calcium, iron, potassium, aluminium and silicon near the outside surface of the used element. All these are major components of filter ashes as can be seen from Table 5.

The presence of these components of filter ash except aluminium and silicon is virtually negligible in the middle section of the used element (about 5 mm from outer surface). It is, however, difficult to evaluate the amount of aluminium and silicon in dust particles penetrated into the filter because the original filter material contains both aluminium and silicon. The SEM photograph of the used element looks visually similar to that of the clean element (Figure 6).

In the DC-test the strength of the used element after 340 hours of exposure in PFBC conditions has been reduced by about 10 % compared with the strength of the unused element (Table 7).

TABLE 7
Strengths of the used and unused filter elements

	Strength MPa	Upper limit MPa	Lower limit MPa
Unused	2.6	2.8	2.5
Used in PFBC (after 340 hours)	2.3	2.5	2.2

168

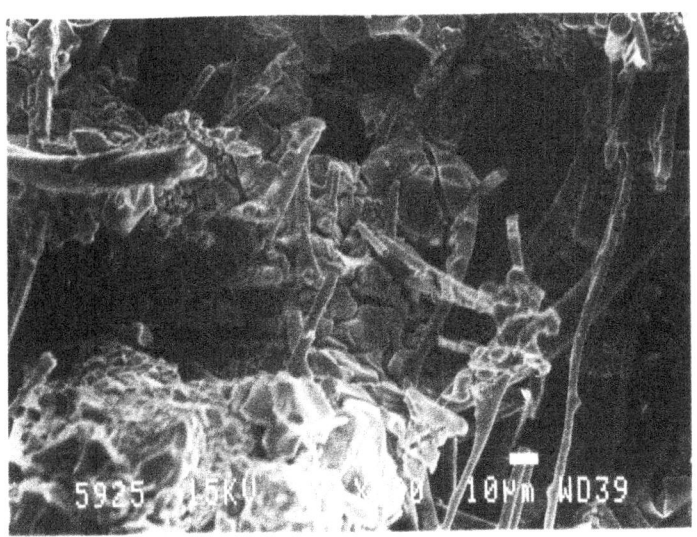

CLEAN

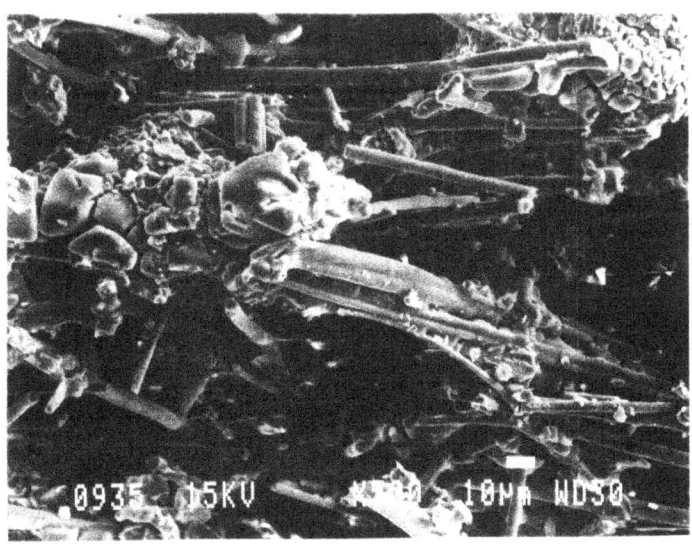

USED for 340 hours (flow direction ---->)

Figure 5. SEM photographs from near the outside surface of the clean and used filter element
(500 X)

169

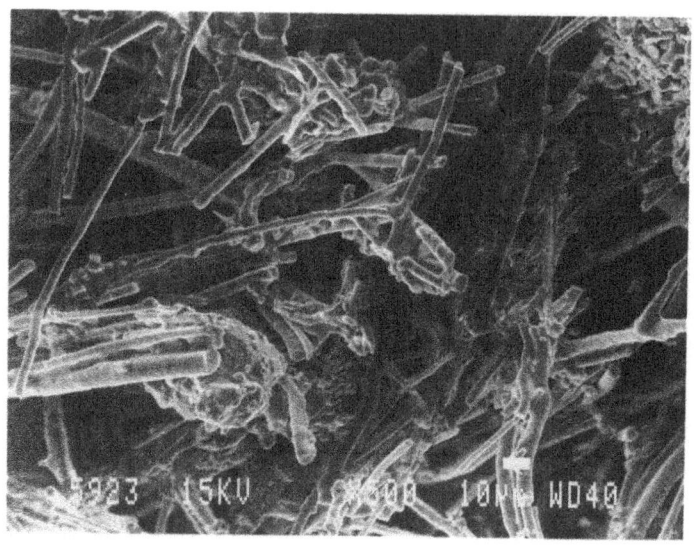

CLEAN

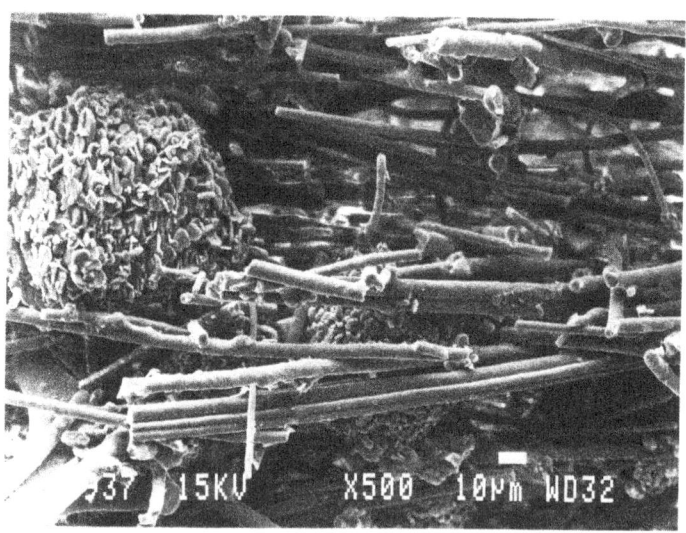

USED for 340 hours (flow direction ---->)

Figure 6. SEM photographs from the middle section of the clean and used filter element
(500 X)

CONCLUSIONS

The following conclusions can be made from the results presented in this paper:
1. The collection efficiency of the studied filters was very high, typically higher than 99.9 %, which indicates clearly that these elements will meet particle removal requirements of most gas turbines and environmental emission limits.
2. The permeance of the filter behaved as expected: it first dropped quickly and then it seemed to stabilize to a constant level or about 15 % of the original permeance of the clean element. This suggests that the filter elements probably would not get blocked even for longer exposure.
3. The filter elements performed satisfactorily during the tests and only minimal dust particle penetration into the outside surface layer of the element occurred.

However, it has to be taken into account that the exposure time of the filter elements (340 hours) was too short for making reliable, long-term conclusions. Further research will be needed to evaluate the long-term durability of this filter type in PFBC conditions.

ACKNOWLEDGEMENTS

The work presented in this paper is a part of the Finnish Combustion Research Program "LIEKKI". It was financed by the Ministry of Trade and Industry of Finland, A. Ahlstrom Oy, Imatran Voima Oy and Helsinki University of Technology. The authors gratefully acknowledge the staff of Laboratory of Fuel and Process Technology of Technical Research Centre of Finland, as well as the personnel of the authors´ laboratory.

REFERENCES

Brown, P.K. (1992). Analysis of three dust samples and a used Cerafil HS element from a PFBC application at Helsinki University of Technology. Development Report. Foseco International Limited, Materials Group.

Hulkkonen, S., Jahkola, A. & Kurkela, E. (1989). The Otaniemi Pressurized Fluidized Bed Test Facility and Research Project. Proceedings of the 10th International Conference on Fluidized Bed Combustion. ASME, pp. 233-237.

Jahkola, A., Jalovaara, J. & Hulkkonen, S. (1991). Testing of Ceramic Candle Filters at Otaniemi PFBC Test Rig. Proceedings of the 11th International Conference on Fluidized Bed Combustion. ASME, pp. 947-951.

Jalovaara, J. (1992). Hot Gas Dust Filtration from Combustion Systems. Proceedings of Second EPRI Workshop on Filtration of Dust from Coal-Derived Reducing and Combustion Gases at High Temperature. EPRI, presentation number 23.

Jalovaara, J., Hippinen, I. & Jahkola, A. (1993). Particle Removal in PFBC Using a Fibrous Ceramic Candle Filter. Proceedings of the 1993 International Conference on Fluidized Bed Combustion. ASME, pp. 713-718.

Kurkela, E., Ståhlberg, P., Laatikainen, J. & Nieminen, M. (1991). Removal of particulates and alkali metals from the product gas of a pressurized fluidized-bed gasifier. Proceedings Filtech Conference. The Filtration Society, pp. 449-467.

Withers, C.J., West, A.A., Twigg, A.N., Courtney, R.S., Seville, J.P.K. & Clift, R. (1990). Improvements in the Performance of Ceramic Media for Filtration of Hot Gases. Filtration & Separation. Elsevier Advanced Technology, January/February 1990, pp. 32-37.

APPLICATIONS OF LOW DENSITY CERAMIC FILTERS FOR GAS CLEANING AT HIGH TEMPERATURES

C J C Beattie and C J Withers
Cerel Ltd
Catherine House, Coventry Road, Hinckley, Leicestershire

ABSTRACT

Low density ceramic filters are now available for incorporation into gas filtration units. Trials have been undertaken with such units on a variety of duties. These duties have mainly concerned the cleaning of gases from high temperature processes prior to their release to atmosphere. Other duties have used high temperature filtration as an integral unit operation, simplifying operating processes.

The paper presents the experiences of the authors with installations in a number of different applications in a variety of industries. The paper discusses the properties and performance of low density ceramic media in general terms and relates these properties to their performance in each of these applications. It identifies specific strengths and weaknesses of this type of medium and proposes projects in other industries which would benefit from the use of this technique.

The paper presents detailed, quantified case studies of the use of low density ceramic media in several industrial processes which are subject to the Environmental Protection Act.

INTRODUCTION

Many industrial processes involve the generation of hot waste gases, which can be contaminated with either solid, liquid or gaseous pollutants. With the advent of the Environmental Protection Act 1990, these processes must now meet tough emission standards for these pollutants. How rigorous these limits are depends on the nature and size of the process. Although the legislation covers a range of airborne pollutants, in practice application of the Act focuses on the removal of the particulate emissions, ie. the most visible sign of pollution. However, until recently, cleaning up hot gas streams presented considerable problems.

The options available for cleaning a hot gas stream are limited. Essentially, these are bag filter technology incorporating fabric filter media, electrostatic precipitators, wet scrubbers and cyclones. With fabric filters and scrubbers, cooling of the gas stream is essential and maximum efficiency is thus linked to maintaining the gas temperature within a narrow band. This can make the technology sensitive to process upsets such as unexpected temperature surges. Traditional hot gas technology such as ESPs and cyclones are becoming less attractive with the tightening of emission limits under the EPA, which favours the more efficient barrier filtration techniques. With this background, development of hot gas clean up technologies has tended towards producing a barrier type filter, capable of meeting the ever decreasing limits for particulate emissions and free from the restriction of a temperature excursion vulnerability.

Effective and efficient hot gas filtration is now a reality (Seville *et al*, 1989; Anon, 1991; Withers, 1992) with the availability of several low density ceramic filter products on the UK market and the practical application of low density ceramic filtration in numerous industrial sectors (Anon, 1992; Elliott, 1992). Ceramic candles are available from such companies as BWF, Heimbach, Didier and Cerel. This paper dicusses the author's experience with a number of processes incorporating the ceramic filter products of Cerel. While all the data may not be exactly the same for the different products, the problems and benefits encountered will certainly be universal in their occurrence. Frequently, full and open statements of the filter performance cannot be made because to do so would breach customer confidence.

CHARACTERISTICS OF CERAMIC FILTER CANDLES

The standard form of low density ceramic filters is a candle, closed at one end and open at the other. The open end is usually flanged to allow the mounting of the candle on a header plate separating the dirty and clean gas flow. Dirty gas impinges on the outer surface of the candle and clean gas is carried away down the hollow centre of the candle. The particulates are captured on the outer face of the candle and cleaned off with pulses of compressed air which are directed down the centre of the candle, much as in a reverse pulse jet filter. Captured particulates then fall down from the filters and are collected in a hopper.

The candles are formed from ceramic fibres bonded with both organic and inorganic binders. Consequently, the filters are stable to high temperatures and are not damaged by sparks or incandescent particles; it is claimed that the filters can operate continuously at temperatures of up to 1000°C. A further characteristic of the ceramic structure is that the filters are inert to attack from aggresive chemicals, such as steam and acid gases.

As with fabric filters, the bulk of the filtration is achieved by the build up of a porous cake of particulates on the surface of the filter, which prevents further penetration of the media (Koch *et al*, 1992). In some situations, this barrier of dust is artificially added (pre-conditioning) to improve the filter's performance. This barrier, or conditioned, layer of particulates gives an increase in pressure drop over the virgin media. However, it also gives rise to excellent filtration efficiencies. Due to the fact that this conditioned layer is permanent, the rigid ceramic elements do not clean down completely when pulse cleaned as fabric filters do. This results in virtually no penetration of the filter and performances which can result in less than 5 mg/m^3 of particulates in the emissions.

For a given application, the filtration velocity can be higher than that used for fabric filters, as the ceramic filter is permanently protected from dust penetration by the conditioned layer. With a fabric filter, pulse cleaning periodically strips away the protective layer, allowing particulate penetration of the media.

The combination of the filter's temperature resistance and its ability to withstand surges in volume means that ceramic filters are less likely to be susceptible to damage caused by variable conditions and process upsets.

PRACTICAL CONSIDERATIONS

While the ceramic filters are more durable than their fabric equivalents, their use must be tempered with common sense. Although ceramic candles have excellent filtration characteristics, and an improved resistance to process upsets, they are by no means the answer to all problems.

High filtration Velocities
Opening the door to higher filtration velocities is not free from hazards. It is not widely appreciated that operating at high filtration velocities not only increases the system pressure drop, but can also generate other problems.

A situation which can develop, and which can be made worse by high filtration velocities is dust re-entrainment. When this occurs, particulates cleaned off by the reverse pulse do not immediately fall down into the hopper, but are captured again by the filters. Consequently, with a poorly designed filter, the dust loading the filter sees can be artificially high and high pressure drops result.

The point at which this becomes a serious problem will of course depend on a number of factors, such as the particle size range of the dust, the dust loading and the path the particles must follow to escape the filtration zone. With certain dusts, careful consideration must be given as to whether off-line cleaning is a requirement on the filter plant.

An additional potential difficulty to be aware of is the fact that some dust cakes are compressible, and that the pressure drop across the filter in these cases will not rise proportionally with the filtration velocity. Substantially higher pressure drops can result which are difficult to predict without pilot plant work.

Operational Temperature

Although the upper temperature limit of low density ceramic filters is often high - 900 to 1000 °C - there are technical difficulties in operating at these limits. Special consideration must be given to the design and construction of the filter housing, and often the dusts being filtered have different characteristics at these elevated temperatures. Certain types ash can sinter at temperatures as low as 700°C. Patently, a sintered or melted material on a rigid media could present significant difficulties when pulse cleaning.

Higher temperatures also mean higher pressure drops. The pressure drop across a filter is proportional to the gas viscosity. For example, with air, the gas viscosity at 500°C is roughly 150% that at 170°C. For filters operating at the same filtration velocity, and seeing the same dust loading, the pressure drop of a filter operating at the higher temperature would be 50% higher than that of an equivalent filter operating at the lower temperature.

Media Strength

Low density ceramic candles, although possessing sufficient strength to withstand repeated pulse cleaning, do not have any more capability than bag filters to survive direct impingement by abrasive particles. Filter designs must appreciate this and sensible engineering design can eliminate the threat from "shot blasting" of the elements.

Maintenance

As with fabric filters, proper maintenance is essential. Failure of a ceramic element will lead to the failure of others if left unattended. Much as in fabric filters, holes can develop in the elements if particulates leaking into the clean side are pulsed into the elements at the high pressures used for pulse cleaning.

Ceramics

In certain situations the ceramic filters can also be prone to flux attack. If a substantial amount of combustible material and flux is allowed to build up on a filter element, then sintering and weakening of the element can occur. Usually, this build up is due to a design problem with the filter or the failure of an ancillary service.

APPLICATIONS OF CERAMIC FILTRATION MEDIA

To a great extent this has been determined by two factors. One is the state of the current legislation, which provides the major driving force, while the other is the need to find practical clean up solutions for industries traditionally seen as the problem children of gas clean up.

Incineration

Ceramic filters have been employed in two of the principal areas of incineration, clinical and chemical. Each presents its own difficulties and problems.

Case 1: Chemical Waste Incineration

TABLE 1
Performance data for a pilot plant on a chemical waste incinerator

Duty:	405.9 Nm3/hr (238.4 scfm)
Filtration velocity:	0.026 m/s (5.12 fpm)
No of ceramic candles:	36
Filtration Temperature:	150 - 300 C
Pressure Drop (max):	280 mm Water Gauge
Filter Resistance (x10^9)	4.6 /m

(For a definition of filter resistance, see Appendix no.1)

Solvents and other chemical wastes are incinerated in this process and the hot gases used to generate steam. The gases leave the boiler at around 400°C and contain ash, and various acid gases.

The particulates in the flue gas are there by virtue of the atomization process which injects the waste chemicals into the incineration chamber. They tend to be of a uniform size, and are also sub-micron: SEM work revealed that the bulk of the particles are actually in the order of 0.1 to 0.2 microns (see figure 1). Dust of this order of size would present problems to most fabric filter media, which would allow a the sub-micron portion of the dust to pass straight through. The ceramic filter appeared to stop all the particulates as there was no visible plume from the filter, but at the price of a high pressure drop due to the close packed dust cake.

Figure 1. SEM photograph showing agglomerate of sub-micron particulates

The waste gas also has variable amounts of acid gases, including HCl and SO_2. Both attacked the metal work of the filter plant, although the ceramic filter elements were uneffected.

The gas temperature was fairly stable for individual solvent feedstocks to the incinerator, but did vary significantly from day to day. The plant was also shut down on a fairly frequent basis for cleaning of the heat exchanger, and as and when feedstocks were exhausted.

Case 2: Clinical Waste Incineration

Ceramic filters were used in a pilot scale trial on the effects of lime injection to remove acid gases such as hydrogen chloride and sulphur dioxide (Gang & Loffler, 1990). The trial involved a filter plant containing 36 1m candles, handling gases from a clinical waste incinerator. Lime was injected into the gas flow directly before the filter plant, and the effects of operating temperature, lime dosing rate and pulse cleaning interval on the acid gas removal rate were studied. The results of this study will be summarised in a brochure by ETSU, who part funded the work, and are discussed in a paper by A J Startin (Startin, 1993).

TABLE 2
Performance data for a pilot plant on a clinical waste incinerator

Duty:	1153.8 **Am3/hr** (678 acfm)
Filtration velocity:	0.047 **m/s** (9.25 fpm)
No of ceramic candles:	36
Filtration Temperature:	180-210 **C**
Dust Loading (max):	5.72 **g/Am3 (mostly lime)**
Pressure Drop (max):	295 **mm Water Gauge**
Filter Resistance (x10^9)	2.67 **/m**

Ceramic filters are particularly suited for this type of application. Clinical waste incinerators typically handle individual bags of waste which can contain a wide variety of clinical waste; anything from glassware to nappies. The variability of the feed material can lead to process fluctuations which are not easily handled by fabric filters, with surges in temperature, acid gas content and volume flow.

As the work was primarily concerned with the acid gas removal, no measurements were taken of the particulate level in the exhaust gases, although no visible emissions from the filter could be seen.

Foundry applications

The filtering of gases from the various furnaces in this industry presents some real problems for enduser and the filter manufacturer. The chief cause of grief is the multitude of greases, oils and paints which adhere to aluminium scrap, and which can blind, clog and otherwise render ineffective filtration media.

The filters described below have to handle a wide range of temperatures, dew point excursions and sticky dusts. Additionally, the media must be tough enough not to be rendered ineffective by fires or incandescent particles reaching the filter.

Case 3: Aluminium Scrap Melting - Clean Scrap

Case 3 concerns a filter installed on an indirectly fired furnace melting what would normally be classified as clean aluminium scrap. This consisted mainly of sprues from diecasting, or partially painted scrap. The furnace had approximately half a tonne capacity.

TABLE 3
Performance data for a filter on a furnace melting aluminium scrap

Duty:	2380 Nm3/hr	(1400 scfm)
Filtration velocity:	0.029 m/s	(5.8 fpm)
No of ceramic candles:	180	
Filtration Temperature:	140 - 300 C	

Previously there was no gas clean-up system, and waste gases were vented directly to atmosphere. However, once a ceramic filter was installed, the problem of visible emissions did not altogether disappear. While fumes and soot particulates were captured very effectively, so that no emissions were visible from the filter, at certain, seemingly random, parts of the melt cycle, emissions became visible from the flue stack.

The filters showed no sign of damage and no leaks could be found. However, on monitoring the temperatures throughout the filter system, it became clear that the appearance of the plume was in fact temperature related. When the temperature in the filter reached a certain point *smoke* became visible, and when the temperature dropped beneath the critical point, the emissions disappeared.

It became apparent that the emissions were oil vapours which were passing thorugh the filter as gas and condensing on hitting the atmosphere. In this case the problem was to be solved by the addition of an afterburner before the filter.

Case 4: Aluminium Scrap Melting - Contaminated Scrap

TABLE 4

Performance data for a filter on a furnace melting heavily contaminated aluminium scrap

Duty:	8080 Nm3/hr	(4752 scfm)
Filtration velocity:	0.05 m/s	(10 fpm)
No of ceramic candles:	416	
Filtration Temperature:	200 C	
Pressure Drop (max):	275 mm Water Gauge	
Filter Resistance (x10^9)	2.2 /m	
Emissions:	0.6 mg/m3	Isokinetic

The filter handles the exhaust gases from four indirectly fired furnaces, melting scrap which is heavily contaminated with oil, grease and plastic. The furnaces run 24 hours a day and are regularly dosed with fluxes.

This filter is interesting for two reasons. Firstly, the candles are vertically mounted, and secondly, a high face velocity is used without an excessive pressure drop. Isokinetic sampling has established that the filtration efficiency of the filters is not compromised. The filter experienced some initial problems with the build up of solids near the header plate. As this build-up had a significant carbon content, it was prone to catch fire and burn intensely, allowing flux attack of the ceramic structure to occur. This lead to embrittlement and loss of filtration efficiency for several of the candles. The problem was overcome by a small redesign of the filter assembly.

Case 5: Aluminium Scrap Melting - Rotary Furnaces

TABLE 5

Performance data of the test filter on a rotary aluminium scrap melting furnace

Duty:	max 250 Am3/hr	(147 cfm)
Filtration velocity:	0.01 - 0.04 m/s	(2 - 8 fpm)
No of ceramic candles:	9	
Filtration Temperature:	max 200 C	
Dust Loading:	1 g/Am3	
Pressure Drop (max):	500 mm Water Gauge	
Filter Resistance (x10^9)	up to 6.0 /m	

This filtration exercise was intended to provide a test platform to characterise ceramic filters. The filter contained 9 1m elements, and was drawing fumes and gases from three rotary furnaces (Callis, 1992). One of the furnaces uses oxygen rather than air, which leads to high burner flame temperatures and consequently a higher proportion of fines in the emissions due to vapourised metal oxides.

The filter exhibited a problem not before encountered with ceramic elements by the author. The resistance of the filter was found to increase with increasing filtration velocity. This effect is most likely due to the dust layer being compressible.

Coal Processing

Case 6: Smokeless Fuel Production

TABLE 6
Performance data for a filter on a smokeless fuel process

Duty:	50276 **Nm3/hr** (29540 scfm)
Filtration velocity:	0.03 **m/s** (6 fpm)
No of ceramic candles:	4800
Filtration Temperature:	270 **C**
Dust Loading:	1.0 - 35.0 **g/Am3**
Pressure Drop (max):	220 **mm Water Gauge**
Filter Resistance (x10^9)	2.4 /m
Emissions:	less than 1 **mg/m3** Isokinetic

This is the largest single application of ceramic filters in the country, if not the world, and has had its share of difficulties (Rogers & Jones, 1993). The installation consists of two 50,000 m3/hr ceramic filters, incorporating a total of 4800 ceramic candles in horizontal mounting. The filters are filtering gases from a smokeless coal process.

Coal is heated in a reducing atmosphere to drive off the low temperature volatiles which would normally cause the coal to smoke when burnt. The gas produced by this process has a significant calorific value, and is burnt in two boilers to produce steam. Problems arise with the particulates in the gas. The gas has typically anything from 1 to 35 g/m3 of dust, which is approximately 20-40% w/w carbon, although this figure can be as high as 60%.

The filter candles were mounted in modules of 36 elements, with a **cell** of candles containing 6 modules, arranged 2 modules wide and 3 modules high. Several of the modules in each cell have 4 candles removed to form blanked off inspection ports.

The ceramic filters of two manufacturers were tested extensively in pilot plant trials. The ceramic filter candle finally selected successfully filtered the gas over a period of 10 months, acquiring 4000 hours of operation. However, on installation of the main plants, problems began to occur almost immediately.

A frequent occurrence in both the pilot plant and the main filter was the dust catching fire. If air hits the hot filters, the carbon ignites, glowing red hot. Ash in the dust then sinters to form a porous cake.

In the pilot trials, no difficulties arising from fires were observed. In the main plant, the occurrence of a fire, which was frequently due to burner flameouts, would lead to blocking of the filter cells and consequent breaking of the filters due to the solids build up. Interestingly, some candles removed from these cells clean down satisfactorily and display no increase in resistance to flow over what would be expected from a conditioned filter.

Why the cells block is not clear. It may be that sintered material bridges between the ceramic candles, and that once this occurs the dust cannot escape from the filtration zone and gradually blocks up the filter cell. The horizontal mounting configuration means that any material cleaned off the candles can have a distance of 2m to fall before clearing the filtration zone. This, combined with the property of the dust to sinter and gather on any horizontal surface contributes to exacerbate the problem.

At the present time, the only solution found for the problem is to detect a fire as it starts and then to flood the filter with nitrogen and steam. While not ideal this now prevents the fires which are clearly linked to the cells blocking.

General Applications

Case 7: Furnace Melting Lead Solder

This filter was the subject of a study by the consultancy group Gibb Environmental Sciences. To comply with the legislation on emission levels of lead, the filter had to meet some very tough requirements and a very low level of emissions was required. This was complicated by the presence of ammonium chloride in the gas as a result of the smelting process. The filter achieved 0.34 mg/m^3 at a relatively high face velocity.

TABLE 7

Performance data for a pilot plant filter on a furnace melting lead solder

Duty:	1272.8 **Nm3/hr** (748 scfm)
Filtration velocity:	0.057 **m/s** (11.3 fpm)
No of ceramic candles:	36
Filtration Temperature:	29 **C**
Dust Loading:	101.4 **mg/Am3**
Pressure Drop (max):	300 **mm Water Gauge**
Filter Resistance (x10^9)	5.8 **/m**
Emissions:	0.34 **mg/m3** Personal Sampler

Although a certain level of particulate was found on the filter used to monitor emissions, this was found to be mainly ammonium chloride. This is used in the recovery process and was probably captured after subliming from the furnace and passing through the ceramic filter in vapour phase.

Case 8: Gasification of Biomass Waste

TABLE 8

Perfomance data for a filter on a gasification process

Duty:	337 **Nm3/hr** (198 scfm)
Filtration velocity:	0.064 **m/s** (12.7 fpm)
No of ceramic candles:	24
Filtration Temperature:	580 **C**
Pressure Drop (max):	280 **mm Water Gauge**
Filter Resistance (x10^9)	1.27 **/m**

The ceramic candles are incorporated into a "Waste to Energy" gasifier unit. The core of this process is a small gasification unit which will accept a variety of organic waste material as fuel, such as sewage sludge, wood waste, poultry litter, straw and coal. The fuel is gasified to give a dirty producer gas which is cleaned by a succession of equipment before being combusted to generate electricity in a biogas internal combustion engine.

The gas clean-up includes a ceramic filter operating at high temperatures and face velocity, followed by a water scrubbing stage. The operator noticed a considerable difference in the quality of the gas when the original cyclone was replaced

184

by the ceramic filter unit. There were no visible particulates in the gas leaving the ceramic filter.

The whole process is skid mounted, and enclosed in a container. The figures quoted above are from a mid range sized unit, incorporating 24 candles. The unit has performed satisfactorily without operational difficulties of any description.

Case 9: Machine Tool Dust Extraction

In this particular application, ceramic filters were adopted over an wet scrubbing system. The principal reason for this was that by using a dry filtration route, biological degradation of the leather and the subsequent effluent handling problems were avoided. Standard fabric filters were unsuitable because of the tendency of the leather to catch fire.

TABLE 9

Performance data for filter on a leather grinding machine

Duty:	1522 Nm3/hr (895 scfm)
Filtration velocity:	0.1 m/s (19.6 fpm)
No of ceramic candles:	24
Filtration Temperature:	20 C
Pressure Drop (max):	285 mm Water Gauge
Filter Resistance (x10^9)	1.58 /m

The only problem encountered with this filter was due to the leather catching fire. The machine operators had to be persuaded not to attempt to put out the fires with either buckets of water or brooms: fire extinguishers containing carbon dioxide were considered to be more appropriate. After this initial teething problem, the filters performed as can be seen above.

The application is also interesting from the point of view of the extremely high filtration velocity used. This was possible due to the coarse nature of the dust.

Case 10: Secondary Aluminium Processing

This process recovers the aluminium from foil products such as margarine wrappers or packaging for cigarettes. The foil has a paper backing, which is burned off in a furnace. The products of combustion and particles of aluminium are first passed through a multi-cyclone and then to the ceramic filter. The multi-cyclone knocks out the fraction of the particulates of greater than 50 μm size, which contains 82% by mass of recoverable metallic aluminium.

TABLE 10
Performance data for a filter on waste gases from a furnace

Duty:	16275 Nm3/hr	(9575 scfm)
Filtration velocity:	0.032 m/s	(6.3 fpm)
No of ceramic candles:	1296	
Filtration Temperature:	200 - 250 C	
Dust Loading:	500 mg/Am3	
Pressure Drop (max):	203 mm Water Gauge	
Filter Resistance (x10^9)	2.5 /m	
Emissions:	0.3 mg/m3	Isokinetic

Ceramic elements were selected over fabric filters for two main reasons. As the filter could operate at elevated temperatures, the exhaust gases did not require dilution before filtration in the ceramic filter. Thus the size of the filter was substantially reduced over the fabric filter equivalent. Additionally, the high level of aluminium fines in the dust constitutes a fire risk, and the ceramic elements had already demonstrated their ability to withstand a fire in the filter during pilot plant trials.

Case 11: Zirconia Production

TABLE 11
Performance data for a filter on an arc furnace melting zirconia

Duty:	562 Nm3/hr	(330 scfm)
Filtration velocity:	0.028 m/s	(5.6 fpm)
No of ceramic candles:	36	
Filtration Temperature:	max 60 C	
Dust Loading:	330 mg/Am3	
Pressure Drop (max):	460+ mm Water Gauge	
Filter Resistance (x10^9)	8.3 /m	
Emissions:	max 0.166 mg/m3	Isokinetic - 5 tests

This process melts zirconia using an arc furnace. The resulting hot vapours were drawn off and filtered in a 36 element pilot plant. Due to the high temperatures required for the process, silica is actually vaporised during the melt and it is this condensing which forms the bulk of the particulate captured on the filter.

As the material being melted was slightly radioactive, emission limits of 0.3 mg/m^3 were required. Fabric filters could not achieve this level and, although the ceramic filters achieved the desired limit the cost in terms of pressure drop was high.

DISCUSSION

The case studies above illustrate the great flexibility of the low density ceramic candle. The combined strengths of its wide band of operational temperature, its high efficiency and chemical resistance make it the ideal choice for many applications. The examples also show its use in applications which do not necessarily require a high temperature filter, but which benefit from the filter's other characteristics

The barriers of cost and confidence will doubtless begin to crumble as the product is more widely accepted in the marketplace, and more rigorous demands are placed on the polluter. Efficiency is becoming a key issue, to the extent that endusers are looking to technology which will give them security for the future as well as compliance with todays emission targets.

As the EPA gathers pace and with successive reviews, the ceramic filter will find a place in such sectors of industry as cement, mineral processing and power generation. Already, ceramic filters are quoted as BATNEEC in HMIP guidelines (HMIP, 1992), and their versatility will guarantee their place in the gas clean up systems of the coming decade.

REFERENCES

Anon (1991). Editorial of The Chemical Engineer. No 505, 10 Oct 1991.

Anon (1992). Cerafil Filters Commissioned at Mountstar Metals. Filtration and Separation. 29, no. 2.

Callis, R. (1992). Integration of Processes and Abatement Techniques to meet EPA Requirements in the Metallurgical Industries. Presented to a Meeting of the Filtration Society, UCL, September.

Elliott, G. K. (1992). Ceramic Filtration Accepted as Particulate Removal Technique. Pollution Prevention. June 1992.

Gang, P. & Loffler, F. (1990). Combined Separation and Retainment of Particulate and Gaseous Matter with Cleanable Filters. Presented at 5th World Filtration Conference, Nice 1990.

HMIP (1992). Chief Inspector's Guidance to Inspectors. Waste Disposal and Recycling. Process Guidance Note IPR 5/1. Merchant and In House Chemical Waste Incineration. pp 20-21.

Koch, D., Cheung, W., Seville, J. P. K., & Clift, R. (1992). Effects of Dust Properties on Gas Cleaning using Rigid Ceramic Filters. Filtration and Separation. 29, no.4.

Rogers, P. & Jones, M. (1993). The Evaluation of Low Density Ceramic Filters for Gas Cleaning in a Boiler Flue Gas Clean-up System. To be presented to Second International Symposium on Gas Cleaning at High Temperatures, 27-29 September, University of surrey. Publ. Elsevier.

Seville, J. P. K., Clift, R., Withers, C. J. & Keidel, W. (1989). Rigid Ceramic Media for Filtering Hot Gases. Filtration & Separation, July/August.

Startin, A J (1993). Acid gas treatment at a Cerafil Pilot Plant. To be presented to the Second International Symposium on Gas Cleaning at High Temperatures, 27-29 September, University of Surrey. Publ. Elsevier.

Withers, C. J. (1992). Considerations in the Specification of Hot Gas Filters. Presented to a Meeting of the Filtration Society, UCL, Sepember.

APPENDIX 1

Filter Resistance

The pressure drop of a filter can be expressed as a function of the filtration velocity, the gas viscosity and the cake and filter resistance to flow:

$$\Delta P = K \, \mu \, v$$

where	ΔP	$=$	pressure drop across the filter and cake, Nm^{-2}
μ	$=$	the gas viscosity, Nsm^{-2}
v	$=$	the filtration velocity, ms^{-1}
K	$=$	the resistance of the filter and cake, m^{-1}

This simple expression provides a basis for comparison of the different filter applications, as theoretically it is independent of the filtration temperature and face velocity. Therefore the expression characterises the medium and the associated filter cake. It is likely that this expression is too simple to encompass all filtration duties. Other versions of this equation exist which draw in such factors as the dust loading and the time between cleaning pulses. However, they require substantial amounts of information to be of use and do not necessarily give a basis for comparison between applications.

THE EVALUATION OF LOW-DENSITY CERAMIC FILTERS IN A BOILER FLUE GAS CLEAN-UP SYSTEM

MICHAEL JONES and PAUL ROGERS
Coal Products Ltd.
Coventry Homefire Works
Keresley
Coventry
West Midlands

ABSTRACT

The use of small multi-element pilot filter units are an essential stage in determining the suitability of low-density ceramic filters for large scale gas clean-up projects. The analysis of the data obtained during prolonged testing of such filters can give valuable insights into the behaviour of the materials being filtered and provides information necessary for the finalisation of the design of full size units. Even so, problems not encountered even during extensive pilot filter trials, can arise during the operation of full stream units causing possible radical re-design and unexpected expense.

INTRODUCTION

Solid smokeless fuel is manufactured at Coal Products Coventry Homefire Works by a process involving the low-temperature fluidised bed carbonisation of crushed coal. The waste gas produced by this process, amounting to some $17,000nm^3/Hr$, and containing large quantities of tar and fine carbonised coal particles (char), was previously incinerated and its products of combustion discharged to the atmosphere via a 65m high chimney, resulting in a heavily ash laden emission. Over recent years increasing pressure from the media, local action groups, Her Majesty's Inspectorate of Pollution and the realisation by the Company that the particulate emissions from the incinerator chimney had to be reduced, led to the introduction, in mid 1990, of a large scale clean-up programme at the Works.

Of the two options considered for the clean-up of the process waste gas, its continued incineration and the subsequent cleaning of the flue gases was the most favoured method, although not without its difficulties. The second option, that of cleaning the gas prior to incineration, was expected to be more problematic due to the high gas temperature (400°C) and is tar laden nature. Incineration of the lean 'cleaned' gas was also necessary as it possessed a very strong odour. The extracted tars and char would also be difficult to dispose of.

The main problem encountered in cleaning the flue gases following incineration of the waste gas was the availability of suitable equipment. With an incinerator exhaust temperature in the order of 700°C and the small size of the particulate matter discharged, conventional gas cleaning techniques were considered inappropriate. Although some 40% of the waste process gas was already being used to fire coal dryers and boilers, replacing oil as the primary fuel and considerably reducing operating costs, it was envisaged at this stage of the project that the cleaning plant still had to be capable of dealing with the flue gases from incineration of the total waste gas production, to make allowances for plant failures.

Notwithstanding the fact that low-density ceramic filtration had not been employed in any major flue gas cleaning scheme and little practical information was available, the Company considered the technique worth evaluating in the absence of more conventional solutions. In late 1990, a pilot ceramic filter was purchased with the aim of cleaning a side stream of flue gas from the incinerator chimney. This filter contained thirty-six, horizontally mounted, one metre long Cerafil S ceramic elements and was fitted with reverse pulse cleaning controls. The flue gas was drawn from the incinerator chimney via an insulated duct, by a small fan situated downstream of the clean side of the filter. Once in operation it soon became apparent that combustion conditions within the incinerator varied considerably, preventing stable filtration from being achieved. High moisture levels in the filter inlet gases and the probability of tar vapours being drawn through the filter elements when incinerator combustion was poor led to the conclusion that this was a far from ideal installation on which to carry out the first evaluation of the filter.

As more stable conditions were obviously required to successfully evaluate the filter's performance it was agreed that the unit should be transfered to the outlet of one of the Works watertube boilers which was burning process waste gas. Although the boiler flue gas temperature was considerably lower than that in the incinerator chimney it was heavily particulate laden and was discharged into the incinerator chimney base adding to the total solids emission.

PILOT CERAMIC FILTER TRIALS

The Cerafil S pilot filter was set up to clean a side stream of heavily particulate laden flue gas from the outlet of the most frequently used of the two Works boiler. The dust burden in the flue duct was generally in the order of $7.0g/m^3$, increasing significantly during boiler sootblowing. The unit was initially set to operate at a filter element face velocity of 3.0cm/s, giving typical values of pressure drop prior to cleaning of 190mm H_2O and 135mm H_2O following cleaning. Due to poor thermal insulation of the boiler flue duct the inlet temperature to the filter remained between 200 and 260°C although the boiler outlet gases ranged between 270 and 360°C. An on-line pulse cleaning regime was adopted, with the extraction fan continuing to draw gas through the filter during cleaning. The elements were cleaned once per hour at a pulse pressure of 6.5 bar and the pulse duration was set at 120mS throughout the trials. Filter pressure drop and inlet gas temperature were monitored and recorded continuously from the Works control room.

Trial 1

The pilot unit was operated in the on-line mode for some 800 hours of almost continuous duty during which time the pressure drop after cleaning, the base line pressure drop, remained at a level of 135 to 145mm H_2O. The quantity of dust collected averaged 20kg per day. These samples were analysed for particle size distribution (Table 1) and carbon content. It was found that the carbon content of the ash varied between 25% and 45% depending upon the firing rate of the boiler. As the boiler combustion chamber was not specifically designed for burning what was in effect a pulverised solid fuel, poor combustion of the particulate matter due to the lack of residence time within the boiler appeared to be the reason for the high carbon carry-over. During this trial the filter outlet dust concentration was measured by the Works laboratory and found to be very encouraging, with levels in the order of 2.3mg/m^3 being recorded. Owing to the very low mass of solids collected and the potential for experimental error being significant at these levels it was determined that any outlet solids concentrations could be regarded as being less than 20mg/m^3.

Trial 2

The filter was then operated in the off-line cleaning mode for a further 300 hours in order to investigate the advantages, if any, over the on-line mode. It was thought that off-line cleaning, with no flow through the filter during the cleaning period, would achieve more efficient cleaning of the ceramic elements and result in a lower base line pressure drop. At a face velocity of 3.0cm/s. dust re-entrainment, the rapid re-capture of some of the dust removed from the element surfaces during on-line cleaning, was expected to be minimal and produce only a very small increase in base line pressure drop. Comparison of the differential pressure traces obtained during on and off-line cleaning confirmed that this was in fact the case, with the values of pressure drop soon after cleaning being almost identical. (Figure 1).

After approximately 1,100 hours of operation, two of the ceramic elements were removed for permeability tests and scanning electron microscope (SEM) inspection by the manufacturers. They were replaced in the filter by new elements. These tests indicated that no significant dust penetration of the elements had in fact occurred.

Trial 3

The pilot unit was again returned to duty, with the face velocity raised to 3.6cm/s. for some 80 hours in each of the pulse cleaning modes. With a filter inlet temperature of 220°C the pressure drop stabilised in the range 230 to 180mm H_2O and remained so throughout the trial. The test was repeated for similar durations, at a nominal face velocity of 4.1cm/s. and at a gas inlet temperature averaging 230°C. This gave a filter pressure drop in the range 265 to 215mm H_2O. Some evidence of dust re-entrainment was observed at both velocities during the on-line cleaning phase. Small but very rapid rises in filter pressure drop were observed immediately following the cleaning of each row of elements. This gave rise to a slightly higher base-line pressure drop than with off-line cleaning.

TABLE 1
Typical Boiler Flue Ash Size Analysis.

Size (Microns)	Weight Cumulative % Undersize.
53.5	100.0
37.6	99.5
28.1	98.8
21.5	96.1
16.7	90.2
13.0	81.0
10.1	70.2
7.9	57.5
6.2	42.6
4.8	29.0
3.8	20.0
3.0	13.2
2.4	8.2
1.9	5.0

Determined By Malvern Diffraction Size Analysis

Testing continued at a face velocity of 4.6cm/s. for a further 48 hours, with limited success. This resulted in a pressure drop in the range 280 to 230mm H_2O, but due to the lack of fan suction the gas flow through the filter decreased slightly as maximum differential pressure was attained prior to cleaning.

It was concluded that for stable long term evaluation, the filter should be operated at a nominal face velocity of 3.6cm/s. with an on-line cleaning regime. On-line cleaning was chosen because changes in base-line pressure drop could be more easily recognised, on completion of pulse cleaning, by regular inspection of the differential pressure traces. Testing continued in this mode for a further 1,200 hours with the filter pressure drop remaining in the range 230 to 180mm H_2O throughout. For the final 100 hours of the trial the interval between reverse pulse cleaning sequences was reduced from 60 minutes to approximately 40 minutes in order to investigate whether further improvements in cleaning resulted. In fact very little change in base-line pressure drop was observed.

The stability of the relationship between filter face velocity and base-line pressure drop throughout the trials was very encouraging in

193

terms of potential ceramic element life (Figure 2). Dust penetration of the ceramic medium appeared to be negligable as the base-line pressure drop had remained constant for long periods. The inspection by the manufacturers and other agencies of all eight elements removed during the trials had revealed that very little dust had penetrated more than 200 microns from the elements surfaces.

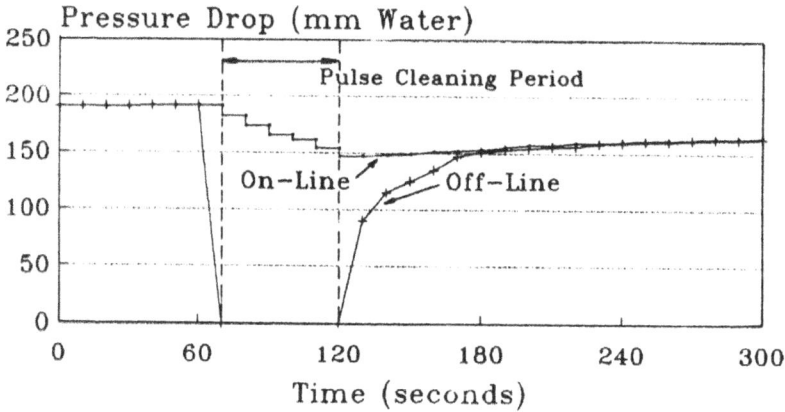

Figure 1. Comparison of Pressure Drop Traces During On and Off-Line Cleaning. Cerafil S Filter at Face Velocity of 3.0cm/s

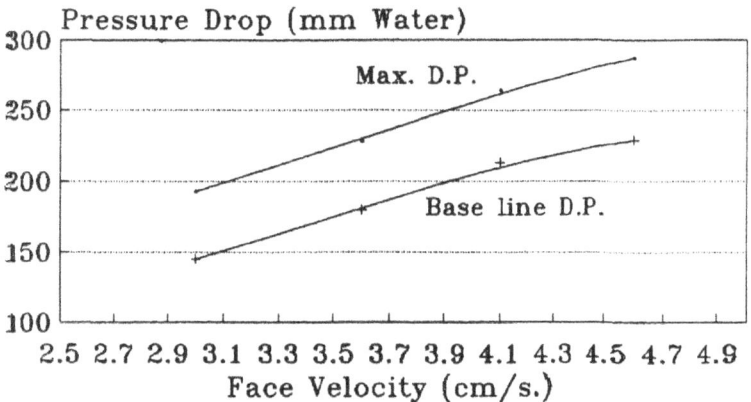

Figure 2. Face velocity/ Base-line pressure drop relationship at an average temperature of 220°C. Cerafil pilot filter

B.W.F. Ceramic Test Filter

A second ceramic pilot filter was also supplied for evaluation at the Works. This unit was capable of operation at higher temperatures than the Cerafil unit (700°C.) and was equipped with nine horizontally mounted 1.5m long B.W.F. low-density ceramic elements. The suppliers claimed that these elements were capable of filtering at much higher face velocities for given pressure drops compared with those of Cerafil S. The filter was installed adjacent to the Cerafil unit drawing flue gas from the same boiler flue duct. Initial problems due to the lack of effective thermal insulation of the filter inlet duct led to the cooling of the flue gas below its dew point, resulting in wetting and softening of the ceramic elements. Once oven dried the elements returned to their original state without any apparent effect on filtration performance.

Testing commenced in April 1991 with the filter operating at a nominal face velocity of 8.0cm/s. Pulse cleaning pressure was 4.5 bar with on-line cleaning initiated by pressure drop and set at 380mm H_2O. Base-line pressure drop stabilised at approximately 115mm H_2O but within 24 hours had risen to 215mm H_2O and after a further 24 hours had reached 240mm H_2O. At this point the face velocity was lowered to 6.0cm/s. and the cleaning point reduced to 305mm H_2O. The pulse air pressure was also raised to 5.2 bar to give more thorough cleaning.

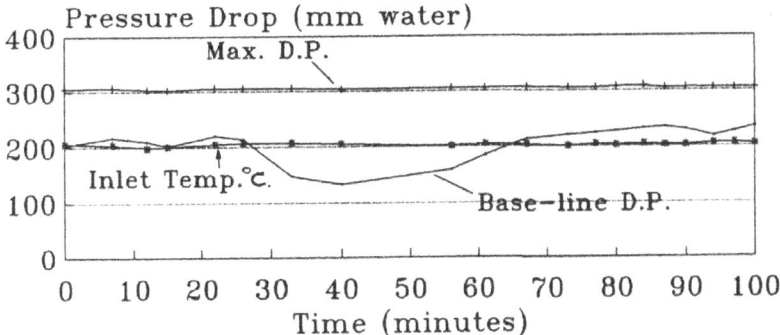

Figure 3. Base-line Pressure Drop Variation, B.W.F. 1.5m Ceramic Filter at Face Velocity of 6.0cm/s.

Testing continued with the base-line pressure stabilising briefly at 200mm H_2O with a gas inlet temperature of 205°C. Due to the large surface area of the filter house higher temperatures could not be achieved even when fully insulated. The base-line pressure drop then started to drift necessitating a further reduction in face velocity to 5.0cm/s. and a lowering of the cleaning point to 250mm H_2O. Studies of the differential pressure traces at 6.0cm/s. revealed that base-line pressure drop remained reasonably constant for a period of time and then after drifting downwards would slowly returning to a stable level but at a higher level than the previous stable period. (Figure 3).

195

At a nominal face velocity of 5.0cm/s, the filter operated successfully with the base-line pressure drop stable at 190mm H$_2$O for over 550 hours of continuous operation. Evidence of dust re-entrainment on the elements was observed at the higher face velocities. Again, rapid rises in pressure drop following the cleaning of each row of elements were observed but were absent at a face velocity of 5cm/s. (Figure 4).

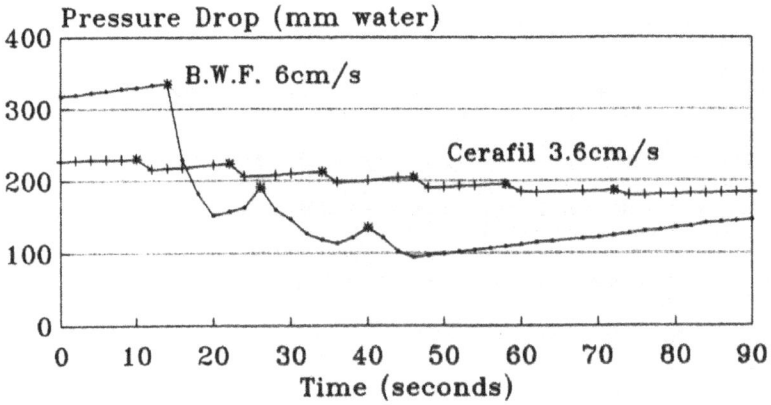

Figure 4. Dust Re-entrainment on Ceramic Elements During On-line Pulse Cleaning. B.W.F. and Cerafil Pilot Filters.

Filter Fires and Ash Sintering

On several occasions during the operation of the Cerafil filter, combustion of the collected dust occurred within the filter chamber. These principally involved the burning of dust accumulations in the collection hopper but combustion of dust on the ceramic element surfaces was also observed. As a result of two particularly intense fires, sintering of the ash on the ceramic elements took place. Following the first incident the layer of sintered ash was very light, being only about 2mm thick. An attempt was made to remove this layer by repeated reverse pulse cleaning but without success, so the filter was returned to operation in order to examine what effect the sintered layer had on the ceramic elements performance. After at least 500 hours of operation the pressure drop characteristic appeared little changed from that observed prior to the sintering incident.

The second incident was far more serious, resulting in a crisp sintered 'crust' up to 8mm thick in places, on all thirty six elements. Observations of the base-line pressure drop indicated that the sintered layer had appreciably reduced the ability of reverse pulse cleaning to remove dust from the surface of the elements and a total blockage of the filter was thought likely if it continued in operation. Although the base-line pressure drop had risen by about 25mm H$_2$O and was continuing to rise, the unit appeared to filter quite normally. The ceramic elements were removed and an attempt made to clean off the sintered ash layer by hand brushing. This revealed that the ash had not adhered strongly to the ceramic medium but was attached only loosely to the surface fibres.

Cleaning did not prove difficult but due to the very fragile nature of the elements breakages were inevitable. These were replaced by new elements, the undamaged ones refitted and the filter put in operation at a face velocity of 3.6cm/s. for a further 800 hours without incident. Filter pressure drops in the range 230 to 180mm H_2O were again experienced.

As the ash sintering phenomenon had occurred during boiler sootblowing and at higher than normal firing rates, it was concluded that a combination of extremely high dust burdens with a high carbon content, higher than average gas temperatures and a rapid rise in excess air in the boiler flue were the cause of the ignitions. At this point the boiler combustion controls were manually operated and it was the practice to increase draught through the boiler furnace chamber during sootblowing to increase flame stability. The resulting increase in combustion air to the furnace caused the flame to be drawn through the boiler creating a significant rise in gas outlet temperature and flue gas oxygen levels.

Conclusions From Pilot Filter Trials

The trials confirmed that ceramic filter units could be operated successfully for considerable lengths of time in this application without serious difficulty. There had been no failures of ceramic elements in service and all inspections carried out on sample elements had revealed that no significant dust penetration or mechanical damage of the ceramic media had occurred. The Cerafil pilot unit had performed wholly within the modest limits specified by the suppliers throughout some 3,200 hours of testing, while the second, B.W.F. equipped unit, had operated satisfactorily at somewhat lower face velocities than those claimed by the suppliers for almost 600 hours. The ash combustion and sintering incidents within the Cerafil filter had not seriously damaged the ceramic elements and as the boiler operating procedure was to be modified to prevent the ash sintering conditions from re-occurring, it was considered that potential problems had for the large part been overcome. Solids loadings in the outlets of both filters were very encouraging with levels as low as 2.0mg/m^3 being recorded. It was therefore concluded that low-density ceramic filtration was a viable method for the clean-up of flue gases from the two Works boilers, so recommendations were made to proceed with the design of a full sized filter. By now the original concept of a total clean-up of the incinerator flue-gases had been abandoned in favour of the filtration of the exhaust gases from the Works boilers. By operating both boilers at near maximum capacity, about 65% of the total waste gas could be burnt and the products of combustion filtered. Much of the remaining waste gas was already being burnt in the coal dryers and their discharges to the incinerator chimney were to be cleaned by other means.

Boiler Ceramic Filter Units

The imposition, in June 1991, of a two part Improvement Order by Her Majesty's Inspectorate of Pollution on the particulate emissions from the incinerator chimney required the fitting of suitable filters to the flue gas ducts of both Works boilers. The first part of the Order required one of the boilers to be so equipped by November 1991, along with other plant modifications. The second part required the remaining boiler to be compliant by July 1992.

The first of the boiler filters, that for No.1 boiler, was designed with the full co-operation of Coal Products staff in the light of experience gained in pilot filter operations. This filter was capable of handling 50,000Am^3/Hr of boiler flue gas at a normal operating temperature of 270°C and was designed to operate at a maximum temperature of 400°C. The expected operating pressure drop was 180 to 250mm H_2O. The filter house contained 2,592 Cerafil S elements mounted in seventy-two, thirty-six element, horizontal cassettes and was divided into six pairs of cells, each pair having a common dirty gas inlet and individual clean gas outlets for on-line maintenance. Each cell was fitted with six removable element cassettes in two vertical rows of three.

On and off-line cleaning facilities initiated by filter pressure drop and controlled by a PLC unit were incorporated. The reverse pulse pipes were arranged to clean each vertical row of 18 elements and operated in pairs in each cell. Pulse frequency and duration were regulated by individual control packages on each cell. An automatically controlled damper was fitted to the clean outlet of each cell to enable off-line cleaning.

Operation of No. 1 Boiler Filter Unit

In October 1991, No. 1 boiler filter was commissioned in conjunction with improved combustion controls and oxygen monitoring fitted to the boiler. The filter was operated in the on-line cleaning mode, with cleaning point set at 230mm H_2O and reverse pulse air pressure of 4.5 bar, for five weeks during which time outlet gas solids concentrations varied between 5 and 11mg/m^3. Flue gas inlet temperatures ranged from 280 to 330°C with occasional excursions to 380°C. The differential pressure was in the range 250 to 180mm H_2O. During a boiler shutdown for maintenance, the clean chambers were opened and inspected as part of a routine check on filter performance revealing that there had been no observable dust slippage into the clean side of the filter.

The boiler then ran for a further six weeks before failure of several of the pulse cleaning diaphragm valves occurred. Acidic flue gases circulating in the reverse pulse pipework had condensed on the cool valves causing the corrosion of diaphragm components, leading to valve failure. Inspection of the ceramic elements revealed that a large amount of ash sintering had taken place and in several cells the spaces between the elements were completely blocked with ash. The degree of sintering in some of the element cassettes was so great that breakage of a large percentage of the elements had resulted. The sintering was believed to have been largely due to the failure of the pulse cleaning system allowing large masses of carbon rich dust to accumulate on and between the ceramic elements. The constant passage of compressed air through the failed diaphragm valves was then thought to have supplied the necessary oxygen for combustion to take place, as flue gas oxygen levels remained in the order of 6.0% during operation of the boiler. Examination of failed elements revealed that once a single element had failed within a cell, dust slipping through to the clean chamber was likely to be entrained in the pulse cleaning air jet causing dust erosion of the fragile material of other elements. A 'domino effect' of failure was then considered inevitable.

The valve diaphragms were replaced with corrosion resistant components and all ceramic elements were removed. Broken and damaged elements were replaced and lightly sintered ones brushed clean, although this action certainly caused failures to occur later. One cell was completely re-equipped with new elements and thermocouples were strapped to the surfaces of six of the elements. These thermocouples were connected to a high-speed survey recorder in order to continuously monitor the surface temperatures of the elements. The accepted practice of allowing the formation of a thick dust cake on the elements surfaces to aid filtration was abandoned in this cell, in favour of frequent cleaning in order to minimise the amount of resident dust and thus prevent ash sintering. The pulse cleaning regime was changed to timed on-line cleaning with each row of elements being cleaned every 90 seconds with the same end in mind. Pulse air pressure was also raised to 5.0 bar to aid cleaning. Once in operation the temperature recorder registered a very rapid rise in all monitored elements surface temperatures at every boiler flame failure. (Figure 5). Due to the instability of the waste gas supply to the boiler, flame failures were likely to occur without warning. At this time only the fuel valves were isolated at flame failure, the air dampers remaining open to purge the boiler. After a week of operation, inspection of the elements in this cell revealed that the high carbon content ash had burnt on the elements surfaces, but had not sintered.

Following the discovery of ash combustion immediately following flame failure, a phenomenon which had not been observed during pilot filter trials, the boiler air dampers were arranged to close on flame failure in order to reduce air passage through the filter. The fact that it was necessary to air purge the boiler combustion chamber prior to re-ignition somewhat reduced the effectiveness of this operation although if purging was delayed to allow the filter to cool, ash combustion was considered less likely.

After a further six weeks of duty, failure of the filter occurred again with heavy sintering and element breakages found throughout the cells, apart from the thermocouple equipped cell. The intensity of the combustion around some of the ceramic elements was so great that several of the steel pulse cleaning venturis had melted or were heat distorted. A total replacement of elements was undertaken along with a modification of the cleaning regime to that of timed on-line cleaning with each row of elements being cleaned every 90 seconds as in the single cell already modified. An off-line cleaning programme was also initiated with each cell being taken out of service and cleaned every 45 minutes. It was considered that by employing both on-line and off-line cleaning modes the most effective cleaning of the ceramic elements could be achieved.

Two further failures have occurred in No.1 boiler filter due to the combustion and sintering of high carbon dust at boiler flame failure, requiring the complete replacement of the ceramic elements. In both cases very large amounts of sintered ash had adhered to many of the elements and caused their breakage, the broken elements then falling to cause further damage to less affected elements.

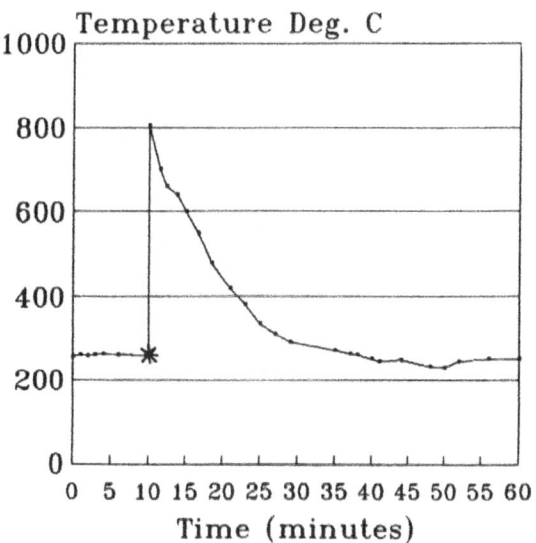

Temperature Deg. C

Time (minutes)

*   Boiler Flame Failure

Figure 5.   Boiler Ceramic Filter Element Surface Temperature
At Flame Failure.

Single Element Test Rig

A single ceramic element test rig was designed and constructed in order to determine the conditions required for combustion of solids in the boiler outlet gases.   This involved the drawing of dust laden boiler flue gas through an insulated and heated chamber containing a single ceramic filter element and measuring the rise in the filter's surface temperature with increasing oxygen levels and gas temperatures.   The rig also had the facility to 'dope' the flue gas with extra char particles in order to increase the carbon content and hence the volatility of the dust sample.   Following test rig experiments the results were analysed and recommendations made (Carroll & Perry, 1992) as follows:-

a) A sustained reduction in both temperature and oxygen content of the flue gas.

b) The isolation of the filter unit immediately following boiler flame failure and the blanketing of the unit with nitrogen.

c) A daily survey of dust loading in the clean gas leaving each filter cell in order to give early warning of element failure.

Implementation of Recommendations

In July 1992, both boiler filters were modified with the addition of inlet and outlet isolating dampers and filter by-pass systems allowing the boilers to be fired on oil only without using the filters. (Figure 6). Low pressure steam was introduced into each filter inlet duct immediately on boiler flame failure to create an initial inert 'buffer' preventing air ingress into the filter while the isolating dampers closed. Once the filter was isolated the steam was replaced by a nitrogen purge directly into the filter. The test rig work had indicated that combustion did not occur in the dust at temperatures below 250°C irrespective of oxygen levels or carbon contents, and

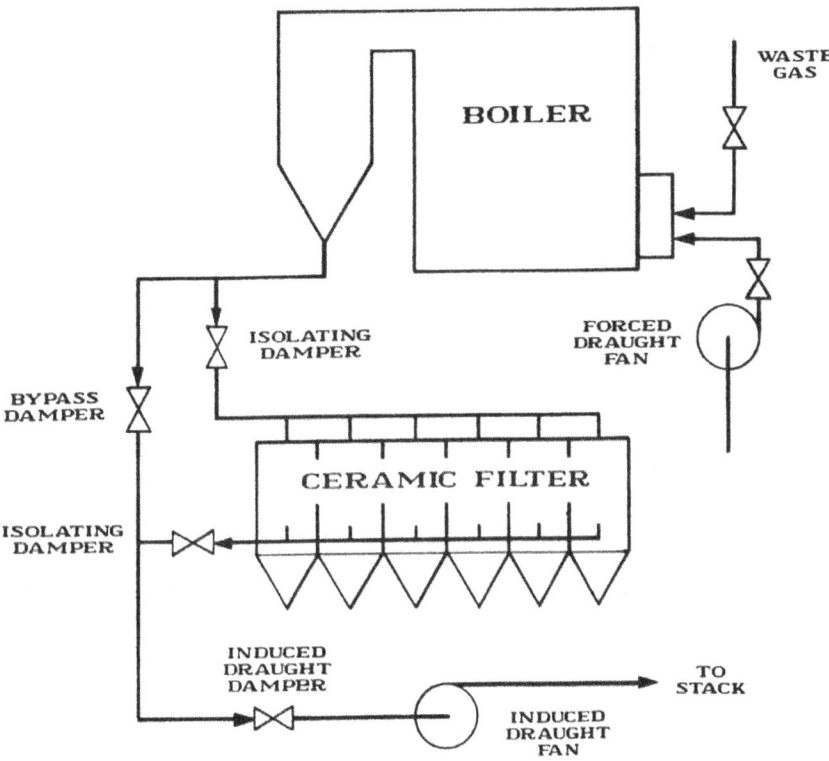

Figure 6. General Arrangement of Boiler Filter Isolation and By-pass System. June 1992.

at temperatures above 250°C, combustion was not observed at oxygen levels of less that 16.0%. Although No.1 filter had operated with inlet temperatures sometimes in excess of 250°C with flue gas oxygen levels of 6.5%, even a slow smouldering of the ash was unlikely to have occurred on line as no high temperatures had been recorded on the surfaces of the monitored elements while the boiler was in operation. Following the installation of isolating dampers and a nitrogen purge, ash combustion and sintering continued to take place on occasions in No.1 filter following boiler flame failure. Although great efforts had been made to maintain the boiler outlet temperature at below 250°C, these incidents coincided with periods of operation at outlet temperatures in excess of 250°C, due to plant difficulties and waste gas supply problems. This resulted in the overloading of the combustion chamber with high levels of volatile char particles leading to a rise in the boiler outlet temperature.

No.2 Boiler Ceramic Filter

The second boiler filter was commissioned in August 1992 and has operated successfully to date for some 6,800 hours of almost continuous duty. The filter inlet temperature was maintained at below 250°C throughout this period, aided by the fitting of a larger, and more efficient economiser to the boiler. As a result of these measures, no ash sintering on the ceramic elements has taken place. Regular inspections of both clean and dirty sides of each cell has also revealed that no element failures have taken place. Clean gas dust loadings have been less than 2mg/m^3. although much higher levels have been recorded as a result of dust slippage past the by-pass damper, which was found to be not fully closed.

A similar programme of inspection has been initiated on No. 1 boiler filter with each pair of cells being periodically taken off line and an examination of ceramic elements and clean chambers undertaken in order to provide early warning of catastrophic failure. No. 1 boiler has also been operated in a similar manner to No. 2 boiler since February 1993, and to date the filter unit has shown no signs of ash combustion on the ceramic elements.

REFERENCES

Carroll. G. & Perry. E.R. (1992). The Design and Operation of a Single Filter Test Rig. Coal Products Ltd. Scientific Department Internal Report.

Jones. M. & Rogers. P. (1991). A Practical Application of Rigid Ceramic Filters for Flue Gas Cleaning. Paper presented to The Coke Oven Managers Association. Cardiff.

ASPECTS OF PULSE-JET CLEANING OF CERAMIC FILTER ELEMENTS

S. LAUX
B. GIERNOTH
H. BULAK
U. RENZ
Aachen University of Technology, Germany
Lehrstuhl für Wärmeübertragung und Klimatechnik,
Eilfschornsteinstraße 18, D-52056 Aachen, Germany

ABSTRACT

Rigid ceramic filter elements for high temperature and pressure gas filtration are of high interest for hot gas clean-up of combustion and gasification gases. To assess filtration and cleaning performance, long term experiments over more than 13,000 hours were carried out during the past seven years at Aachen University of Technology, Germany. A pilot filter with six ceramic elements was operated with PFBC combustion gas and heated air with injected dust at temperatures ranging from 250 to 850 °C. A detailed report on these experiments can be found elsewhere (Laux *et al.*, 1991, 1992a/b, 1993 and EPRI, 1989, 1992).

In addition to the investigations on filtration and cleaning performance at this technical scale, various laboratory experiments are carried out to address the mechanisms of pulse-jet cleaning and dust reentrainment. Measurements of transient pressure and video observation during pulse-jet cleaning in a cold filter unit are supported by finite-difference flow calculations to identify the influences on flow and pressure distributions during element cleaning.

Since filter pressure drop during operation depends on fly ash characteristics, dust permeability measurements are performed as batch samples in the laboratory to provide assistance for the scale-up to larger filter units.

FLY ASH CHARACTERISTICS

The total pressure drop of a filter element is the sum of the pressure drops across the filter element, the residual dust layer, and the temporary dust layer, see Figure 1 (Schiffer, 1990). The pressure drop caused by the residual layer and the filter element structure is called baseline pressure drop. This is the pressure drop after all elements are cleaned and it depends on a large set of parameters, mainly the interaction of the dust cake with the filter element surface and the residual dust layer permeability.

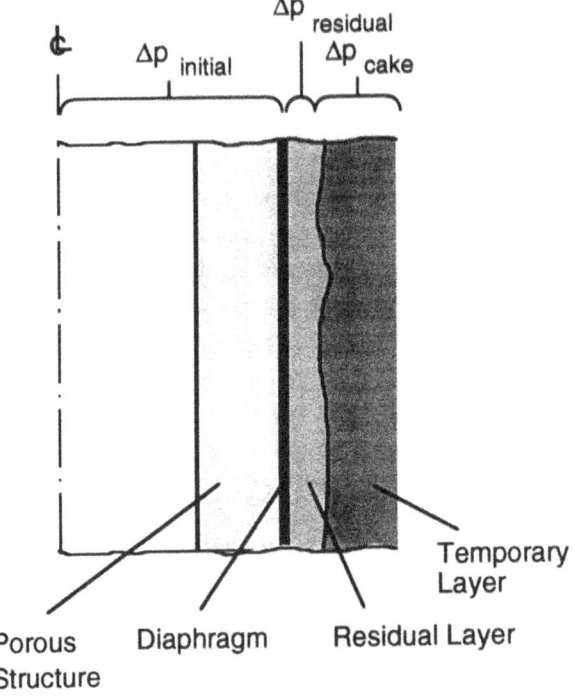

Figure 1. Filtration Model

Since these influences depend on operational conditions, it is impossible to precisely predict the baseline pressure drop from laboratory experiments. However, the dust permeability which determines the pressure drop of the temporary dust layer can be measured in a laboratory set-up.

A scheme of the permeability test facility is shown in Figure 2. The facility can be operated either with dried air at low temperatures (up to 300 °C) or with nitrogen at high temperatures (up to 850 °C). A ceramic filter disk at the lower end of a vertical tube serves as a filter specimen. Fly ash is dispersed into the laminar gas flow and filtered at the disk with a face velocity typical for hot gas filter applications (0.01 to 0.06 m/s).

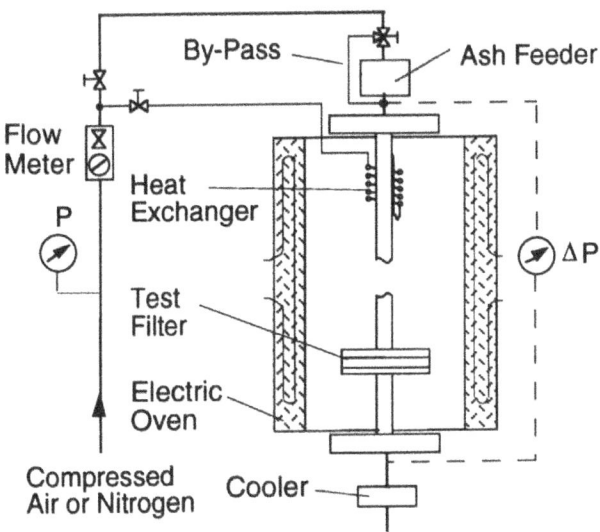

Figure 2. Permeability Test Facility

The increase of the pressure drop (Δp_A) due to the dust layer is recorded. With the known filter face velocity (v) and the gas viscosity (μ) as a function of temperature, the permeability of the ash layer (k_A) can be determined using Darcy's law:

$$k_A = \frac{v\,\mu}{\Delta p_A}$$

The permeability (k_A) has the unit [m] and depends on the thickness of the dust layer (δ). The product of permeability and layer thickness yields a layer permeability (k^*), which is nearly independent of the layer thickness and can be regarded as an ash property.

$$k^* = k_A \, \delta$$

The layer permeability has the unit [m^2]. After the measurement, the filter and the supporting ring are weighed to determine the mass of the filtered ash. Finally, the thickness of the ash layer is measured with a microscope. The dust layer density is calculated as an additional dust parameter from the layer thickness and mass. Thus, two important properties, permeability and density, of the dust layer generated by the filtration process are determined in a single lab-scale test.

Figure 3 shows the dust permeability and density as a function of filtration velocity at ambient temperature for a gasification ash. For the results shown, the filtration velocity was kept constant at 4 cm/s during ash feeding. In a real filtration process however, this is hardly the case. Shortly after a cleaning sequence, the group of elements just cleaned has a low resistance in comparison to the remainder of the filter elements. As a consequence, the filter face velocity increases substantially for individual filter elements resulting in a denser sub-layer of low permeability. Since the temporarily increased filtration velocity is present after each cleaning, all elements exhibit a residual dust layer with low permeability and high density after the conditioning phase. However, the permeability of the residual dust layer can not be determined without experimental data, because it strongly depends on the filter cleaning strategy.

The result of a permeability measurement at elevated temperature is shown in Figure 4. After generating a dust layer on the filter plate, the whole facility is heated stepwise, while the filter face velocity is kept constant at 4 cm/s. The permeability of the dust increases indicating that the influence of temperature for gases is not included in Darcy's law. This result was also found by Pulkrabek and Ibele (Pulkrabek & Ibele, 1987).

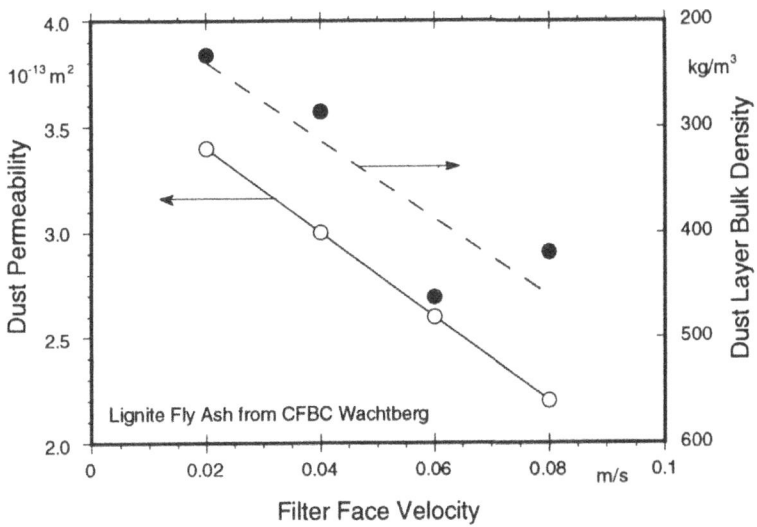

gure 3. Variation of Cake Permeability and Density with Face Velocity

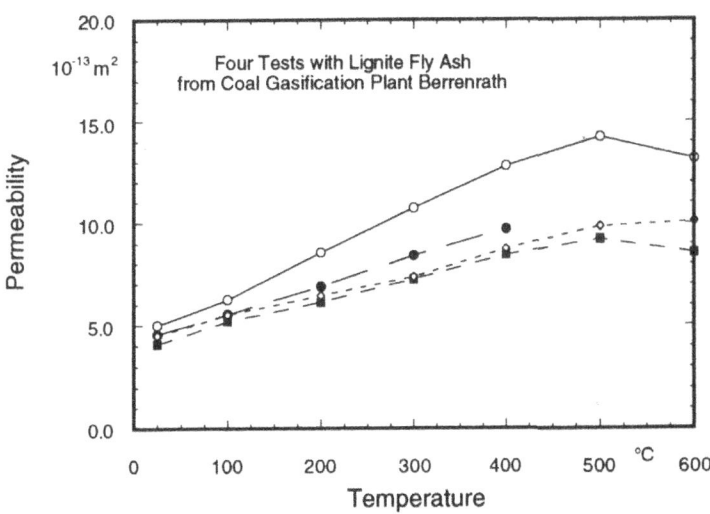

Figure 4. Hot Permeability Test

STABILITY OF FILTRATION PROCESS

Pulse cleaning of rigid ceramic filter elements requires that the dust cake be lifted from the outside surface. To apply an outward radial force to the dust cake, the pressure drop during filtration has to be neutralized. Internal stress is applied to the dust layer and a pressure drop across the filter and dust cake is built up by a sudden reversed flow. It has been observed that the dislodgment of the temporary dust layer does not occur directly at the interface between filter element surface and dust cake, but somewhere inside the dust layer itself.

Stable pulse cleaning results in the well-known pressure plot. During filtration, the pressure drop increases linearly. With a sufficiently strong pulse the entire dust layer accumulated between two time-triggered pulses can be removed (see left side of Figure 5). Experiments showed that a specific threshold pulse pressure is necessary to achieve cleaning at certain operating conditions. Pulse pressures below this threshold fail to remove any dust cake. This results in a permanent buildup of the dust layer and subsequently in a shutdown of the facility. The right side of Figure 5 shows a characteristic plot of the corresponding tube sheet pressure drop. The pressure generated by a pulse inside the filter cavity must always exceed the pressure drop of the filter to achieve a local flow reversal.

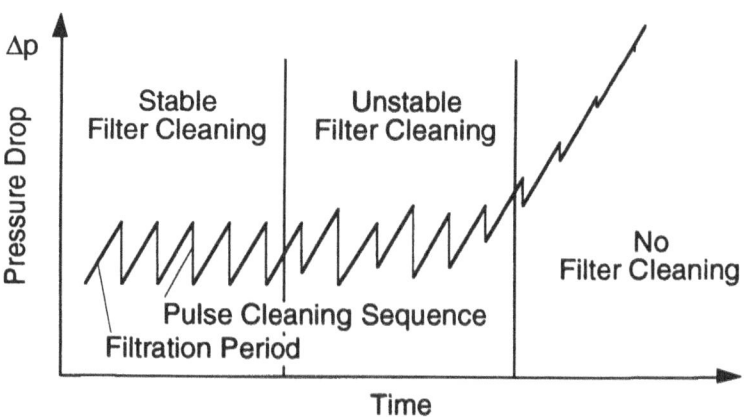

Figure 5. Idealized Filter Pressure Drop

Pulse pressures slightly above the threshold pressure seem to provide sufficient cleaning over a short time period. However, the amount of dust settling on the surface is larger than the dust removed from the surface by the cleaning pulse (middle section of Figure 5). This results in an increasing residual dust layer and higher pulse pressures are necessary to act against an increased pressure drop. If the pulse air pressure is not increased at this point, operation turns into the failure of the pulse cleaning shown in the right hand section of Figure 5. At constant gas flow and dust load, operation with an unstable setting of the pulse pressure is possible for several hours, even days. However, at some point the dust accumulated on the filters starts to increase rapidly. Stable operation can nearly instantly be regained with an increase of the pulse pressure.

Figure 6 shows unstable filtration experiments with a FBC fly ash. The results were obtained at the six filter element hot gas filtration test facility at RWTH Aachen. The permeability defined according to Darcy's law is plotted versus the exposure in the time period from hour 9890 to 10050. Experiments were run at a temperature of 800 °C, a filter face velocity of 6 cm/s, and a pulse pressure of 4 bar. Dust load and filtration cycle length were varied. The decay of permeability during some experiments is substantial. The partial recovery in the time between is due to problems with the dust feeder, i.e. at times no dust was fed and permeability recovered or stayed constant. Low or no dust load results in low pressure drops before the pulse, making cleaning easier. However, when dust concentration in the gas was constant, the permeability declined constantly. Data analysis showed that maximum pressure drops before cleaning were between 100 and 110 mbar during these high velocity experiments. The numerical modeling of the pulse cleaning event presented later in this paper showed that the pressure inside the cavity generated by a pulse of 4 bar at a filtration temperature of 850 °C is only approximately 130 mbar. Thus, the effective pressure differential for cleaning was only approximately 20 mbar. Although this value still dislodges dust, it is not sufficient to clean all dust from the elements that got to the element during the previous filtration period. A similar unstable operation of the hot gas filter is described in an ASME-Paper (Laux *et al.*, 1993). Filter operation at 4 cm/s and 4 bar pulse pressure was always stable. A pulse pressure setting of 3 bar led to unstable operation at this face velocity. Future work will focus on the amount of over-pressure in the filter cavity necessary for stable filter operation.

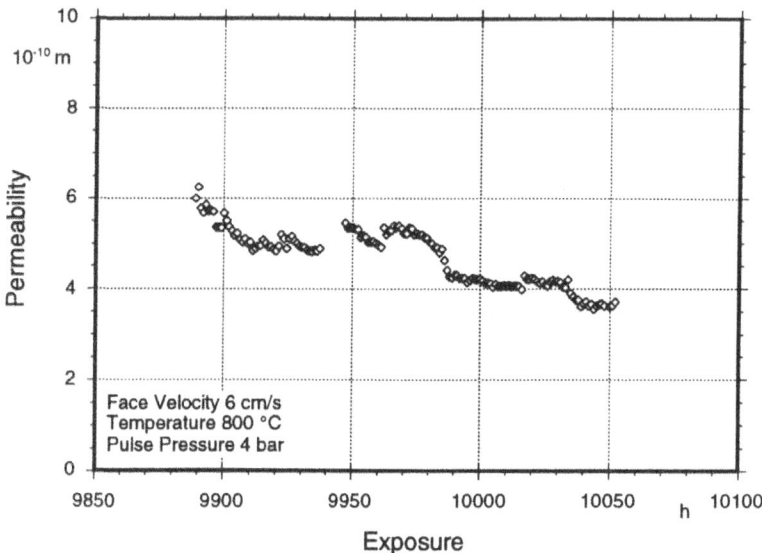

Figure 6. Unstable Filtration Experiments at High Face Velocity

PERFORMANCE OF THE PULSE-JET SYSTEM

Key element of any pulse-jet cleaning system is the discharge of a high momentum gas jet from a nozzle into a filter element cavity of arbitrary shape. The momentum of the jet is transformed into a pressure increase inside the filter element cavity.

In the available literature the pulse pressure in the reservoir is used quite often to show the performance of the cleaning system. However, the value of the pulse pressure alone is not suitable as a measure of the performance of the system. High momentum jets can be achieved either with a large pipe diameter and low pulse pressure or with a small pipe diameter and high pulse pressure. Therefore, pulse-jet performance data from different filtration facilities has to be judged in conjunction with the specific set-up of the pulse cleaning system. How much of the momentum is converted into an increase of the pressure inside the cavity depends on the configuration of the nozzle-cavity arrangement.

210

In general, mass flow exiting from the pulse-jet system is a measure of the pressure increase. This mass flow depends mainly on the pressure ratio between reservoir and nozzle exit and the geometry of the pulse cleaning system. In high pressure filtration, it is particularly important to improve the pulse jet system to keep reservoir pressures and costs for gas compression as low as possible. Frictional losses in pipes, valves, and bends reduce the mass flow through the cleaning system and should be reduced as much as possible.

Transient Pressure Measurements

Transient pressure measurements were performed on a cold filter in the laboratory of the Institute. The test facility is shown in Figure 7. It can be operated with a controlled blower and a dust feeder. Three filter elements (1 m long, 60 mm o.d.) are mounted with bayonet rings from beneath the tube sheet. A large window allows observation of the cleaning process as well as easy access to the filter elements. Pressure transducers in the pulse cleaning system and at the inner wall of one filter element measure the transients during pulse cleaning.

Each filter element has its own cleaning system. It consists of a solenoid valve attached to a pulse air reservoir, a conical contraction to reduce the pipe diameter from 20 to 8 mm i. d., and a lance to convey the pulse air to a location directly above the tube sheet opening. The pulse-jet exits from the lance shortly above the tube sheet.

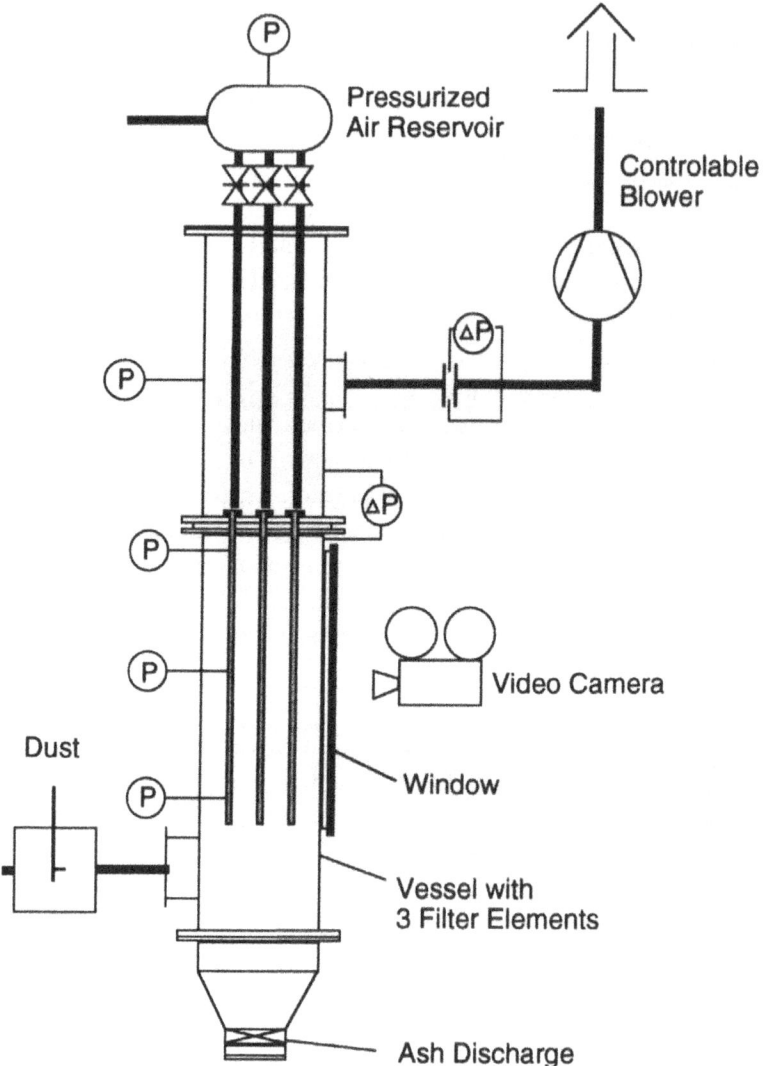

Figure 7. Filter Cleaning Test Facility

Figure 8 shows the pressure downstream of the solenoid valve and at the exit of the pulse cleaning system for a reservoir pressure of 4 bar and a duration of the electric signal to the valve of 100 ms. The valve displays a delay of 40 ms before it opens, although the pressure increase to the maximum pressure is achieved within 10 ms. After the valve is fully open, the pressure plot shows essentially steady-state conditions with a slight decrease of the pressure due to insufficient pulse reservoir volume. The delay between electric signal shut-off and the start of the closing action is substantial (120 ms). Although the valve used is quite fast due to its size (20 mm i.d.), it shows in general the same transients as larger valves. The experiments conducted at the hot gas filtration test facility in the boiler house of the RWTH heat and power plant showed that valves of different sizes and manufacturers exhibit similar behavior. Larger valves show a slower initial pressure increase to the maximum pressure and a much longer opening period. However, no influence of the initial gradient or the duration of the pulse-jet on the cleaning success could be detected. It can be concluded from the pressure measurements that cleaning is essentially achieved by the steady-state pressure in the filter element cavity.

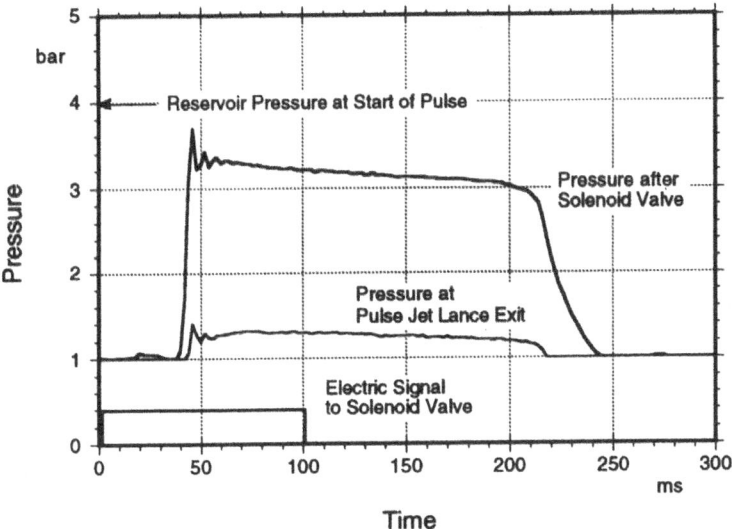

Figure 8. Pressures in Pulse-Jet System

The pressure at the valve is approximately 0.7 bar less than in the reservoir, which accounts for the losses in the valve and for the acceleration of the gas. The pressure decrease between valve and exit of the pulse-jet system is quite large due to the small internal diameter of the lance and the substantial acceleration to sonic velocity. When the pressure after the valve exceeds 2.5 bar, the pressure at the exit of the pulse-jet lance rises above ambient pressure, indicating that sonic velocity is present in the exit. Thus, pressures below 2.5 bar at the valve (less than approximately 3 bar in the reservoir) result in subsonic conditions. Since the sonic velocity places an upper limit on the exit velocity, any reservoir pressures above 3 bar result in an increased exit pressure. This high velocity, high pressure jet expands shortly after the nozzle and is subsequently decelerated in the filter element cavity. This process was modeled with the finite difference software FLUENT.

NUMERICAL MODELING

Simulations of the pulse cleaning event were performed with the finite difference software FLUENT (Fluent, 1990), in order to verify the pressure measurements and to further study flow phenomena. A single filter element and part of the surrounding vessel shown in Figure 9 were modeled on a two-dimensional Cartesian computational grid using version 3.02 of the program. The equations of continuity, momentum and energy were solved and a standard k-ε-model for turbulence modeling was used. The pressure at the exit of the pulse-jet lance was varied with time according to pressure measurements described above.

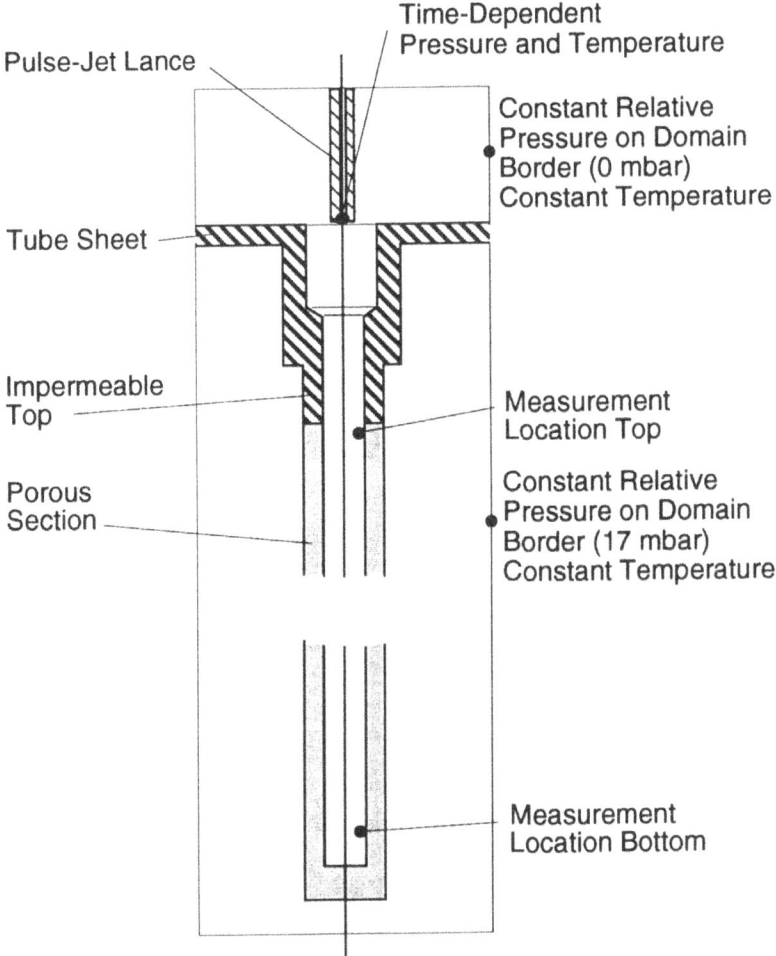

Figure 9. Computational Domain

Figure 10 shows the differential pressure across the filter material as a function of time for a 4 bar cleaning pulse. The first 40 milliseconds of the measurement and of the FLUENT calculations are shown at three different locations in the filter cavity. The starting point of the pressure increase of measurement and calculations were adjusted accordingly to allow easy comparison. Before the onset of the pulse, the pressure drop across the element during continuos filtration causes negative differential pressures. The prediction of the transient pressures are in good agreement with the measurements at the bottom and center of the candle. The slight differences can either be a result of the damping influence of the pressure transducers and the interconnecting plastic hoses or due to deficits of the numerical model to fully describe the reality. However, the steady-state differential pressure at the topmost position is calculated to be 120 mbar, whereas the measurement only yields approximately 80 mbar. The reasons for this are not fully understood. Since the gas jet is not fully mixed and local velocity and turbulence at this location are quite large, the pressure measurement is problematic anyway. These probably also influence the gradient of the measured pressure signal, which is measured less steep than the calculated gradient.

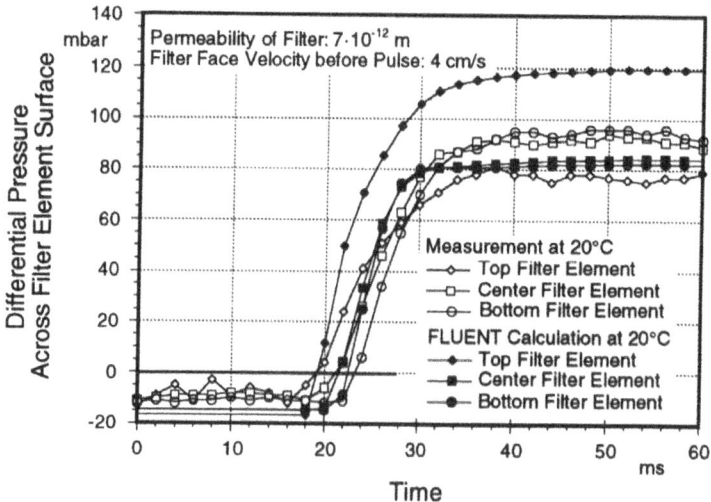

Figure 10. Calculated and Measured Transient Pressures in Filter Element

The mass flow at the lance exit and at the inlet cross-section of the filter cavity during the beginning of the pulse cleaning event is plotted in Figure 11. With the conditions modeled, the continuos flow from the filter is reversed immediately after the start of the pulse. Approximately one third of the mass entering the candle is entrained from the clean gas side at steady state. However, this result is different at high temperatures. Calculations on

216

the same grid at temperatures of 850 °C show that the pulse-jet mass flow is always larger than the mass flow into the filter element. The increase of gas viscosity with temperature in Darcy's Law results in a larger flow resistance at higher temperatures. Subsequently, the mass flow entering element cavity is smaller. More information about the flow characteristics during pulse-jet cleaning can be found in (EPRI Vol 3, 1992).

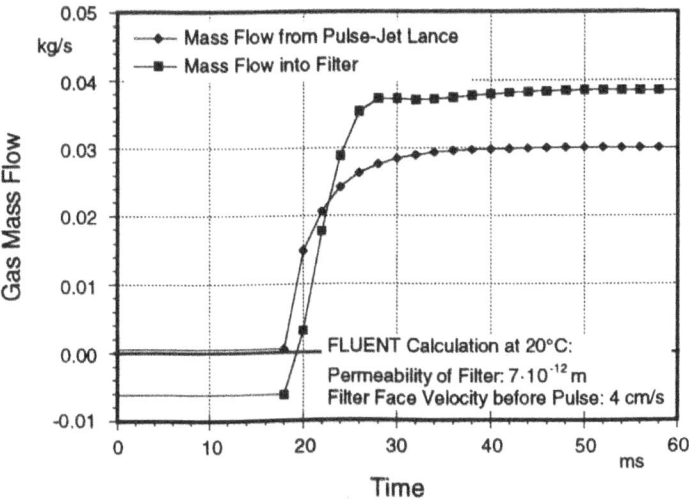

Figure 11: Calculated Mass Flow into Filter Element

OBSERVATION OF THE PULSE CLEANING EVENT

By observing the pulse cleaning event and simultaneously measuring the pressure inside the filter element cavity, important information about the cleaning process could be obtained. The dislodgment of the dust layer from the ceramic filter element surface was observed with a commercially available video-camera, while the pressure differential between the inside and the outside of the filter element was measured near the observed point. The camera captures a frame every 20 ms, which was stored on a video-recorder. Although the beginning of the trigger signal to the solenoid valve was indicated in the video images by a light emitting diode, the exact time of visible cake break-up can only be determined with an error of 20 ms.

However, from the various sequences taped it can be concluded that when a complete cake break-up is visible the full steady-state pressure level was achieved. The individual frames were loaded into a personal computer and processed with imaging software on a personal computer (Image, 1992).

Figures 12 to 15 show the beginning of a cleaning sequence for a pulse pressure of 4 bar. The right hand half of the 60 mm element at a location approximately at the center is shown (1 m element). The sequence shows that the dust layer is dislodged from the filter element surface as large longitudinal flakes. All flakes move immediately and fall down along the filter element surface, and finally settle in the ash hopper. Disintegration after the flakes become detached is small. After approximately 300 ms only a small amount of dust consisting of small agglomerates is still in the settling phase. This observation at ambient temperature is consistent with results reported earlier for high temperatures (Laux *et al.*, 1991 and EPRI Vol 2, 1992).

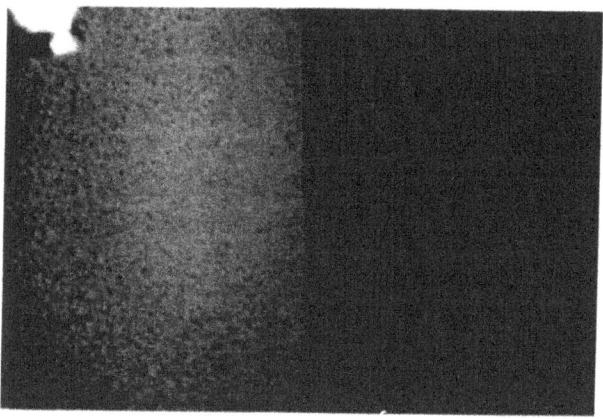

Figure 12. Filter Element Section before Cleaning Pulse

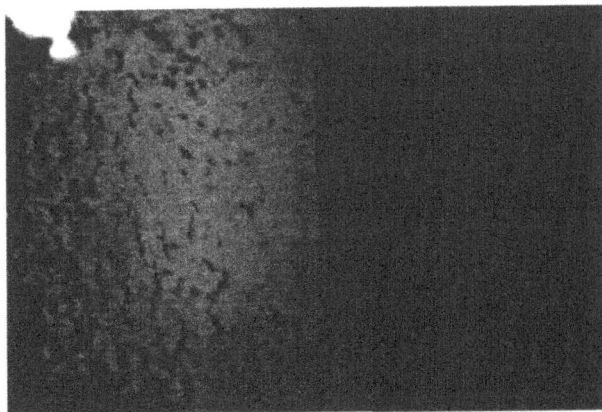

Figure 13. 4 bar Pulse Approximately 20 ms after Cleaning Started

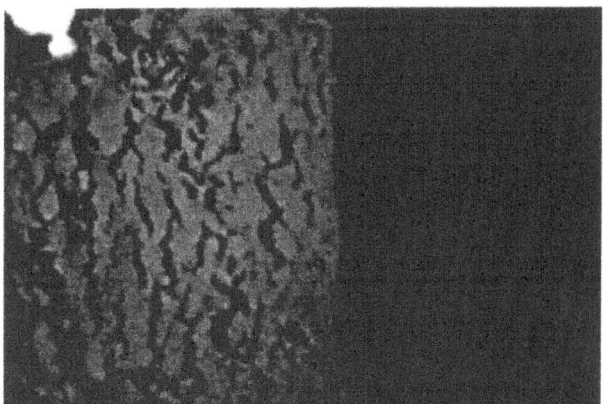

Figure 14. 4 bar Pulse Approximately 40 ms after Cleaning Started

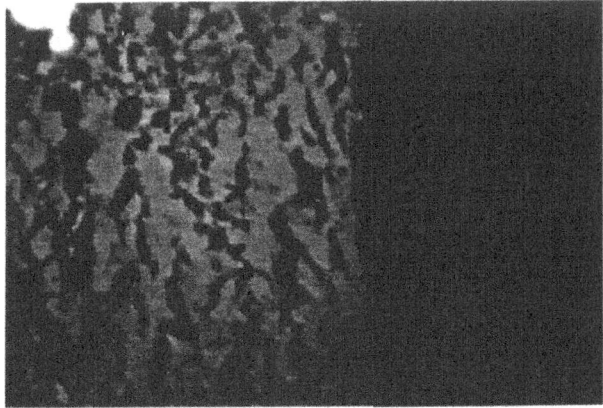

Figure 15. 4 bar Pulse Approximately 60 ms after Cleaning Started

Figures 16 through 18 show frames of cleaning sequences with 2, 4, and 6 bar pulse pressure, respectively. The frames were taken approximately 40 ms after the pressure in the filter element cavity started to increase. Dust layer thickness was 0.64 mm. The frames show much finer flakes for increased pulse pressure. The frame at 2 bar shows that the dust is barely lifted from the surface. The dust cake in the 6 bar image seems to have regions of very fine agglomerates. However, the difference in flake size between 4 and 6 bar is relatively small, although the cleaning process itself is much faster at 6 bar as indicated in Table 1 by the radial cake velocity. In addition, the steady-state differential pressure across the filter element surface is given in the table as well.

TABLE 1
Cake Characteristics during Cleaning Pulse

Pulse Pressure	max. Steady State Pressure inside Element	Average Flake Width	Average Radial Flake Velocity after 40 ms
bar	mbar	mm	m/s
2	35	12	0.06
4	90	5	0.13
6	170	4	0.25

The various sequences analyzed showed that the gaps in the dust cake develop at preferred locations on the surface. The relatively coarse gap pattern in the 2 bar sequence can be found at the same locations in the sequences at larger pulse pressures. In addition, new gaps appear and break down the dust flakes into smaller pieces. The filter structure has probably preferred passages for the cleaning gas due to inhomogeneities in the filter material.

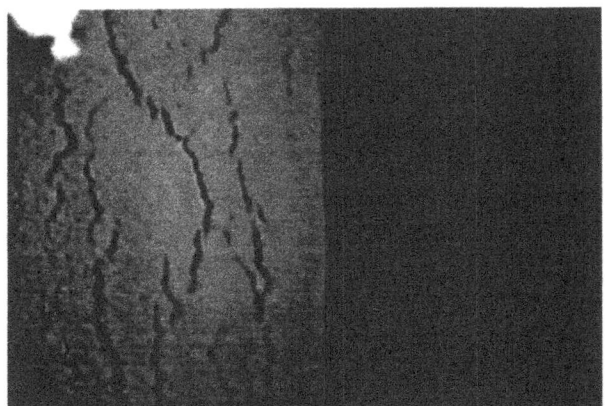

Figure 16. 2 bar Cleaning Pulse, Pressure in Element 35 mbar

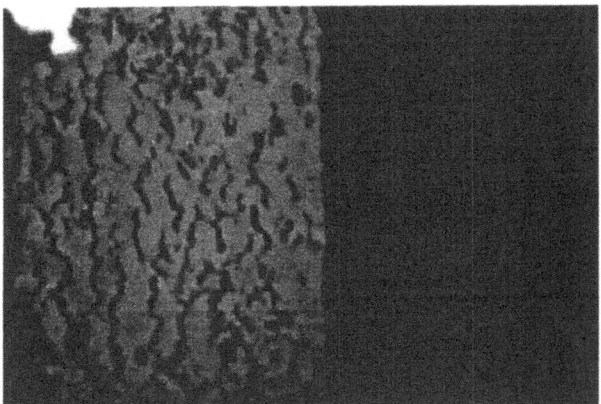

Figure 17. 4 bar Cleaning Pulse, Pressure in Element 90 mbar

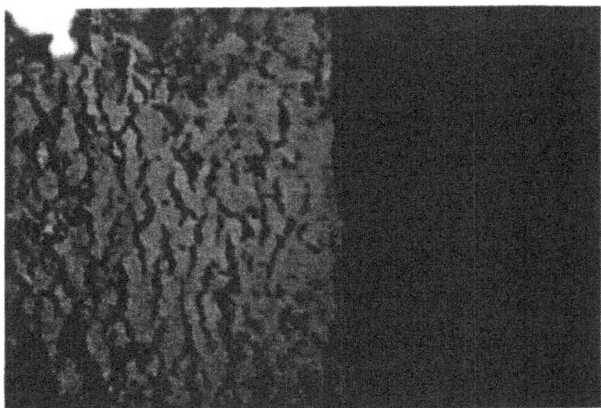

Figure 18. 6 bar Cleaning Pulse, Pressure in Element 170 mbar

The transient pressure values at the center of the element are shown in Figure 19. The pressure drop before the pulse was 10 mbar in all cases. The transients are similar to those measured at the solenoid valve (Figure 8). All pulse pressure settings reverse the pressure differential across the filter. However a 2 bar pulse generates only little over-pressure. In case of high tube sheet pressure drops this pressure might not be sufficient to achieve total cake dislodgment, thus leading to the unstable conditions discussed above.

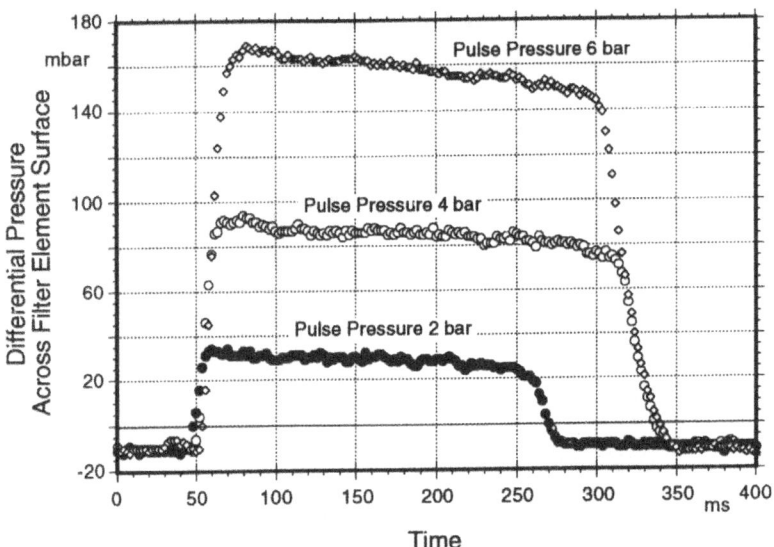

Figure 19. Transient Pressures across Filter Element during Pulse

CONCLUSIONS

Laboratory scale experiments were conducted to support the results of the filtration experiments on rigid ceramic filter elements at the RWTH Aachen hot gas filtration test facility. The permeability of fly ash samples was measured at temperatures up to 850 °C to provide information about ash characteristics for filtration process scale-up. A strong influence of filtration velocity on dust layer permeability was found. Additional experiments focused on the performance of the pulse-jet cleaning system. The stability of the filtration process was related to the pressure generated by the pulse-jet inside the filter element cavity. The results of a numerical analysis of the cleaning process agreed well with pressure measurements at a cold filter test facility. By observing the pulse cleaning event and simultaneously measuring the pressure inside the filter element cavity, important information about the cleaning mechanisms was obtained. It was found, that the dust layer dislodges at the beginning of the pulse as flakes from the filter element surface. Both the size of the flakes and their radial velocity are a function of the pulse pressure.

ACKNOWLEDGEMENTS

The project was jointly funded by the Federal Ministry for Research and Technology (BMFT), Deutsche Babcock Energie- und Umwelttechnik AG, Rheinbraun AG, RWE Energie AG, and Schumacher Umwelt- und Trenntechnik GmbH (Ref. No: 0326752A/B). The responsibility for the contents of this paper lies solely with the authors.

The support of the organizations which have co-operated in the project and the participation of many individuals is gratefully acknowledged. The authors express their appreciation to the many contributions and suggestions given by our project partners. Special recognition is extended to all student workers for their excellent support.

REFERENCES

EPRI, Electric Power Research Institute (1989). Design and Commissioning of a Filter Module for High Temperature and Pressure. EPRI GS-6489, Project Report 1336-7.

EPRI, Electric Power Research Institute (1992). High Temperature Gas Filtration. EPRI GS-6489, Final Report EPRI-Research Project 1336-7, Volumes 2 to 4.

Fluent Inc. (1990). Fluent User's Manual, Version 3.02, Lebanon, NH.

Image (1992). Public domain image processing and analysis software (Macintosh), National Institutes of Health, Research NIMH.

Laux, S., Giernoth, B., Bulak, H., Renz, U. (1992a). Besonderheiten beim Betrieb von keramischen Heißgasfiltern mit unterschiedlichen Stäuben. Proceedings VGB Conference "Wirbelschichtsysteme", Cologne, Germany, September 2 - 3.

Laux, S., Giernoth, B., Bulak, H., Renz, U.(1992b). Betriebsverhalten von Filterelementen und Anlagenteilen zur Heißgasfiltration unter realen Rauchgasbedingungen bei hohen Temperaturen im Dauerversuch", Final Report BMFT Research Grant, 1990-1992, Reference: 0326714A.

Laux, S., Giernoth, B., Bulak, H., Renz, U. (1993). Hot Gas Filtration with Ceramic Filter Elements. Proceedings of the 12th International Conference on Fluidized-Bed Combustion, San Diego, USA, May 8 - 13.

Laux, S., Schiffer, H.-P., Renz, U. (1991). Performance of Ceramic Filter Elements for Combined Cycle Power Plant High Temperature Gas Clean-Up. Proceedings of the 11th International Conference on Fluidized-Bed Combustion, Vol. 2, Montreal, Canada, April 21 - 25.

Pulkrabek, W. & Ibele, W. (1987). The Effect of Temperature on the Permeability of a Porous Material. Int. J. Heat Mass Transfer, Vol. 30, No. 6, pp 1103-1109.

Schiffer, H.-P. (1990). Entstaubung heißer Rauchgase mit keramischen Filterelementen. Ph.D-Thesis RWTH Aachen.

PULSE JET CLEANING OF RIGID FILTER ELEMENTS AT HIGH TEMPERATURES

Stefan Berbner and Friedrich Löffler
Institut für Mechanische Verfahrenstechnik und Mechanik,
Universität Karlsruhe (TH), 7500 Karlsruhe, Germany

ABSTRACT

This investigation concerns the influence of high temperatures on the cleanability (cake detachment) of hot gas filter media and the time dependence of the residual pressure drop.

The objective of this project is to define the conditions for operation and cleanability of the filter to assure a steady operation even under the difficult conditions of hot gas cleaning. A steady operation includes reliable cake removal and constant residual pressure drop during operation of the filter. This should lead to an improved design and construction of filters and pulse jet cleaning systems for hot gas filtration and an extended range of their applications.

The application of cleanable filter media at high temperatures is important in the fields of waste gas cleaning and product recycling. It is the basis for the further development of advanced environmental technologies, e. g. in power plant technology to protect plant components (e.g. gas turbines) and for recycling of catalyst dusts or in waste incinerators and pyrolysis plants.

INTRODUCTION

At the Institut für Mechanische Verfahrenstechnik und Mechanik of the University of Karlsruhe there are experimental and theoretical investigations conducted to improve the construction and cleanability of rigid filter elements at high temperatures.

During surface filtration the filter medium acts as a barrier to the solids, so that a dust cake is built up on the upstream surface, with relatively little penetration into the medium itself. In general, the rigid media behave in this way even under hot gas conditions. In order to maintain an acceptable total pressure drop across the filter, the cake must be removed periodically, either by reverse pulse or reverse flow. It is clearly important to be able to assess the magnitude of the required cleaning action (1-17). In pulse jet cleaning processes the pressure increase inside the candle must be applied to overcome adhesion of the cake to the medium and be applied to overcome the cohesion bet- ween the particles. In order to build good filters for hot gas cleaning processes the influence of the high temperatures on the kinetics of residual pressure drop and cake detachment forces has to be known.

The aim of these first investigations is to get basic information about the pulse jet cleaning behaviour of rigid ceramic filter elements at ambient conditions. In a later step there will be experiments under real hot gas conditions to get information about the influence of the high temperature, the gas atmosphere and the dust composition on pulse jet cleaning.

THE EXPERIMENTAL APPARATUS

Areal Dust Density Profiles

The measuring arrangement was developed by Klingel (2) to determine areal dust density profiles on fabric fibre filter bags. Figure 1 shows the schematic of the measuring device.

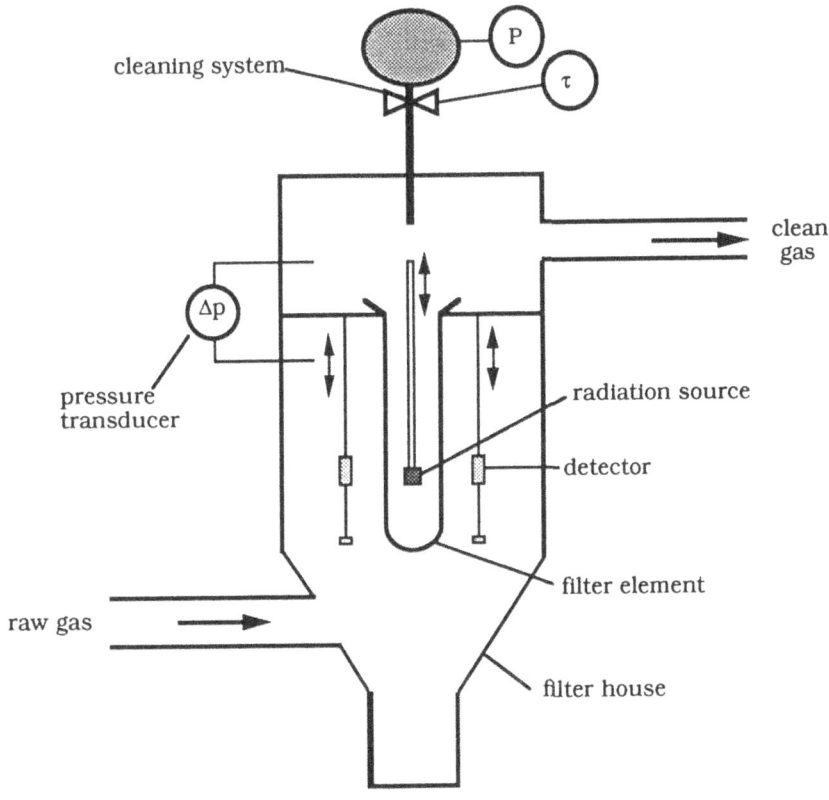

Figure 1. Measuring device to determine areal dust density profiles

The measuring principle is based on the absorption of x-ray fluorescence radiation. The radiation source is located at the centre of the filter element and is movable along the candle simultaneously with three detectors located at the outside of the element. Data collection and processing are performed with a computer equipped with adequate peripheral devices. The results are the simultaneous "in situ" measurements of three areal dust density profiles at one height of the filter elements. The x-rays radiated from the source serve to determine the mass of particles accumulated in and on the fabric. During pulse jet cleaning, the device is removed from the filter candle. A detailed description of the measuring technique is given in additional literature (2, 4, 6, 7).

Pressure Measurement inside the Candle

A schematic of the experimental apparatus used for the investigations is shown in figure 2.

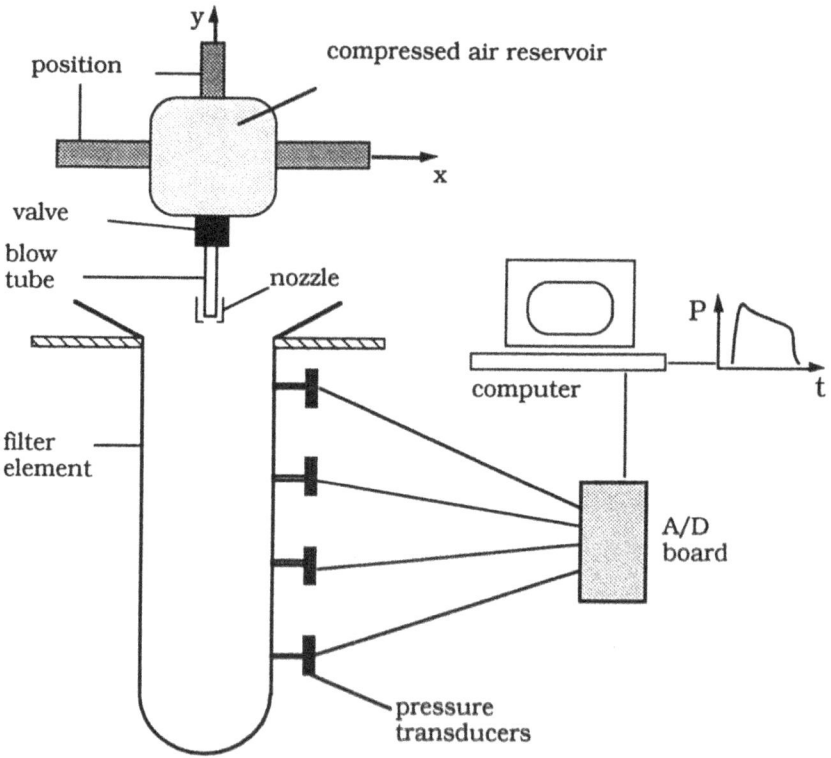

Figure 2. Experimental apparatus to measure the pressure increase inside the filter element during the cleaning process

The pulse jet system consists basically of a compressed air reservoir, a diaphragm and solenoid valve system with a timer, a blow tube with one nozzle and one filter candle. The filter elements hang freely. Different types of cleanable filters can be used with an adapter system. The blow tube can be raised and lowered with the aid of a special frame construction. Thus it is possible to vary the distance between the blow tube and the filter candles.

During the experiments blow tubes with different nozzles were used, in order to vary the outlet diameter. To introduce a pressure pulse into the filter candle, the diaphragm valve is opened for a short duration (0,05 to 0,5 seconds) by means of the solenoid valve and the timer, allowing compressed air to flow from the reservoir through the blow tube and the nozzle into the candle. During the investigations, the pressure drop in the reservoir, the pressure increase at 3 positions in the blow tube and the pressure increase at several locations along the length of various candles are measured. All pressure transducers record the pressure against atmosphere.

Hot Gas Experimental Set Up

A schematic of the experimental set up for the investigations under hot gas conditions is shown in figure 3.
A 300 kW natural gas burner produces exhaust gases. A continuously working dust feeder provides a particle laden raw gas stream, which enters the filter through the raw gas inlet. Directly in front of the raw gas inlet the dust concentration is continuously monitored by a photometric device measuring the extinction of a laser beam. The filter house itself is heated and contains two types of filters: In the centre hangs a technical filter element as is used in many industrial applications of hot gas cleaning. At one wall of the filter house is a filter disc attached, made out of the same material as the technical filter element in the centre. Therefore, both filters are exposed to the same raw gas conditions. Both filters have their own pulse jet cleaning systems. The reasons for using a filter disc to predict the behaviour of a technical filter element are described in detail in (8) and (12). In surface filtration at ambient conditions this experimental technique has been developed as a VDI standard (VDI 3926). One goal of the investigations at high temperatures with this experimental set up is to establish such a standard for high temperatures filter testing..
At the opposite side of the filter disc are high temperature resistant windows to enable high speed photographs to be taken during the cleaning process.

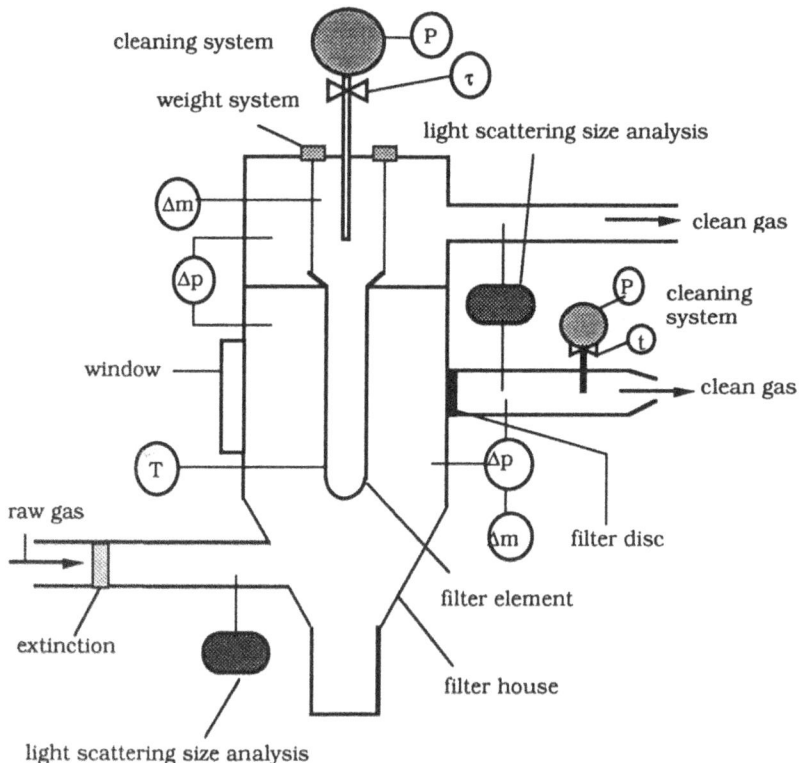

cleaning system

weight system

light scattering size analysis

clean gas

window

cleaning system

clean gas

raw gas

filter disc

filter element

extinction

filter house

light scattering size analysis

Figure 3. Hot gas experimental set up

At the top of the filter house is a system installed to measure the weight of the filter element using force transducers before and after pulse jet cleaning. Hence together with the information from the pressure transducers (residual pressure drop) it is possible to estimate the success of the cleaning effort.

The filter house is built of SiC ceramic blocks. With the heating appliance a maximum temperature of about 1000 °C can be maintained during the experiments.

EXPERIMENTAL RESULTS

Areal Dust Density Profiles

The ratio of the removed to the remaining dust fraction allows the calculation of the incremental (between two successive pressure pulses) cleaning efficiency R:

$$R = \frac{W_{E,i} - W_{R,i}}{W_{E,i}} = 1 - \frac{W_{E,i}}{W_{R,i}} \tag{1}$$

$W_{E,i}$: Incremental areal dust density just before the cleaning process
$W_{R,i}$: Incremental areal dust density just after the cleaning process
i: sample location on the candle

For a variety of experiments the cleaning efficiency R is given as a function of the reservoir pressure P_T, the valve opening time τ and the areal dust density $W_{E,i}$. If cleaning is very thorough, it is possible that R>1. If no cleaning occurs or if dust is redeposited on the filter during the cleaning experiment R<0. In the case of steady filtration conditions $W_{R,i}$=0, the cleaning efficiency is always R=1. This state was not reached in the course of these experiments.
For all experiments rigid ceramic fibre filters are used. The test dust is limestone. The cleaning process is always off line.

In figure 4 the cleaning efficiency along the length of the filter element is given as a function of the reservoir pressure P_T. The pressure P_T is given against atmosphere and varied between 2 and 6 bar. The electrical valve opening time τ = 0,08 s. The diameter of the outlet nozzle of the blow tube d = 6,5 mm. The incremental areal dust density before cleaning $W_{E,i}$ = 100 g/m². The dust concentration in the raw gas stream c = 5 g/m³. The filtration velocity v = 0,05 m/s. The length L = 0 m is at the top of the hanging filter element; the length L = 1,38 m is close to the bottom of the filter element.

Figure 4 shows that the cleaning efficiency increases strongly with an increasing reservoir pressure. More important than the reservoir pressure P_T is the pressure reached in the blow tube just in front of the outlet nozzle, which will be discussed later. Figure 4 also shows that for reservoir pressures lower than 3 bar the pressure in the cleaning system is not high enough to get a satisfactory cleaning. For P_T = 5 bar a very good cleaning efficiency all along the filter candle is reached, and operation approaches the steady state filtration. For reservoir pressures above 5 bar the candle is over-cleaned. The results show that there are high cleaning efficiencies even at the top of the filter element. This means that the distance between the outlet nozzle of the blow tube and the top of the filter candle is long enough to get a free pulse jet with secondary entrained air. The pulse jet strikes the filter medium at the top of the filter candle. There is no further air entrainment by the pulse jet inside the filter element. An over pressure prevails all along the filter during the pulse jet cleaning process. This is an important fact, because in pulse jet cleaning of rigid hot gas filter elements no movement of the fabric occurs (fabric acceleration/decceleration) to enhance the cleaning. All forces acting upon the dust accumulations to overcome the adhesion forces between the filter medium and the dust cake must be produced only by the static over-pressure inside the filter element.

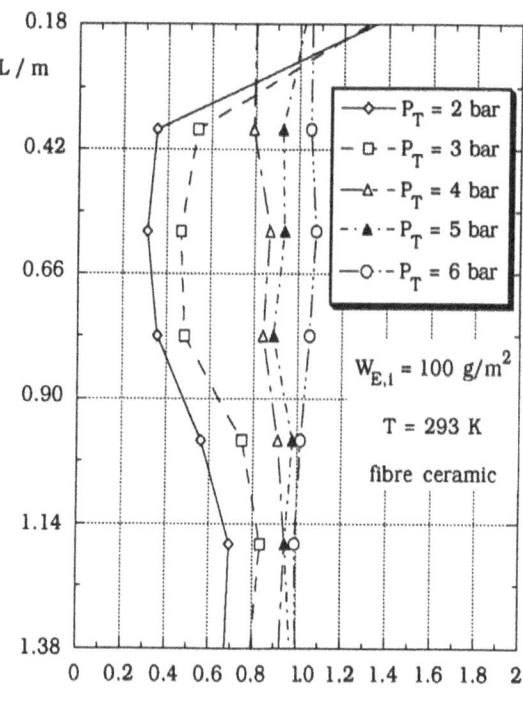

Figure 4. Cleaning efficiency as a function of the reservoir pressure P_T along the length of the filter element

Figures 5 and 6 show the cleaning efficiency R as a function of the valve opening time τ along the length of the filter element at different incremental areal dust densities. The electrical opening time τ is varied between 0,05 and 0,2 s. The reservoir pressure P_T = 5 bar. All other settings are as above mentioned.

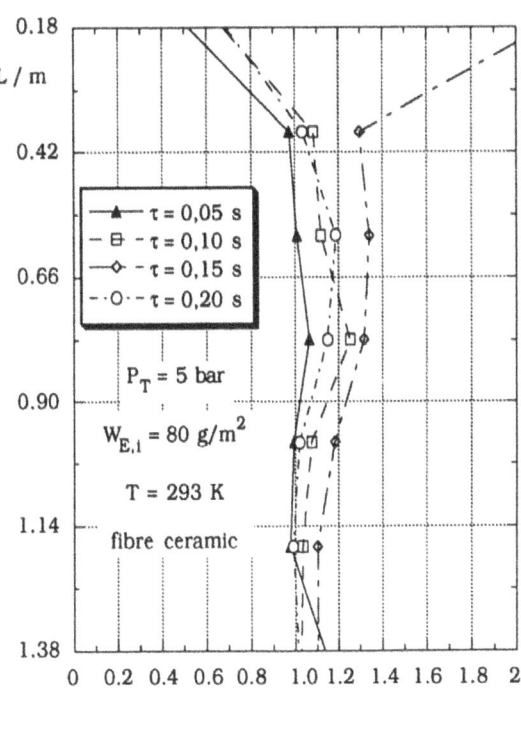

Figure 5. Cleaning efficiency as a function of the valve opening time τ along the length of the filter element, $W_{E,i}$ = 80 g/m²

Figures 5 and 6 show there is almost no influence of the valve opening time τ on the cleaning efficiency even at small areal dust densities. Previous work has shown that, during on-line pulse jet cleaning, the valve opening time is more important in order to reduce that part of the dust, which is redeposited on the filter surface directly after cleaning. The results here support the theory that a very rapid pressure increase within the filter element is necessary to reach high cleaning efficiencies. The duration of the pulse jet (duration of the pressure signal) itself is of less importance under these conditions.

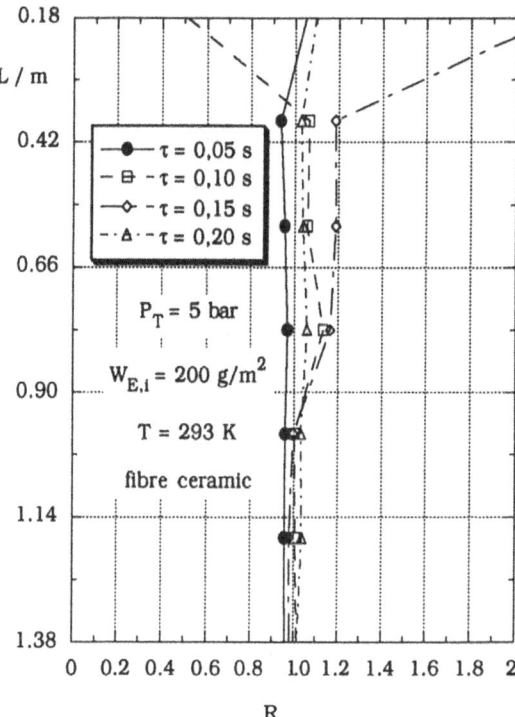

Figure 6. Cleaning efficiency as a function of the valve opening time τ along the length of the filter element, $W_{E,i} = 200$ g/m²

Pulse Pressure inside the Filter Element

The highest velocity a gas can reach upon being emitted from a cylindrical orifice is sonic velocity. This can be accomplished if the ratio of the pressure outside the blow tube to that within falls below the critical value of 0,528 (17). With the assumption that the environmental air pressure outside the blow tube is 10^5 Pa (1 bar), this velocity can be reached by air emitted from the nozzle at an over pressure of $P^* > 1,89$ 10^5 Pa .

Figure 7 shows the pressure in the blow tube as a function of time and of the nozzle diameter d. The reservoir pressure P_T=5 bar and the electrical valve opening time $\tau = 0,05$ s. In this diagram are compared the pressures measured at the transducer next to the outlet nozzle of the blow tube.

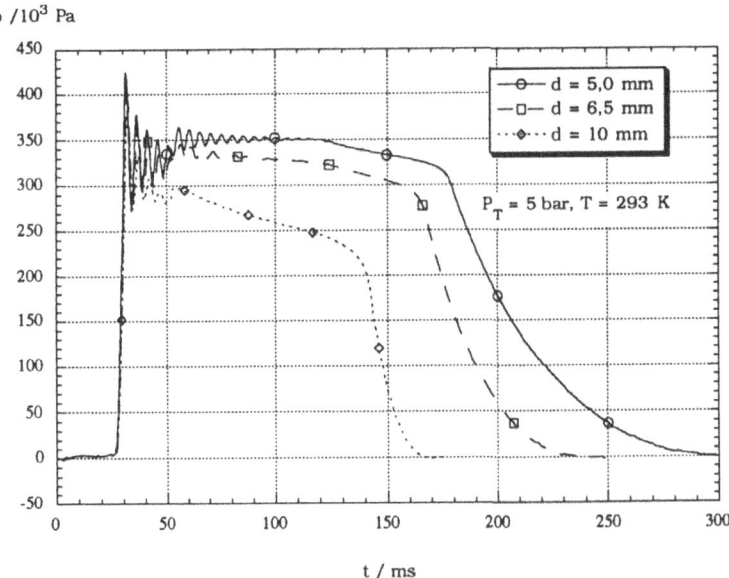

Figure 7. Blow tube pressure as a function of time and the nozzle diameter d

It shows that for all nozzle diameters used the blow tube pressure is higher than the critical pressure P^* to reach sonic velocity. The

235

duration of the over pressure inside the blow tube decreases with increasing nozzle diameter d. Other investigations have shown that the valve opening time τ does not influence the pressure inside the blow tube significantly. Sonic velocity is always reached with reservoir pressures higher than 3 bar. In the cleaning system used for the following investigations no significant pressure loss appears in the blow tube.

Typical traces of the pressure signals obtaining at various positions along the filter candle are shown in Figure 8. The reservoir pressure $P_T = 5$ bar. The electrical valve opening time τ = 0,06 s and the outlet diameter of the nozzle d = 10 mm. The distance H between the nozzle and the top of the filter candle is 80 mm.

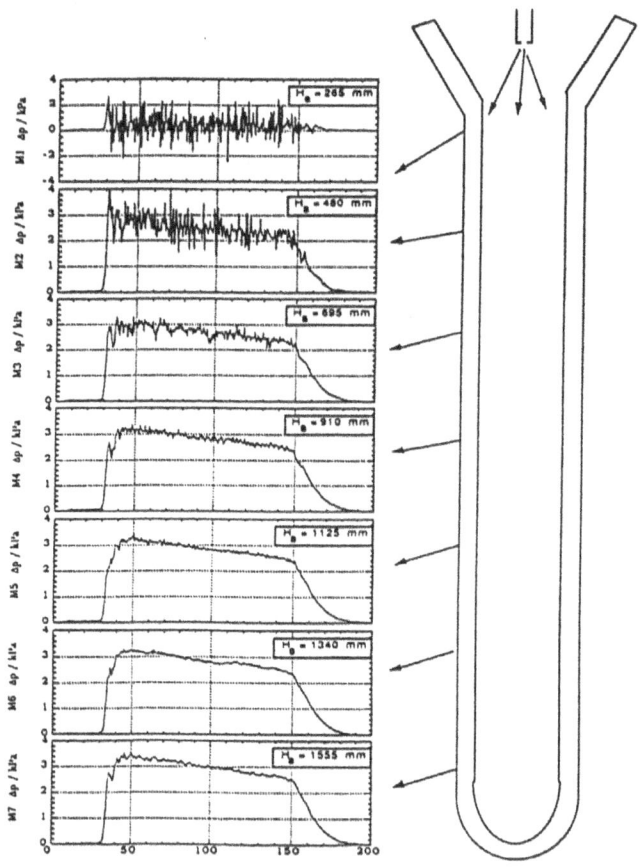

Figure 8. Typical pressure traces at various locations along the filter (fibre ceramic)

236

It should be noted from these pressure traces that despite the high reservoir pressure P_T only low static pressures are built up in the filter candle. This follows from the fact that the pressure increase in the candle depends on the discharge rate of the nozzle, the amount of secondary entrained air at the top of the filter element and the permeability of the filter medium. In order to compare different pulse jet cleaning systems, one should therefore consider the pressure inside the element and not that of the compressed air reservoir.

Observing the measured pressure signals along the length of the filter candle, it can be seen that nearly the same peak pressure is reached at every position along the candle. In contrast to measurements at long fabric fibre filter bags (18) no significant variation in the pressure increase time occurs for this relatively short ceramic filter candle. According to the permeability of the filter medium, air escapes through the filter medium in the upper regions. The cleaning pulse jet therefore progressively loses energy as it travels down the candle. This leads to the effect that the peak pressure occurs at the bottom of the candle after a certain lag of time. The duration depends on the length of the filter element.

Additionally figure 8 shows that pressure oscillations exist in the upper candle regions. Large negative pressures are reached sucking environmental air into the candle. Positive pressures appear after the pulse jet strikes the wall of the filter element. The height where these positive pressures appear first depends on the reservoir pressure P_T, the outlet nozzle diameter d and the distance between the blow tube and the filter candle H.

Figure 9 shows the influence of the distance between blow tube and filter candle H on the pressure in the upper region of the filter element. All experimental settings are the same as mentioned above. The distance H is changed from 10 to 70 mm. In figure 9 are shown the signals recorded by the pressure transducer located next to the top of the filter element.

It is shown that with increasing distance H the positive part of the pressure signal increases. The pressure peak is not as high as in the lower region of the filter candle. On the one hand the positive part of the pressure in the upper region increases with increasing distance H; on the other hand the pressure peaks in the lower region decrease with increasing H. For each cleaning system the geometry has to be optimized.

These results are important especially with regard to hot gas filtration. The temperature of the pulse jet is very often much lower than the temperature of the filter element and the clean gas. If the pulse jet striking the wall of the filter candle is too cold it could cause strong thermal stresses in the ceramic and damage the element.

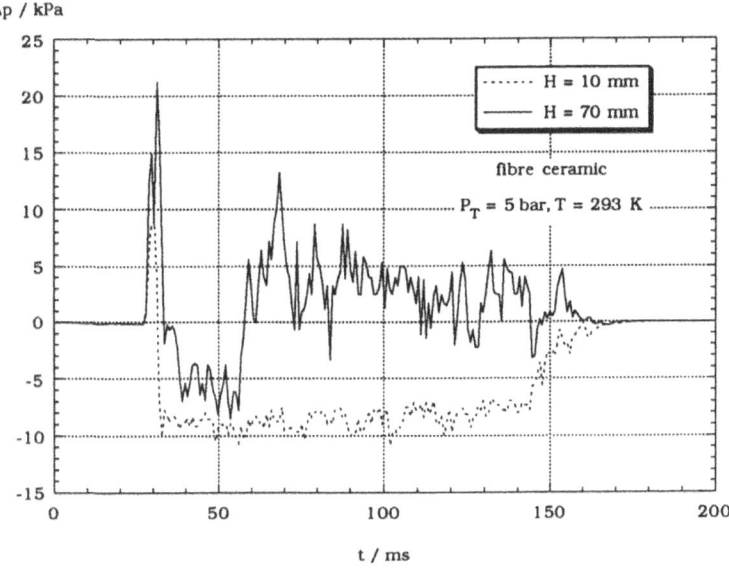

Figure 9. Influence of the distance H between the blow tube and the filter candle on pressure in the upper candle region for sintered ceramic

The durability of the ceramic filter elements under hot gas conditions is still one of the major problems. Due to this fact the cold pulse jet must entrain enough environmental hot gas to reduce the thermal shock to the filter elements during the cleaning process.

Figure 10 shows the influence of the pulse jet duration on the pressures reached in the candle. The electrical valve opening time τ is varied between 0,05 and 0,2 s. The reservoir pressure P_T = 5 bar. In the diagram are compared the signals of one pressure transducer in the middle of the filter element. It is shown that the valve opening time τ (the pulse jet duration) has no significant influence on the pressure peaks in the filter element. These results support the theory mentioned earlier.

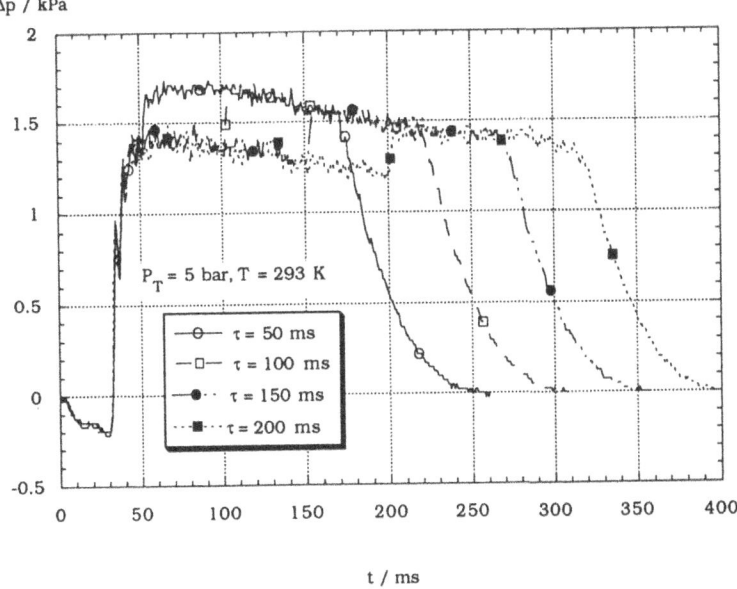

Figure 10. Influence of the pulse jet duration on the pressure peaks
inside the filter element

Another important influence on the pressures reached in the rigid filter
elements is the permeability of the filter medium. In figure 11 are
compared the pressure signal of one transducer located in the middle of
different filter candles. The candles nearly have the same geometry, but
one is made of sintered ceramic pellets, while the other is made of
ceramic fibres with inorganic reinforcement. The experimental settings
are the same as mentioned above. Figure 11 shows that in the filter
element with the lower permeability, that made of sintered ceramic
pellets, the pressures are much higher than in the ceramic fibre
element. This is due to the higher pressure drop of the sintered
ceramic element. The air of the pulse jet has to overcome a higher
friction passing through the wall. So more mass of the pulse jet reaches
the lower regions of the candle and results in a higher pressure on the
transducer. This does not mean that the separation forces to detach the
filter cake from the outside surface are stronger. The reverse flow loses
more energy in the wall of the filter element. So one further aspect of
investigations in hot gas cleaning processes should concern the

relationship between the pressure inside the element, the energy loss through the wall and the local separation forces reached on the outside surface to detach the filter cake.

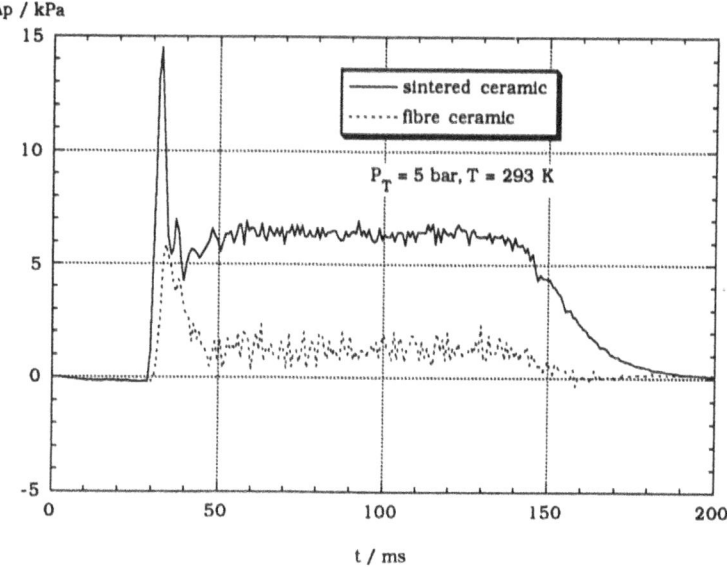

Figure 11. Influence of the permeability of the filter medium on the pressures

CONCLUSIONS

These investigations were carried out to get basic information about the pulse jet cleaning behaviour of technical rigid ceramic filter elements.

It is shown that to get a good cleaning efficiency (steady state filtration) one has to know the separation forces needed locally, on every part of the filter element, to overcome the binding forces between cake and medium.

Based on these investigations under ambient conditions new experiments with rigid ceramic filters are necessary. They have to include the influence of the temperature on the binding forces of the filter cake for different fly ashes under real hot gas conditions. Additionally they have to take into consideration the real gas atmosphere and possible chemical reactions in the filter cake as they occur in hot gas processes.

Further experiments have to investigate whether it is possible to predict the behaviour of technical filter elements by the behaviour of a filter disc under the same rough hot gas circumstances.

Furthermore it is necessary to improve the design of hot gas filter elements, optimising (among other things) durability, pressure drop and cleanability.

ACKNOWLEDGEMENTS

This work is sponsored by the Projektträger Biologie, Energie, Ökologie des Bundesministers für Forschung und Technologie, Forschungs-zentrum Jülich GmbH, Contract No. 0326741A. The authors would like to thank this organisation for the support of this work.

REFERENCES

1. Löffler, F., 1988, Staubabscheiden, Thieme Verlag

2. Klingel, R., 1983, Untersuchung der Partikelabscheidung aus Gasen an einem Schlauchfilter mit Druckstoßabreinigung, Dissertation Universität Karlsruhe

3. Klingel, R. & Löffler, F., 1981, Influence of selected parameters on pulse jet fabric filter efficiency, Filtech Conference, London, GB, 111-120

4. Klingel, R. & Löffler, F., 1982, Studies on dust collection and cleaning efficiency of a pulse jet fabric filter, World Filtration Congress III, Downingstown, PA, USA, I, 35-42

5. Klingel, R. & Löffler, F.,1983, Influence of cleaning conditions on outlet dust concentration of a pulse jet fabric filter, Second Conference in Fabric Filter Technologie for Coal Fired Power Plants,Denver,Colorado, USA

6. Klingel, R. & Löffler, F., 1983, Influence of cleaning intensity on pressure drop and residual dust areal densitiy in a pulse jet fabric filter, Filtech Conference, London, GB, 306-314

7. Klingel, R. & Löffler, F., 1983, On line - Ermittlung von Staubmassenprofilen.an einem druckstoßabgereinigten Filter-schlauch mit Hilfe eines radiometrischen Absorptions-meßverfahrens, Staub-Reinhaltung der Luft 43, 5, 179-185

8. Sievert, J., 1988, Physikalische Vorgänge bei der Regenerierung des Filtermediums in Schlauchfiltern mit Druckstoßabreini-gung, Dissertation Universität Karlsruhe

9. Sievert, J. & Löffler, F., 1985, The effect of cleaning system parameters on the pressure pulse in a pulse jet filter, 16th Annual Meeting of the Fine Particle Society, Miami Beach, Florida, USA

10. Sievert, J. & Löffler, F., 1986, Dust dislodgement in pulse jet filters, Partec, Nürnberg, IV, 111-126

11. Gäng, P., 1990, Kombinierte Abscheidung von Partikeln und Gasen mit Abreinigungsfiltern bei hohen Temperaturen, Dissertation Universität Karlsruhe

12. Gäng, P. & Löffler,F.,1992, F.New procedure and test-rig for the characterisation of cleanable filter media, <u>Clean Air Congress Montreal</u>, Canada

13. Schmidt, E., 1991, Elektrische Beeinflussung der Partikelabscheidung in Oberflächenfiltern, Dissertation Universität Karlsruhe

14. Bower, C. J., Twigg, A.N, Cheung,W., Seville, J. P. K. & Clift,R.,1988, Filtration and cleaning characteristics of ceramic media, <u>International Conference on Fluidised Bed Combustion</u>, London, GB, II/9/1-14

15. Cheung, W., Seville, J. P. K. & Clift, R.,1989, A patchy cleaning interpretation of dust cake release from non woven fabrics <u>Filtration and Separation</u>, May/June,187-190

16. Keidel, W., Withers, L. J., Seville, J. P. K. & Clift, R., 1989, Rigid ceramic media for filtering hot gases, <u>Filtration and Separation</u>, July/August, 265-271

17. ter Kuile, J., Seville, J. P. K. & Clift, R., 1990, Aerodynamic considerations in design and cleaning of rigid ceramic filters for hot gases, <u>5th World Filtration Congress</u>, Nice, France, 530-537

18. Zierep, J. Grundzüge der Strömungslehre,<u>Verlag G. Braun</u>, Karlsruhe

REGENERATION OF RIGID CERAMIC FILTERS

D. KOCH[1], K. SCHULZ[2], J.P.K. SEVILLE AND R. CLIFT
Department of Chemical & Process Engineering
University of Surrey, Guildford, GU2 5XH, U.K.

ABSTRACT

The important issues to be addressed in specification and operation of ceramic filters at high temperatures are the long term conditioning behaviour and the filter "cleanability". These two aspects of filtration behaviour are inextricably linked and, at the present state of knowledge, very much system-specific, so that experimental work using the dust of interest is essential. In the past, the authors have introduced the concept of "cake removal stress" to characterise the adhesion of the filter cake to the filtration medium and have developed a coupon test method to measure both conditioning and cake removal behaviour. Coupon filter tests are described in which a low-density fibrous ceramic medium has been used to filter calcium carbonate dusts, and the conditioning behaviour and cake removal stress distributions have been measured over a range of particle size and cake areal dust loadings. For reverse flow cleaning, the cake removal stress decreases strongly as the cake areal mass increases from 100 to 500 g/m^2, and less strongly thereafter. Cake removal stresses were also measured using centrifugal acceleration; the results are consistently lower than those for the reverse flow test and nearly independent of cake loading. The difference between the two tests is thought to be due principally to cake "hinging" to the medium after effective detachment. At cleaning pressures which are typical of industrial practice, cake detachment occurs by patch separation and a fracture mechanics interpretation of this phenomenon is advanced.

1.   Present address: Nestec Ltd., Research Centre, Vers-Chez-Les-Blanc. P.O. Box 44, CH-1000 Lausanne 26, Switzerland.

2.   Diploma student from Institüt für Mechanische Verfahrenstechnik und Mechanik, Universität Karlsruhe, Germany.
      Present address: Schumacher Umwelt- und Trenntechnik, Zur Flügelau 70, Postfach 1562, D-7180 Crailsheim, Germany.

NOMENCLATURE

a	patch size radius (Fig. 9)	(m)
d_P	particle size	(m)
$E^{\bullet}$	elastic modulus of the cake	(Pa)
f	fraction of the cake removed	(-)
h	cake thickness	(m)
K	function of dust and cake properties (eq. 6)	$(N^{\frac{1}{2}}m^{\frac{1}{2}})$
k_C, k_M	modified cake and medium resistances (eq. 2 and 3)	$(Pa\ s\ m^{-1})$
N	number of filtration/cleaning cycles	(-)
P_C	excess pressure required to expand patch	(Pa)
p_{PP}	pressure difference across coupon produced by reverse pulse	(Pa)
t_{PP}	duration of reverse pulse	(s)
W_A	areal cake loading	(gm^{-2})
γ	particle surface energy	(Jm^{-2})
γ_a	adhesive fracture energy of cake to medium	(Jm^{-2})
$\Delta P_C, P_M$	pressure difference across cake and medium during reverse flow	(Pa)
ΔP_T	total pressure difference during reverse flow	(Pa)
ϵ	void fraction of cake	(-)
$\nu^{\bullet}$	Poisson's ratio of cake	(-)
σ_c	cake detachment stress	(Pa)
σ_{50}	stress required for 50% cake detachment	(Pa)

INTRODUCTION

The development of ceramic materials capable of filtering micron-sized particles from process gases at temperatures up to 1000°C is one of the most significant recent advances in gas filtration technology. There are two generic types of filter medium (Seville, 1993): a granular bonded or sintered material consisting typically of silicon carbide, and a fibrous bonded material consisting usually of aluminosilicates in various proportions. The former type was developed in response to the need for hot gas filters for power generation plant; the latter has been developed more recently (Seville et al., 1989b) and has the advantage of a very high void fraction, typically 0.9-0.95, and therefore low

245

resistance to flow and low weight. It is therefore more suitable for many process industry applications.

In use, the filter medium is formed into long hollow "candles" with one closed end, conventionally suspended vertically from a tube plate, with the dirty gas flowing radially inwards, along the inner clean side of the candle and out of the plenum chamber. The medium is designed to act as a surface filter so that dust is deposited on the outer candle surface and builds up into a cake, which is periodically detached by a reverse pulse of some suitable cleaning gas. The usefulness of the filter in practice does not depend significantly on the efficiency with which it removes incoming particles; this is very close to 100% for all but the most demanding of applications. It depends to a much greater extent on its long term pressure-drop history. After each cleaning pulse, the resistance to flow of the candle decreases to its "residual" value which, during the initial "conditioning" period, increases from cycle to cycle. There is evidence that during this period a "residual layer" of dust builds up on the surface of the medium, which may have a different composition or void fraction from the rest of the filter cake (Seville et al., 1991) and may not be removed by the cleaning pulse. When cleaning occurs, the filter cake therefore detaches from a layer of deposited dust rather than the medium itself. While the residual layer is probably essential to prevent penetration of dust into the medium, it is important that it does not build up so much as to cause intolerably high operating pressure drops. There is therefore a practical problem of designing a filter cleaning system which is sufficiently effective to remove the filter cake but not to remove all of the residual layer, or to redisperse the dust cake so that it simply goes back onto the filter after cleaning.

Ideally, it should be possible to predict the cleaning conditions directly from the gas and particle properties, including particle size and surface energy, but this is not realisable at the present state of knowledge. As an immediate objective, it is desirable to have a practical test for cake detachment which could be used to select the cleaning conditions, and it is the results from such a test which form the subject matter of this paper. The importance of prior testing is underlined by the highly system-specific nature of conditioning and cleaning behaviour. In the trials of silicon carbide filters at the Grimethorpe PFBC Establishment (Burnard et al., 1993), for example, major differences between the behaviour of dusts from similar feedstocks were recorded, even with a change in sulphur sorbent alone. If such differences in behaviour can be detected by means of small-scale tests such as that described here, there are likely to be major savings in the cost and complexity of full scale hot filter trials.

Operating temperature can have a strong effect on cake detachment because changes in temperature alter the surface forces which bind the cake together. The authors have previously studied the effect of temperature on filter conditioning and cake removal using a slip-stream from a gasifier (Cheung et al., 1988) and more recent results are presented by Cahill et al. (1993). However, there is no doubt that conditioning and cake detachment occur by the same mechanisms at all temperatures so the more fundamental study presented here is concerned only with ambient temperatures.

FILTER CAKE DETACHMENT

Several analyses of the problem of cake detachment from flexible fabric filters have been presented (Koch, 1993) but the common assumption in all of them is simply that the dust cake detaches from the filter medium when it experiences a tensile stress sufficient to overcome either the strength of the adhesive bond between the cake and the medium (or a residual dust layer) or the internal cohesion of the cake. In theory, as soon as the strength of this adhesive or cohesive bond is exceeded (by whatever cleaning mechanism) the cake detaches everywhere simultaneously. In practice, however, neither the adhesive/cohesive cake strength nor the applied stress is entirely uniform across the filter surface so that "patchy" cleaning results, i.e. cake is completely detached from some areas of the filter and completely retained in others. This sort of behaviour is to be distinguished from progressive removal of dust layers, which is not observed.

In a conventional bag filter it is usually assumed that the required tensile cleaning stress is set up primarily by the movement caused by the cleaning pulse or, in the case of mechanically-cleaned filters, the shaking action. Pulse cleaning displaces the fabric outwards. When it becomes taut, it decelerates sharply, normally at many times gravitational acceleration. The cake then experiences a tensile stress which depends on its areal density and on the deceleration. If the stress is sufficient, fracture occurs so that

247

the cake is thrown clear of the medium. Rigid media such as the ceramics considered here show no displacement on cleaning. The tensile stress is therefore entirely the result of the pressure drop imposed across the cake due to reverse-flow of cleaning gas, as shown below.

Analysis of Cleaning by Reverse Flow or Pulse

Consider first the case of a filter medium on which a uniform cake has been laid down. A cleaning flow is now imposed in the opposite direction to the filtration direction, as shown in Fig. 1. During reverse-flow cleaning a pressure difference will be set up across the filter, consisting of contributions from the cake and the medium.

$$\Delta P_T = \Delta P_C + \Delta P_M \tag{1}$$

The gas viscosity is effectively constant and the cake and medium thicknesses can be incorporated into modified resistances, k_C and k_M, so that

$$\Delta P_M = k_M U \tag{2}$$

and

$$\Delta P_C = k_C U \tag{3}$$

combining equations (1), (2) and (3):

$$\Delta P_C = \Delta P_T [\frac{k_C}{k_C + k_M}] \tag{4}$$

This is the pressure drop across the cake itself and also, as shown in Fig. 1, the tensile stress acting at the cake/medium interface. It is therefore this quantity which is of prime interest when investigating the cake removal characteristics of a given dust/medium combination.

FIG.1

Pressure Distribution
in Medium and Cake
During Reverse Flow

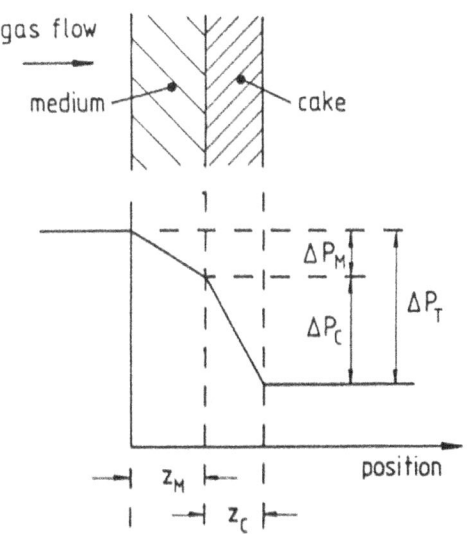

Equation (4) was developed for a uniform cake. However, it is also theoretically applicable to a partially cleaned filter (Seville et al, 1989a). Because of its inhomogeneous resistance to flow, a patchily cleaned filter will show regions of preferential gas flow. However, in the uncleaned areas the total pressure drop across the filter must still be distributed across the medium plus cake as shown in equation (4), provided that the flow is rectilinear; i.e. in the uncleaned areas the gas velocity is the same in the medium and in the cake. This approximation is valid provided that the undetached cake patches are large compared with the cake thickness.

From the analysis above it is clear that when comparing the cleaning behaviour of different dust/medium combinations it is ΔP_C, the pressure drop across the cake alone, which should be considered and not ΔP_T, the total pressure drop. Indeed, comparison of values for ΔP_T necessary to detach the cake may be misleading, because they depend on the cake loading and medium resistance. It is sometimes asserted that thick cakes are easier to clean from filter media than thin ones. The foregoing analysis provides one reason why this may appear to be so. If the total pressure drop across the cake required for detachment is constant, a thicker cake requires a smaller pressure gradient for removal and therefore less cleaning gas flow and less pressure drop across the medium itself.

The approach outlined above considers only steady reverse-flow cleaning, but it can also be applied to pulse cleaning, since the maximum stress to which the cake is subjected corresponds to the steady flow cleaning value from equation (4) (Cheung, 1989). In many industrial applications, the pressure rise associated with the cleaning pulse is, in any case, relatively slow so that the process is better considered as steady reverse flow cleaning.

EXPERIMENTAL METHODS

The intensity of cleaning action delivered to a filter candle varies with position so that this geometry does not lend itself to more fundamental test work. Experiments were therefore carried out on 60 mm diameter flat "coupons" of filter medium using a test method originally adopted by Cheung et al (1988). The filter medium consisted of bonded alumino-silicate fibres and had a void fraction of approximately 90%; it was supplied by Cerel Ltd., under the name "Cerafil-S-1000".

FIG.2

Coupon Test Rig
- Schematic

1. compressed air

2. air filter

3. pressure reservoir

4. dust feeder

5. dispersion chamber

6. filter coupon

7. absolute filter

8. exit

9. bag filter

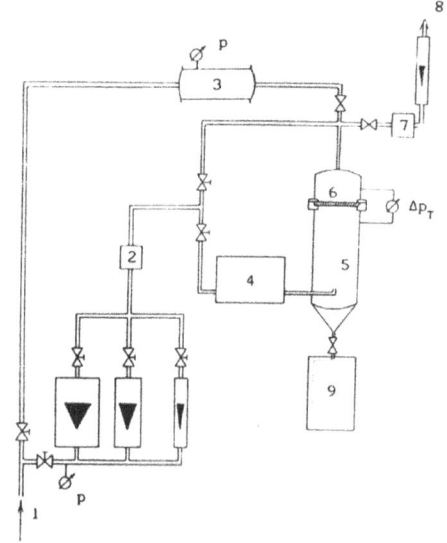

The test rig is shown in Fig. 2; it allows a metered flow of dust-laden gas to pass upwards through the filter coupon during the filtration cycle, depositing the filter cake on the lower surface. The associated rise in flow resistance is measured by pressure transducers and logged continuously. Cleaning is carried out at either a preset time interval or a prescribed differential pressure by administering a reverse pulse of gas to the chamber above the coupon. Two fast-response strain-gauge pressure transducers are used to measure the differential pressure transient across coupon and cake during cleaning. The rig operation is fully automatic since filter conditioning may last as long as 300 cycles of, typically, 2.5 minutes duration. After conditioning, the "cake removal stress" curve is determined by cutting off the main filtration flow and applying a controlled reverse flow (rather than pulse) of cleaning gas. The reverse flow is increased in small increments, recording the corresponding total pressure difference, ΔP_T, across the medium plus cake at each point and weighing the amount of cake removed. The cake removal stress can be calculated from equation (4), since the resistance of the cleaned medium, R_M, and the medium plus complete cake prior to cleaning, $(R_M + R_C)$, are known. As previously discussed, the cake is observed to detach in patches as ΔP_C is increased, i.e. some areas of the filter appear to clean down to the residual layer while others remain uncleaned. This results in a "cake removal stress curve" relating the fraction of cake removed to cleaning stress, as shown in the experimental results presented later in this paper.

The test dust used in all of the experiments described here was a fine limestone, dispersed by means of the "Wright dust feeder" (Wright, 1950). The original as-supplied size distribution (referred to here as "mixture") was further sub-divided into a "fine" and an "intermediate" fraction. It is well known that in redispersion of dust some classification usually occurs in the pipework of the apparatus. The extent of this "drop-out" was carefully monitored in this case and size distributions of the three dust fractions were obtained from the deposited dust cakes, not the original powder, using a Coulter LS130 diffraction sizer. The dust mixture was approximately log-normal with a mass median diameter of 2.0 μm and a geometric standard deviation of 2.6. The fine fraction was also nearly log-normal with a mass median diameter of 1.8 μm and a geometric standard deviation of 1.9. The intermediate fraction was coarser, with a mass median diameter of 3.5 μm and strongly cut off at both extremities of the distribution.

RESULTS

Conditioning

Figure 3 shows conditioning results over 250 cycles for a face velocity of 19 cm/s and three different pulse cleaning conditions (defined by maximum differential pressure across coupon plus cake, and pulse duration). It is evident that, as expected, the intensity of the cleaning pulse has an effect on the long-term conditioning behaviour: a higher differential pulse pressure leads to quicker conditioning and lower equilibrium pressure loss. For most of the subsequent experiments reported here, 23.1 kPa was adopted as the cleaning pressure during conditioning. Reduction in the pulse duration from 1.3 to 0.3 s was found to have no effect on residual pressure loss development and conditioning. This result is consistent with observations of the effect of pulse duration on cleaning of a full candle by Cheung et al. (1989) and Berbner and Löffler (1993).

FIG. 3   Filter Conditioning - Dust Mixture

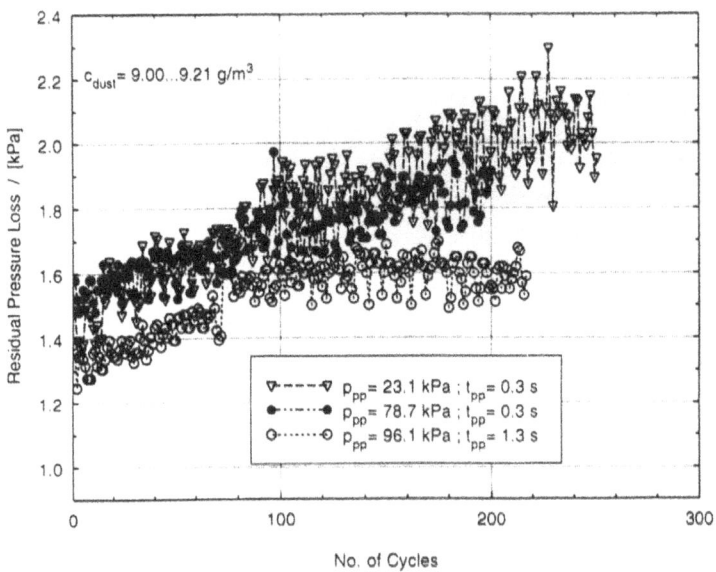

As expected, the extent of conditioning has an effect on the measured cake removal stress (Schulz, 1992). For cake areal loadings of 1000 g/m², the least adhesive cakes were found with a "virgin" unconditioned coupon. For conditioned coupons, the stress required to remove 50% of the cake, σ_{50}, was typically doubled by conditioning with a cleaning pressure of 23.1 kPa rather than 96.1 kPa. It is apparent, therefore, that less efficient cleaning has a cumulative effect; if the cleaning pressure applied is low initially, the cleaning pressure required may then increase in subsequent cycles.

Cake Detachment

A typical set of reverse-flow cake detachment curves is presented in Fig. 4, showing the reproducibility to be expected from replicate measurements. Two features are particularly noticeable; the long "tail", the stress required for 90% cake removal being typically an order of magnitude larger than that required for 50% removal, and the strong effect of cake areal loading, W_A. The shape of the curves is undoubtedly connected with the difficulty of removing the remaining small patches of cake at the end of the cleaning process (see later). Further cake detachment curves for a range of cake areal loadings are presented in Figs. 5(a) and (b) for the dust mixture and the fine fraction, and the stress required for 50% cake removal, σ_{50}, is plotted against cake loading, for all size fractions, in Fig. 6. In all cases the detachment stress decreases with increase in cake loading, the decrease from 100 to 500 g/m² being particularly marked. One factor here may be the continuity of the cake. The surface of the filter medium is rough so that 100 g/m² is barely enough to form a continuous cake over the filter surface. As expected, the cake removal stress for the courser dust is less than that for the two finer fractions, a doubling of the mass median diameter resulting in a decrease in the detachment stress of about 50%.

FIG. 4   Cake Removal Curves - Reverse Flow Detachment - Dust Mixture

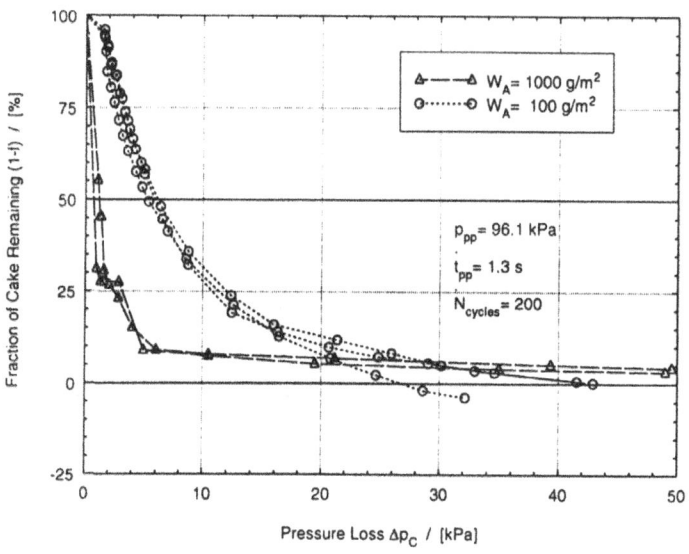

G. 5(a)   Cake Removal Curves - Reverse Flow Cleaning - Dust Mixtur

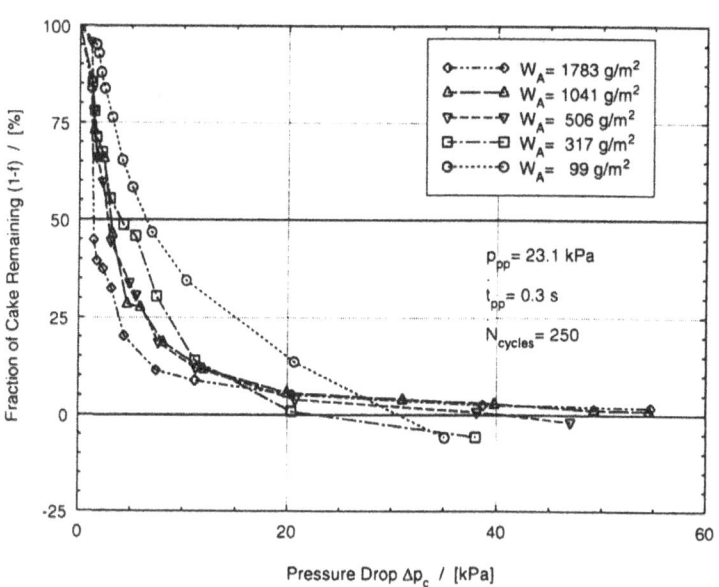

254

FIG. 5(b)　　　Cake Removal Curves - Reverse Flow Cleaning - Fine Fraction

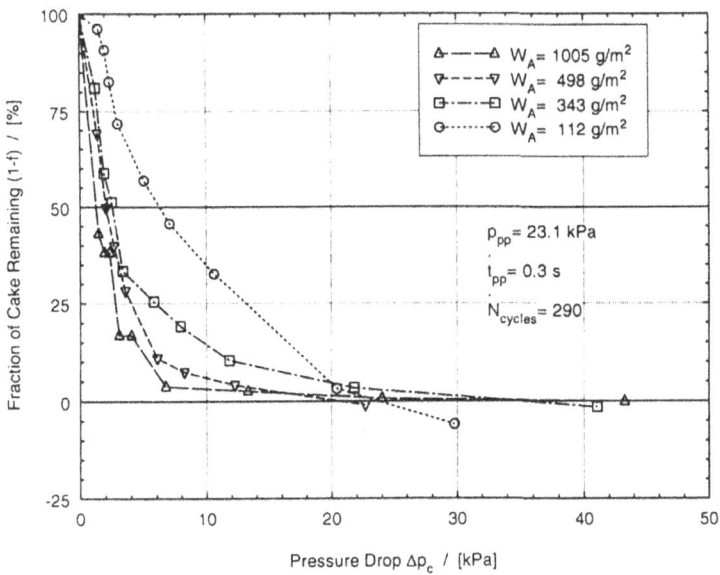

FIG. 6　Reverse Flow Cleaning - Effect of Cake Loading on Detachment Stress, σ_{50}

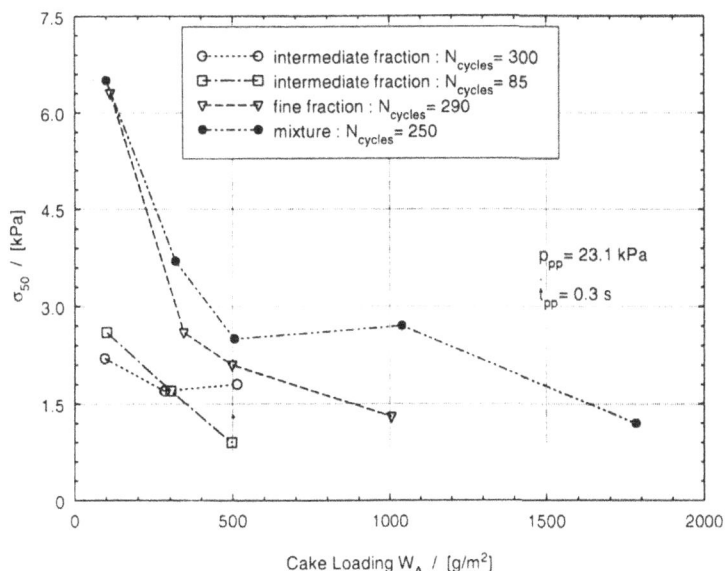

One possible reason for the observed dependence of cake removal stress on cake loading is that the void fraction of the deposited cake depends on its areal mass. However, this seems unlikely in this case since the measured cake resistances to flow vary linearly with cake loading (Koch 1993). It is known that compression of filter cakes can occur (Cheung, 1989; Schmidt and Löffler, 1991), but this would tend to enhance the cake removal stress for thicker cakes, the opposite of the effect observed here. Further cake detachment tests were therefore carried out at different cake loadings, using two alternative methods of removal (Koch, 1993).

Firstly, a standard laboratory centrifuge (M.S.E. Mistral 2000) was modified to accept a conditioned filter coupon, and cake removal was observed and recorded for increasing steps in rotation speed. The applied detachment stress is simply calculated from the rotation speed and the areal dust loading. Secondly, a reverse flow "burst pressure" test was adopted, in which the reverse flow rate was gradually increased until the first major cake detachment occurred. The detachment pressure was recorded at this point, irrespective of the fractional cake detachment achieved.

FIG. 7   Cake Removal Curves - Acceleration Cleaning - Fine Fraction

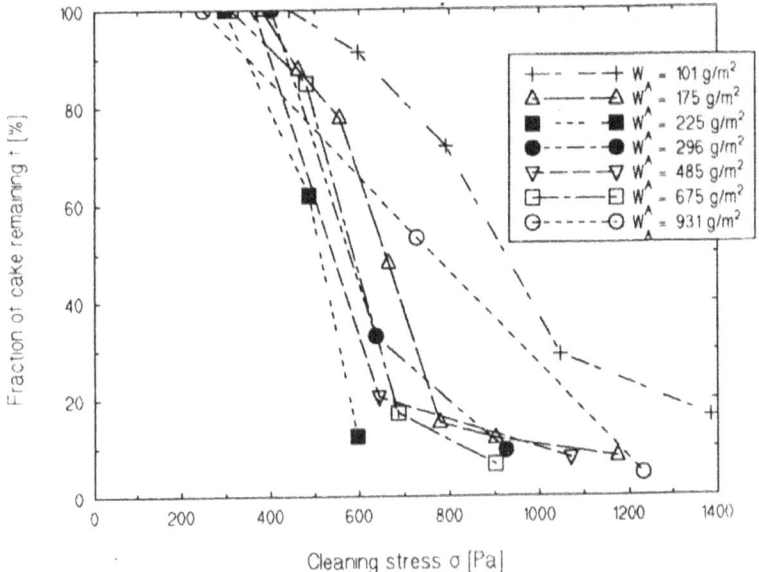

FIG. 8   Effect of Dust Loading on Median Cake Removal Stress

Dashed line indicates "burst pressure test"; other symbols indicate acceleration tests, both for fine fraction.

1.   Reverse flow - fine fraction
2.   Reverse flow - 4.5 μm limestone on singed polyester needle-felt (Sievert, 1988)
3.   As 2, but using acceleration
4.   As 3, but fabric chemically stiffened

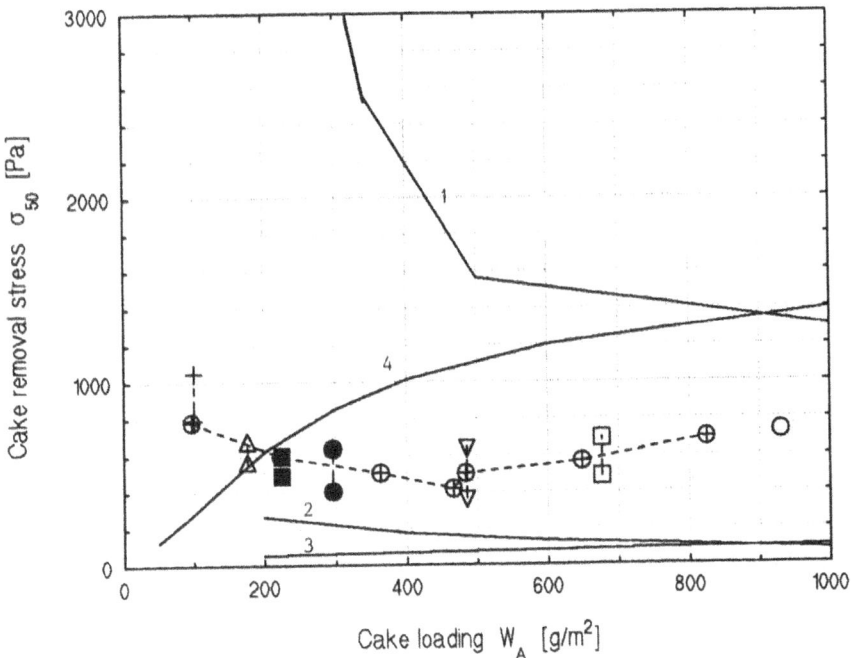

257

The results of the acceleration test are shown in Fig. 7, for the fine fraction only, and the dependence of median cake removal stress on cake loading for all methods is summarised in Fig. 8, again for the fine fraction only. The cake removal stresses measured by acceleration and by the burst pressure test coincide almost exactly with each other and are consistently lower than those from the reverse flow test and nearly independent of cake loading. The results of all three methods converge at high cake loadings ($>$ 1000 g/m^2). Closer observation of partially-cleaned filter coupons after all three cleaning methods reveals that the difference between the original reverse flow cleaning method and the others is principally due to cake "hinging", whereby patches of cake remain loosely attached to the filter surface after reverse flow cleaning, even though their true detachment stress has been exceeded. They are not then included in the measured detached dust mass. No such effect can occur in the centrifugal acceleration test, which is therefore a true measure of the adhesion of cake to conditioned medium. However, since in practice cake removal occurs due to reverse flow, the reverse flow test method remains the most appropriate for estimation of the cleaning flow requirement.

In Fig. 8, the combined detachment stress results for the fine fraction from this work are compared with results obtained by Sievert (1988) for the dependence of cake detachment stress on areal loading using limestone dust of mass median diameter 4.5 μm and a range of fabric media. The first and industrially most significant feature of this comparison is that the cake removal stress is at least one order of magnitude higher for the rigid ceramic media used in this work than for the non-woven flexible fabric used by Sievert. However, in one significant series of experiments Sievert used chemically-stiffened fabric, causing an increase in cake removal stress to values which are very similar to those obtained in the present work. It is highly probable, therefore, that the reason for the much lower cake detachment stresses measured using fabrics is their flexibility, particularly their ability to stretch in tension and thus crack the cake before removal.

PREDICTION OF THE CAKE DETACHMENT STRESS

At present it is not possible to predict *a priori* the stresses which must be imposed to clean the filter medium. One approach is to sum the interparticle forces at each of the contact points across the failure surface, taking due account of their directions, as in the classic agglomerate strength model due to Rumpf (1962). This leads to an expression for the detachment stress, σ_c; for a cake of monosized particles, as follows:

$$\sigma_c = \pi \frac{1-\epsilon}{\epsilon} \frac{\gamma}{d_p} \tag{5}$$

where γ is the surface energy of the particles and d_p is their diameter. Recent work by Aguiar and Coury (1993) for detachment of phosphate rock dust from polyester fabric has shown roughly the dependence on ϵ and d_p predicted by equation (5), albeit for a fitted value of γ. This approach, however, assumes simultaneous failure of all particle-particle contacts across the failure surface, and implies that there should be no influence of cake thickness on σ_c. Morris et al. (1987) have advanced a theory based on patch detachment which shows an increasing cake detachment stress for a thicker cake, while Sievert (1988) has suggested two mechanisms by which, at a constant filtration gas flow rate, the increased pressure drop across a thicker cake enhances the attachment stress and so also causes an increased value of σ_c for a thicker cake. These arguments are discussed in full by Koch (1993). None of the previous approaches has taken into account the presence of flaws or weak points in the cake/medium interface. It is reasonable to assume that filter cake adhesion is poor in some areas, thus giving rise to flaws in the interface which will act as points of stress concentration during cleaning. This is analogous to the problem of "blistering" in the adhesion of coatings to surfaces, which has been the subject of fracture mechanics analysis (e.g. Williams and Anderson, 1977). Using this analysis as a starting point, it is possible to advance a criterion for patch detachment from a filter surface. The patch shown schematically in Fig. 9 stores energy in bending. Expansion of the patch will occur when the rate (with respect to patch radius) at which stored energy is released equals the energy required to create the new surface, which is the adhesive fracture energy. The stored energy is simply found from thin plate bending theory, from which it can be shown that the critical excess pressure required to cause the patch to expand is given by:

$$P_C = \frac{K\gamma_a^{1/2}}{(2a)^2} \qquad\qquad (6)$$

where

$$K = [\frac{512E^{*}h^3}{3(1-\nu^{*2})}]^{1/2}$$

γ_a is the adhesive fracture energy of the cake to the medium

a   is the patch radius

h   is the cake thickness (proportional to cake loading)

E^{*} & ν^{*} are the elastic modulus and Poisson's ratio for the cake

Insertion of values for E^{*} and γ_a deduced from deformation and fracture experiments performed on dust compacts (Abdel-Ghani et al., 1991), and approximate values of the other unknowns, yields values for P_C of order 1 kPa, which is in agreement with the values of σ_{50} shown in Fig. 8. However, there are difficulties remaining in attempting to use equation (6) to predict values for σ_{50} from the fundamental particle and cake properties. Most significantly, the patch size is not determined by this analysis, and since this is observed to increase with cake loading (Figs. 10 and 11) this would reduce the sensitivity of P_C to variation in h, in accordance with the acceleration results shown in Fig. 8. Since the dependence of P_C on both γ_a and E^{*} is relatively weak (Abdel-Ghani et al., 1991), this implies a relatively weak dependence of P_C on particle size and this is again in accordance with observations.

Though the thinking behind equation (6) is somewhat speculative, the measure of agreement between P_C and σ_{50} is encouraging and warrants further investigation. The crucial difficulty lies in predicting a value for the patch size, a, and further experimental and theoretical work is proceeding in order to try to solve this problem.

FIG. 9   Cake Patch Detachment - Schematic

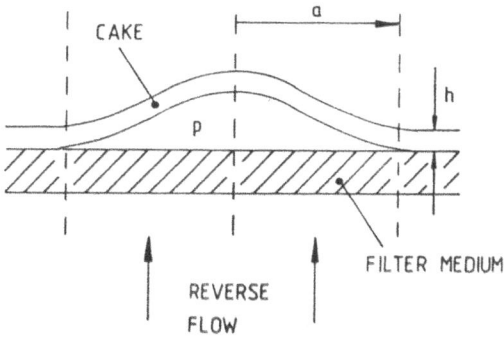

FIG. 10   Effect of Dust Loading on Patch Detachment

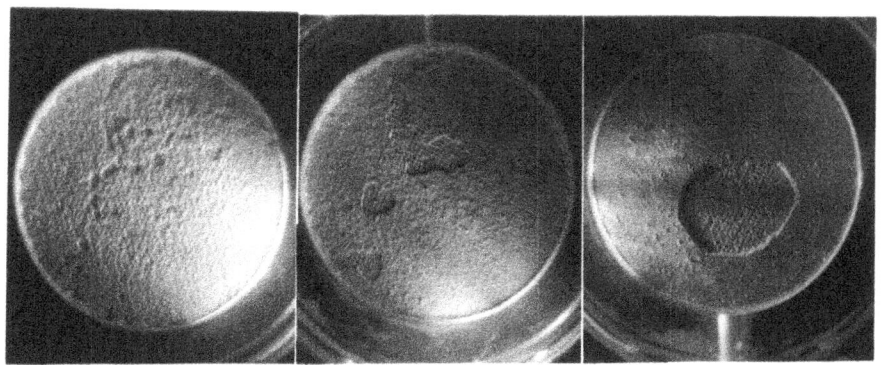

101 g/m²   175 g/m²   675 g/m²

FIG. 11   Effect of Dust Loading on Measured Patch Diameter (Fine Fraction)

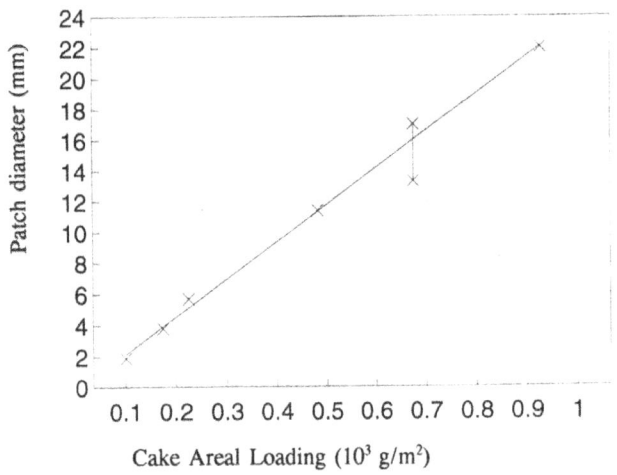

Cake Areal Loading (10^3 g/m²)

CONCLUSIONS

One of the major issues to be addressed in the development of rigid ceramic filters is the maintenance of an acceptable resistance to flow over a long period of operation. This in turn depends on the ability of the filter cleaning system to remove the deposited dust cake. A significant advance in understanding was the recognition that dust cake detachment occurs entirely as a result of the pressure difference across it imposed by the reverse flow of cleaning gas. Since the cake removal stress cannot be predicted *a priori*, it must be determined by experiment, most conveniently by tests using small coupons of filter medium. The results of the coupon tests presented here using reverse flow to detach the cake suggest that the median cake detachment stress decreases strongly as the areal cake loading is increased from very low values (typically in the range 100-500 g/m²) and less strongly thereafter (> 500 g/m²). The effect of cleaning efficiency is cumulative: if the cleaning pressure applied is low during the early filter life, the required cleaning pressure may subsequently be higher.

Cake removal stresses measured by applying a centrifugal acceleration are consistently lower than those from the reverse flow test although there is some evidence that the results of the two methods converge at high cake loadings (> 1000 g/m^2). Measurements of the stress required for the first major cake removal event during reverse flow cleaning, the "burst pressure" test, coincide with those obtained by acceleration. The difference between the original reverse flow test and the others may be attributable to the fact that some patches of cake remain hinged to the filter surface after cleaning, even though their intrinsic cake detachment stress has been exceeded. While the acceleration and burst pressure tests may give a truer representation of the intrinsic cake adhesion to the filter, the reverse flow test remains the most appropriate for estimation of the cleaning flow requirement. At cleaning pressures typical of those employed in practice, cake removal occurs by detachment of patches, whose dimensions increase with increase in areal cake loading. While no wholly satisfactory theory of patch detachment has yet been developed, it seems likely that patches are able to extend by fracture along the cake-medium interface, a process known as "blistering", which can be described by a fracture mechanics analysis. Future work will seek to apply this analysis in order to predict cake removal stresses from independent measurements of cake properties.

Acknowledgement

The work described here was carried out with the financial support of the Science and Engineering Research Council (Specially Promoted Programme in Particulate Technology) under grant number GR/E78227. The authors would like to acknowledge the help of Cerel Ltd., a Burmah Castrol Company, for the generous supply of ceramic filter media; the staff and students of the Institüt für Mechanische Verfahrenstechnik und Mechanik of the University of Karlsruhe for many helpful discussions; the British Council for an award under the British-German Academic Research Collaboration Programme for collaboration between the Universities of Surrey and Karlsruhe; and to Mr Bryan Underwood for his assistance with the centrifuge experiments.

REFERENCES

Abdel-Ghani, M., Petrie, J.G., Seville, J.P.K., Clift, R. and Adams, M.J. (1991). Mechanical Properties of Cohesive Particulate Solids, Powder Technol. 65, 113-123.

Aguiar, M.L. and Coury, J.R. (1993). Air Filtration in Fabric Filters : Cake-Cloth Adhesion Force, submitted to Solid-Fluid Sep. J.

Berbner, S. and Löffler, F. (1993). Pulse-Jet Cleaning of Rigid Filter Elements at High Temperature, these Proceedings.

Burnard, G.K., Leitch, A.J., Stringer, J., Clarke, R.K. and Holbrow, P (1993). Operation and Performance of the E.P.R.I. Hot Gas Filter at Grimethorpe P.F.B.C. Establishment : 1987-1992, these Proceedings.

Cahill, P., Rasmussen, G., Tustin, M. and Robertson, D. (1993). Filtration Properties of Dust from Fluidised Bed Gasification Systems, these Proceedings.

Cheung, W., (1989). Filtration and Cleaning Characteristics of Ceramic Media. PhD Thesis, University of Surrey.

Cheung, W., Seville, J.P.K., and Clift, R. (1988). Filtration and Cleaning Characteristics of Ceramic Media. Proc. 4th Int. Conf. on Fluidised Bed Combustion. Inst. of Energy, London, II/9/1-14.

Koch, D. (1993). Characterisation of the Regeneration Performance of Rigid Ceramic Filters, PhD Thesis, University of Surrey.

Morris, K., Allen, R.W.K. and Clift, R. (1987). Adhesion of Cakes to Filter Media, Proc. Filt. Soc. (Filtn. and Sepn.) 24(2), 41-45.

Rumpf, H. (1962). The Strength of Granules and Agglomerates. Int. Symp. Agglomeration, Interscience, London, 379-418.

Schmidt, E. and Löffler, F. (1991). The Analysis of Dust Cake Structures, Part.Part. Syst. Charact. 8(2), 105-109.

Schulz, K. (1992). Orientierende Experimentelle Untersuchungen zum Betriebsverhalten einer Laborfilteranlage für Starre Keramische Filtermedien, Diplomarbeit thesis, Universität Karlsruhe.

Seville, J.P.K., (1993). Rigid Ceramic Filters for Hot Gas Cleaning, KONA 11, to be published.

Seville, J.P.K., Cheung, W. and Clift, R. (1989a). A Patchy-Cleaning Interpretation of Dust Cake Release from Non-Woven Fabrics. Proc. Filt. Soc. (Filt. and Sepn.) 26(3), 187-190.

Seville, J.P.K., Clift, R., Withers, C.J. and Keidel, W. (1989b). Rigid Ceramic Media for Filtering Hot Gases. Proc. Filt. Soc. (Filt. and Sepn.) 26(4), 265-271.

Seville, J.P.K., Legros, R., Brereton, C.M.H., Lim, C.J. and Grace, J.R. (1991). Performance of Rigid Ceramic Filters for C.F.B.C. Gas Cleaning. 11th Int. Conf. on Fluidized Bed Combustion, A.S.M.E. Montrèal.

Sievert, J. (1988). Physikalische Vorgänge bei der Regenerierung des Filtermediums in Schlauchfiltern mit Druckstoßabreinigung. Dissertation Universität Karlsruhe Fortschr. Ber. VDI Reihe 3, Nr. 161.

Williams, M.L. and Anderson, G.P. (1977). Adhesive Fracture Mechanics, Proc. 4th Int. Conf. Fracture, "Fracture 77", Vol. 1, 643-661.

Wright, B.M. (1950). A New Dust Feed Mechanism, J. Sci. Instr. 27, 12.

PULSE JET CLEANING AND INTERNAL FLOW IN A LARGE CERAMIC TUBE FILTER

SHIGEO ITO

Advanced Coal Technology Section, Advanced Energy Department,

Central Research Institute of Electric Power Industries (CRIEPI)

2-6-1 Nagasaka, Yokosuka, Kanagawa, 240-01 Japan

ABSTRACT

For high temperature dust elimination in IGCC power generation in Japan, operating at 693K and 2.55MPa, CRIEPI has selected a large ceramic tube filter. The filter is made of silicon carbide and consists of a surface film with fine pore and a coarse pore base tube. In a pressure vessel several elements are connected and pulled down as a module. A one-dimensional analysis of the filter tube flow and a numerical model of pulse jet cleaning led to identification of a suitable filter size and appropriate design of the cleaning system. The cleaning velocity through the filter wall varied somewhat along the filter axis, depending on volume flow rate of the cleaning gas, filter size, filter permeability, etc. The total pressure of the pulse jet needed to be 5MPa to acquire enough cleaning flow. This investigation evaluated the effect of these factors in the pulse jet system.

NOTATION LIST

A	: Cross section area of nozzle and filter	$[m^2]$
a	: Sonic velocity	$[m/s]$
c_p	: Heat capacity at constant pressure, per unit mass	$[J/(kg \cdot K)]$
c_v	: Heat capacity at constant volume, per unit mass	$[J/(kg \cdot K)]$
D	: Internal diameter of filter tube	$[m]$
e	: Total fluid energy	$[J/kg]$
f	: Friction factor at internal surface	$[-]$
L	: Filter length	$[m]$
M	: Rate of momentum entering diffuser throat	$[kg \cdot m/s^2]$
P	: Pressure	$[Pa]$
r_0, r_1	: Internal and external radius of filter	$[m]$
t	: time	$[s]$
T	: Temperature	$[K]$
u	: Axial velocity in internal filter tube	$[m/s]$
u_r	: Cleaning velocity at external surface	$[m/s]$
v_w	: Cleaning velocity at internal surface	$[m/s]$
w,W	: Mass flow rate	$[kg/s]$
x	: Distance from filter inlet	$[m]$
ΔPr	: Cleaning pressure	$[Pa]$
μ	: Entrainment ratio $= w_e/w_n$	$[-]$
κ	: c_p/c_v	$[-]$
κ_f	: Permeability of filter	$[m^2]$
η	: Gas viscosity	$[kg/(m \cdot s)]$
ρ	: Gas density	$[kg/m^3]$

Suffix

0	: Reservoir of pulse jet or nozzle inlet
e	: Entrained gas
n	: Nozzle exit
o	: Rude gas side
t	: Diffuser throat

INTRODUCTION

An advanced coal burning electricity generating system, known as the integrated coal gasification combined cycle (IGCC) power generation system, is being developed in Japan. It consists of an air–blown entrained bed gasifier, a hot gas clean–up operated at 420°C and 25atm, and a 1300°C gas turbine. CRIEPI has opted for a porous ceramic filter for dust removal in the hot gas clean–up. The filter has a fine–pore surface film and a coarse–pore silicon carbide base tube. Dust particles collect on the external surface of the filter, and the dust layer is removed by on–line pulse jet cleaning. One of the most important factors in developing this filtration system is the scaling–up of the facility. To obtain a compact filtration facility we jointed several tubular elements to produce a long filter which can be pulled down from tube sheet with the aid of a metal rig. This filter system will be tested in 1993, with the 20T/D fixed bed hot gas clean–up pilot plant, a project set up by the electric power industries of Japan. The pilot plant has been installed as 1/10 of the slip stream of the 200T/D IGCC pilot plant in the Nakoso power station. These large tubes are being used to investigate whether filtration and cleaning flow become uniform along the filter axis. The cleaning system is also important. To reduce the thermal shock during filter cleaning, the pulse jet should have the highest entrainment ratio while retaining sufficient cleaning velocity. Pulse pressure, diffusers (sometimes called venturi) and pulse nozzles are the main elements of the pulse jet system.

The numerical simulation is a one dimensional steady–state calculation based on the balance of mass, momentum, and energy. An axisymmetrical or three–dimensional simulation requires some unconfirmed models such as the k-ε turbulence model, distribution of permeability and heat transfer in porous media, all of which may cause inappropriate results. We investigated a simple model and confirmed the results with some experiments, and estimated the required operating condition in the hot gas filtration.

PULSE JET FLOW EXPERIMENT

In pulse jet cleaning the cleaning pressure can change suddenly. Figure 1 illustrates the test apparatus used to examine the change of cleaning pressure. Four elements (①②③④ of Table 1) were connected to make a long filter with a 7cm external diameter, 4cm internal diameter and 4m length. There is a diffuser at the inlet and the other end is closed with an impermeable cap. A straight tube is placed at the diffuser inlet to act as a pulse

Table 1

Experimental condition of pulse jet flow

Press. and Temp.	Pulse gas: 0.20–0.74MPa, 283K
	Entrained gas: 0.10MPa, 293K
Filter size	External diameter: 0.07m
	Internal diameter: 0.04m
	Length: 1, 4m (1m×4)
Permeability	① $2.91 \times 10^{-12} \text{m}^2$
	② $4.64 \times 10^{-12} \text{m}^2$
	③,④ $5.29 \times 10^{-12} \text{m}^2$
	⑤ $2.14 \times 10^{-11} \text{m}^2$
Friction factor	0.01
Pulse tube	Diameter: 0.006m
	Length: 0.5m
Diffuser	Throat(Dt): 0.01–0.04m
	Throat length: 4×Dt
	Friction factor: Blasius's Equation

nozzle. The nozzle jet entrains the surrounding gas, which flows into the diffuser, and the mixed gas acquires the pressure required for cleaning in the diverging cone. The diffuser cone is set at an angle 8° to both inlet and exit.

There is a suggestion that filter cleaning performance is controlled by the pressure difference in the dust cake (Ciliberti et al., 1986). This was not confirmed in our experiments, in which pressure loss recovery improved with cleaning velocity. Therefore we assumed filter permeability and set the pulse pressure to obtain a particular cleaning velocity.

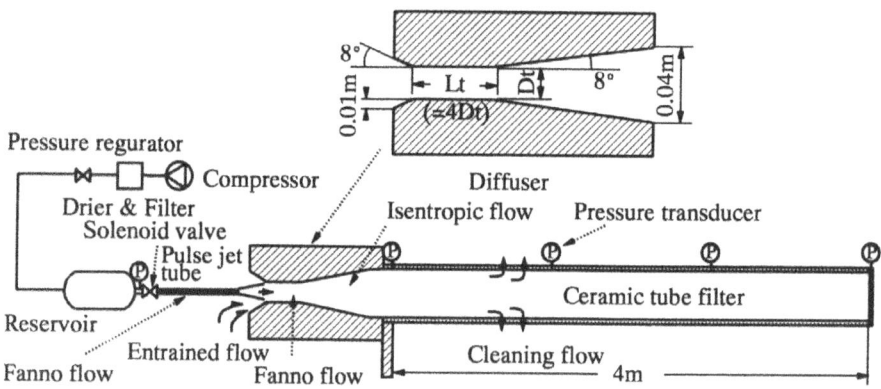

Figure 1. Test apparatus for pulse jet flow in porous tube filter

NUMERICAL CALCULATION OF JET AND ENTRAINED FLOW

The calculations relative to the pulse jet system include the nozzle jet, the entrained flow at the diffuser inlet, flow in the diffuser throat and diverging cone and the cleaning flow through the filter wall.

Nozzle Discharge Rate

The nozzle jet can be Fanno or adiabatic, depending on the nozzle type. If the nozzle is a straight pipe, the flow is Fanno. The exit pressure at the nozzle is then approximately the cleansed gas pressure.

Diffuser Inlet

The mass, momentum and energy balance are considered between nozzle exit and diffuser throat.

(Mass)

$$w_n + w_e = W \tag{1}$$

(Momentum)

$$w_n u_n + P_n A_n + P_e (A_t - A_n) = W u_t + P_t A_t \tag{2}$$
$$= M$$

(Energy)

$$w_n e_n + w_e e_e = We$$
$$e_n = c_p T_n + \frac{1}{2} u_n^2$$
$$e_e = c_p T_e \quad (\mu > 0), \quad e_n \quad (\mu < 0) \tag{3}$$
$$e = \frac{\kappa}{\kappa - 1} \frac{P_t}{\rho_t} + \frac{1}{2} u_t^2$$

Here W_n, P_n, P_e, T_e, M, e_n, e_e are known. Assuming the entrainment ratio, the throat inlet velocity is given as Eq.(4). Pressure and gas density at the throat inlet are calculated using Eq.(2) and (3).

$$u_t = \frac{\kappa}{\kappa + 1} \frac{M}{(1 + \mu) w_n} [1 - \sqrt{1 - 2 \frac{\kappa^2 - 1}{\kappa^2} \frac{w_n^2}{M^2} (1 + \mu)(e_n + \mu e_e)}] \tag{4}$$

Diffuser Throat and Diverging Cone

The flow in the diffuser throat can be regarded as Fanno, and as adiabatic in the diverging cone.

Filter Tube

The one–dimensional equations of continuity, motion, and energy in the filter tube are given as Eqs.(5)–(7). Here the friction factor has a significant effect. If the internal surface of the filter tube is smooth, the friction factor has to be corrected with something relate the film theory (Mizushina et al., 1971). In this case, however, the internal surface is so rough that the friction factor is set at the value of zero flow through the filter wall.

(*Continuity*)

$$\frac{\partial \rho}{\partial t} + \frac{\partial \rho u}{\partial x} = -\frac{4}{D} \rho v_w \tag{5}$$

(*Motion*)

$$\frac{\partial \rho u}{\partial t} + \frac{\partial}{\partial x}(\rho u^2 + P) = -\frac{4}{D} f \frac{1}{2} \rho u^2 \tag{6}$$

(*Energy*)

$$\frac{\partial}{\partial t}(\rho c_v T + \frac{1}{2}\rho u^2) + \frac{\partial}{\partial x}(\rho c_v T + \frac{1}{2}\rho u^2 + P)u = -\frac{4}{D}(\rho c_v T + \frac{1}{2}\rho v_w^2 + P)v_w \tag{7}$$

In a steady–state condition these equations are reduced to the following.

(*Density*)

$$\frac{d\rho}{dx} = -\frac{1}{u}(\rho \frac{du}{dx} + \frac{4}{D} \rho v_w) \tag{8}$$

(*Pressure*)

$$\frac{dP}{dx} = -\rho u \frac{du}{dx} + \frac{4}{D}(\rho u v_w - \frac{f}{2}\rho u^2) \tag{9}$$

272

(Axial velocity)

$$\frac{du}{dx} = -\frac{4}{D}\frac{1}{a^2-u^2}[(a^2+\frac{\kappa+1}{2}u^2+\frac{\kappa-1}{2}v_w^2)v_w - \frac{f}{2}\kappa u^3] \qquad (10)$$

The flow in porous media is calculated using Darcy's law. If the flow is isothermal, the relationship between filtration velocity and pressure is given as Eq.(11). However the flow through the filter wall is neither isothermal, adiabatic, nor incompressible. Moreover permeability is not constant through the filter wall. It can change suddenly near the collection surface. Therefore we use Eq.(11) as the first approximation.

(Darcy's Law)

$$\vec{v} = -\frac{\kappa_f}{\eta}\nabla P$$

$$v_w = \frac{\kappa_f}{2\eta r_0 \ln(r_1/r_0)}\frac{P^2-P_o^2}{P} \quad , \quad u_r = \frac{\kappa_f}{2\eta r_1 \ln(r_1/rr_0)}\frac{P^2-P_o^2}{P_o} \qquad (11)$$

Since the throat inlet conditions are known, we can calculate the condition in the throat, in the diverging cone, and at the filter inlet, the boundary conditions of Eqs.(8)–(11). Integrating those equations from the filter inlet enables us to obtain the axial velocity at the filter end. The entrainment ratio is then changed until $u|_{x=L}$ drops to zero.

DISCUSSION

Transition of cleaning pressure

The transition of the cleaning pressure is shown in Figure 2 and Figure 3 in which the pulse duration time is 0.2s. The cleaning pressure in the filter approaches steady state within 50ms in this type of long filter, which means that the balance of mass, momentum, and energy can be calculated as steady state. Figure 2 also shows the distribution of the cleaning pressure without the diffuser, in which there is no cleaning pressure near the cleaning gas inlet. It is necessary to fit the mixing region at the filter inlet with a length of impermeable material or a diffuser. Figure 3 shows the cleaning pressure calculated, which agrees fairly well with the experimental result.

The distance between the pulse nozzle and the diffuser inlet has little effect on the

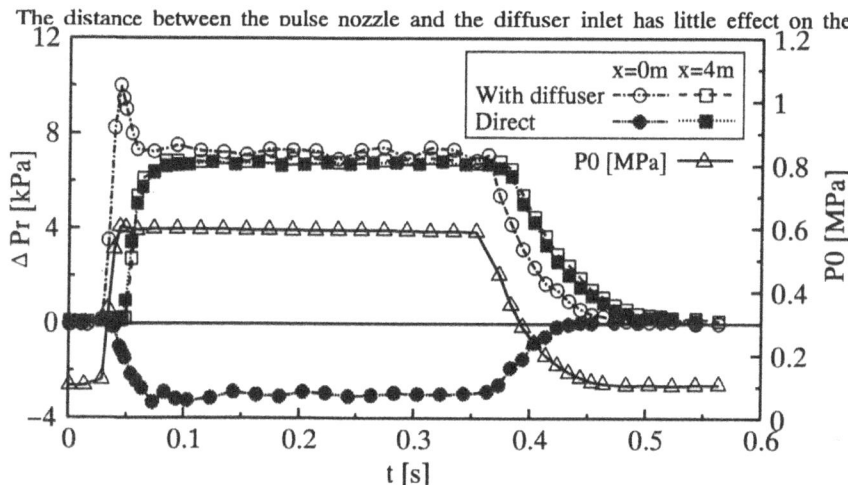

Figure 2. Transition of cleaning pressure in a porous tube filter with and without diffuser

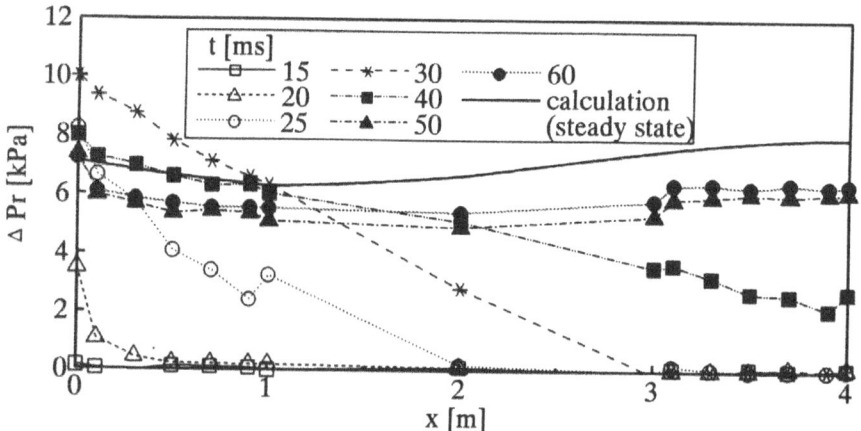

Figure 3. Propagation of cleaning pressure in a porous tube filter

cleaning pressure provided the distance is short. A distance of 2cm showed the highest cleaning pressure in these tests.

Flow in filter tube

Having adequate boundary conditions of Eqs.(8)–(11), this simulation shows good agreement with experiments, as shown in Figure 4. With external filtration, the filtration velocity increases slightly near the cleansed gas exit, but there are only small variations as long as filtration velocity is low. The variations become more evident during cleaning because the flow rate is much higher

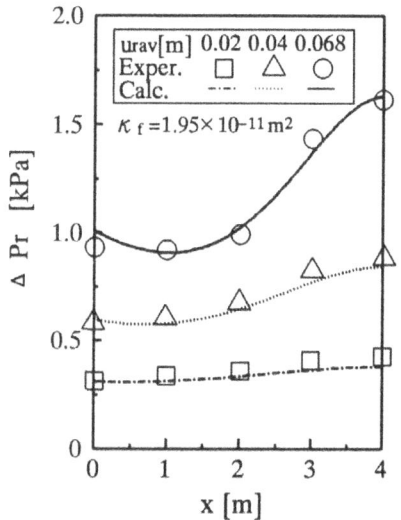

Figure 4. Distribution of cleaning
pressure

than during filtration. The cleaning pressure becomes higher at the filter end but lower near the cleaning gas inlet, since the axial velocity decreases and the momentum is recovered as static pressure. Such variation is more evident in long and small diameter filters, high permeability filters, and at high cleaning pressure. If suitable filter sizes are selected, such variations become small and can be ignored.

Diffuser effect and entrainment flow

To evaluate the calculation we compared the result with experiments. Figure 5 shows the effect of the diffuser, which means that there is an effective throat size for the highest cleaning pressure. Figure 6 shows the effect of permeability. The high permeability filter shows lower cleaning pressure, but the cleaning velocity rises, and it would therefore be better to evaluate cleaning performance with the cleaning velocity. As shown in these figures, this simulation gives good agreement with experiments. The cleaning system was evaluated at hot gas clean-up values of 693K and 2.55MPa with this method.

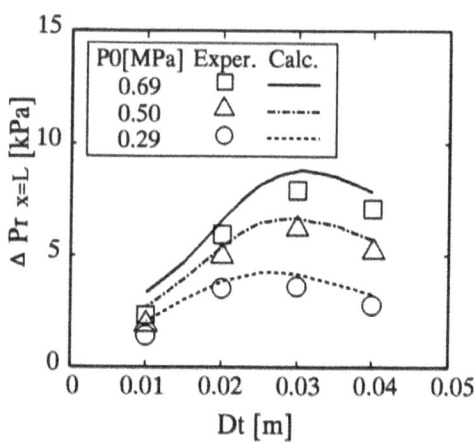

Figure 5. Effect of diffuser on cleaning pressure

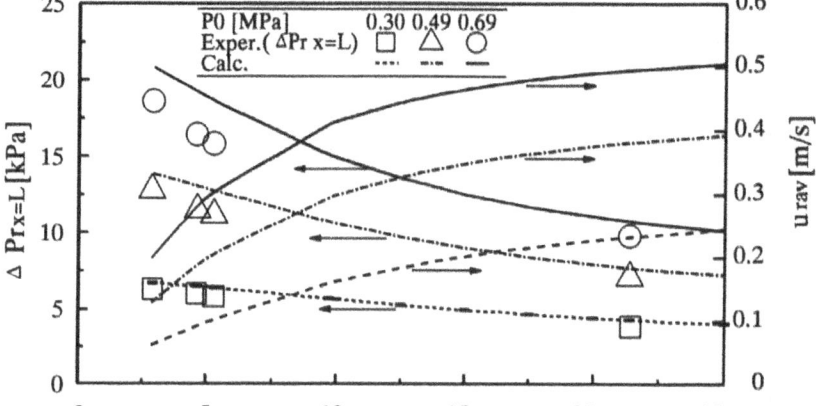

Figure 6. Effect of filter permeability on cleaning pressure

Hot gas clean-up condition

Filter characteristics and simulation values of the hot gas filtration are listed in

276

Table 2. Three large elements are connected with a metal rig to make a large tube module. The pulse tube is placed at the diffuser inlet of each module.

Figure 7 and 8 show the effect of pulse pressure on cleaning velocity and entrainment ratio. Since the cleaning velocity has to be higher than 10cm/s in a 2T/D hot coal gas filtration test (Ito

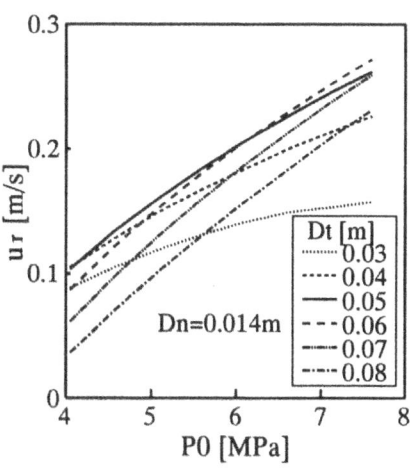

Figure 7. Effect of pulse pressure
on cleaning velocity

et al., 1990), the pulse pressure should be as high as 5MPa. The entrained flow becomes twice that of the jet with the 5cm diffuser and the cleaning gas temperature is higher than 600K, as shown in Figure 9. The temperature change is less than 100K, and therefore thermal shock will be negligible.

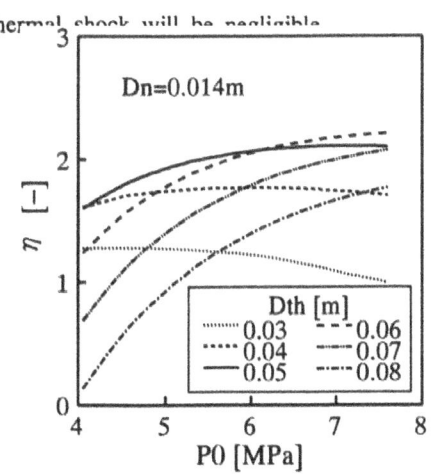

Figure 8. Effect of pulse pressure
on entrainment ratio

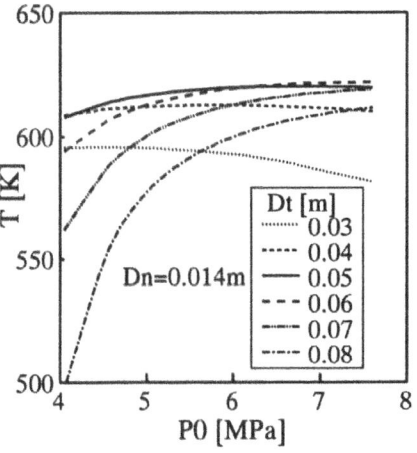

Figure 9. Effect of pulse pressure on
cleaning gas temperature

Table 2

Simulation condition of hot gas filtration

Filter	Internal diameter 8cm, External diameter 11cm
	Length 4.5m(1.5m×3), Effective length 4m
	Permeability: $1.4 \times 10^{-12} m^2$
	Friction factor: 0.01
Filtration	Temp.: 693K, Press.: 2.55MPa
	Pressure loss before cleaning: 29.4kPa
Cleaning	
Pulse gas	Total Temp.: 473K, Press: Variable
Pulse tube	Diameter: Variable, Length: 2m (Fanno flow)

Figure 10 and 11 show the effect of the nozzle size. A large nozzle raises the cleaning velocity but decrease the entrainment ratio. The distribution of the cleaning velocity is shown in Figure 12. With this filter size the filtration velocity and the cleaning velocity show only small variations.

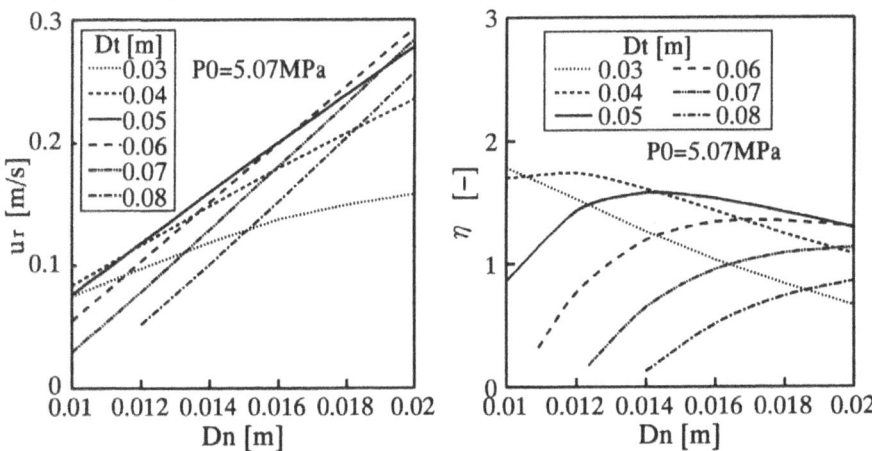

Figure 10. Effect of nozzle diameter on cleaning velocity

Figure 11. Effect of nozzle diameter on entrainment ratio

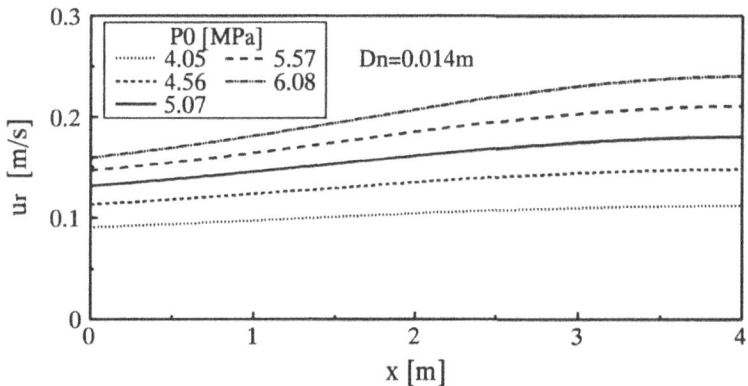

Figure 12. Cleaning velocity along filter axis

CONCLUSION

Since the pulse jet flow reached steady state within 50ms, we can regard the cleaning flow as steady state. One-dimensional analysis of the cleaning flow showed good agreement with experiments and yielded the following results.

The cleaning velocity of the filter shows some variations along the filter axis, but the pilot plant filter showed almost uniform cleaning velocity. The diffuser with the 5cm throat is effective in improving cleaning for the pilot filter. The pulse pressure needs to be as high as 5MPa to give an effective cleaning performance.

The 20T/D pilot plant test will confirm these results in 1993.

REFERENCES

Ciliberti, D.F. & Lippert, T.E. (1986), IChemE SYMPOSIUM SERIES No.99, pp. 193
Mizushina, T., Takeshita, S. & Unno, G. (1971), J. Chem. Eng. Japan, 4, pp. 135
Ito, S., Tanaka, T., & Kawamura, S. (1990), 5th World Filtration Congress, 2, pp. 524

INVESTIGATION INTO THE CLEANING OF FIBRE CERAMIC FILTER ELEMENTS IN A HIGH PRESSURE HOT GAS DEDUSTING PILOT PLANT

RAINER SKROCH, GERNOT MAYER-SCHWINNING
Lurgi AG, Research and Development Division
W-6000 Frankfurt/Main, Germany

UWE MORGENSTERN, EKKEHARD WEBER
Institute of Environmental Engineering,
University of Essen, W-4300 Essen, Germany

ABSTRACT

In a pilot plant for dust collection from hot flue gas at pressures of up to 15 bar, systematic tests were carried out on rigid ceramic fibre filter elements to study the effectiveness of pulse jet cleaning. After a brief description of the pilot plant and the filter medium employed, the relationships between pulse pressures during cleaning, maximum differential pressures during reverse flow, reverse flow duration and cleaning gas rate will be outlined.

INTRODUCTION

Since the demand for electrical energy has been increasing world- wide and coal has been constantly gaining in significance for power generation, it has become a major objective to optimize power generation efficiencies. This can be achieved by combined cycle power plants, where, apart from the traditional steam turbine process, coal is burnt or gasified at high temperatures and the resulting flue or fuel gas fed into a gas turbine to generate additional power. Before entering the gas turbine, however, the gas needs to be cleaned, i.e. any dust or other impurities have to be removed. Most turbine manufacturers restrict the maximum allowable dust load at the turbine inlet to 5 mg/m^3.

Dust collection systems in such applications typically have to work at 350 to 1000 °C and 10 to 25 bar, depending on the type of process applied upstream. Therefore, only high efficiency dust collectors like electrostatic precipitators or filtration separators can be employed for this purpose.

Possible applications for high temperature dedusting can also be found in metallurgical processes, beneficiation of raw materials and glass production. Here, however, dust collection takes place at atmospheric pressure.

The tests for this study were carried out in a pilot plant designed for dust collection downstream a combined cycle power plant based on pressurized coal combustion /1/.

FILTER MATERIAL

The filter medium examined is a commercially available, low density, ceramic fibre filter material/2, 3, 4, 5/. This type of filter element comes in modular form and is fabricated from aluminium silicate ceramic fibres which are applied to a mould by a filtering method.

The filter element, cf. figure 1, consists of a downward tapering upper part, a cylindrical middle part and a semispherical bottom part. The essential dimensions are given in the table. The filter elements used in the tests were of type KE 85/150, with a diameter of 150 mm, a material thickness of 20 mm and a length of 1,5 m.

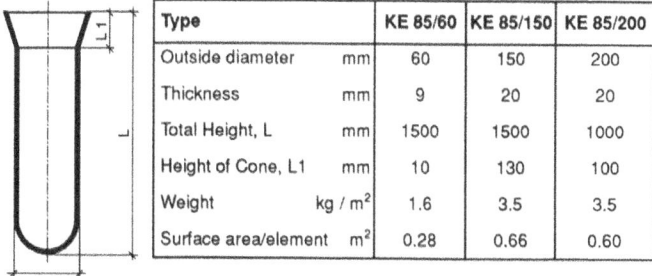

Type		KE 85/60	KE 85/150	KE 85/200
Outside diameter	mm	60	150	200
Thickness	mm	9	20	20
Total Height, L	mm	1500	1500	1000
Height of Cone, L1	mm	10	130	100
Weight	kg / m²	1.6	3.5	3.5
Surface area/element	m²	0.28	0.66	0.60

Figure 1. Form and technical data of fibre ceramic filter material type KE 85.

Ceramic fibre filter elements can withstand temperatures up to 900 °C at a mimimum and do not require any support structure.

DESCRIPTION OF THE PILOT PLANT

The essential parts of the pilot plant are a pressurized combustion chamber and the downstream dust collection system as shown in the flow sheet (fig. 2). The flue gas produced in the combustion chamber at 1300 to 1600 °C is dedusted either by two filter separators working in parallel or by a single electrostatic precipitator. To support dust collection, a partial flow can be drawn off the filter bottom and cleaned in an additional filter. The results of such an arrangement have been reported in another publication /6/.

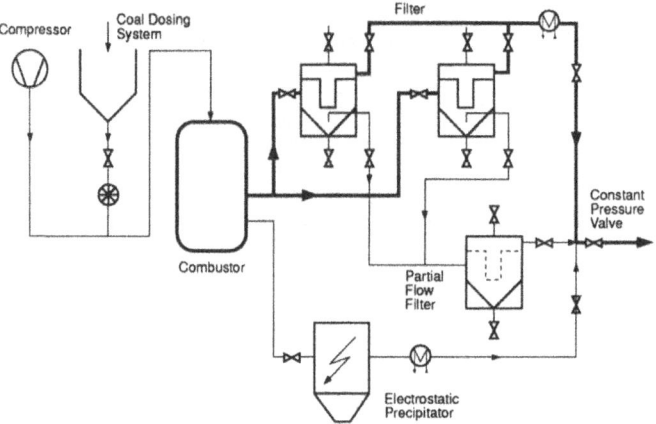

Figure 2. Flow sheet of pilot plant for HPHT-Dedusting
The characteristic data of the plant are summarized in figure 3.

Coal feed	100	kg/h
Thermal power	0,8	MW
Flue gas flow rate	1200	m³/h (NTP)
Pressure	20	bar
Flue gas temperature	700-1000	°C
Filtration area	4	m²
Filtration velocity	2-6	cm/s

Figure 3. Data of pilot plant.

The two parallel-connected filter separators are pressure
vessels with internal insulation and house 3 filter elements each. The cleaning valves are arranged on lances
whose orifices are directly above the filter elements
(Fig. 4).

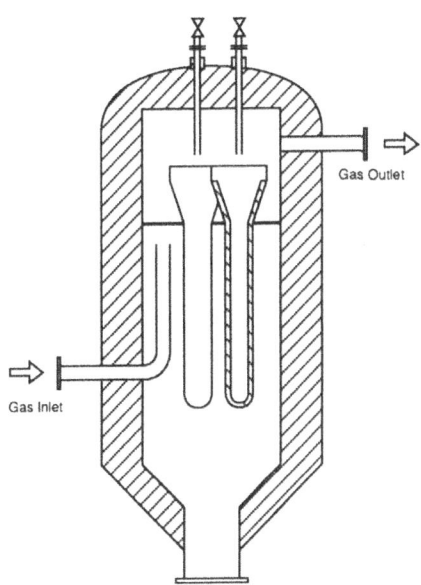

Figure 4. HPHT-filter with ceramic fibre elements

Figure 5 shows how the filter elements are fastened to
the cell plate. The downward tapered top of the element
is simply held by an equally shaped counterpart in the
cell plate opening.

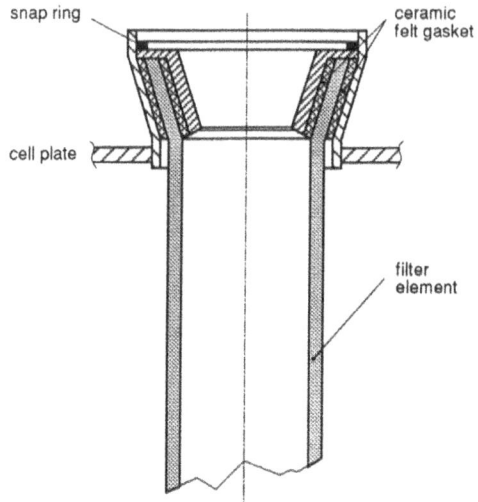

snap ring

ceramic
felt gasket

cell plate

filter
element

Figure 5. Fixing of filter element in cell plate

PULSE-JET CLEANING

There is a fundamental difference between bag filter and
rigid filter cleaning. Pulse-jet cleaning of bag filters
is accomplished by sending a temporary reverse flow of
the clean gas across the filter bag, the cleaning process
being promoted by a certain inertia effect resulting from
the sudden flexing of the bag. This inertia effect cannot
occur in the case of rigid filter elements. However, depending on the specific application, bag flexing may also
have some disadvantages, because, residual dust deposits
are further compressed and dust particles penetrate the
filter medium, when the filter bag collapses.

Another point to be taken into account in the context of
hot gas filtration is that the cleaning process may be
seriously hampered by the poorer agglomeration and settling characteristics of the separated dust. This can be
seen in figure 6 that shows the pressure drop curves
across a single-element filter at 300 °C and 500 °C. The
raw gas dust load was 9 g/m^3 in both tests. At the lower
temperature, a residual pressure drop of 10 mbar and a
cycle time of 40 minutes were determined. During these
40 minutes the pressure drop rose to 25 mbar. At 500 °C
and a cycle time of 15 minutes, an increase in the pressure drop from 25 to 40 mbar was observed, i.e. the
values were notably higher.

284

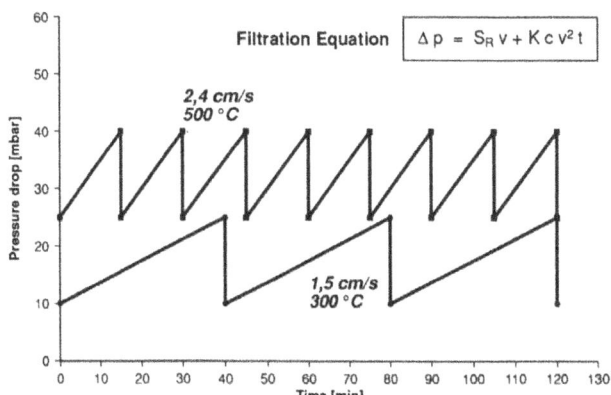

Figure 6. Typical pressure drop curves of filter elements

While the dust load, as mentioned above, was the same for
both tests, the filtration velocity was different. Therefore, the parameter values of the filtration equation /7/
represented in figure 6 were determined. They also show a
clear upward trend with increasing temperatures.

The residual resistance of the filter medium depends to a
high degree on the condition of the filter medium itself.
The residual resistance of a new filter element, for
instance, is about 17.000 Pa/m/s, whereas that of a used
filter element exhibiting a residual deposit of 1 mm to
1.5 mm thickness is approximately 77.000 Pa/m/s (fig. 7).

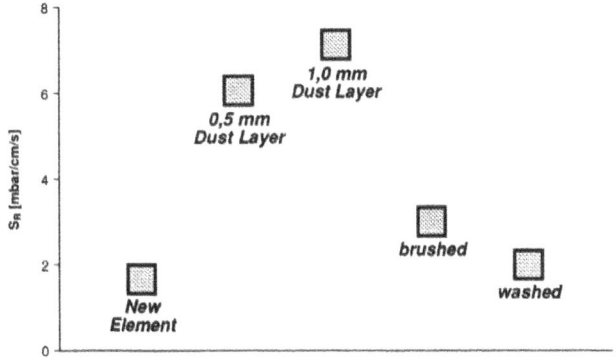

Figure 7. Residual resistance S_R of filter element.

The latter value corresponds to the steady state condition with fly ash after 500 hours of operation. Some specimen filter elements were cleaned and washed manually, but even then it was impossible to reach the original residual resistance value.

PULSE PRESSURE

During filtration, a descending pressure profile across the filter cake, the residual deposit and the filter medium can be observed. A qualitative representation of this profile is provided in figure 8. During cleaning, this profile is reversed due to the temporary pressure rise inside the filter element. It is obvious that ΔP_{eff}, the essential parameter for the cleaning process, is a function of this inside pressure.

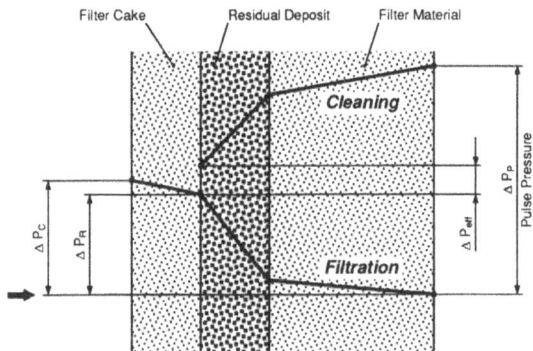

Figure 8. Pressure drop across filter material and filter cake.

In order to measure the pressure inside the filter elements during cleaning, the measuring equipment shown in figure 9 was added to the pilot plant. Lances were installed in two positions along height of the filter element and connected to a pressure transmitter. This arrangement combined with a transient recorder allowed us both to determine the pulse pressure, being the amplitude of the absolute internal pressure, and to record the oscillation of the differential pressure during cleaning.

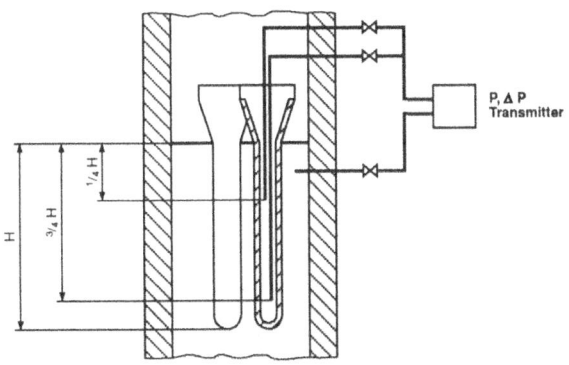

Figure 9. Measuring arrangement for pulse pressure and
differential pressure during cleaning.

Figure 10 represents the pressure inside a filter element
as a function of the reservoir to system pressure ratio.
It can be noted that there is not a big difference between
the values of the upper and the lower measuring point.
The pressure inside the element rises from 20 mbar at a
pressure ratio of 3 to about 90 mbar at a pressure ratio
of 8. This test is to be considered an orientative test,
since it was carried out at atmospheric conditions, using
cold gas and zero filtration velocity.

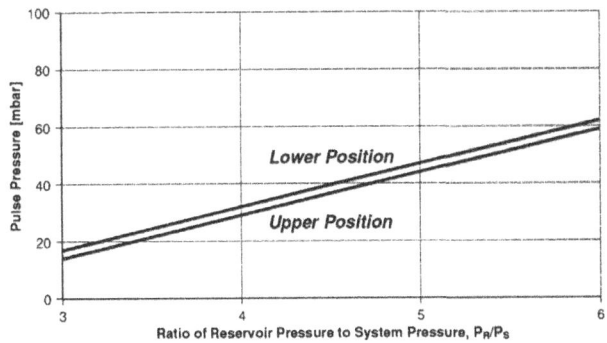

Figure 10. Pulse pressure inside filter element vs. ratio
of reservoir pressure to system pressure.

In the following figure 11, the pulse pressure is shown as a function of the filter temperature at a system pressure of 5 bar and a pressure ratio of 2.6. The pulse pressure clearly rises with increasing temperatures, because the actual cleaning gas volume released due to the jet effect increases with rising temperatures.

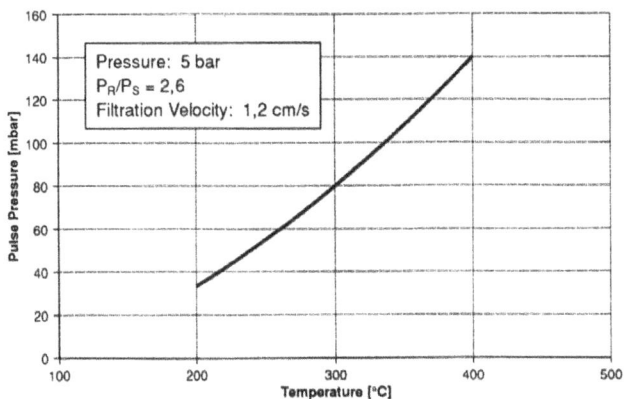

Figure 11. Pulse pressure vs. temperature.

DIFFERENTIAL PRESSURE DURING CLEANING

Depending on the opening time of the valve and on the pressure ratio, the pressure oscillation across the filter element varies in degree. A qualitative representation of the possible pressure oscillation profile is shown in figure 12.

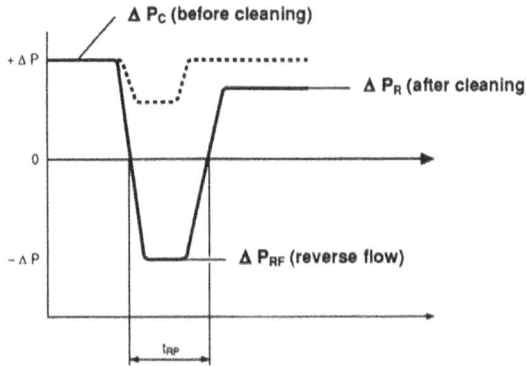

Figure 12. Pressure drop during pulse jet cleaning.

The decisive criterion is a sufficiently long reverse flow caused by a sufficiently big reverse flow pressure drop. The differential pressure amplitudes observed at the filter element during cleaning are plotted in Figure 13 as a function of the pressure ratio at various temperatures. Above 0, there cannot be any cleaning effect whatsoever, but the farther the differential pressure moves into the negative range the better is the cleaning effectiveness. Again, it is obvious that cleaning effectiveness improves with rising temperatures and increasing pressure ratio.

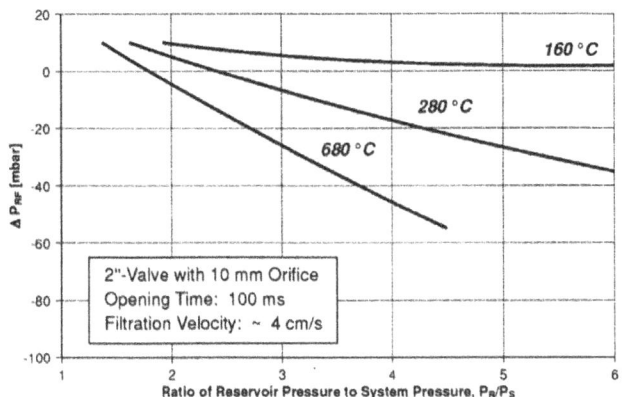

Figure 13. Reverse flow pressure drop vs. reservoir to system pressure ratio

In the tests underlying figure 14, two different filtration velocities were applied at a constant pressure ratio of 4. The graph shows the differential pressure across the filter element as a function of temperature. This is another proof of the strong inter-relationship between temperature and cleaning effectiveness.

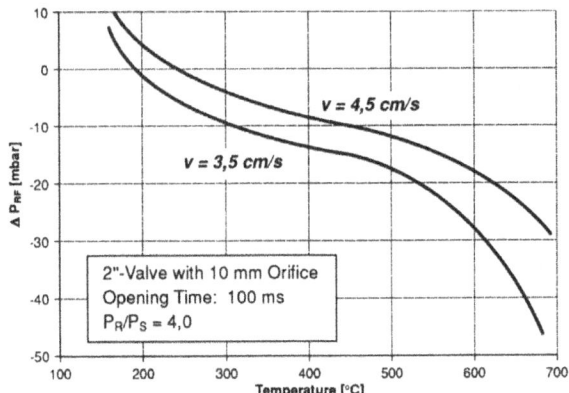

Figure 14. Reverse flow pressure drop vs. temperature.

REVERSE FLOW DURATION

In addition to a sufficiently negative differential pressure the cleaning effectiveness is governed by the duration of the reverse flow. This parameter is represented in Figure 15 as a function of the pressure ratio at various temperatures.

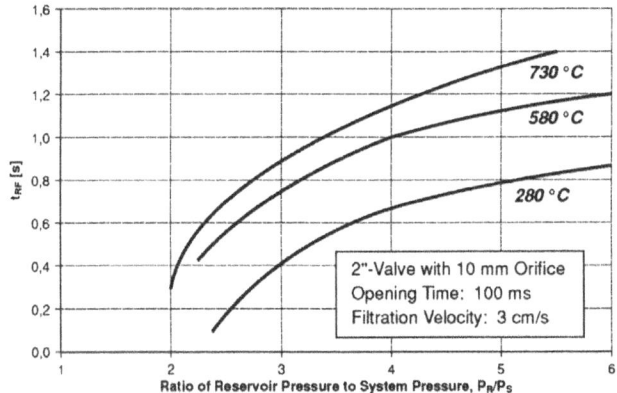

Figure 15. Duration of reverse flow as a function of pressure ratio.

At a pressure ratio of 3, for example, and a temperature of 280 °C the reverse flow time is 400 ms. At 730 °C with all other conditions remaining unchanged, the aforementioned increase in the actual cleaning gas rate resulting from temperature rises leads to a reverse flow time of 900 ms. Reverse flow duration as a function of the pressure inside the filter element is depicted in Figure 16. Here it should be noted that even at a pressure amplitude of only 20 mbar inside the filter element reverse flow times between 800 and 1100 ms were observed, depending on the temperature.

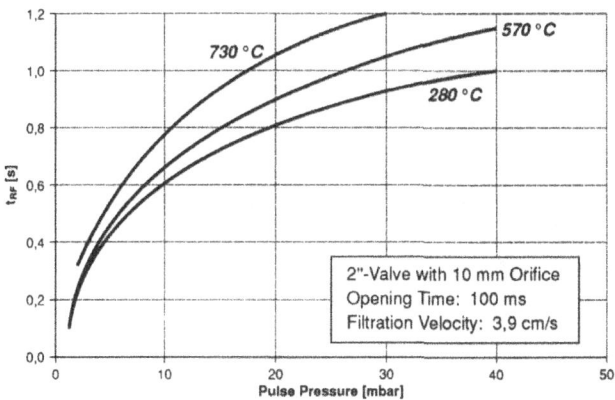

Figure 16. Duration of reverse flow vs. pulse pressure.

CLEANING GAS RATE

The gas rate required for the cleaning process is of vital importance for the operation of pulse-jet filters. The pilot tests carried out with 3/4" and 2" cleaning gas valves showed the correlations represented in Figure 17. At opening times of 100 and 200 ms, the 3/4" valve injected between 10 and 40 l of cleaning gas (NTP), a rate range known from bag filters operating at atmospheric pressure. The rates injected by the 2" valve were considerably higher, i.e. between 50 and 120 l.

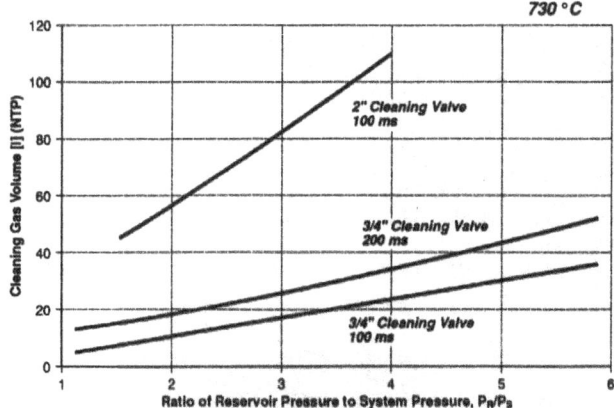

Figure 17. Quantity of cleaning gas for one pulse as a function of pressure ratio.

The following considerations demonstrate that the observed consumption data are realistic. Tests on bag filters operating at atmospheric pressure have shown that for successful bag cleaning a gas volume of abt. 10 l is required per bag and cleaning cycle. When translating this value to high pressure high temperature filters, the correlation shown in Figure 18 becomes evident. With rising temperature and constant pressure, cleaning gas consumption actually drops. When increasing the temperature, e.g., from 200 °C to 800 °C, consumption is only half as high. However, one must not forget that consumption increases as a linear function of system pressure rise. Therefore, high system pressures result in rising cleaning gas consumptions, which may be well above 100 l (NTP) as it was measured for the 2" valve of the pilot plant.

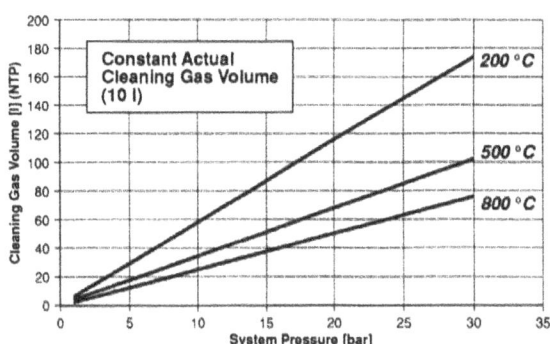

Figure 18. Cleaning gas volume for one pulse jet as a function of temperature and pressure.

292

CONCLUSION

The tests carried out in the pilot-scale high pressure high temperature dedusting system produced a number of quantitative correlations for various parameters, which describe the cleaning process. In addition, during the tests which were carried out over a two year-period, the soft ceramic filter elements in use were found to be very resilient and furnished us with design data for commercial high pressure high temperature filters.

At present, additional tests are to be carried out in order to determine, on the basis of the findings so far, the impact of such parameters as pulse pressure, reverse flow duration and reverse flow pressure drop amplitude on the cleaning effectiveness.

The authors wish to express their gratitude to the German Minister of Research and Technology who sponsored the research project upon which this paper is based.

REFERENCES

/1/ Weber, E.; Schulz, R.; Bender, J.: Bau und Versuchs-betrieb einer Pilotanlage für die Hochtemperaturgas-reinigung mit Filtrationsabscheidern. BMFT-Bericht 03E-1259-A, Universität Essen, 1986

/2/ Schmidt, D.; Schulz, R.; Weber, E.: Long Term Tests with Ceramic Dust Separators at Gas Temperatures between 800 °C and 1000 °C. IChemE Symp. Ser. 99, Gas Cleaning at High Temperatures, Surrey, Sept. 1986, S. 245 - 262

/3/ Product Information of BWF, Offingen, Germany

/4/ Schulz, R.: Untersuchungen zur Staubabscheidung aus Gasen mit Filtrationsabscheidern bei hohen Tempera-turen und hohen Drücken. Fortschritt-Berichte VDI Reihe 3, Nr. 156, Düsseldorf 1988

/5/ Weber, C.: Advances in Hot Gas Filtration Technology. Filtr. & Sep 25 (1988), 3/4, 100-103

/6/ Mayer-Schwinning, G.; Skroch, R.; Weber, E.; Schulz, R.: Progress in Hot Gas Dedusting by Fiber Ceramics. Proceedings of PARTEC, 2. European Symposium "Separa-tion of Particles from Gases", Nürnberg 1992

/7/ Leith, D.; Ellenbecker, J.: Theory for Pressure Drop in a Pulse-Jet cleaned Fabric Filter. Atmospheric Environment 14 (1989), 845/852

THE WORLD MARKET FOR HOT-GAS MEDIA FILTRATION:
CURRENT STATUS AND STATE-OF-THE-ART

Lutz Bergmann
FILTER MEDIA CONSULTING, INC.
P.O. Box 2189, LaGrange, Georgia 30241, U.S.A.

ABSTRACT

Hot-gas media filtration for temperatures above 250° and
mainly up to 500° C is an evolving technology becoming more
and more popular in different industry segments. This paper
discusses the different filter materials and major
applications. The status of most current systems worldwide
is discussed as well as the state-of-the-art technology.
Worldwide hot-gas systems in 1992 amount to approximately
$70 - 75 million. By the year 1996 - 1997 that market should
have grown to approximately $170 - 180 million. Filter
media may represent 20% - 30% of this total , for flange-to-
flange systems.

INTRODUCTION

This author reported about the subject of media hot-gas
filtration as early as November, 1982 [1]. Hot-gas systems
ten years ago were almost unknown with the exception of a
few specialized applications. During those last ten years,
however, a number of very challenging developments have lead
to a much wider recognition of a new potential filtration
area basically dealing with temperatures above 250°C, in
general between 850°C - 1000°C.

Depending on the application, different filter materials
are currently being used to accommodate high temperatures,
sometimes pressure, and most importantly performance
characteristics like acceptable differential pressure and
efficiency.

The following is a summary of the current status of
media used in hot-gas filtration worldwide.

294

FILTER MEDIA - SUMMARY

Sintered Stainless Steel Semi-Rigid Filter Elements (BEKAERT, MEMTEC, Others)

Sintered Stainless Steel Pleated Filter Cartridge (Numerous)

Nested Fiber Filter (NFF) - (BATTELLE)

Sintered Porous Powder Metal (PALL, MOTT, NEWMET, KREBSÖGE, FUJI, Others)

Ceramic Woven Nextel™ 312 Fabric (3M COMPANY)

Ceramic Fiber Based Pulp-Type Cartridge KE-85 (BWF)

Sintered Plastics/Oxides Ceramics - SINTHERAMIC (HERDING, DCE, Others)

Rigid Ceramic Candle Based On Oxide and Non-Oxide Materials (SCHUMACHER, CEREL, UNIVERSAL POROSICS, Others)

Ceramic Cross-Flow Filter (COORS)

New Ceramic Cordierite Monolith (CERAMEM CORP.)

Sintered Stainless Steel Semi-Rigid Filter Elements

Such materials are based on stainless steel fiber metal vlies which is sintered and temperature resistant (depending on the alloy used) up to 550°C - 600°C. Randomly orientated fiber metal medium is normally a stainless steel fiber matrix - created by a random air layer - of selected diameters and length of a specified depth. The material is then compacted for mechanical strength and specific filter characteristics. Fibers bonded together into a much stronger medium by calendering and sintering, are generally selected for specific applications and high efficiency requirements.

MEMTEC has developed Fibermet[R] fiber metal by forming a non-woven structure of very fine metal fibers. A web is formed of loose random metal fibers and then sintered by varying the fiber diameters and the method of web construction. Fibermet[R] filter media can have particle removal ratings as fine as 0.3 micron in gas applications. There are numerous systems in operation worldwide.

Sintered Stainless Steel Media Pleated Filter Cartridge

These filter configurations, normally in pleated cartridge form, were originally developed for polymer melt applications in liquid filtration. Very similar configurations, however, are being currently used for so-called "blow back" systems in catalysts and precious metal recovery systems. This is one of the largest hot-gas applications today. Many different companies worldwide offer such systems; among them are: PALL, MEMTEC, MOTT, FAIRY, FUJI and many others.

Nested Fiber Filter (NFF) - BATTELLE

This concept based on a nest of needle-like fibers is made from an alloy suited for a given application. The filter is a structure in which fibers are formed in a dendritic chain-like structure and used for hot-gas conditions. This material has not yet been tested commercially.

Sintered Porous Powder Metal

One of the oldest metal media configurations is sintered powder metal filter elements being used predominantly in catalyst and precious metal recovery systems. Powder metals can be produced from almost all materials. Shape, size and distribution of the powder particles are important parameters which effect the properties of high porosity sintered materials. Sintered metal powder is manufactured from titanium, aluminum and many other alloys. Stainless steel and bronze are the most commonly used materials for these elements. Sintered powder metal elements are predominantly produced by gravity sintering techniques. Metal powders requiring higher sintering temperatures are normally formed by compacting in a die and the parts are sintered individually.

Ceramic Woven Nextel™312 Fabric

The 3M COMPANY has produced such materials for at least 10 years and manufactured into filter bags temperature resistant up to 1400°F (760°C). The fiber composition consists of alumina, boria and silica. The weight of the fabric is approximately 13 oz./sq.yd. with an air permeability of approximately 16 cfm. Ceramic Nextel™ filter bags are currently operating in a number of hot-gas installations.

Ceramic Fiber Based Pulp-Type Cartridge KE-85

Developed at the University of Essen and commercially produced and sold by BWF in Germany, an impressive number of installations employ this relatively new ceramic fiber based material. The filter bag configuration does not require a cage or support structure, and is temperature resistant, according to the company, up to 850°C. Most installations today, however, are being operated up to 500°C. Applications include separation of calcium carbide, aluminum and zinc dust, atmospheric and pressurized fluidized bed combustion and waste incineration.

The emission rates indicate a superb efficiency of such materials often better than .004 gr./ft.^{2} at the outlet. Since these filter elements can be operated at air-to-cloth ratios, of up to 5:1, total systems can operate very economically.

Sintered Plastic/Oxides Ceramic - SINTHERAMIC

The inventor of sintered plastic filter elements, HERDING in Germany has developed a hot-gas media with temperatures up to 500°C. Applied to a large pore sintered core made of ceramic oxide materials with a thickness of about 4mm and mean pore size diameter of 60 micron, a thin coating of 15 - 40 micron is applied resulting in a mean pore size diameter of 1.5 micron. Since the core and the coating

consist of the same material and they are bonded by
sintering, a homogeneous filter material is obtained with an
optimum performance characteristic. The filter material is
composed of aluminum oxide Al_2CO_3.

Rigid Ceramic Candle Based on Oxide and Non-Oxide Materials

There are numerous manufacturers of such filter elements
and they have been known for decades. Typical companies in
this field are: CEREL, LTD.

> DIDIER
>
> SCHUMACHER
>
> UNIVERSAL POROSICS, INC.
>
> Others

Since these materials are the subject of numerous
conference presentations, they are listed here only for
completion.

Ceramic Cross-Flow Filter - COORS

The cross-flow filter is unique and the cross-flow
geometry is formed from alternate layers of two types of
porous ceramic tiles. The two component tiles consist of
flat porous ceramic sheets with channels separated by ribs.
The channels are oriented at 90° to each other in successive
layers so that thin porous "floors" of each channel become
the actual filter surface. The tiles can either be machined
into pressed unfired plates or extruded through dies. Many
alternating layers of plates are stacked onto each other and
this assembly is then fired in a kiln, resulting in a
unitized monolithic ceramic structure. This filter concept
has been tested extensively over the last 10 years and has
been used at temperatures up to 1500°F (860°C). The material
is manufactured by COORS and marketed by WESTINGHOUSE
ELECTRIC CORP.

New Ceramic Cordierite Monolith - CERAMEM

This relatively new unique proprietary ceramic gas filter is currently being tested in a number of applications, but has not been commercially used. This filter is based on the use of the porous cordierite monolith of the type widely used in automotive catalytic converters. The monolith contains a large number of parallel passageways, which extend from one end face to an opposing end face. The passageways are plugged at the monolith end face to provide a "dead-end" flow configuration. Passageways open at the inlet end face are plugged at the outlet end face, and vice versa.

A thin - approximately 50 micron - ceramic microfiltration membrane coating is applied to the passageway's walls to obtain a composite filter structure. The pore size of the membrane coating (0.2 - 0.5 micron) is about 100-fold finer than that of the porous monolith material.

Applications

General Comments

Since during this conference, many of the following applications are discussed in much detail, only a brief summary of the most current applications for hot-gas media systems follows:

> Catalyst/Precious Material Recovery
> Waste Incineration
> Pressurized Fluidized Bed Combustion
> Gasification
> Calcination
> Catalytic Cracking/Refining Industry
> Chemical Process Industry
> Other Applications

Catalyst/Precious Material Recovery

This application is not necessarily recognized as hot-gas media filtration, but is one of the largest current

299

market segments and one of the best established and experienced hot-gas filter systems. Typical applications include: precious catalysts, nickel, platinum, polyethylene fines, silicon, silica alumina materials, pharmaceutical products and other organic and inorganic materials often found in the specialty chemical industry. Some of these products are very expensive and, in extreme cases, may cost as much as $20 - $100/lb. The recovery of such materials obviously pays handsomely for the hot-gas media system. There are probably several hundred systems worldwide in different sizes. Some applications are very unique, for instance, radioactive strontium collected in a hot-cell operation. The high level radiation means that the product could only be handled remotely. The valuable product, radioactive strontium, is encapsulated as purified strontium fluoride. Later the recovered precipitate is sintered in a furnace to drive off all volatiles.

One of the largest suppliers of these systems [2] has listed some unique applications: deductor, hydrofluorinator - off gas treatment, aniline recovery, fluid cracking catalyst regenerator, nuclear fuel cycle, removal of transuranic element oxides from spray calciner, uranium recovery, catalyst fines from fluid bed reactor, polyethylene recovery, polycrystalline - fumed silica removal, fly ash - shale oil, fluorinator off gas treatment, phthalic anhydride process catalyst recovery, low density polyethylene production, proprietary fluid bed reactor off gas, nuclear enrichment plant off gas, nuclear fuel production, and numerous catalyst recovery systems.

Waste Incineration

Although more or less a great opportunity for hot-gas filters, most of today's incineration plants are controlled with so-called conventional baghouse technology.

There are a number of examples, however, where hot-gas

media filters have performed very well. A Dutch company developed a thermal decomposition plant and has operated soil incineration successfully since 1986. The out gases are controlled with a hot-gas filter currently using rigid ceramic filter elements [3].

There are several medical waste incinerators operated under hot-gas conditions, some of them are proprietary. One U.S. manufacturer [4] reported in 1992 about a successful installation with rigid ceramic candle filter elements. A German equipment manufacturer [5] has supplied a number of hot-gas filters for incineration all based on rigid ceramic filter elements.

Pressurized Fluidized Bed Combustion

This application is probably one of the most promising for hot-gas media systems and is subject to a number of presentations at this conference. Two projects, however, should be mentioned briefly. The single largest project is the large scale demonstration unit conducted at the TIDD POWER PLANT in Brilliant, Ohio. The AMERICAN ELECTRIC POWER SERVICE CORP. plans to evaluate integrated engineering design of a commercial scale filter using 1/7th flow slip stream at the 70 MW plant.

A $20 million hot-gas clean up system is funded by the U.S. Department of Energy. WESTINGHOUSE ELECTRIC CORP. is going to build a hot-gas filter unit at the Tidd Power Plant in Ohio.

This filter concept is based on rigid ceramic candles, and the filter unit consists of 384 cylinder-shaped silicon carbide filter elements. Individual elements will be assembled on 3 cluster assemblies. The clusters then will be placed into the 10 X 40 ft. cylindrical hot-gas vessel. The system will handle about 1/7 of the exhaust gas discharged by the PFBC combuster. Ash-laden gases will enter the side of the cylindrical vessel depositing the ash on the outer surface of the ceramic elements.

301

Once every 30 minutes a 600 psi air pulse will flow into the elements and remove the collected ash which will be cooled and conveyed to the fly ash removal system.

At the R. E. BURGER PLANT of OHIO POWER the BABCOCK AND WILCOX SO_x-NO_x-RO_x-BO_x-Technology is currently being tested at a 5 MW demonstration project. This process involves an SCR unit for NO_x control, injection of either calcium or sodium sorbents for SO_2 control, and a hot-gas temperature pulse jet fabric filter for fly ash removal. This pilot unit has been operating since March of 1992. Field tests today demonstrate that the design objectives of 70 - 90% SO_2 removal, 90% NO_x removal, and 99% particulate removal can be exceeded.

Selective catalytic reduction (SCR) is the most widely used technology for post combustion NO_x control. The application of the SCR coal-fired units, however, presents some unique problems. High concentrations of sulfur oxide and particulates in the flue gas can cause sulfur poisoning of the catalyst along with erosion and fouling by fly ash.

The BABCOCK & WILCOX process for the combined removal of SO_2, NO_x and particulates is a unique method using a special catalyst from NORTON designated NC-300. The NC-300 catalyst delivers many benefits to the SNRB process. This includes the ability of operate within a wide range of temperatures (from 600° - 1000° F), low ammonia slip, very low conversion of residual SO_2 to SO_3, and high resistance to poisoning by flue gas components. There was previous successful pilot testing at the BABCOCK & WILCOX Alliance facility in Ohio, and the Department of Energy selected the 5MW demonstration project for this process.

Gasification

The integrated coal gasification combined cycle climbed out of obscurity to take a prominent place among new power generation options in the 1990's.

There are numerous systems which will be discussed in much detail at this conference.

Calcination

The application of calcination of magnesium oxide is a somewhat unique process. Raw material which is supplied in different qualities to the plant is crushed and separated by sieving. Coarser material is recycled. The material is then fed into the calciner. Preheated off-gases are led to a primary cyclone and back into the main combustion area. Eventually off-gases are led to a secondary cyclone. The collected material from the hot-gas filter is recycled. Hot-gases treated in the fabric filter are also fed into the heat exchanger and then recycled into the system. This hot-gas filter is currently operating between 500° - 550° C.

The entire system handles 7 t/hr of material. Approximately 150 tons of material is processed per day. A total of 3,000 kg/h enters into the collection system. Fifty percent of this is collected in the primary cyclone. Of the remaining 1,500 kg/h approximately 750 kg/h is fed into the hot-gas filter. The material varies depending on the different blends processed in the operation. The company manufactures up to 15 - 20 different grades in size and consistency blended for different end-uses and customer specifications. Bulk density is 400 g/l; however, that again varies depending on the material being processed.

Total air flow of the filter system is 6,000 m^3/h which is converted into 14,000 m^3/h at approximately 520°C. Actual operating temperatures vary between 480° - 520° C with maximum temperatures occasionally up to 550°C.

The total filter consists of 2 compartments, each with 80 filter bags 150 mm in diameter and 3,450 mm long, with a total filter area of 256 m^2.

The cleaning is accomplished with compressed air at pressure of 6 bar on-line, but permanently 24 hours to maintain a differential pressure which varies between 120 and 160 mm (4.7 - 6.2 WS).

SUMMARY OF INSTALLATION

Application:	Magnesium Oxide
Location:	MAGINDAG, Oberdorf/St. Katherin Lamig Austria
Manufacturer:	RESEARCH-COTTRELL GMBH Germany
Filter System:	Pulse Jet
Total Air Volume:	14,000 cbm/hr. at 520°C
Filter:	2 units - 80 filter bags ea.
Filter Bag Size:	3450 mm long X 150 mm dia. Total 256 m^2 filter area (2816 sq.ft.)
Filter Media:	Bekipor^R ST 25 GFX Inconel Stainless steel metal fiber sinter vlies
A/C Ratio:	3:1 (0.9 cbm/m^2/min.)
Cleaning:	Compressed air - 6 bar "Lanzen" system 24 hours continuously
Start-up:	Originally 1985 Operating since May 1986 without any problems

Catalytic Cracking/Refining Industry

There are some proprietary installations which operate successfully but are not available for any presentation. In general, however, the power recovery of a fluid bed catalytic cracking unit is crucial. Temperature conditions can be as high 1000° - 1200°F (538° - 650°C). Sometimes these units operate under pressure. The catalyst is degenerated in the cat cracker regenerator where carbon is burned off releasing particles in a fluidized bed. The hot, dirty off gas is sent through multi-stage high efficiency cyclones to remove large particles. Efficiency of removal of particles smaller than 5 microns is non-existing; therefore, a hot-gas filter has tremendous advantages. The application deals with off gases from a storage tank of highly abrasive catalysts. In producing catalysts, heat is recovered by a turbine after cyclone cleaning. Once cooled, the gas is filtered in a conventional baghouse down stream. The storage tank application could very well become a necessity for many refineries. The added turbine protection, however, by replacing a cyclone with a hot-gas filter is also very important.

Chemical Process Industry

Unfortunately many of the systems are proprietary, and little information is available. Potential for hot-gas systems, however, is in the following industry segments:

> Electronic Chemicals
> Titanium Dioxide
> Sulfur Recovery
> Nitric Acid Production

In the application of polyphosphate production in the early '80's a hot-gas filter with 20,000 Nm^3/h at 390°C operated satisfactorily in a pulse jet baghouse system with a total filter area of 2,803 sq.ft. The idea was to replace a wet scrubber at an annual energy savings of about $250,000.00. There are many different examples of systems mainly based on metal media filter element configurations.

Other Applications

Obviously there are many applications for hot gas systems which are not covered in the above examples. Many are small; others are very unique. Companies in the glass, ceramic and brick industries have evaluated hot-gas systems as BEKAERT demonstrated several years ago at a fractional condensation process in connection with the filtration of

AS-Sb oxides. This particular filter operated at the time at approximately 400°C and replaced an electrostatic precipitator. The hot-gas media filter offered better efficiency, but it is not known if filter systems for this application are being widely used.

Market

Worldwide hot-gas systems in 1992 represented approximately \$70 - \$75 million. By the year 1996/1997 a total world market for such systems is expected to represent between \$170 - \$180 million. Filter media may represent 20% - 30% of this total, for flange-to-flange systems, perhaps as much as 50% depending entirely on applications.

Finally, it should be stated that hot-gas media systems cannot be seen as isolated pollution control or product recovery units. They must be seen as a system offering to the user an incentive since these systems are generally more expensive than conventional pollution control equipment.

REFERENCES

(1) November 1982, 70th Annual Convention IFAI, Las Vegas,NV "New Fibers in Hot-Gas Filtration"

(2) PALL CORPORATION, Annual Report 1990

(3) CEREL, LTD. (formerly FOSECO) Press Release 1987

(4) UNIVERSAL POROSICS, INC., A Report on Novel Materials of Construction After More than a Year of Operation in a Hot Gas Filter

(5) BHS WOLFF, Herne/Germany

HIGH TEMPERATURE CERAMIC FIBER FILTER BAGS

Timm J. Gennrich
Ceramic Materials Department
3M Center
St. Paul, MN 55144-1000

ABSTRACT

Conventional fabric filters are limited to operating temperatures lower than 500° F (260° C). Filtration at higher temperature can offer several advantages: it can eliminate the need and expense for equipment to cool a gas stream before filtration and to cleanup dilution streams, it can reduce maintenance costs and extend the lifetime of process equipment by avoiding dew points of corrosive species, and it can increase overall operating efficiency by recovering energy or valuable by-products.

Nextel™ ceramic fiber filter bags have demonstrated operation up to 1400° F (760° C). The bags are made from continuous polycrystalline metal oxide fibers typically 10-12 microns in diameter. The fibers are produced from a chemical solution thereby providing well controlled composition and properties. They are processed into engineered yarns and converted into a variety of woven and braided forms for industrial applications at temperatures up to 2500° F (1370° C).

The high temperature filter bags are woven as seamless tubing and further processed to enhance their filtration properties. First tested in 1981, they have undergone thousands of hours of field evaluation. These evaluations, mechanical and filtration properties, and performance in a variety of commercial applications are reviewed.

INTRODUCTION

Particulate control systems based on baghouse technology are restricted by the performance characteristics of the fabric filters. Despite advances in fabric design, durability, and filtration efficiency in recent years, conventional filter fabrics are limited to operating temperatures below 500˚ F (260˚ C).

At the same time, passage of clean air legislation and development of advanced combustion technology has increased interest in the removal of pollutants from hot gas streams, particularly in coal-fired power generation processes. Beyond the utility industry, filtration above 500˚ F can offer significant advantages for other industrial processes.

High temperature filtration can eliminate the need and expense for process equipment to cool a gas stream before filtering and the additional expense to clean up dilution streams. A simplified process can reduce capital, installation, and maintenance costs and space requirements. The lifetime of equipment can be extended by avoiding dew points of corrosive species or difficult to filter hydrocarbon tars. Overall operating efficiency can be increased by recovering the energy of hot gas streams or collecting valuable by-products.

Development of new materials and filtration media has provided more choices for hot gas filtration, such as ceramic and metal fibers, wire cloths, rigid ceramics, ceramic granular beds, and hot electrostatic precipitators.

Baghouse filtration is a simple technology that has matured over the years through sophisticated application of basic filtration and mechanical principles. The successful demonstration of Nextel™ high temperature filter bags enables the application of baghouse collectors to temperatures up to 1400° F (760° C).

NEXTEL™ CERAMIC FIBERS

3M has been a leader in ceramic fiber technology since the early 1970's. This technology is based on a "sol-gel" process whereby a chemical sol is extruded through a spinneret and fired to form individual continuous ceramic filaments. The process provides excellent control of composition, fiber diameter, and physical properties. The resulting metal oxide fibers (62% Al_2O_3, 24% SiO_2, 14% B_2O_3) are polycrystalline and non-porous which gives them good strength, flexibility, and chemical resistance (Tompkins, 1992). They differ from glass based fibers which are drawn from a melt of inorganic materials to form an amorphous glassy fiber. An important result is that the polycrystalline Nextel™ fibers are comparatively abrasive resistant. Fiberglass and silica based fibers are sensitive to abrasion and lose strength quickly if protective coatings such as polytetrafluoroethylene (PTFE) or silicone are not present.

Applications. Individual filaments are combined to form rovings, engineered yarns, and sewing threads. The continuous nature of the fibers allows converting them into many textile forms such as braided sleevings, woven fabrics, and filament windings using conventional equipment. The combination of processability, abrasion resistance, excellent tensile strength, and high temperature stability provides utility for a variety of industrial applications at temperatures up to 2500° F (1370° C). These include aircraft firewalls, aerospace thermal protection, high temperature electrical insulation, furnace curtains and zone dividers, thermal insulating blankets, and various gaskets, seals, and barriers for industrial thermal protection. The common aspect of these applications is the need to maintain strength and flexibility during and after exposure to high temperatures.

Properties. Figure 1 shows the tensile strength of individual filaments of Nextel™ 312 after four hours of exposure at temperatures up to 2370° F (1300° C) in both oxidizing and reducing atmospheres (Holtz & Grether, 1987). In air, the tensile strength changed slightly up to 2190° F (1200° C) and had 67% retention after 1300° C. In hydrogen, strength decreased above 1800° F (1000° C) reaching 50% of initial strength after 1200° C.

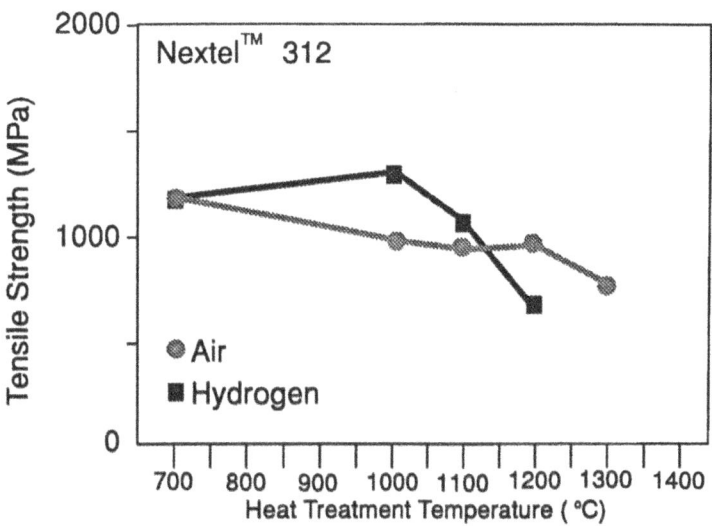

Figure 1. Single Filament Tensile Strength After Four Hours of Heat Treatment

Nextel™ fibers have been shown to have excellent chemical resistance. Table 1 shows that exposure to ten percent solutions of various acids and bases has little effect on fiber strength. One exception is hydrofluoric acid which dissolves the SiO_2 from the fibers with a corresponding loss of strength. Oxides of alkali metals will attack the fibers at temperatures above 1400° F (760° C). They melt and form a glass which fuses the fibers together and embrittles them. Other metal oxides may form low melting eutectics causing similar damage, but usually not at temperatures below 1830° F (1000° C).

Chemical	Chemical/Name	Concentration	Strength Retention
Acids			
HNO_3	Nitric Acid	10%	95%
HCl	Hydrochloric Acid	10%	98%
H_2SO_4	Sulfuric Acid	10%	65%
H_3PO_4	Phosphoric Acid	10%	100%
Bases			
KOH	Potassium Hydroxide	10%	86%
NaOH	Sodium Hydroxide	10%	78%
NH_4OH	Ammonium Hydroxide	10%	77%
CaO	Calcium Oxide	Saturated	99%

Table 1. Chemical Exposure Effects

FILTER BAG DEVELOPMENT

The application of Nextel™ fabrics in industrial markets resulted in many different fabric weave designs. The inherent permeability of these fabrics coupled with the proven strength, durability, and temperature resistance of the fibers led to the development of a filtration fabric in the early 1980's. This work was based on applying existing knowledge of woven fabric filter design to ceramic fiber fabric weaving (White et.al., 1986). The intent was not to improve fabric filtration properties but to extend that capability to temperatures above the useful operating range of existing fabrics. Figure 2 gives a comparison of the recommended continuous operating temperature of various commercially available fabrics.

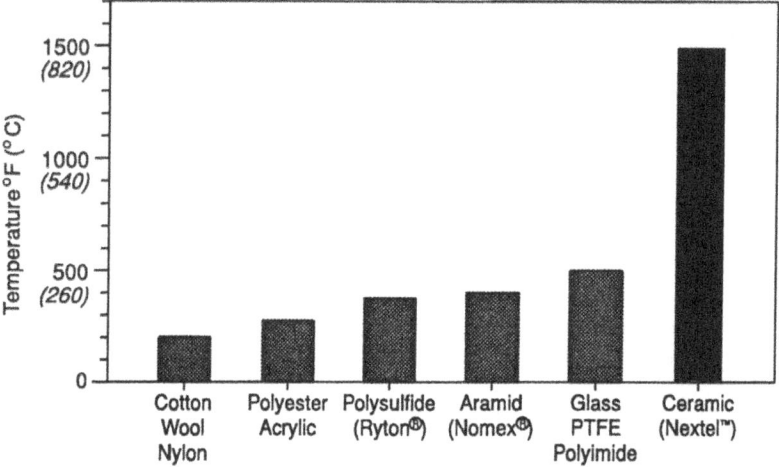

Figure 2. Filter Fabric Temperature Limits

Early Test Results

In 1981, Furlong and Shevlin (Furlong & Shelvin, 1981) reported technical feasibility of fabric filters made from ceramic fibers at temperatures up to 1500° F (815° C) using a pulverized coal-fired combustor at ambient pressure. Filtering efficiency up to 97.9% was achieved over 20 hours of operation. They also reported having tested a bag for 100,000 reverse air cleaning cycles at an air temperature of 1000° F (540° C) (no dust loading) with no damage to the bag.

About the same time, Ciliberti and Lippert (Ciliberti & Lippert, 1981) tested ceramic fiber filters at high temperature and high pressure (HTHP) for application in pressurized fluidized-bed combustion (PFBC) (Chang & Lips, 1985). The tests were conducted by Westinghouse under sponsorship of the Electric Power Research Institute (EPRI). 19 filter bags, 6 inch (15 cm.) diameter by 4.5 (1.4 m) feet long, were tested at 800° F (427° C) to 1500° F (815° C) and 165 psia (1140 kPa). Efficiencies exceeded 99.5% on a test dust of reinjected PFBC ash and limestone. Face velocity was varied from 2.8 to 5.6 ft/min (1.4 to 2.8 cm/sec.) and dust loadings ranged from 2-6 gr/scf (5 to 14 g/m³). Total operating time at temperature was about 50 hours. Bags were cleaned on line using a short-duration high pressure pulse or series of pulses. Effective cleaning was acheived when the pulse to test pressure ratio was about three or greater.

The fabric for both of the above tests was an 8 harness satin weave made with Nextel™ 312 fiber. The bags had a stitched seam with quartz thread. During the Westinghouse test a seam failed in one of the bags. Failure was attributed to poor seam design. Other threads and seam designs were tried with similar results. Finally, based on work by White, et al, a seamless bag was developed to eliminate the problem of seam failure (White et.al., 1986). To improve fabric performance and reduce penetration during pulse cleaning the weave was changed to a 5-harness pattern. The new bag design was tested at Acurex (Chang & Lips, 1985) and Westinghouse (Lippert et.al., 1986) in work supported by the DOE and EPRI. The tests confirmed the improved bag was capable of high collection efficiency (>99.9%) at HTHP conditions (1500° F, 815° C, 11 bar) and face velocities from 3 to 6 ft/min (1.5 to 2 cm/sec.).

The design changes made in 1985 have also eliminated subsequent occurrence of bag failure. Advances in handling and bag processing have been made to improve product quality, but the basic 5-harness, seamless tube construction of Nextel™ 312 filter bags has remained the same. For installation, the bags are fitted on perforated metal support tubes which have closed bottoms, and are clamped at the top and bottom to provide a tight seal. The perforation pattern of the tube can vary to provide 40-58% open area but does not reduce the effective filtration area of the filter assembly. Typically, the support tubes are flanged on top and attached to the tube sheet with a high temperature sealing gasket. Filter cake solids are removed with reverse pulse cleaning either on or off line.

Filtration Properties. Figure 3 shows filtration curves for Nextel™ fabric and various commercial low temperature filter fabrics. Efficiency is plotted as a function of time. The laboratory test data was generated under more rigorous conditions of small particle size (0.1μ mean particle NaCl aerosol), low mass concentration (100 mg/m^3, .04 gr/ft^3), and high face velocity (10 ft/min, 5 cm/sec). The test conditions result in lower initial filtration efficiency and longer time to reach high efficiency, however the advantage is better resolution of data. It is expected that in a baghouse greater efficiencies would be reached in a shorter period of time, since larger particles would be less likely to penetrate through the fabric.

As shown in Figure 3 the Nextel™ fabric has a lower starting efficiency than the other fabrics. But efficiency increases much faster with time. Felted products have higher initial efficiency than woven fabrics. All fabrics were able to reach greater than 99.9% efficiency within a short period of time. With proper conditioning and cleaning all would filter effectively in a pulse-jet baghouse. In other recent work, Helfritch and Feldman (Helfritch & Feldrman, 1990) concluded that the filtration performance characteristics of Nextel™ filter bag media was very similar to that of conventional woven fiberglass.

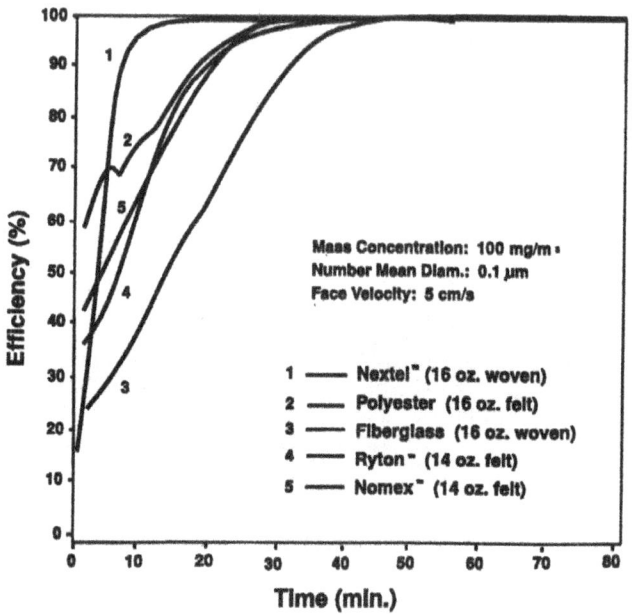

Figure 3. Fabric Filtration Efficiency

APPLICATIONS

Demonstration Projects

University of North Dakota. From October 1985 to June 1989, EPRI sponsored a two phase program to evaluate the performance and durability of Nextel™ fabric filter bags (Weber & Schelkoph, 1990b). The project was conducted at the Energy and Environmental Research Center (EERC) on a pilot-scale pulse-jet baghouse filtering a slipstream of flue gas from a coal fired steam plant. Results have been reported previously and are summarized in Table 2 (Weber & Schelkoph, 1990a). Phase I was scheduled for a total of 1000 operating hours. At the conclusion of the test, a second set of nine filter bags (6 inch [155mm] diameter x 8 ft. [2.4m] long) was installed. Phase II was planned for 6,000 hours to allow broader ranges for system operating conditions but was extended to accumulate additional operating time.

	Phase I	Phase II
Total Test Time	1022 Hours	16877 Hours (5827 Total Pulses)
Particulate	Fly Ash	Fly Ash
Operating Temperature	427° to 510°C (800° to 950°F)	427° to 538°C (800° to 1000°F)
Face Velocity	0.9 m/min (3.0 ft/min)	1.8 m/min (6.0 ft/min)
Mass Median Diameter	2.3 μm	2.6 - 14.8 μm
Inlet Mass Concentration	>3.4 g/m³ (>1.5 grain/ft³)	5.7 g/m³ (2.5gr/ft³)
Efficiency	99.19 - 99.96%	99.66 - 99.99%

Table 2. Test Results from University of North Dakota

Total operating hours and cleaning cycles at shut-down were 16,877 and 3,441 respectively. Actual pulse cycles totaled 5,827 due to multiple pulse cycles and cold-flow testing. Baghouse particulate collection efficiency ranged from 99.66 to 99.99%. Air-to-cloth ration was nominally 4.6 - 6.0 ft/min (2.3 - 3.0 cm/sec) with excursions as high as 7.9 ft/min (4.0 cm/sec). Baghouse temperature ranged from 600 - 930° F (350 - 500° C) with excursions up to 1000° F (540° C). Particulate mass loading at the baghouse inlet ranged from 1.3 to 4.4 gr/scf (3 to 10 g/m³) and consisted of both flue gas particulate and supplemental injected fly ash.

To determine bag durability, bags were removed and replaced at intervals during the operation. Single bags were removed after about 1900, 4400, and 8600 hours. Tests were conducted on these and all remaining bags at the conclusion to determine fabric tensile strength in the warp and fill directions. As shown in Figure 4, the bags retained 85% of their initial strength.

The Nextel™ fabric filter bags met or exceeded all performance expectations. No bag failures of any kind were observed, differential pressure was relatively low and easily controlled, and collection efficiency was sufficient to meet requirements for regulatory emission standards and for protecting downstream equipment.

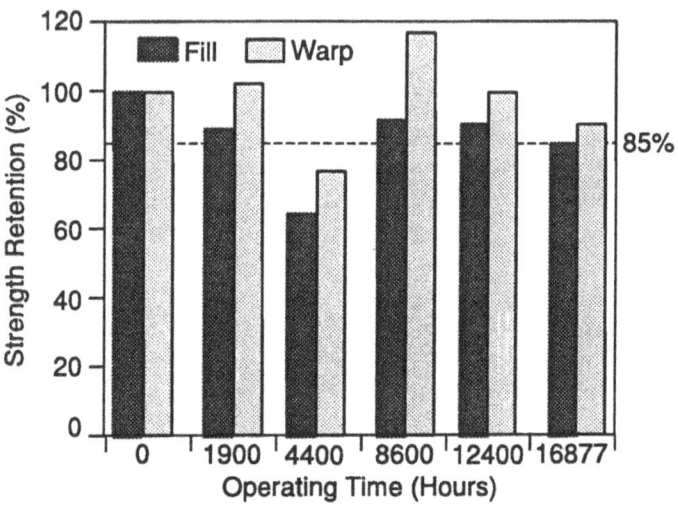

Figure 4. Fabric Durability - Tensile Strength Retention

Babcock & Wilcox SNRB Project. As part of the US DOE Clean Coal II Program, B&W is conducting a demonstration project of their patented SNRB (SO$_x$, NO$_x$, Rox, Box) process for the combined removal of sulfurous oxides (SO$_x$), nitrous oxides (NO$_x$), and particulate emissions from a coal-fired flue gas. The key to the SNRB process is a high temperature pulse-jet baghouse in which simultaneous removal occurs (Figure 5) (Wilkinson etal., 1991). Ammonia and a dry sorbent such as hydrated lime or sodium bicarbonate are injected into the flue gas upstream of the baghouse at 900 - 1100° F (480 - 590° C). The sulfated sorbent products and flyash are collected on the filterbags. In the clean side of the baghouse particle-free gas at 700 - 850° F (370 - 455° C), passes over an SCR catalyst where the ammonia reduces NO$_x$ to N$_2$ and water. The design greatly reduces the potential for SO$_2$ poisoning of the catalyst and erosion and fouling of the catalyst by flyash. Since the removal of all three pollutants are integrated into a single unit operation, the SNRB process offers lower capital and operating costs, simplicity of operation, and reduced space requirements.

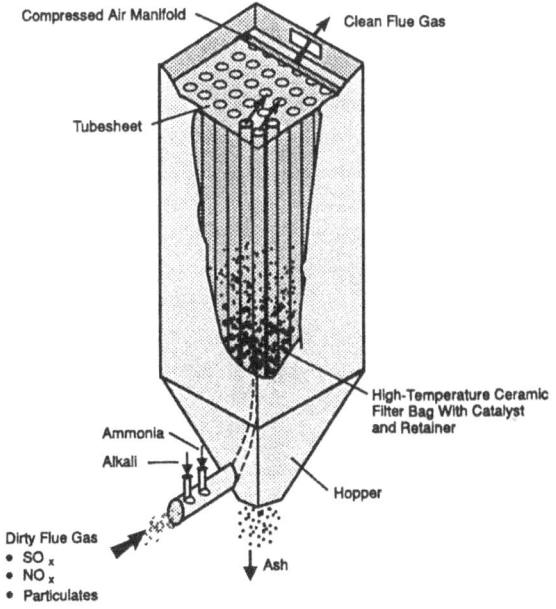

Figure 5. Schematic of the SO_x-NO_x-RO_x Box™ (SNRB) Baghouse

A pilot-scale facility was operated at B&W's Alliance Research center in 1990-1. The high temperature pulse-jet baghouse contained 12 commercial-size bag/catalyst assemblies. Each bag was 6 1/8 inch (155mm) diameter and 20 feet (6.1m) long, for a total collection area of 385 ft^2 (36m^2). The baghouse was operated at temperatures up to 850° F (455° C). An optimum temperature of 800 - 850° F was required to achieve objectives of 70 - 90% SO_2 removal and 90% NO_x removal. An additional objective was to demonstrate particulate emission levels less than the New Source Performance Standards (NSPS) requirement of 0.03 lb/million Btu. Figure 6 shows emissions sampled downstream of the baghouse were consistently less than 0.03 lb/million Btu (13 g/10^9 joule). The baghouse was operated at an air-to-cloth ratio of 3.6.

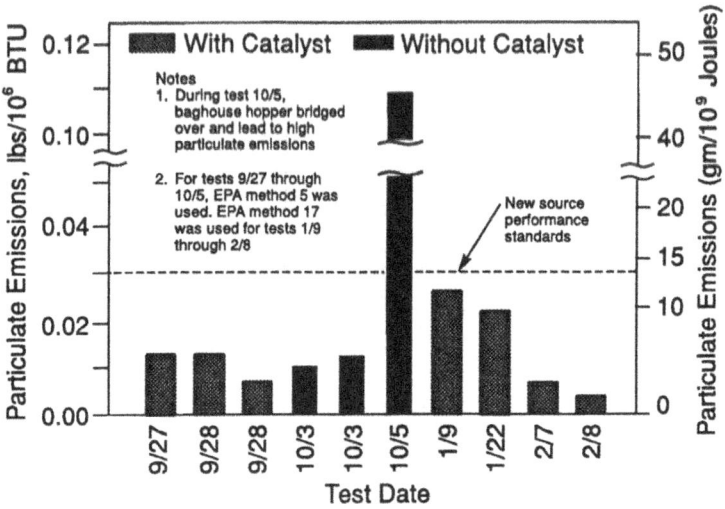

Figure 6. SNRB Pilot Facility Particulate Emissions

At the conclusion of the pilot test, the filter bags were removed and fabric tensile strength measured. These results are reported in Table 3 and compared to results for the bags from the North Dakota study as discussed above. Also for comparison, results are reported for bags which were pulsed 100,000 times in a laboratory pulse-jet baghouse at ambient temperature without dust loading.

Sample	Warp	Fill	Edge
UND - Clean	51(231)	79(358)	-
- Dirty	54(245)	80(363)	-
SNRB - Clean	57(258)	63(286)	37(168)
- Dirty	55(249)	89(404)	47(213)
Lab - New	-	73(331)	42(191)
100,000 pulses	-	45(204)	21(95)

Peak load in pounds (N) for 0.5 (12.7 mm) inch wide specimen

Table 3. Filter Bag Tensile Strength

316

The demonstration phase of the SNRB Clean Coal Program was a 5-MWe facility located at Ohio Edison's R.E. Burger Plant. The 23,000 acfm slipstream unit was designed similarly to the pilot-scale process. The test program was planned to optimize the baghouse operating conditions and demonstrate process feasibility on a commercial scale. The baghouse consisted of six modules and a total of 252 Nextel™ fabric filter bags (6 1/8 inch diameter x 20 ft. long) for a total filter area of 8200 ft^2 (760 m^2).

The facilityran from June, 1992 through May, 1993 with a total of about 2200 hours of operation at temperature. The baghouse was operated at an air-to-cloth ration of 4 with outlet dust loadings averaging 0.012 lb/million Btu or 0.0062 gr/acfm. Operating data correlates well with the previous experience at the Alliance pilot-scale facility (Holmes et. al., 1992).

Other Applications

In addition to the coal-fired utility demonstration projects described above, Nextel™ fabric filter bags have been used in several other industrial hot gas filtration applications. These are summarized in Table 4.

Refinery Catalytic Cracker. A southern California refinery installed a high temperature baghouse to recover catalyst fines from hot gases exiting the catalyst regenerator (Baldwin & White, 1987). Previously, in order to comply with particulate removal standards, the refinery had to use electrostatic precipitators with high initial and maintenance costs or install additional equipment to provide a cooling air stream prior to a standard baghouse collector and risk condensing corrosive acid from the gases. MikroPul Environmental Systems designed and built a reverse pulse-jet baghouse to withstand temperatures up to 1300° F (700° C). 64 bags (4.5 inch. [117mm] diameter x 10 feet [3m] long) were used to provide 755 ft^2 (70 m^2) of filter area. E&L Engineering installed the baghouse which has been operating and meeting compliance standards since 1988.

Fumigant Scrubber. Sur-Lite Corp. designed a high temperature baghouse to remove limestone dusts from hot exhaust gas exiting a fumigant scrubbing system. The compact unit was built to withstand 650° F (350° C). Methyl bromide gas used to fumigate cotton is incinerated in a gas fired kiln. Kiln exhaust is passed through a bed of limestone to scrub the bromide prior to venting to the atmosphere. The unit came on-line in late 1991, and has met emission standards for both acid and particulate content.

Metal Smelting. Bench and pilot scale studies have been conducted on a variety of metal smelting processes. Nextel™ fabric filter bags have been used successfully to separate and recover arsenic, antimony, gold and other metals from hot process gases.

Thermal Remediation. Several projects are ongoing to filter hot gas from incinerators and thermal oxidizers used in the treatment of waste, spill cleanups, and contaminated soils. In most cases, the need for modular or transportable equipment places a premium on process size and simplicity. Proper application of high temperature baghouse collection can minimize space requirements by eliminating the need for dilution streams. The ability to filter hot kiln gases also allows more flexibility for proper incineration to control creation of toxic products of combustion.

Application	Catalyst Cracker	Coal Fired Boiler	Coal Fired Boiler	Fumigant Scrubber
Temperature (F)	860 - 1300	825	800	650
(C)	460-700	440	420	350
Gas Stream	Air	Flue gas	Flue gas	Combustion gas
Solids	Catalyst dust	Flyash	Flyash, limestone	lime, CaCO3, CaBr2
Loading (gr/acf)	<30	2.5	2 - 3	Variable
(g/m*)	<70	6	5 - 7	
Particle size - MMD (μ)	40	8	3	--
Face velocity (fpm)	3.2	5.6	2.8	3
(cm/s)	1.6	2.8	1.4	1.5
Pressure drop (in)	4	4	2	1 - 4
(mbar)	10	10	5	2.5 - 10
Efficiency (gr/acf)	<.01	>99.95%	>99.9%	<.01
(g/m*)	<.02			<.02

Table 4. Nextel™ Filter Bags Test History

SUMMARY

Nextel™ ceramic fibers have sufficient strength and chemical resistance to be used in a variety of industrial textile applications at temperatures up to 2500° F (1370° C). The fibers are resistant to oxidizing and reducing atmospheres at the temperatures of interest for most fossil fuel combustion processes. The fibers' strength and flexibility led to the successful development of a filtration fabric for pulse-jet baghouses.

Early development tests revealed weaknesses in fabric design and seam construction, but demonstrated effective filtration capability in a variety of conditions including temperature to 1400° F (760° C), pressure to 165 psia (1140 kPa), and face velocity to 6 ft/min (3 cm/sec). The fabric design was improved in 1985 to the current seamless construction.

Since that time, Nextel™ ceramic filter bags have been shown to filter effectively for thousands of hours of operation at temperatures much higher than conventional filter fabrics can withstand. They have proven efficiency and durability in a variety of commercial applications.

REFERENCES

1. Baldwin, J.R. & White, L.R. (1987), Refinery Application of Ceramic Fiber in High Temperature Bag Filter, 1000°-1300° F," in Proceedings of the American Institute of Cemical Engineers Technical Meeting on Expanding Frontiers, Anaheim, CA,April 21.

2. Chang, R. & Lips, H. (1985), "Ceramic Fabric Material Testing" . Work performed under DOE Contract No. DE-AC21-83MC20110, Mountain View, CA.

3. Ciliberti, D.F. & Lippert, T.E. (1981), "Evaluation of Ceramic Fiber Filters for Hot Gas Cleanup in Pressurized Fluidized-Bed Combustion Power Plants." EPRI CS-1846, Topical Report, May.

4. Furlong, D.A. & Shevlin, T.S. (1981), "Fabric Filtration at High Temperature," Chem. Eng. Prog., January, p. 89.

5. Helfritch, D.J. & Feldman, P.A. (1990), "High Temperature Filter Media Evaluation," presented at the Eighth Symposium on the Transfer and Utilization of Particulate Control Technology, San Diego, CA, March 20-23.

6. Holmes, A.R., Redinger, K.E., Evans, A.P., McCoury, J.M., Johnson, H. and Bolli, R.E. (1992), SOx-NOx-Rox Box™ (SNRB) Process Development and Demonstration," Paper presented at 9th Annual Pittsburgh Coal Conference, Pittsburgh, PA, October 12-16.

7. Holtz, A.R. & Grether, M.F. (1987), "High Temperature Properties of Three Nextel™ Ceramic Fibers," 32nd Internaitonal SAMPE Symposium, Anaheim, CA, April 6-9.

8. Lippert, T.E., Ciliberti, D.F., Tassicker, O.J., & Drenker, S.G. (1986), "Test and Development of Woven Ceramic Bag and Ceramic Candle Filters for HTHP Application," in IChemE Symposium Series No. 99, pp. 215-231.

9. Tompkins, T.L. (1992), "Nextel™ Ceramic Fibers, Fabrics, and Applications," Presented at Hi-Tech Textiles 1992 International Exhibition and Conference, Greenville, SC, June 9-11.

10. Weber, G.F. & Schelkoph, G.L. (1990a), "Performance and Durability Evaluation of 3M High Temperature Nextel™ Filter Bags," presented at the Eighth Symposium on the Transfer and Utilization of Particulate Control Technology, San Diego, CA, March 20-23.

11. Weber, G.F. & Schelkoph, G.L. (1990b), "Performance and Durability Evaluation of 3M High Temperature Nextel™ Filter Bags," EPRI GS-7055, Project 1336-16, Final Report.

12. White, L.R., Forester, R.J., O'Brien, D.L. & Schmitt, G.A. (1986), Ceramic Fabrics for Filtration at High Temperatures, 290-820° C, in IChemE Symposium Series No. 99, pp. 263-281.

13. Wilkinson, J.M., Chu, P., K.E. Redinger, K.E., Gennrich, T.J., Hsieh, K.C. (1991), "5-MWe SNRB Demonstration Project," Presented at the EPRI Ninth Particulate Control Symposium, Williamsburg, VA.

CERAMIC HONEYCOMB FILTER FOR HOT GAS CLEANING

Yasuo Akitsu, Hideyuki Masaki and Osamu Kyo
Ceramics Apparatus Department,
NGK Insulators, LTD.
1 Maegata-cho, Handa city, Aichi pref. 475 Japan

ABSTRACT

A new ceramic honeycomb filter is currently under development. This filter is applicable to hot gas cleaning and dust collection, and possesses a high filtering area per volume. The filter is made of mullite or cordierite and employs a honeycomb structure. It has external dimensions of 150 mm square x 500 mm in length, a 9.9 mm square cell, porous cell wall thickness of 1 mm, mean pore diameter of 15 μm, and filtering area density of 145 m^2/m^3.

The cleaning apparatus is composed of packs that contain the filters, a pulse dedusting device and a housing. The apparatuses have been successfully used for stationary diesel engines, waste incinerators and chemical process powder collection. In a diesel engine exhaust application, the collection efficiency exceeded 99% under conditions of an unburned carbon soot load of 0.5 g/Nm^3, filtering velocity of 1.0 m/min and maximum gas temperature of 400 ℃. Under these conditions, a stable pressure drop was maintained at 4 kPa after 5000 hours of operation.

INTRODUCTION

This report describes the configuration, performance and field operation examples of a ceramic honeycomb filter used for hot gas cleaning. The need for a high-performance cleaning device for hot gases has been growing in recent years. Many development activities in this field have been conducted with ceramic candle filters and tube-shaped filters. Because of their relatively small filtering area per volume, they require large vessels for containing the filters and large installation space, resulting in increased apparatus cost.

The ceramic filter presented in this report has the shape of a honeycomb, possessing a fairly large filtering area per volume. This unique property makes the gas cleaning device compact.

Development activities concerning a ceramic cross flow filter with a honeycomb and/or monolithic shape have already been reported by Westinghouse Electric Corp. (Reference 1) and Asahi Glass Corp. (Reference 2).

We have proceeded with the development of the ceramic honeycomb filter in view of the following advantages over ceramic cross flow filters:

(1) Ease of ceramic fabrication,

(2) Low pressure drop due to a thin wall for the filtrate, and

(3) Ease of assembling and disassembling the filter packs for maintenance.

We initially started development activities on a filter for gases below 500℃ , and so far ten or more apparatuses are now being operated in various hot gas cleaning applications below 500 ℃ , such as for stationary diesel engine exhaust gas. We are now proceeding with the development of a ceramic honeycomb filter applicable to hot gases above 500℃ like PFBC, molten reducing iron manufacturing and other industrial processes.

This report describes a ceramic honeycomb filter focusing primarily on gases below 500℃ , on which the majority of our testing was conducted, and substantial field data has been obtained.

FILTER CONFIGURATION AND CHARACTERISTICS

Honeycomb filters are made of either mullite ($3Al_2O_3-2SiO_2$) or cordierite ($2MgO-2Al_2O_3-5SiO_2$). The material to be used is primarily determined by the temperature and corrosiveness of the gas. More specifically, mullite is used preferentially in the case of corrosive gases, while cordierite is used in the case of hot gases due to its high resistance to thermal shock. The corrosion rates of both materials were measured by the liquid immersion method using small honeycomb blocks. The blocks were immersed in solutions of 10% HCl, 10% H_2SO_4 and 10% NO_3, respectively, at 90℃ for 100 hours, and then the weight loss of each block was measured. Taking the weight loss of mullite 1 as the index, the weight loss of cordierite was 160, 170 and 95, respectively. Although the corrosion conditions in this case may be much more severe than conditions actually encountered for hot gases, the results indicate that mullite is the preferable material in acid gas applications, especially at low and medium temperatures, due to possible condensation. A thermal shock test was conducted by repeating pulse jetting of ambient temperature air on a honeycomb filter (150 mm square x 500 mm in length) heated in an electric furnace. After 50 cycles of pulse jetting at each temperature (50℃ increments), the filter was visually inspected for the presence of any cracks or damage. As a result, mullite one was damaged at a temperature difference of 550-600℃, while cordierite one was not damaged even up to a temperature difference of 900℃. Based on these results, we generally use mullite for temperatures of 500℃ and below, and cordierite at temperatures of 500℃ and above. Unless specified otherwise, the following explanation refers to the use of a mullite filter.

Fig. 1 shows the configuration of the honeycomb filter and Fig. 2 shows a photograph of that filter. The standard filter dimensions are 150 mm square and 500 mm in length for the outside configuration. The square cells that form the honeycomb pattern have a 1 mm wall thickness and 9.9 mm cell pitch. All cells are alternatively sealed at either end and the cell walls serve as the filtration surface.

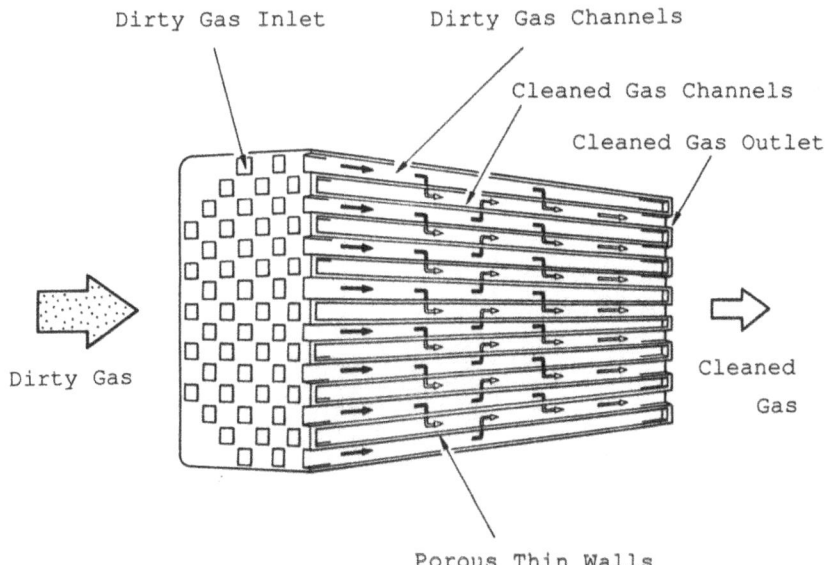

Fig. 1   Schematic Representation of Ceramic Honeycomb Filter

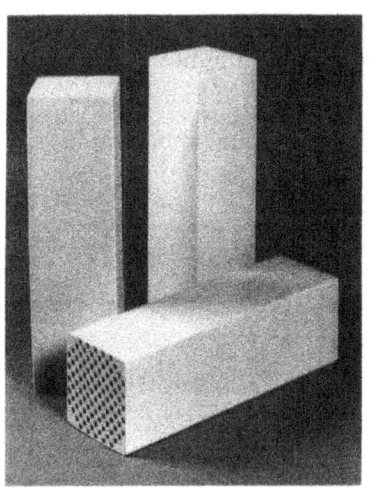

Fig. 2   Photograph of Ceramic Honeycomb Filter

The pore size distribution of the cell wall, measured by mercury injection (using a porocimeter), is shown in Fig. 3. Mean pore size, 50% as cumulative volume, is approximately 15 μm for both mullite and cordierite, and maximum pore size is 150 to 170 μm.

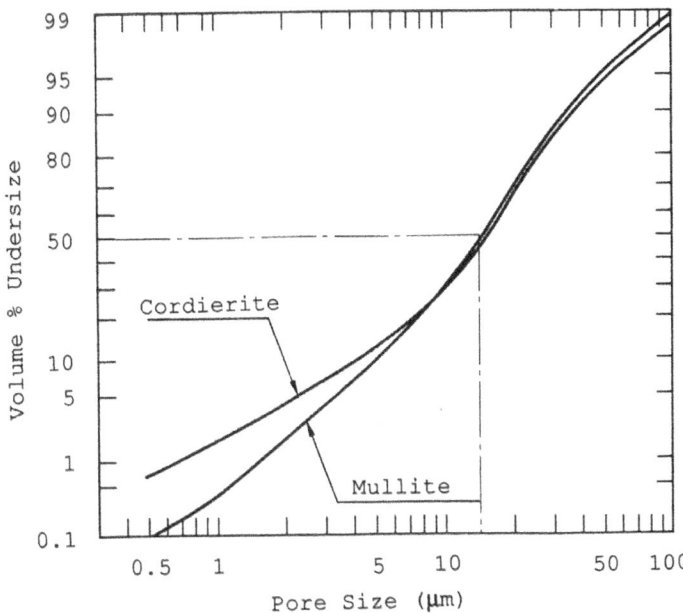

Fig. 3 Pore Size Distribution of Ceramic Honeycomb Filter

The effective filtration area of a standard filter (150 mm square x 500 mm in length, cell pitch of 9.9 mm) is 1.63 m², and the effective filtration area per volume is 145 m²/m³.

Table 1 shows the representative physical properties of the filter. The thermal shock resistances shown in the table indicate the test results for honeycomb filter of the standard size.

Table 1   Standard Characteristics of Ceramic Honeycomb Filter

Item		Characteristics		
Material		Mullite Cordierite		
Size	Total Size (mm)	150□ x 500 ℓ		
	Cell Pitch (mm)	7.4	9.9	15
	Cell Wall Thickness (mm)	0.8	1.0	1.2
Filtration Area (m²/m³)		192	145	101
Average Pore Size (μ m)		Mullite Cordierite		15 15
Porosity (%)		40		
Thermal Expansion Coefficient 40~800 ℃ (x 10^{-6}/℃)		Mullite Cordierite		4.5 1.2
Softening Temperature (℃)		Mullite Cordierite		1800 1400
Thermal Resistance (℃) at Pulse-Jet-Cleaning		Mullite Cordierite		550 $\geqq$ 900
Compressive Strength (MPa)	Axial	40		
	Transverse	15		

Fig. 4 shows the air flow characteristics of the fresh filter
when using clean air.   The pressure drop is proportional to the
surface velocity, or filtering velocity, in the range of 2 m/min
or less.   The deviation of pressure drop fell within 12% maximum
when 220 filters were measured.

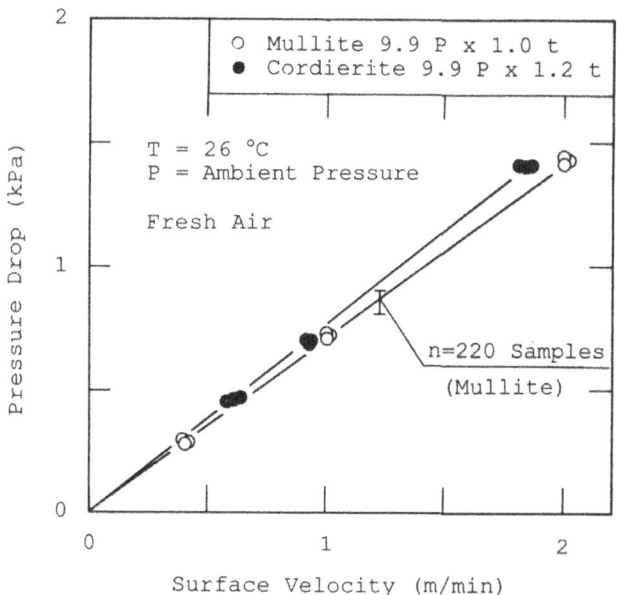

Fig. 4 Surface Velocity vs Pressure Drop of Ceramic Honeycomb
Filters

FILTER ASSEMBLY

The schematic construction of the filter unit using ceramic
honeycomb filters is shown in Fig. 5. The unit comprises a pack
integrating filters with a metal frame, a pulse-jet cleaning
device, a unit housing and a dust box. Dust-containing gas enters
the unit from the dust shoot portion of the housing and is
cleaned by the filters. As dust accumulates on the filter
surface, the pressure drop through the filters increases
gradually. Therefore, the filters are periodically cleaned by
pulse-jet cleaning with compressed gas. The dust that is blown
off from the filters by pulse-jet cleaning, then falls down into
the dust box.

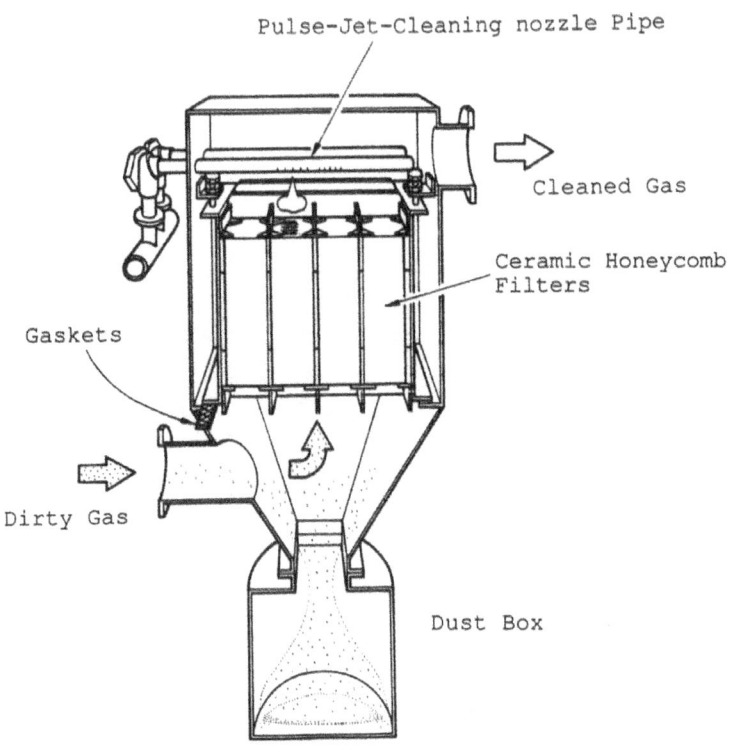

Fig. 5 Schematic Construction of the Filtration Unit Used
Ceramic Honeycomb Filters
(Type of Pulse-Jet-Cleaning)

The pack, bundling a multiple number of filters, is fastened
to the flange of the housing with gaskets. Normally, a single
pack contains either 16 or 25 filters. The pack configuration for
low temperature use is shown in Fig. 6.

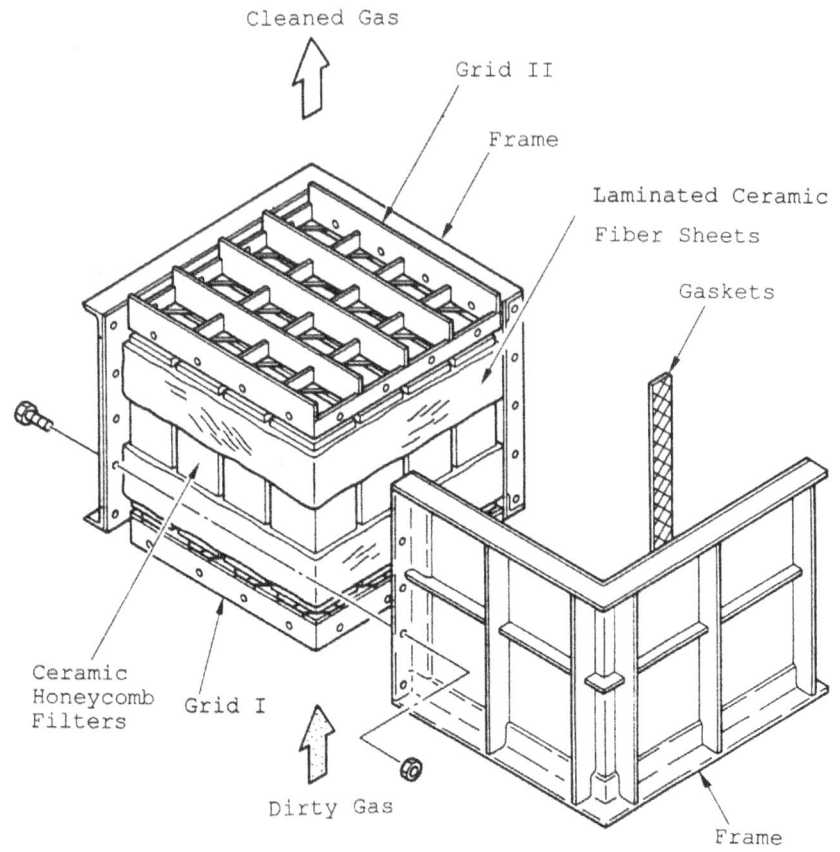

Fig. 6 Schematic Construction of the Pack for Low Temperature Gas

Laminated ceramic fiber sheets (in which Al_2O_3-SiO_2 fibers are laminated in the form of blankets) are adhered around the periphery of each filter with an inorganic adhesive. After arranging the filters within a pair of L-shaped metal frames, the metal frames are tightened to the specified dimensions and an adequate pressure is applied to the ceramic fiber sheet between filters to ensure proper dust sealing. The amount of gas leakage

329

at these portions is initially 0.5% or less per filter at a differential pressure of 4.9 kPa. When applied to diesel engine exhaust gas, there was no leakage of dust during one year of operation, consisting of daily starting and stopping, at 300℃ and at a pressure drop of 4.9 kPa. The laminated ceramic fiber sheets inserted between the filters were selected from the absorptivity of the filter manufacturing tolerance, the relief of stress on the filter (shock during handling and transport, vibrations during operation, deformation of the metal frame and so on), and the long-term dust sealing characteristics.

The configuration of a pack used for hot gases at 500℃ and above is shown in Fig. 7. Due to the high thermal expansion of the metal against the ceramics, and the large decrease of elasticity of the ceramic fiber serving as the sealing material, the filters are sustained individually in the pack and sealed with flat gaskets made of SiO_2 fibers. Each filter is provided with a flange at the end of the filter, which is sintered into a single structure with the ceramic honeycomb. Laminated ceramic fiber sheets are inserted around the periphery of the filters. With this pack configuration, the reliability of the flat gasket is very important to ensure dust sealing performance over an extended period of time. Thus, we are still continuing with evaluation of various gasket materials and seal structures.

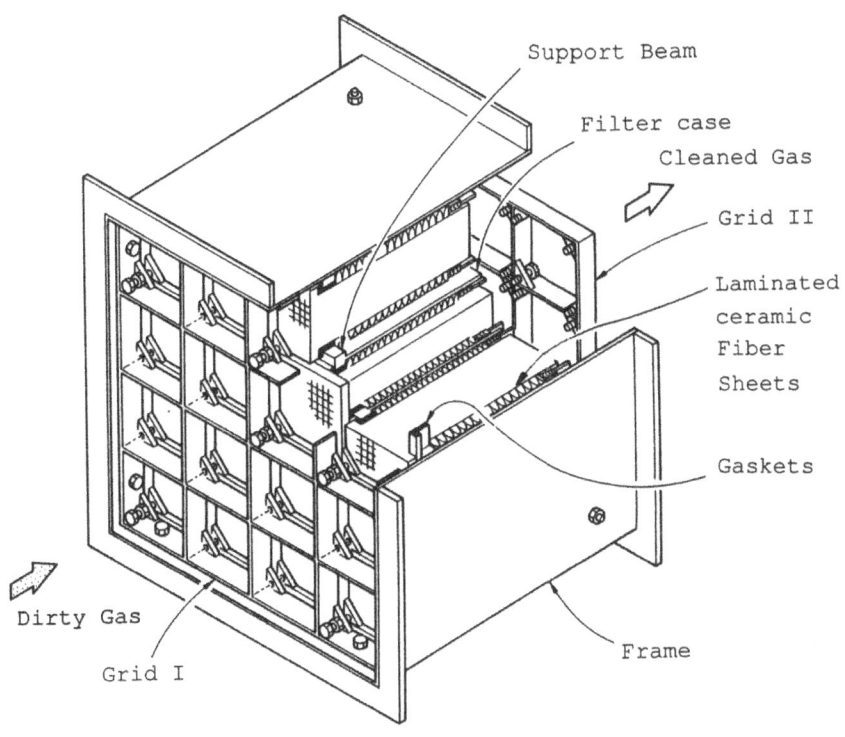

Support Beam

Filter case

Cleaned Gas

Grid II

Laminated ceramic Fiber Sheets

Gaskets

Dirty Gas

Grid I

Frame

Fig. 7 Schematic Construction of the Pack for high Temperature Gas

The thermal shock resistance of the flanged filters was equal to that of the filters without flange in the thermal shock test of repeated pulse-jet cleaning. Grids, in the form of metal lattice frames provided with support pieces in four corners for each filter, are mounted on both sides of the pack to secure the filters against differential pressure, impact force resulting from pulse-jet cleaning, as well as flat gasket sealing pressure.

These grids also serve to reinforce the pack.

The impact force resulting from pulse-jet-cleaning is the largest mechanical load that acts on the filter. Thus, the filter configuration and its strength were designed to bear the load of pulse-jet cleaning, taking into account any possible movement of the filter, the supporting area and the filter material strength.

We have applied two types of cleaning devices in actual installations. These consist of pulse-jet and mobile continuous cleaning. As shown in Fig. 5, pulse-jet cleaning is performed on the downstream side of the filters by compressed gas introduced through nozzle pipes connected to a compressed gas reservoir. Pulse-jet gas is injected through the slit openings of the nozzle pipes into the filters, and pulse-jet cycle and jet duration are controlled by solenoid valves.

The mobile continuous cleaning device is shown in Fig. 8. The cleaning nozzle pipe, provided with injection holes, moves axially during the cleaning cycle by a gear motor drive as illustrated, or by a pneumatic cylinder. This device is suitable for dedusting of adhered dust because of its powerful cleaning force. It is also suitable for minimizing the effects of pressure on the upstream side during the cleaning period. On the other hand, the pulse-jet device has strong advantages as a cleaning device in terms of long-term reliability because of the simplicity of the system and the absence of moving parts. Consequently, this device is particularly suitable for high temperature applications.

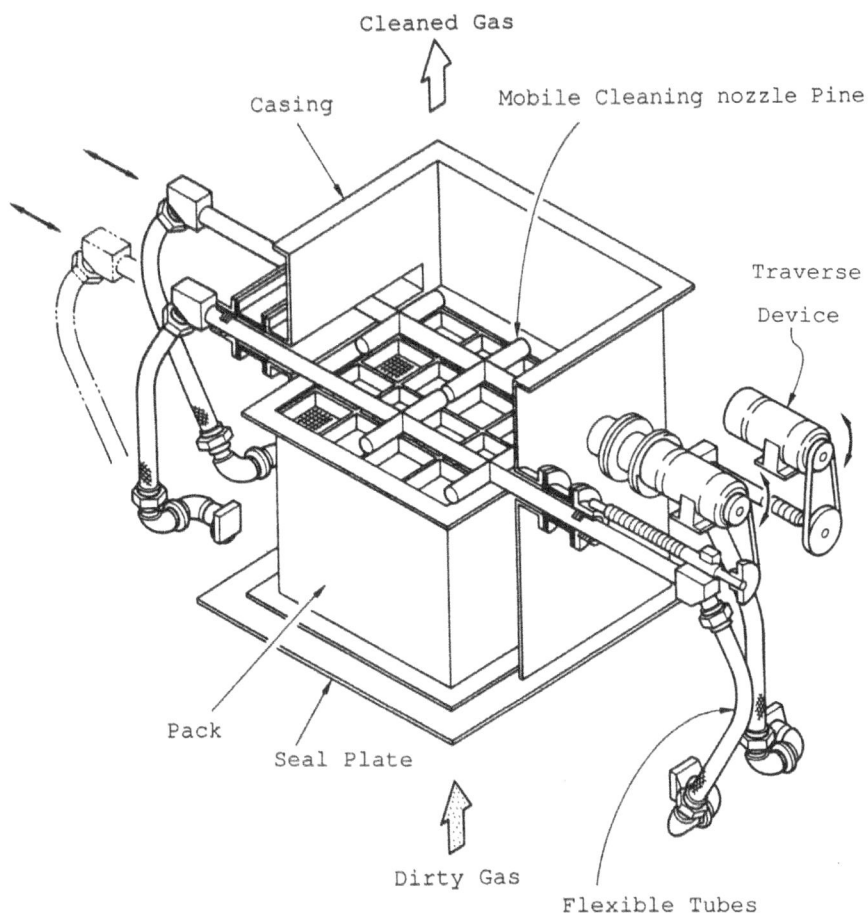

Fig. 8 Schematic Construction of the Filtration Unit Used
Ceramic Honeycomb Filters
(Type of Mobile Continuous Cleaning)

333

FILTER PERFORMANCE

The following provides an explanation of the pulse-jet-cleaning system since it is used most commonly. Since one nozzle pipe of the pulse jet serves several filters, slit openings are provided on the nozzle pipe in the perpendicular direction to convey uniform cleaning force to the filters as shown in Fig. 5. The relationship between pulse jet conditions and residual dust weight was examined for a diesel engine exhaust gas. The amount of residual dust in a filter was confirmed to strongly relate to pulse jet force as shown in Fig. 9. The maximum surface velocity during pulse jet cleaning, which is equivalent to pulse jet force, was calculated from the measured discharge velocity at the end of the filters.

The results of an endurance test are shown in Fig. 10, in which one filter was tested by pulse jet cleaning of dust collected from diesel engine exhaust gas with a bag filter.

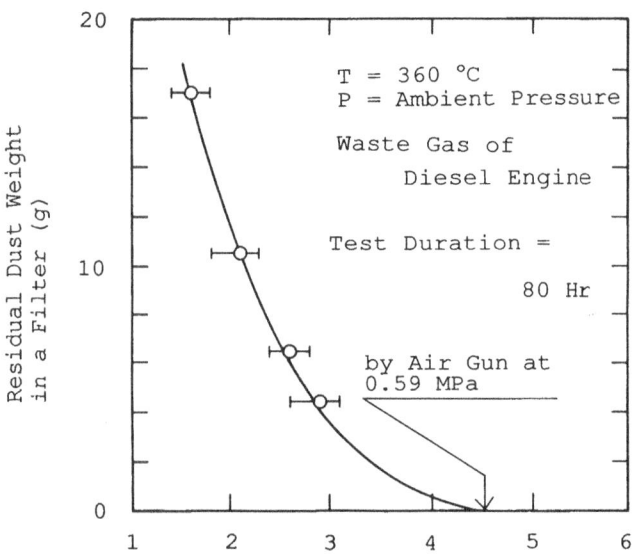

Maximum Surface Velocity at pulse-Jet-Cleaning (m/min)

Fig. 9 Residual Dust Weight in a Filter as a function of
Maximum Surface Velocity at pulse-Jet-Cleaning

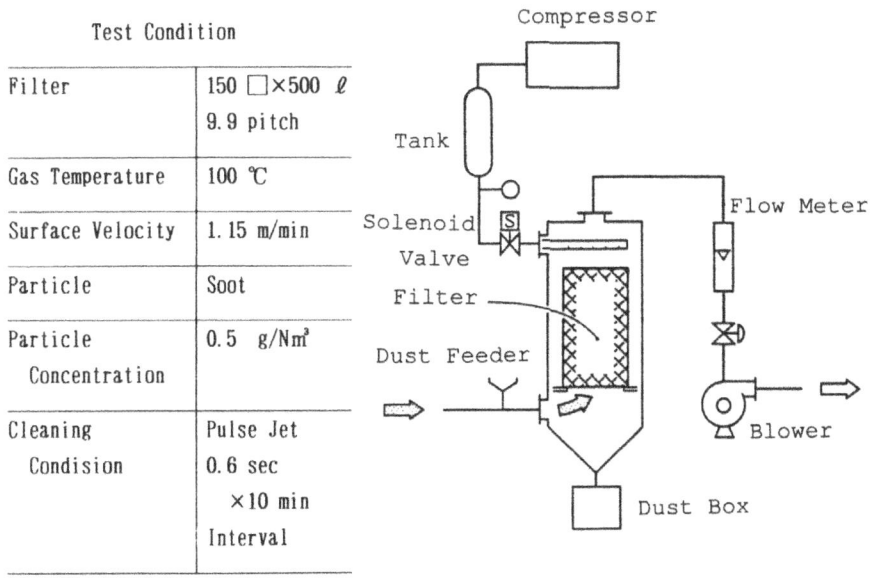

Test Condition

Filter	150 □×500 ℓ 9.9 pitch
Gas Temperature	100 ℃
Surface Velocity	1.15 m/min
Particle	Soot
Particle Concentration	0.5 g/Nm³
Cleaning Condision	Pulse Jet 0.6 sec ×10 min Interval

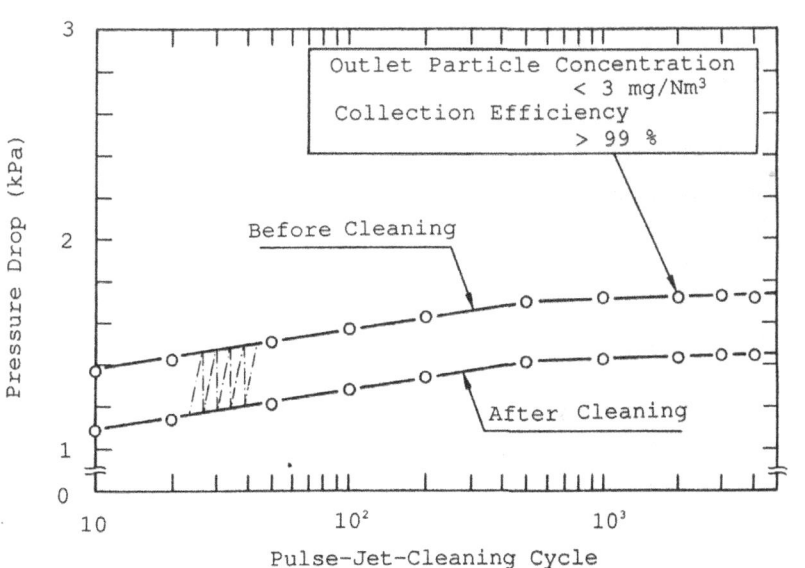

Fig. 10   Pressure Drop of a Filter as a Function of Pulse-Jet-Cleaning Cycle

335

The mean particle size was 10 μm after drying as shown in Fig. 11.

The particles were observed microscopically to be agglomerated. The particles suspended in the exhaust gas were estimated to be one-tenth of the measured size. This endurance test was conducted with a pack containing 16 filters, and the test conditions and results are shown in Fig. 12. After the test, the filters were sliced open to observe the extent of dust adherence to the cell wall. In addition to observation of uniform adherence, a small amount of dust was observed to have accumulated at the bottom end of the cell hole. This accumulating tendency was found, in the other tests, to be promoted when pulse jet force becomes weaker. We believe that this accumulation is caused by weak scavenging gas flow at the bottom end of the cell hole.

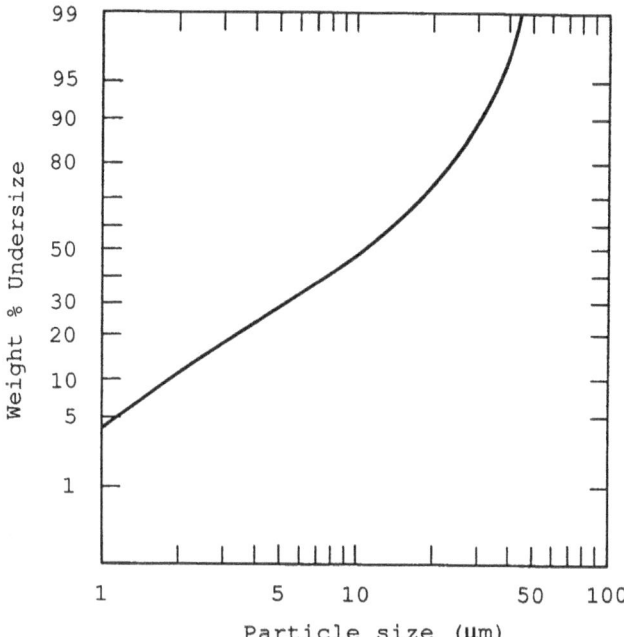

Fig. 11   Particle Size Distribution of the soot used in Test

INDUSTRIAL APPLICATION AND PERFORMANCE

A dust collector with the ceramic honeycomb filters was installed to clean hot exhaust gas from a diesel engine laboratory. The schematic flow of this system is shown in Fig. 13, and the system specifications are shown in Table 2.

Table 2   Specification for Comercial Dust Collector

Item		Specification
Gas Condition	Gas	Waste Gas of Diesel Engine
	Temperature	Max. 400℃ (50~400 ℃)
	Volume	240 m³/min at 400 ℃
	Pressure	-1 kPa
	Particle	Soot, 0.3~0.5 g/Nm³
Specification of Dust Collector	Filter Element	Ceramic Honeycomb filter Mullite, 150□ x 500 ℓ 9.9P x 1.0t
	Pulse Jet Cleaning	0.59 MPa Air x 0.2 sec 10min interval
	Filtration Area	209 m²
	Particle Collection Efficiency	90% over
	Pressure Drop	4.9 kPa under

Exhaust gas from diesel engines is introduced into the filtration unit, and the cleaned gas is discharged from the chimney by a blower. The system handles 240 m³/min of exhaust gas at 50-400℃ with the filters providing 209 m² of filtration area.   Each pack consisting of 16 filters (4 rows x 4 rows) is installed in a housing. The system is equipped with a pulse jet cleaning device that periodically injects pulses of compressed air at 0.59 MPa through a solenoid valve and nozzle pipe.   One nozzle pipe cleans four filters.

337

Test Condition

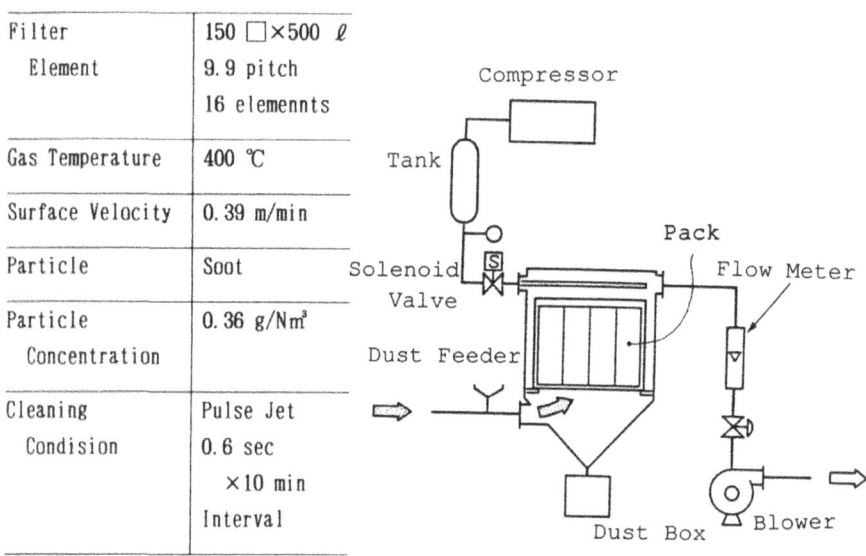

Filter Element	150 □×500 ℓ 9.9 pitch 16 elemennts
Gas Temperature	400 ℃
Surface Velocity	0.39 m/min
Particle	Soot
Particle Concentration	0.36 g/Nm³
Cleaning Condision	Pulse Jet 0.6 sec ×10 min Interval

Compressor

Tank

Pack

Solenoid
Valve

Flow Meter

Dust Feeder

Dust Box

Blower

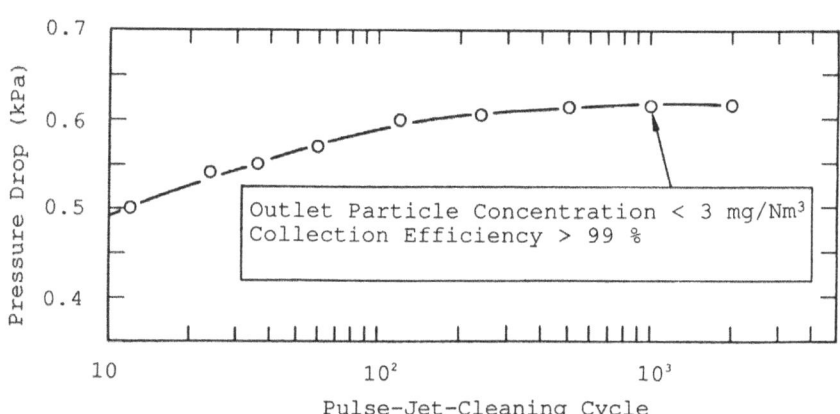

Outlet Particle Concentration < 3 mg/Nm³
Collection Efficiency > 99 %

Pulse-Jet-Cleaning Cycle

Fig. 12   Pressure Drop of a Pack (16 Filters) as a Function of
Pulse-Jet-Cleaning Cycle

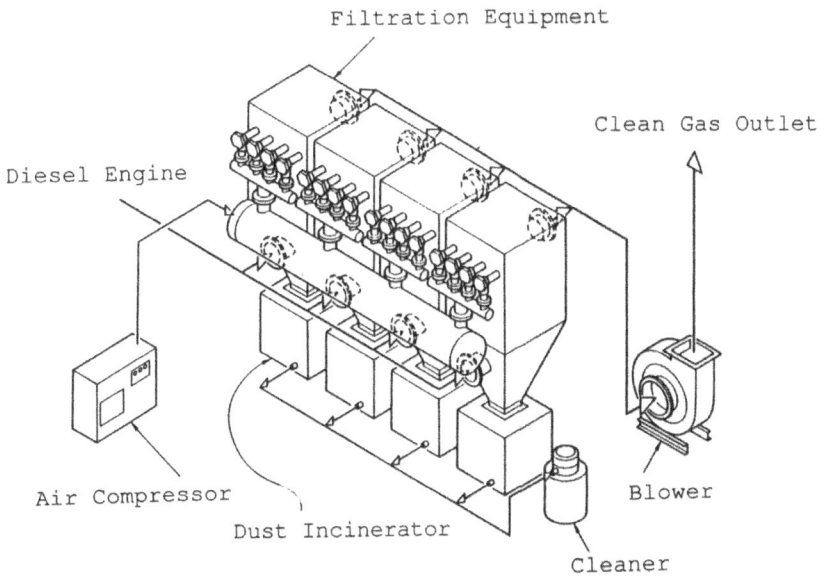

Fig. 13   Schematic Flow of the Dust Collector

A dust incinerator is installed on each housing to burn off the collected unburnt carbon soot. The incinerator is operated during system operation, and the ash remaining is suctioned automatically by a vacuum cleaner at the beginning of operation once a week.

(1) Pulse-Jet-Cleaning Condition

In order to design the dust collector, it is necessary to conduct field tests and determine the necessary or optimal pulse jet condition. However, in the case of the diesel engine laboratory, the necessary pulse jet condition was determined to be a maximum surface velocity of 2.5 m/min or more during pulse-jet-cleaning according to in-house test results.

The maximum surface velocity during pulse-jet-cleaning varies according to the compressed gas pressure, nozzle pipe diameter and opening area of the slits. Some examples of the relationship among maximum surface velocity during pulse-jet-cleaning,

339

compressed gas pressure and duration of pulse-jet-cleaning are shown in Fig. 14. This data was obtained from the system shown in Fig. 5. An increase in pulse time and gas pressure did not result in an increase in maximum surface velocity beyond 2.7 m/min due to system limitations such as nozzle pipe diameter, slit opening area and/or compressed gas reservoir capacity.

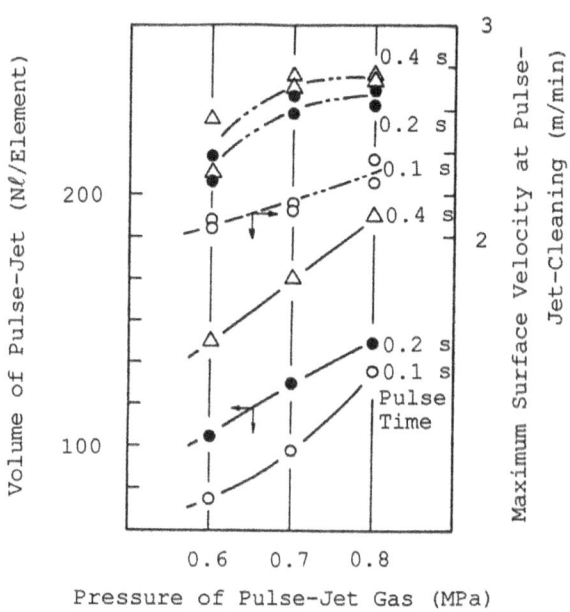

Fig. 14 Condition of Pulse-Jet-Cleaning

Fig. 15 shows the distribution of maximum discharge velocity at the end of the filters during jet-pulse cleaning. Some slight deviation was observed depending on slit location.

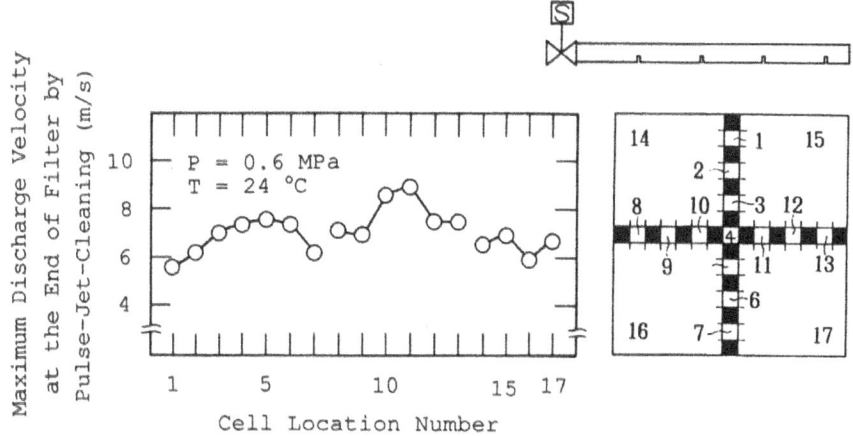

Fig. 15   Distribution of the Discharge Velocity at the end of
Filter by Pulse-Jet-Cleaning

(2) Pressure Drop

When we started operation of the dust collector installed at
the diesel engine laboratory, it was found that the filter
pressure drop was not maintained at a stable level, and that it
exceeded the specified level, being completely different from the
in-house test results.   As a result of examining the filters, the
thickness of the adhered dust layer was far greater than
predicted.   We also recognized that the pulse-jet-cleaning force
was insufficient because of the stronger adhesive nature of the
dust.   The reasons for this discrepancy in dust adhesion between
the in-house and field tests were thought to be: 1) difference in
dust particle size, and 2) water condensation while operated at
low temperatures, especially during daily starting and stopping.
Accordingly, we increased the maximum surface velocity by 30%
from the previous value, changing the compressed air pressure
from 0.49 MPa to 0.59 MPa.   We also added an air purging
operation for a fixed period of time after the system was stopped
in order to prevent condensation.   After these treatments, we
obtained a stable pressure drop as shown in Fig. 16.   The exhaust

341

gas volume, dust concentration and gas temperature varied from hour to hour because of changes in the number of diesel engines operated as well as differences between the engines themselves. In order to accommodate these changes, the pressure drop data presented was corrected to the following conditions.

Gas temperature: 300℃   Conversion of viscosity of air
Filtration rate: 1 m/min   Conversion at proportional
                  relationship for 0.5-1.5 m/min
Dust concentration: 0.3 g/Nm² Plotted over a range of
                  0.3-0.5 g/Nm³

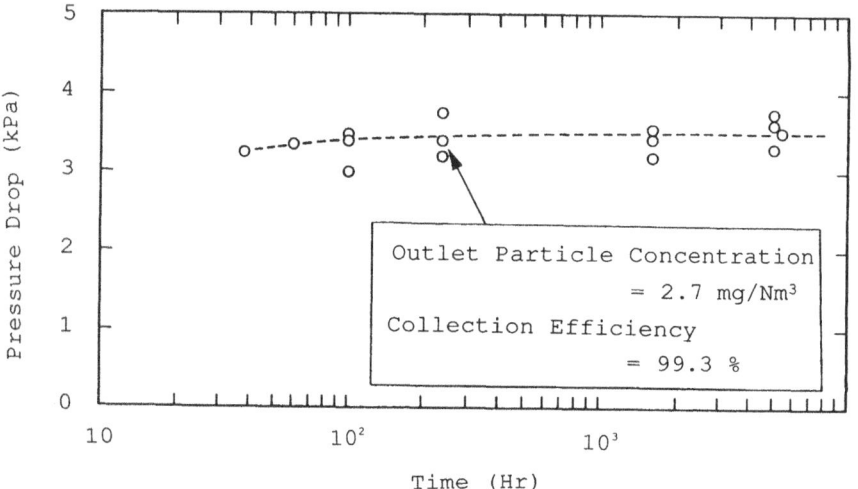

Fig. 16 Pressure Drop at the Dust Collector

After 6 months of operation (approximately 2,400 hours), some filters were removed from the collector for investigation. The thickness of the adhered dust layer was uniform at about 0.1 mm. The accumulation of dust at the bottom end of the cell hole was 20 mm. In addition, slight penetration of dust into the cell wall, in the range of 50 to 70 μm, was observed by SEM and X-ray analysis of the filter cross section. A photograph of this is shown in Fig. 17.

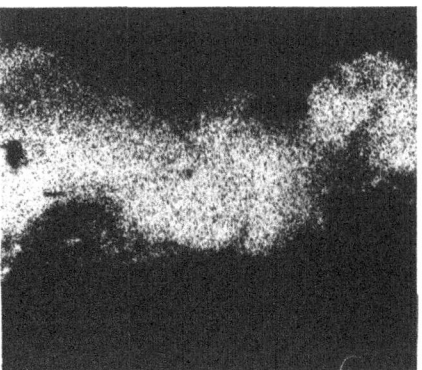

Fig. 17 Photomicrograph of Porous Thin Wall of used Element &
Xray-Photograph of Carbon (Soot) by EPMA (×400)

Moreover, the removed filters had up to 20% higher air flow
resistance than fresh ones, the removed filters having been
cleaned by blowing with compressed air at 0.59 MPa.

So far, ten or more commercial-scale dust collectors with
these honeycomb filters have been installed and are currently
operating satisfactorily. The applications for which these dust
collectors are being used include soot removal from exhaust gas
of stationary diesel engines, product recovery in an activated
carbon manufacturing process, recovery of iron oxide, removal of
soot from incinerator exhaust. In addition, we are also
proceeding with the development of a filter for high temperature
applications at temperatures of, for example, 850℃ in PFBC. The
conceptual construction of this filter is shown in Fig. 18.

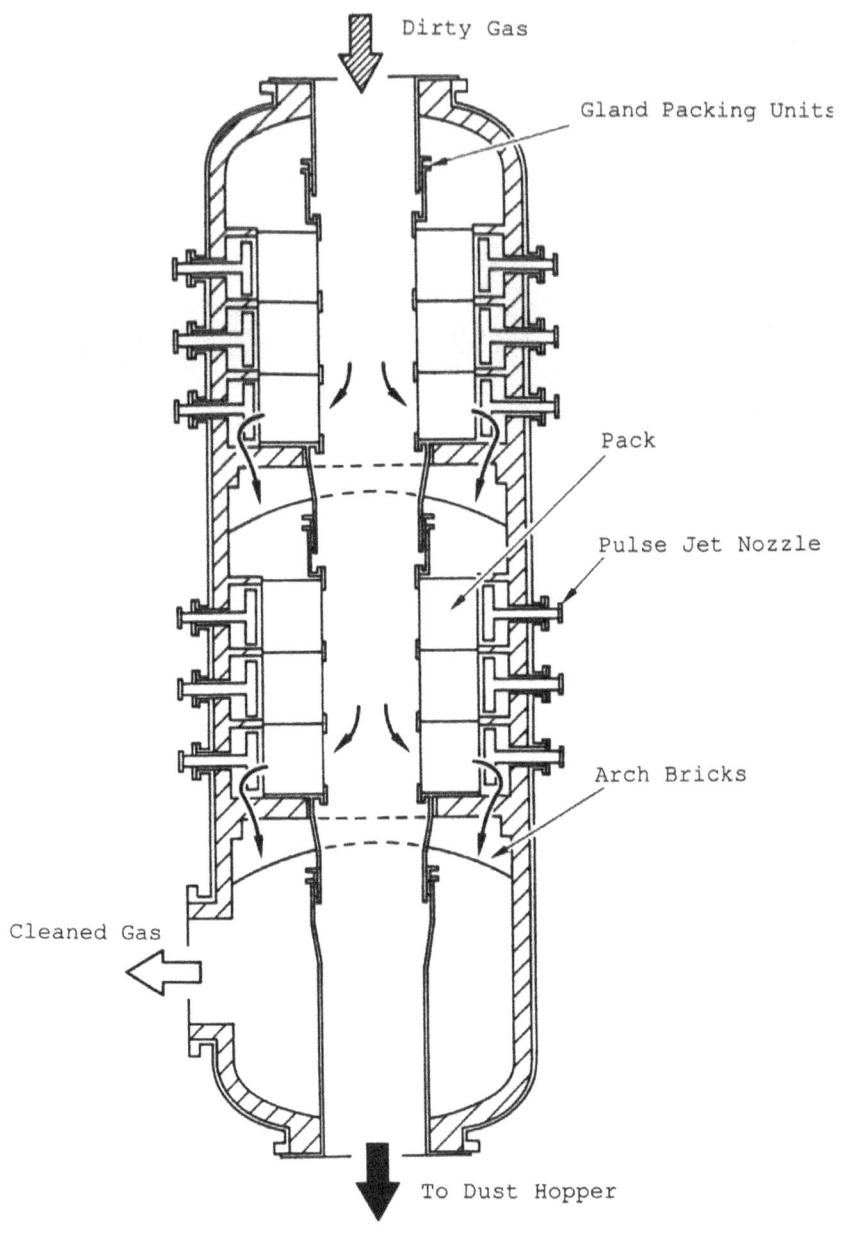

Dirty Gas

Gland Packing Units

Pack

Pulse Jet Nozzle

Arch Bricks

Cleaned Gas

To Dust Hopper

Fig. 18   Conceptual Construction of Filter Equipment for PFBC

CONCLUSION

Ceramic honeycomb filters can be used successfully in low to medium temperature ranges (up to 500℃). Dust collectors for the exhaust gas from stationary diesel engines have demonstrated stable operation for more than 5,000 hours at a maximum gas temperature of 400℃ . The filter offers the advantage of compact system size contributing to reduced system costs. However, in the case of high temperature applications, such as those at 850℃ , technical problems regarding dust sealing and so on remain that will require additional development work in the future.

REFERENCES

1. David F. Ciliberti and Thomas E. Lippert (1986) "Ceramic CrossFlow Filters for Hot Gas Cleaning" IchemE Symposium Series, No99 P193
2. Kazuhiko Takesa, T.U. & S.E. (1991) "Development of Particulate Trap System with Cross Flow Ceramic Filter and Reverse Cleaning Regeneration", SAE paper 910326

COMPACT CERAMIC MEMBRANE GAS FILTER

RICHARD F. ABRAMS
ROBERT L. GOLDSMITH
CeraMem Corporation
12 Clematis Avenue
Waltham, Massachusetts
USA

ABSTRACT

A ceramic membrane filter has been developed by CeraMem Corporation which is extremely compact (540 M^2/M^3) and capable of operation at temperatures exceeding 1270 K. The filter has been tested in a wide range of applications such as removal of flyash from flue gas, diesel exhaust soot filtration, and filtration of hot gases produced in pressurized fluidized bed combustors (PFBC) and gasifiers. All test results have consistently demonstrated complete particulate removal, low pressure drop at typical hot gas filtration face velocities, and complete recovery of clean filter pressure drop upon backpulsing.

INTRODUCTION

This paper describes a novel ceramic filter being developed for hot gas filtration by CeraMem Corporation under U.S. DOE and EPA SBIR grants. The need for hot gas cleanup in the power, advanced coal conversion, process and incineration industries is well documented and extensive development is being undertaken to develop and demonstrate suitable filtration technologies. In general, process conditions include (a) oxidizing or reducing atmospheres, (b) temperatures to 1350 K, (c) pressures to 2,000 kPa, and (d) potentially corrosive components in the gas stream. The most developed technologies entail the use of candle or tube filters, which suffer from fragility, lack of oxidation/ corrosion resistance, and high cost. The ceramic membrane filter described in this paper offers the potential to eliminate these limitations.

FILTER DESCRIPTION

Ceramic Gas Filter Construction

The construction of the CeraMem ceramic filter is based on the use of porous honeycomb ceramic monoliths. These high surface area, low cost materials are widely used as catalyst supports for automotive catalytic converters. The monoliths have a multiplicity of "cells" (passageways) that extend from an inlet end face to an opposing outlet end face. The cell structure can be round, square, or triangular, and the cell "densities" can vary from 4 to 224 cells per cm^2. Porosity of the honeycomb material can be from below 30% to over 50%. The mean pore size can range from about 4 to 50 μm. The superior properties of commercially available honeycomb ceramic monolith materials make them ideally suited for applications requiring high thermal stability, mechanical strength, and corrosion resistance. These rigid ceramics have been used for years as automotive catalyst supports where conditions of high vibration and thermal cycling are encountered in a combustion gas environment. Other applications of these materials as catalyst supports include emission control systems such as catalytic incineration and NO_x selective catalytic reduction (SCR).

The monolith structure used for catalyst support material is readily adapted to function as a filter to remove particulate matter from diesel engine exhaust. Unlike the catalytic convertor application in which automotive exhaust flows in a channel flow mode through the honeycomb cells, the diesel particulate filter (DPF) operates as a wall-flow filter. The carbonaceous soot in the exhaust gas is filtered on and within the cell walls of the monolith. This flow path is achieved by modifying the monolith structure by plugging every other cell at the upstream face of the device (Figure 1) with a high-temperature inorganic cement. Cells which are open at the upstream face of the monolith are plugged at the downstream face. Exhaust gas is thereby constrained to flow through the porous cell walls, and at appropriate intervals, the filter is cleaned by burning off the entrapped soot.

A variety of monolith sizes is available for DPF devices. Typical monolith characteristics are a square cell shape, a cell density of 16 cells/cm^2, a cell wall side of 2.1 mm, and a cell wall thickness of 0.43 mm. DPF devices operate at least in part as depth filters. The pore size of the cell wall material is quite large (20-35μm), and fine particulates enter and plug the cell wall. Deposition inside the walls leads to pore plugging by particulates and makes regeneration by backpulsing difficult, if not impossible.

CeraMem has developed technology for applying thin ceramic membrane coatings to honeycomb monoliths to produce crossflow

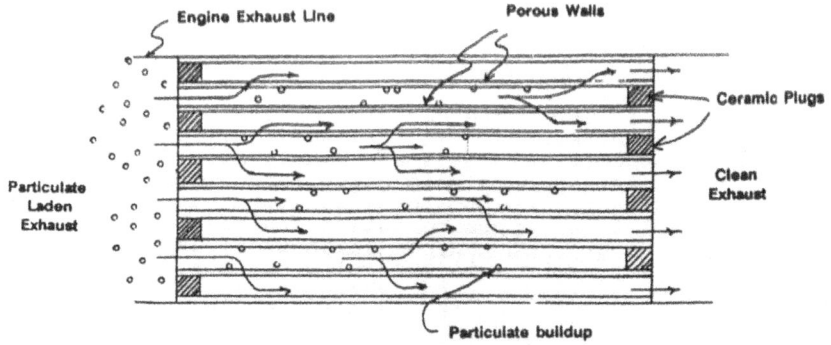

Figure 1: Ceramic Membrane Filter In "Dead-End" Flow Configuration

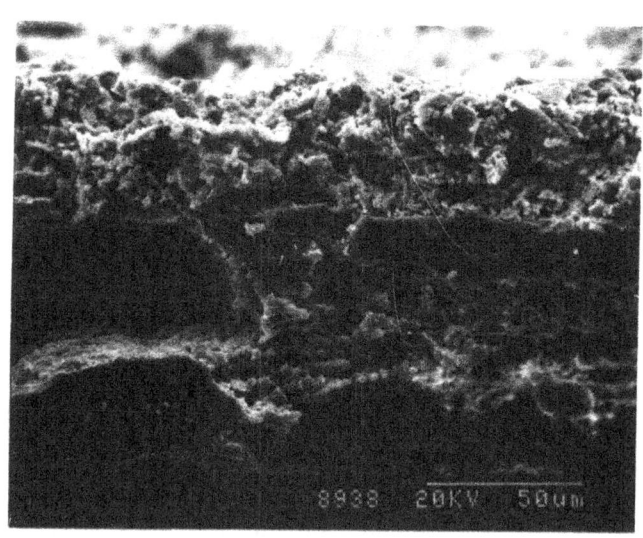

Figure 2. Scanning Electron Micrograph Of Membrane Coating

348

liquid microfiltration (MF) and ultrafiltration (UF) membrane modules. These membrane coatings have pore sizes substantially smaller than the pore size of the monolith support material, and the membrane coatings function as surface filters. Pore sizes of the different CeraMem membrane coatings are in the range from 1.5 μm (the coarsest) to 40 Angstroms. The coating technique involves filling the cells of the monolith with a liquid medium ("slip") containing a mixture of ceramic powders, dispersants, and polymeric binders. The pore structure of the monolith absorbs water from the slip, forming a cake of particles on the walls of the passageways. After a defined absorption time, excess slip is drained from the passageways, and the coated monolith is dried and fired (up to 1600 K) to bond the ceramic membrane particles to themselves and to the cell wall surfaces. Membrane pore size and porosity are determined primarily by the size of the ceramic particles used in the slip.

Ceramic Filter For Hot Gas Particulate Removal

To filter particulates from hot gas, CeraMem modifies the DPF device described above by applying a ceramic MF membrane to the cell wall surfaces. This technique creates a composite filter which can be operated as a backpulsable surface filter. The thin, membrane coating has a pore size approximately 100-fold finer than that of the monolith support (Figure 2). Thus, the retention efficiency of the filter for fine particles is determined by the membrane pore size. By keeping the membrane coating thin ($\approx$50 μm) the gas flow pressure drop is kept acceptably low. Yet, it is possible to use a large-pored, low-resistance support for the membrane. Because the support is coated by the membrane, the pore size of the support does not affect particle retention and the pore structure of the support does not become plugged by particulate matter.

In operation, particle-laden gas flows into the membrane-coated inlet cells. Particulate matter is collected in the inlet cells and the filtered gas exits the module via the downstream cells. As particulate material accumulates, pressure drop increases to a preselected level at which time the filter is cleaned by online backpulsing from the downstream end of the filter.

Properties Of Ceramic Membrane Filters

Ceramic Membrane filters with the sizes shown in Table 1 have been developed. The compactness of the filter relative to other gas filters is evident from the data of Table 2. This compactness leads to small filter housings since the filters can be installed in a closely packed array in the vessel. Both the filters and total filter systems are expected to have costs much lower than those of traditional hot gas particulate filtration systems.

Figure 3 shows a photograph of three of the commercially available filters currently produced. Figure 4 shows a photograph of an assembly of 16 filters in four steel casings with backpulse venturis on top of the filter casings. The filters are sealed in the casing by means of wrapping the filters in a ceramic fiber mat and fitting the filters into the casing. The compressed mat serves to cushion the filters and hold them firmly in place and to provide a particle-tight seal. This array of filters contains approximately 56 m² of filter area in a cube about 0.6m on a side. The assembly could be mounted directly into a tube sheet in a hot gas filter vessel.

TABLE 1
Characteristics Of Ceramic Gas Filters

Dimensions Of Ceramic Gas Filter	Filter Area, m²
144mm Diameter x 152mm Long	0.9
144mm Diameter x 305mm Long	2.0
150mm Side (Square) x 305mm Long	3.4

TABLE 2
Comparison Of Ceramic Membrane Filter With Other Filters

Filter Type	Filter Dimensions	Filter Area (m²)	Area/Volume Ratio (m²/m³)
CeraMem	150mm sq. x 305mm	3.4	155
Fabric Bag	150mmφ x 6.1m long	2.9	8
Candle	60mmφ x 1.5m long	0.26	19
Crossflow	300 x 300 x 100mm	0.77	25

The retention efficiency of the ceramic filter is greatly increased by the addition of the membrane coating. The retention of filters, with and without the membrane coating, has been measured for dilute aqueous suspensions of narrowly-sized alumina particles. The uncoated monolith passes particles with a size of 5 μm almost completely. In contrast, the membrane-coated filter has substantially complete retention for 5 μm particles and >95% retention for 0.5 μm particles. Given the additional particle capture mechanisms in gas filtration, a membrane-coated filter with these liquid retention properties can be expected to have substantially complete retention for submicron particles.

Figure 3: Photograph Of Three Sizes Of Ceramic Membrane Gas Filter

Figure 4: Photograph Of Sixteen Ceramic Filter Assembly With Venturi

Figure 5 shows flow versus pressure drop data for 300mm long filters with and without the membrane coating. The membrane coating increases the resistance over that of an uncoated filter about two- to three-fold. For the uncoated filter, the increase in pressure drop with increasing filtration rate is caused by pressure drop in the filter passageways. For membrane coated filters the primary resistance to gas flow is that of the membrane coating itself.

FILTER TESTING

Thermal Durability Testing

Thermal cycling and shock tests have been undertaken at CeraMem to evaluate the thermal durability of the filters. A 118mm diameter x 127mm long filter was fabricated and tested for air flow/pressure drop and alumina retention before exposure to thermal cycling and shock.

The filter was placed in an electric kiln and thermal cycling tests were performed in which the kiln was fired at the maximum firing rate to heat as quickly as possible to 1200 K. The kiln was held at 1200 K for one hour. Then kiln power was shut off, allowing the kiln to cool to ambient temperature. Each heating cycle took about 45 minutes and the total cycle required about 12 hours. After 50 cycles the filter was removed and inspected. No change was observed in its visual properties.

The filter was then subjected to thermal shock testing in the electric kiln. Pulses of ambient temperature compressed air were discharged onto the face of the filter through a stainless steel tube connected to the compressed air source. A timer/solenoid valve assembly was used to expose the filter to one second pulses of cold air at four minute intervals. During the interval between pulses, the filter inlet face temperature recovered to the kiln temperature of 1200 K after experiencing a drop of several hundred degrees during the cold pulse. The filter was subjected to 1005 pulses and removed from the kiln for evaluation. No visual effects from the pulsing were observed. The thermal cycling and shock described above did not have any effect on filter performance as measured by air flow/pressure drop and retention for 0.5 μm and 5 μm alumina in the alumina suspension filtration tests described above.

Short-Term Field Test Results

To date, field tests of the ceramic filter for particulate removal have been conducted at fourteen sites on a variety of gas streams and under a wide range of test conditions. Results of two hot gas tests are presented below. In general, the following performance characteristics have been observed

353

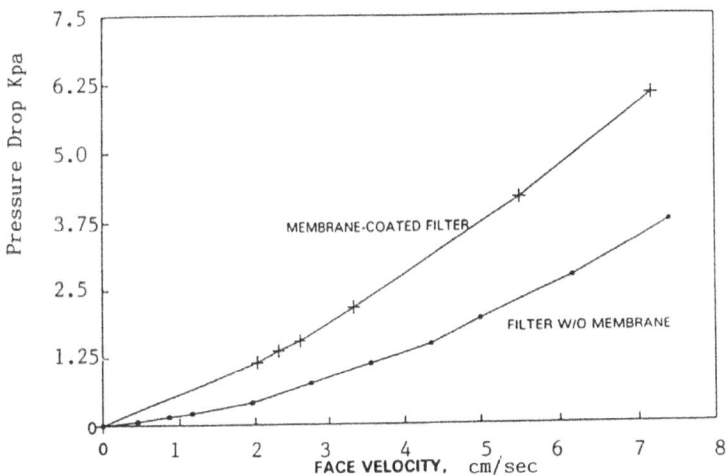

Figure 5.   Pressure Drop Vs. Flow For Ceramic Membrane Gas Filters

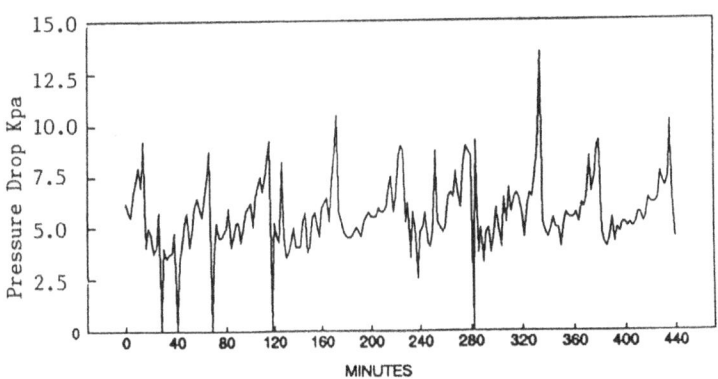

Figure 6:   Pressure Drop Of Ceramic Filter In Tests At EERC

in tests performed to date:

1. Filtration face velocity (equivalent to an "air to cloth ratio") for flue gas tests is comparable to that for pulse jet bags operating at the same pressure drop. (Typical ambient pressure/temperature pressure drop is 1 kPa at 0.02 m/sec) In hot gas tests, flow-pressure drop characteristics have been comparable to those for other ceramic filters.

2. Complete regeneration by a simple backpulse technique is achieved; i.e., no increase in clean filter resistance over repetitive cycles is observed. For example, no increase in clean filter pressure drop was observed at Plant Scholz testing in which over 7000 cleaning cycles were performed.

3. No plugging of the filter passageways by badly caking particulates is observed.

4. Essentially complete particulate removal, including submicron particulate matter, is achieved.

Tests At EERC/University Of North Dakota

In Phase I of a DOE SBIR grant, feasibility tests were conducted at the Energy And Environmental Research Center (EERC) at The University of North Dakota. The reactor system used for testing the filter was EERC's 0.8 kg/sec gasifier. As configured and tested, the reactor product gas was passed through a primary cyclone before introduction into a hot gas cleanup test loop containing a ceramic filter. A heated backpulse system was installed to provide hot backpulse nitrogen. Depending on the discharge pressure and duration of the backpulse cycle, the temperature of the backpulse gas entering the test loop decreased with time. Backpulse frequency and duration were controlled manually to achieve the desired number of pulses and pulse duration.

The fluidized bed calciner was operated in a hydrogen production mode using dolomite as the bed material and Wyodak coal as fuel. Steam and a small amount of oxygen were heated to approximately 1000 K and fed into the bottom of the reactor bed. During the runs, the differential pressure across the filter was monitored continuously and test system operators initiated filter backpulsing when the pressure drop across the filter began to rise at a fairly rapid rate. Backpulsing was initiated at a pressure differential of about 8.72 kPa. Based on the gas flow, ash loading, expected ash bulk density, and filtration cycle time, the passageways of the filter were substantially filled with ash at the time of backpulse regeneration. During the tests, backpulsing was controlled manually, and typically 2 to 4 pulses of 1/2 second duration were used. The backpulsing nitrogen pressure was 517 kPa at the nitrogen cylinders, and less (but not measured) at the ceramic filter.

Test conditions are summarized in Table 4 and representative pressure drop data from 9 cycles of the typical performance observed during the test are shown in Figure 6. Within-cycle pressure fluctuations are associated with gasifier pressure variation caused by a variable coal feed rate. The clean filter baseline differential pressure was maintained through these nine cycles, and over the entire test period.

TABLE 4
Test Conditions For Tests At University Of North Dakota

Filter Dimensions:	118mm dia. x 127mm long
Filter Area:	0.45 m²
Coal Type:	Wyodak
Face Velocity:	0.09 to 0.10 m/sec
Test Temperature:	920 K
Test Pressure:	200 kPa
Ash Content:	2.3 gm/Nm³
Typical Filtration Cycle:	50 minutes
Backpulse Conditions:	Offline Heated Nitrogen
	Four 0.5-second pulses

Although feed ash loadings were determined accurately, attempts to measure the filtered gas ash loading were unsuccessful because of the design of the piping arrangement of the test loop, an arrangement that resulted in ash introduction into the sampling loop during backpulsing. However, it is believed that the ash retention of the filter was substantially complete based on visual observations of the downstream side of the filter and piping system, which were completely clean.

A summary of the conclusions from the EERC tests is as follows:

In 45 filtration/backpulse regeneration cycles in two week test period:

1. No change was observed in the clean filter pressure drop.

2. Ash removal efficiency appeared to be 100%.

3. No degradation was observed in any filter properties after repetitive thermal cycling.

Tests At Westinghouse Science And Technology Center

Short term feasibility tests were conducted at the hot gas test facility of the Westinghouse Science and Technology Center in Pittsburgh in December, 1991. In these tests Grimethorpe ash was injected into combustion gas and a series of filtration cycles were performed over a four-day test period. The general test conditions are given in Table 5. A plot of the typical filter pressure drop over a series of 10 cycles is shown in Figure 7.

TABLE 5
Test Conditions For Tests At Westinghouse

Filter Dimensions:	118mm dia. x 127mm long
Filter Area:	0.45 m²
Fuel Type:	Gas
Face Velocity:	0.04 to 0.08 m/sec
Test Temperature:	920 K
Test Pressure:	690 kPa
Ash Content:	3500 ppm Grimethorpe ash
Typical Filtration Cycle:	15 minutes
Backpulse Conditions:	Online Cold air @ 1030 kPa
	Three 0.5-sec. pulses

Conclusions from the Westinghouse test are as follows:

In 40 filtration/backpulse regeneration cycles in four day test period:

1. No filter plugging was observed.

2. Ash removal efficiency was about 100% (0.5 to 1 ppm ash in filtered gas).

3. Measured pressure drop was two-fold higher than expected, based on extrapolation from ambient air data. The reason for this higher value has not yet been determined.

FUTURE WORK

High Temperature Environment Durability Testing of Hot Gas Filter Materials

The Center for Advanced Ceramic Materials at the Virginia Polytechnic Institute and State University (VPI) is testing CeraMem hot gas filter materials in high temperature environments. The VPI program is funded by the Electric Power Research Institute (EPRI) and the testing of the

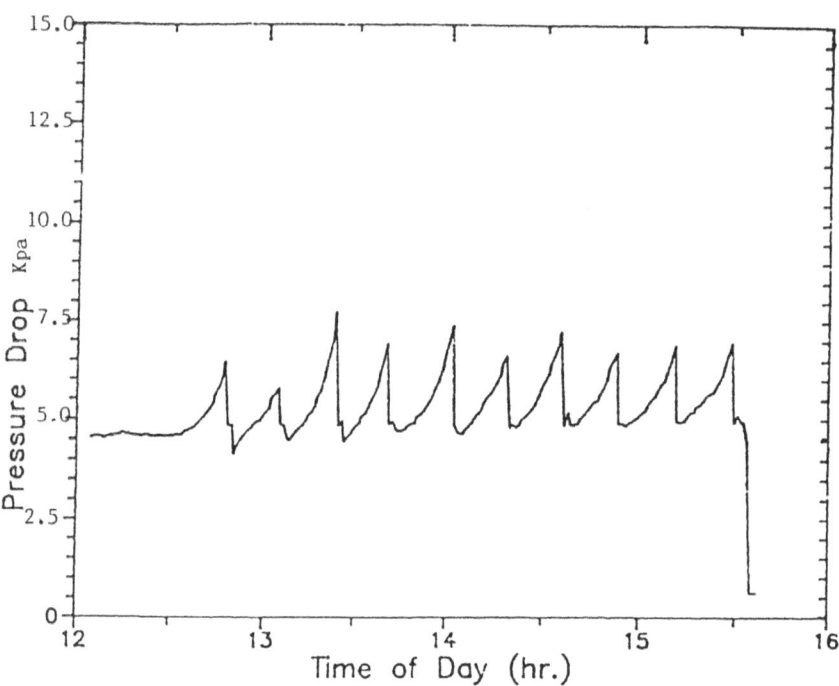

Figure 7:   Pressure Drop Of Ceramic Filter In Westinghouse Tests

CeraMem filter materials is being performed by VPI under an agreement between CeraMem and EPRI. These tests will help determine if the cordierite supports, membrane materials, and/or the plugging cement degrade under simulated hot gas conditions. Examples of degradation could be weakening of the cordierite due to chemical attack and/or microstructural changes, membrane densification causing increased pressure drop, or weakening of plugging cement by chemical attack potentially leading to plug blowout during filter operation. The following describes the materials and experimental procedures for the planned testing.

Samples: All samples are based on cordierite honeycomb produced by Corning, Inc. The cordierite is 50% porous with a 25 μm average pore diameter. The honeycomb has 16 square passageways per square centimeter of frontal area which run parallel from one end face to the opposite end face. All samples are coated with the CeraMem ceramic membrane described above.

Three different types of samples will be exposed to the high temperature environments. The sample types are listed in Table 3. The first sample type is a 25 mm cube with no passageways plugged with cement. These samples will be used for compression strength testing. The second set of samples have a 19mm square cross section and a 50mm length. These samples are also not plugged with cement. After exposure, the passageways will be plugged with silicone adhesive to form the dead-ended passageway filter. These samples will then be used to determine if membrane performance degrades during environmental exposure. The third set of samples are the same as the second except that they are plugged with inorganic cement used to configure membrane-coated monoliths into hot gas filters. These tests will determine if the cement degrades during high temperature exposure.

Exposure Conditions: Samples will be split into three groups containing each sample type for exposure testing. One group will be held as control samples. The other samples will be placed in a furnace containing one of two different high temperature environments. The first set will be exposed to 1120 K static air for 500 hours. The second set will be exposed to an 1120 K static atmosphere of 10 ppm sodium nitrate, 8% steam, and 92% air for 500 hours. The exposure groups are summarized in Table 3.

Sample Testing: After the post-exposure inspection, several tests will be conducted including crush strength, membrane retention, and gas flow versus pressure drop characteristics of the filters. Microstructural and chemical analyses using scanning electron microscopy/energy dispersive x-ray analysis (SEM/EDAX) will also be performed. These tests are summarized in Table 3.

TABLE 3
Summary of Samples, Exposure Conditions, and Testing
Procedures For the VPI Hot Gas Filter Durability Test Project

Exposure	Sample Type	#Samples	Test Procedure
Control	Cube	10	RT Crush; SEM/EDAX
Control	Cube	10	HT Crush; SEM
Control	Cube	10	RT Crush at CeraMem
Control	Cube	10	RT Crush at Corning
Control	50mm; no plug	2	Memb. Perf.; SEM
Control	50mm; w/ plug	5	Memb. Perf.; SEM
HT Air	Cube	10	RT Crush; SEM/EDAX
HT Air	Cube	10	HT Crush; SEM
HT Air	50mm; no plug	2	Memb. Perf.; SEM
HT Air	50mm; w/ plug	5	Memb. Perf.; SEM
HT Na/H_2O	Cube	10	RT Crush; SEM/EDAX
HT Na/H_2O	Cube	10	HT Crush; SEM
HT Na/H_2O	50mm; no plug	2	Memb. Perf.; SEM
HT Na/H_2O	50mm; w/ plug	5	Memb. Perf.; SEM

RT = Room Temperature; HT = High Temperature

The majority of samples will be used for mechanical testing.
Compression strength will be tested at VPI using a Universal
Testing Machine modified to crush samples at temperatures as
high as 1270 K. Samples in lots of ten will be crushed at
either room temperature or 1120 K. The crosshead speed during
testing will be 2.5mm per minute. Additional control samples
will be crushed using similar procedures by both CeraMem and
Corning, Inc. Controls will ensure that the mechanical
testing is reproducible and that CeraMem can correctly
perform crush strength tests on other samples of interest.

Any changes in membrane module performance will be quantified
by measuring membrane retention and gas flow versus pressure
drop characteristics. Membrane retention will be measured by
the alumina retention test described above. Gas pressure
drop of the filter modules will be measured by flowing
nitrogen through the filters at room temperature. Pressure
drop will be typically measured at face velocities of 0.01 to
0.08 m/sec.

Analytical investigation will be conducted by SEM/EDAX.
After mechanical and filter performance testing, samples will
be analyzed for microstructure and chemical changes in the
cordierite support, membrane coating, and cement plugs. If
additional analysis is warranted based on the SEM

investigation, phase analysis by x-ray diffraction and trace chemical analysis by inductively coupled plasma spectroscopy may be performed.

Additional hot gas materials durability testing in a simulated hot gas process will be conducted at the Science and Technology Center of the Westinghouse Electric Company. This test will have the same goals as those at VPI except that fewer samples will be tested and backpulse cleaning at temperature will be simulated.

Future Field Trials

An objective of CeraMem's development program is field demonstration of the filter in longer term tests. Under the CRADA program, the DOE's Morgantown Energy Technology Center (METC) has agreed to work with CeraMem to evaluate CeraMem's ceramic gas filter at METC's AFBC/HGCUP (Atmospheric Fluidized Bed Combustion/Hot Gas Cleanup) facility. METC's evaluation of CeraMem's filter will be in comparison to its existing candle filter data base derived from the same AFBC/HGCUP test facility, to determine its particulate removal performance, residual pressure drop buildup characteristics, regenerability, and resistance to attack under coal-burning AFBC environment. METC's existing AFBC/HGCUP test facility was used in testing and evaluating candle filters at 920 K.

The CRADA program will consist of two phases: Phase I runs of about 400 hours to verify and compare performance of the filter with METC's candle filter data; Phase II for extended operation to a total of 1000 hours. The tests on the filter will be at 169 kPa and temperatures ranging from 920 K to 1255 K.

The test setup will consist of a 3.3 m long vessel in which a specially configured CeraMem multi-filter element will be mounted. The multi-filter will consist of a 144mm dia. x 150mm long filter with cells plugged in a special pattern to produce four small filters. The total filtration area of the unit will be 0.45 m². The multi-filter will be wrapped with MM ceramic fiber mat (3M) and mounted in a 150mm dia. x 300mm long 316 stainless steel pipe. Four fast-acting solenoid valves will be used to backpulse the filters with compressed air when regeneration is needed.

In addition to the DOE sponsored field test, filters will be installed and operated for pilot tests on several other hot gas streams, including:

- 0.28 m^3/sec off gas from a medical waste incinerator, at 922-1255 K

- 0.08 m^3/sec hot gas from lead smelter, at 980 K

- 0.09 m^3/sec shale gasifier hot gas, at 810 K

- Sorbent injection/hot gas filtration for SO_2 removal, at 810 K

- Sorbent injection/hot gas filtration for H_2S removal, at 1030 K

- Simultaneous particulate/VOC removal using catalytic CeraMem filter at 650 K

CONCLUSIONS

A novel ceramic membrane filter has been developed and tested in a variety of hot gas filtration applications. The filter has the attributes of ultra-high particulate removal efficiency, complete regenerability, compactness and low cost. Continued testing of the filter in hot gas environments, for long duration, and to determine its resistance to attack from potentially corrosive components of the gas streams is underway.

ACKNOWLEDGMENTS

The development of this technology is being assisted by Dr. Ramsay Chang and Mr. Richard Brown of EPRI under EPRI Technology License Agreement RP1402-61. The tests at EERC were conducted under the direction of Mr. Jay Haley. The tests at Westinghouse were conducted under the direction of Dr. Thomas Lippert. Finally, the active assistance and support of Dr. Norman Holcombe of DOE-METC is greatly appreciated.

HIGH TEMPERATURE GAS CLEANING – CATALYST RECOVERY

RICHARD WREN
Technical Support Engineer
Scientific and Laboratory Services Department
Pall Europe Limited, Walton Road, Portsmouth, PO6 1TD

ABSTRACT

The modern   petroleum industry has to operate within tight restrictions on refinery effluent discharges. One source of effluent discharge is the flared 'off-gas' from the catalytic cracking of crude oil, which has traditionally contained appreciable levels of catalyst particles. To effectively reduce airborne catalyst emissions, Pall have installed blowback filter units which are capable of removing greater than 99.97% of entrained catalyst particles at 1μm (by weight) from the 600·C 'off gas'.

A regulated blowback of the filter unit enables the porous sintered stainless steel media to be cleaned and regenerated whilst maintaining full forward flow.

In conjunction with existing stages of cyclone units, the Pall blowback filter offers an economic means of reducing the catalyst emission levels to within the present environmental discharge limits.

INTRODUCTION

With ever tightening restrictions on discharge emission levels all process industries are having to review their present operations and incorporate means of reducing to a minimum their effluent discharges (whether released into air, water or land).

This has been very actively pursued in the oil refining industry. In outline a refinery converts crude oil into hydrocarbon based products ranging from liquefied petroleum gases to petroleum coke, by means of physical, physico chemical and chemical processing methods.

Of particular interest is the fluid catalytic cracking (FCC) process, where long chain hydrocarbon oils are converted to gas, liquid petroleum gas (LPG) and blended components for gasoline, gas oil and fuel oil. The catalysts used in general are synthetic zeolites (crystalline alumina silicates), which are usually doped with rare earth and/or precious metal elements.

In the cracking reactor, heavy carbonaceous materials and coke deposit on the surface active catalyst reducing its efficiency. Continuous regeneration of the catalyst is achieved by combustion of the deposits in a fluidised bed regenerator.

During the recirculating cycle from the reactor to the regenerator catalyst fines are carried over in the high temperature flue gas stream (up to 740·C), which prior to passing through power/heat recovery units and eventual discharge to atmosphere (via a 'flare stack') must be stripped of the catalyst.

Most of the catalyst fines entrained in the regenerator flue gas are removed by a number of cyclone stages within the regenerator vessel and returned to the bed. A final stage fines removal unit (typically a third stage cyclone) further cleans the gas before it passes downstream to other equipment. Early experiments with energy recovery equipment in FCC flue gas streams demonstrated that the rotating equipment could not last more than a few hundred hours unless an efficient stage of separation was added to reduce carry over of catalyst fines .

Traditionally, electrostatic precipitators or scrubbers have been employed for final particulate removal prior to exhaust into the environment. However, these methods are limited by their maximum operating temperature and their performance is variable with fluctuating gas flows and catalyst loadings.

Pall have developed a fourth stage high temperature resistant sintered metal filter, which can overcome these operating and performance limitations. The selected grade of filter media is capable of removing the catalyst to an efficiency of greater than 99.97% at 1μm (by weight) and upon reaching pre-determined differential pressure or an elapsed time can regenerate the filter media using a blowback technique.

The development, installation and operating costs of this equipment are intended to effectively reduce the catalyst loadings enabling the refinery to operate within the existing and future emission regulations under the BATNEEC (Best Available Techniques Not Entailing Excessive Costs) principle.

PROCESS DESCRIPTION

Although operating parameters of a FCC unit will vary from one refinery to another, the principle of operation will be the same (see Diagram 1).

To achieve the 'cracking', pre-heated feedstocks (typically gas oils from the vacuum distillation and residuals heated to around 370°C) are contacted with the zeolite catalyst in the feed riser line of the reactor at a temperature of about 725°C. The catalyst is sufficiently hot to vaporise the feed and provide heat for the cracking reaction which takes place within a few seconds. The hydrocarbon/catalyst mixture is separated within the reactor using multi-cyclones to prevent product contamination. The produced hydrocarbons then go to the fractionater at about 525°C, where they are fractionated into overheads, light and heavy cycle gas oils. These fractions are then subsequently processed by further downstream equipment before final products are achieved.

The separated, spent catalyst falls to the stripping section of the reactor where live steam is injected to remove as much residual hydrocarbon as possible. Spent catalyst then feeds on level control to the regenerator, where air for combustion and fluidisation is introduced to burn off carbon deposits on the catalyst so regenerating it. Regenerated catalyst is separated from the flue gas and transferred back to the reactor riser so completing the cycle. Separation usually takes place by two stages of cyclone separators located inside the regenerator.

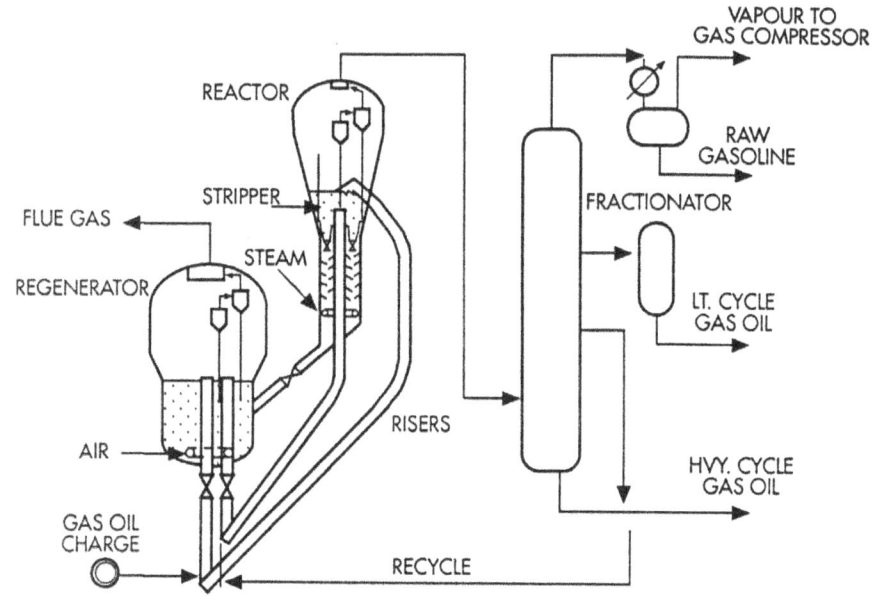

Diagram 1 : Schematic of FCC unit.

The exiting flue gas from the regenerator enters the third stage separator and flows downward and radially outward through a gas distribution grating into swirl veins inserted at the inlet ceramic cyclone tubes. Entrained catalyst is separated from the flue gas in the cyclones. The cleaned flue gas stream then exits the third stage separator through a top outlet nozzle and is directed into an orifice chamber.

The concentrated catalyst exiting the bottom of the third stage separator is entrained in a relatively small flue gas underflow stream. The underflow leaves the bottom of the third stage discharge duct via a critical orifice for final recovery prior to discharge to the atmosphere.

It should be noted that the flue gas circuit surrounding the catalyst regenerator is one of the most hostile environments in the modern petroleum processing industry.

It is in this position that Pall have designed and installed at many locations in the world, a bulk catalyst recovery unit to remove the concentrated catalyst from the flue gas before it is re-combined with the third stage overflow downstream of the orifice chamber. The 'clean' gas then feeds power and heat recovery units before final discharge at the flare stack (see Diagram 2).

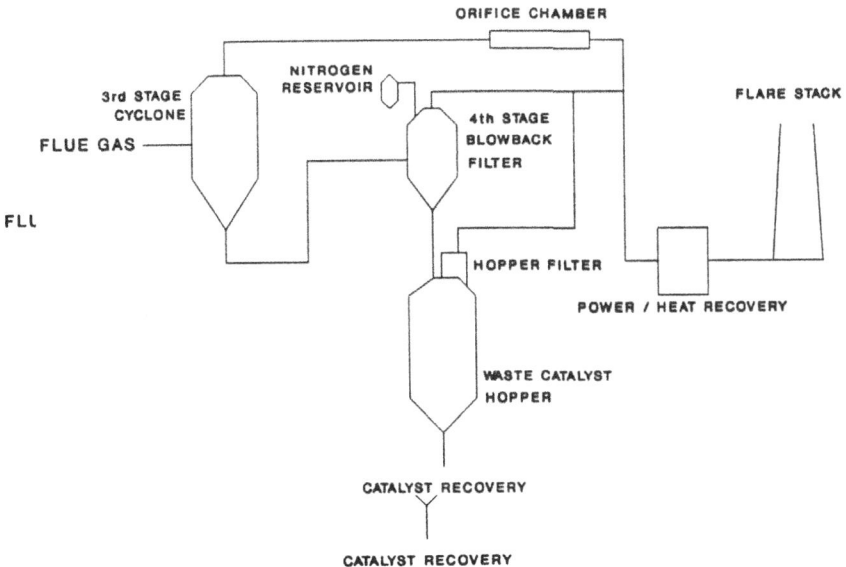

Diagram 2 : Flue gas flow.

PROCESS REGULATIONS

An operating refinery is tightly bound by licence to operate within a set of regulations. The emission specifications are usually stipulated by the country in which the operator is processing, but will compare to the Environmental Protection Act (1990) controlled by Her Majesty's Inspectorate of Pollution (HMIP).

As provided by the Act the current permissible level for release to air of particulate matter arising from process units should not exceed 100 mg/m^3. This is interpreted by the HMIP Chief Inspectors Guidance Notes (IPR1-15).

This then is the maximum limit that can be discharged to air and places a performance criteria for both the third stage cyclone and the Pall blowback filter.

EQUIPMENT DESCRIPTION

The Pall fourth stage blowback filter assembly has been designed to remove the catalyst from the third stage underflow with a 99.97% efficiency by weight.

The filter consists of a number of porous sintered metal cylinders held in a tubesheet arrangement, through which the gas stream passes in an out to in direction. The stainless steel metal filter elements are robust and easily installed. The selected efficiency of the media enables the filters to collect the catalyst fines along the outer surface of the length of the element. At pre-determined parameters, the built up catalyst 'cake' is dislodged by a jet pulse blowback procedure (without stopping forward flow) which effectively regenerates the filter media.

The collected catalyst is then passed into a 'waste catalyst hopper' which has sufficient capacity to allow for the catalyst to cool prior to final discharge into a road tanker.

Filter Elements

These filter elements are manufactured using PALL 'PSS' media. The sintered stainless steel media is produced by an exclusive proprietary sintering process of metal powder. No compression of the powder is used prior to sintering resulting in a highly permeable media with a voids volume of nearly 50%, which gives excellent flow/pressure drop characteristics.

The media can be produced in a range of stainless steel alloys, and a selection of an alloy for the fourth stage blowback application depends on the design temperature of the FCC unit;

TABLE 1

MAXIMUM ALLOWABLE TEMPERATURE (Reducing Environment)	METAL ALLOY
340·C	316L Stainless Steel
620·C	310SC Stainless Steel

The media sheets of the selected alloy are then rolled and welded and made into elements. Each element provides $0.37m^2$ of filter area. The element is closed at one end by a welded end cap. A suitable adapter is welded to the open end for fitting to a tubesheet.

In high temperature systems such as the FCC, elements are bound together in groups of three to form rugged triad clusters. Bands are welded to solid joiner rings at regular intervals between the 'PSS' sections, to provide maximum stiffness and rigidity of the triads.

The number of elements required per unit is a function of the maximum operating flow rate and ensures that the flux rate (ie, flow rate per unit of media area) does not exceed design flux.

Blowback Principles

Catalyst fines removal by the filter media is carried out to either a pre-determined differential pressure or elapsed time, followed by blowback/cleaning of the elements by reverse flow with a pressurised filtered inert gas (such as nitrogen). Various blowback techniques are available, but jet pulse/venturi is preferred for this application due to reduced blowback volume and better solids discharge from the filter surface. For this system, a fitting with a venturi nozzle is provided in the throat of each filter element.

The nozzle is designed to provide the required velocity for minimum gas consumption and efficient cleaning. The blowback gas is directed to the venturi by a small diameter nozzle located just above each filter element, sending a pulse of gas through the filter element in the reverse direction to the process flow.

This high velocity gas pulse releases the catalyst from the outer surface of the filter element. The technique reduces the amount of blowback gas consumed and eliminates the necessity for large inlet and outlet valves, as element isolation from the vessel to system is not required. It also improves the blowback efficiency and reduces the overall differential pressure allowing for further catalyst removal. The time required to provide the pulse is generally less than one second.

This cycle of cake build up and subsequent release is then continuous, whilst maintaining full forward flow to the filter unit (see Figures 1 and 2).

FIGURE 1. HYPOTHETICAL CAKE STRUCTURE

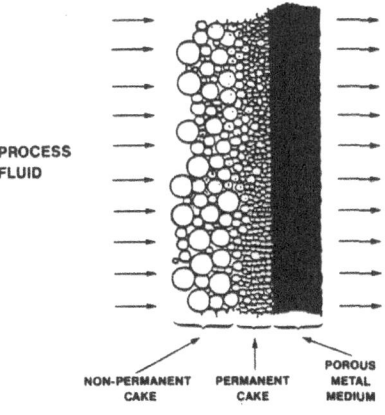

PROCESS FLUID

NON-PERMANENT CAKE   PERMANENT CAKE   POROUS METAL MEDIUM

FIGURE 2. HYPOTHETICAL CAKE RELEASE

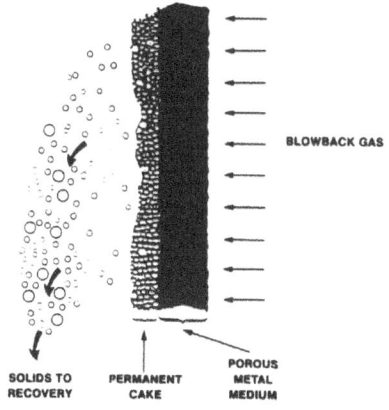

BLOWBACK GAS

SOLIDS TO RECOVERY   PERMANENT CAKE   POROUS METAL MEDIUM

Over a number of cycles a 'saw tooth' plot of differential pressure versus time can be seen, in which the recovery pressure at the beginning of each cycle reaches an equilibrium value (see Figure 3).

371

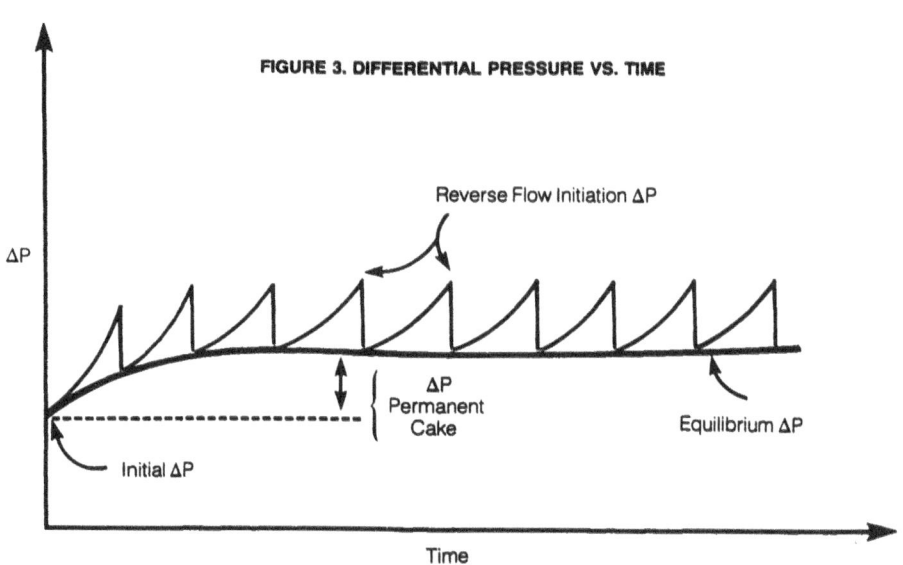

FIGURE 3. DIFFERENTIAL PRESSURE VS. TIME

To assist in the blowback recovery of the catalyst in this application the vessel and filter tubesheets are designed for a downdraft mode of operation. This consists of a central pipe extending upwards from the contaminated gas inlet to below the tubesheet. The inlet gas flow is therefore in the downward direction, assisting released catalyst in settling to the bottom of the vessel (see Figure 4).

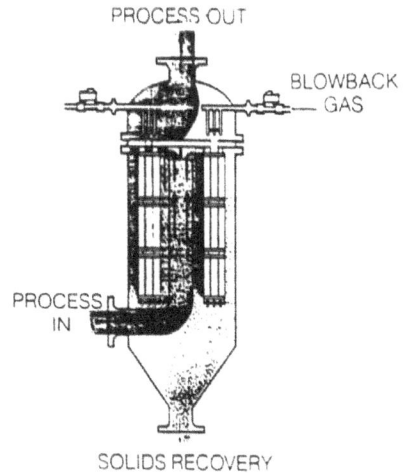

PROCESS OUT

BLOWBACK
GAS

PROCESS
IN

SOLIDS RECOVERY

Figure 4 : Downdraft Design

The tubesheet is divided into segments, which are 'blown back' sequentially at intervals allowing the filter to maintain its full forward flow during blowback operations.

In this application, the blowback gas is nitrogen, which must be dry and filtered to a level equal to or finer than the 'PSS' metallic filter. This precaution will eliminate any contaminants blocking the downstream surface of the blowback filters. The nitrogen is stored at 2-3 times the process pressure in a suitably sized reservoir so as to maintain the required pressure whilst delivering the volume of blowback gas required.

The 'recovered' catalyst from the blowback is then collected in a waste catalyst hopper which acts as a cooling/storage facility. To allow discharge into and venting from the hopper a smaller low flow blowback filter is installed in the top of the hopper vessel. The filtered gas from the hopper filter can then be recombined with the main 'filtrate' line.

The application of Pall 'PSS' blowback filters to FCC off gas has been supported by a continuing programme of test work, in conjunction with refinery operators, over the past 10 years.

373

CASE STUDY

Pall fourth stage catalyst recovery blowback filters have been supplied to Mobil Oil Refineries in Germany, United States of America and England.

Each refinery differs slightly in production capacity and as a result, the size of the filter unit supplied by Pall reflects these variations.

Mobil Coryton is the second largest refinery in the United Kingdom, and at full production processes 60,000 barrels of crude per day. At the design conditions, the catalyst loading of the flue gas from the regenerator is of the order of 170 kg/hr (corresponding to a catalyst concentration of approximately 600 mg/nm³). After separation at the third stage cyclone the catalyst concentration in the underflow line to the fourth stage filter unit has been increased to 30 g/nm³ (ie, a 50 fold increase).

At a catalyst removal efficiency of 99.97% (by weight) this results in a solids loading of 10mg/nm^3 exiting the Pall filter, which when combined with the overflow from the third stage separator results in an overall catalyst discharge level at the stack of less than 50mg/nm^3.

When operating under normal process conditions the filter unit has a 'clean' to 'loaded' differential pressure range of 80 mbar to 110 mbar respectively (this can vary according to catalyst loading and flow rate fluctuations). The frequency of blowback and duration of the pulse can be altered according to the catalyst loading. To obtain an effective blowback of the catalyst Mobil maintain nitrogen pressure at 8 bar g. The size of the pipework on the inlet to the reservoir and the capacity of the reservoir itself ensure the nitrogen pressure sufficiently recovers before the next blowback cycle starts (see Plot below for Blowback Cycles).

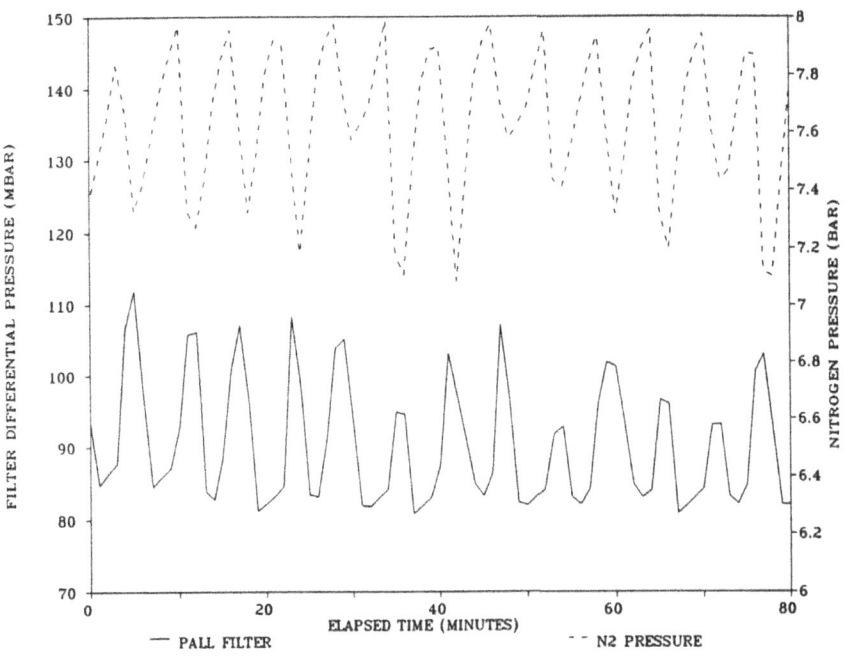

Figure 5 : Mobil Coryton Blowback Cycles.

The unit has been in operation since early April 1992, and by the blowback separation technique has been recovering over 3.5 tons of catalyst per day.

CONCLUSIONS

At present, Pall's fourth stage blowback filter offers an economic means (which are in-line with the BATNEEC principle) of effectively recovering catalyst from an FCC 'off gas' prior to its discharge at the 'flare stack'.

Other large blowback systems using Pall 'PSS' filter media are operating effectively in many applications, both on low and high temperature gas systems. These include CCR Platformers and a range of vent filters

REFERENCES

Hashemi, R., Schuttenberg, K., Weissman, B. (1992). Specifying porous metal blowback filter assemblies for catalyst removal from FCC flue gas. Paper presented at the 23rd Annual Meeting of The Fine Particle Society, Las Vegas, Nevada 13–17 July

HMSO (1992). Process Guidance Note IPR1–15. Petroleum Processes – Crude Oil Refineries.

Pall Corporation (1987). Advanced metal filters for critical gas solid separation problems. Bulletin GSS–1.

Noyes Data Corporation. (1980). Zeolite Technology and applications. Review Nᵒ 170.

HIGH TEMPERATURE DUST COLLECTION USING A UNIQUE METALLIC FIBER FILTER MEDIUM IN BAGHOUSES

Fred K. Pethick
GAS SYSTEMS DIVISION
MEMTEC AMERICA CORPORATION
1750 Memtec Drive
DeLand, FL 32724

ABSTRACT

A metallic bag filter for dust collector baghouse use allows operation at temperatures well beyond the range of traditional fabric bags. The medium was originally developed for high pressure and temperature gas filtration in chemical processes and has a good history there.

For baghouse use the element has been optimized for low pressure high volume service. It will fit easily into most standard collector designs. Details of the design will be presented along with application parameters.

In addition to high temperature capabilities emission levels are much lower than for fabric bags. This is ideal for hazardous dusts and heavy metals. As the medium does not wear the emissions do not increase with age. Emissions testing with various materials will be reported.

Operating data from installations as well as field and laboratory tests will be presented for various materials and process industries. This will include the parameters of pressure drop, flow, temperature, efficiency, solids loading and consistency of performance.

Filtration at high temperatures offers many opportunities for energy and operating cost savings. In some cases the overall process economics can change. Case histories will be discussed in detail.

INTRODUCTION

Baghouse dust collectors are used in many industries. In some cases they are used only for pollution control but they can also be integral to a process collecting products or catalysts. They work well within their temperature and design limits but become maintenance problems if their limits are exceeded.

A metallic fiber filter has been developed which extends these temperature limits, is resistant to abrasive wear and offers finer general filtration levels. It can be used both with pulse-jet and reverse air cleaning much like a fabric bag.

Metal fibers are produced by a complex drawing process using a very pure high grade of stainless steel (or Inconel®) rod. Fibers are manufactured in discrete sizes from 2 to 50 microns diameter. The fibers are then treated like textile fibers. They are carded and laid in a nonwoven web. Each web consists of a specific size fiber. Webs with different size fibers are laid-up together to custom tailor a medium for a specific application.

The lay-up or stack of webs, sometimes including wire mesh for support, is sintered in a furnace to bond the fibers and mesh together. At very near the materials' melting temperature, metallurgical bonds are formed at each contact point so that each fiber is welded to many others along its length. A strong and rigid matrix is formed.

Studying the web shown in Figure 1, you will notice there is more void space than fiber. The void volume is the most important part of the medium because it allows the fluid to pass freely while particles are retained. This is why nonwoven felts make better filters than woven material. Porous materials made from powdered metals or powdered ceramics do not have this open structure. They have more solid than space.

Also, when you consider this picture, you can see that, as the fibers become smaller, the percent of open space where fluid flows becomes greater. Smaller fibers result in more pores per unit area so they pass more fluid with less pressure drop while retaining very small particles.

Many different fiber sizes are manufactured, thus we can tailor a media formulation to the specific requirements of the application for optimum results. The metal fibers used are much smaller than those typically used in fabric bags.

The formulation for the medium used in metallic bags is a thin surface layer of extremely fine fiber supported by coarse fibers and a heavy wire mesh. These layers are sintered together to make up the complete medium. Rolling it to a cylinder and welding the seam makes a "bag" that is rigid and self-supporting. The wire mesh provides the strength so a cage is not required. The complete medium is only 0.6 mm thick and the surface layer, which is the actual filter, is about 0.08 mm thick.

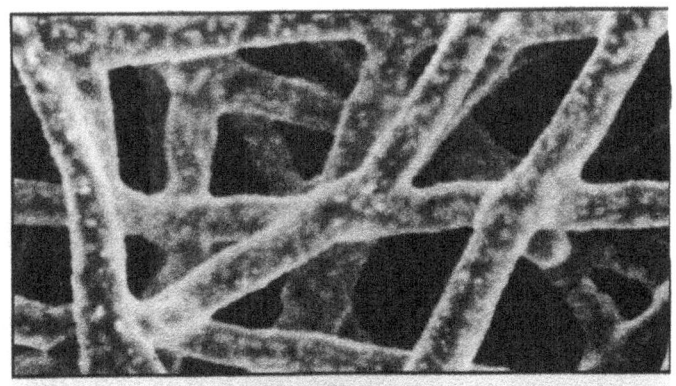

1500X Surface Magnification

FIGURE 1
Cross Section of Web (Magnified 1500 times)

The fine surface layer supports the dust cake on its surface where it is easily removed by pulse cleaning (see Figure 2). A typical fabric bag does not have this extremely fine surface and must be "conditioned". During conditioning, a layer of dust cake builds to provide fine filtration. The ultra fine surface layer in the metal filter eliminates this need for conditioning and provides the advantage of being a fiber filter rather than a compacted powder cake. Higher filtration efficiency is achieved and higher air to cloth ratios can be used. See Table I for test data.

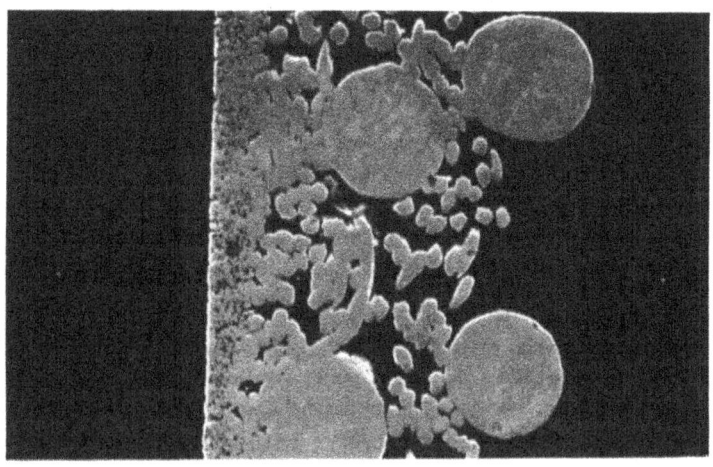

FIGURE 2
Surface Media Cross Section (Magnified 75 times)

With fabric bags much of the cleaning, with either pulse-jet or reverse air, comes from the fabric flexing, stretching and opening up to release the particles. With the metallic medium this does not occur. During cleaning the cake is lifted off the surface by air flowing thru the ultra fine surface virtually everywhere. It then falls to the hopper in the same way as with fabric bags except that the filter cake is less broken up so cleaning is more efficient. Since the metallic filter does not move, wear is eliminated and life of the bag is extremely long. The "puffing", or downstream release of particulate, with each pulse of a fabric bag does not occur.

The surface layer is composed of very fine fibers so it is very efficient. The very thin cross-section keeps the pressure drop low and prevents blinding. The surface is rolled to a smoothness of 20 micro-inches RMS to prevent sticking of the dust cake. The result is similar operation to a fabric bag with higher efficiency and higher flow capacity (air/cloth ratio).

TABLE I
SUMMARY OF TEST DATA

Material	Air/Cloth Ratio cm/sec	Pressure Drop mm WG	Inlet Solids Loading gm/M³a	Outlet Solids Loading gm/M³a	Efficiency %
CHROMIC OXIDE (Very fine powder from a rotary kiln)	4	76	89	.0077	99.991
YELLOW PIGMENT Extremely fine powder	3.5	45	14	.00407	99.97
MANGANESE DIOXIDE Pigment Grade	2.5	35	60	.0031	99.995
CARBON BLACK 99% < 5μ 74% < 1μ	4	150	.76	.000150	99.98
OTTAWA SILICA TEST DUST* 50% < 7μ 10% < 1.5μ	5	NT	2.3	.00046	99.98
SILICA GEL 3% < 20μ 23% < 75μ	2.4	76	NT	NT	NT

NT = Not Tested   *Results supplied by customer

LIMITATIONS OF FABRICS

Fabric filter bags in dust collectors limit the operating temperature of the system. For many processes this means the gas and particulate stream must be cooled safely below this temperature just to do the filtration. Typical fabric temperature limits are polyester - 135°C, Aramid - 200°C, and glass - 260°C. Any application beyond 280°C requires a cooling system for fabric bags to be used.

There are three common cooling methods used: water quench, heat exchange and dilution air. Water quench consists of a water spray directly into the hot gas and particulate stream. It is low in cost but carries the potential for problems. A very short flow interruption will burn the bags or very little excess water will create mud in the system when it mixes with the dust. The valuable heat cannot be recovered.

Heat exchange cooling requires a special heat exchanger design to prevent fouling and abrasion from the particulate. It is costly to build and there is a risk of plugging, fouling or condensation. The heat can be recovered but it is only successful with a limited number of applications.

Dilution air cooling is the most reliable cooling system, is moderately low in first cost but high in operating cost. The heat also cannot be recovered. A dilution air blower adds a large amount of cold ambient air to mix with the hot process gas. The baghouse becomes larger and power is required to pump the ambient air.

The metal fiber filter eliminates the need for cooling and provides many operating advantages.

METALLIC FILTER OPERATION

Elimination of the cooling system eliminates the additional equipment that must be purchased, installed and maintained. The equipment cost of the metallic filter baghouse is about the same, or less, than the equipment cost of a fabric baghouse and cooling system. By the time it is installed the total cost is much less, and a more reliable system results. Emissions are lower and temperature excursions do not result in melted or burnt bags. The careful monitoring of emissions to determine when the bags are worn or broken is mostly eliminated.

In process systems often the hot gas can be recycled or can go on to the next process step without reheating. This saves energy and simplifies the process, the door is also open to improving the overall process design. The collected material can be fed back into the product stream without reprocessing. The high temperature eliminates concern about reaching water and acid dewpoints that could plug the system or cause corrosion. A hot dirty flue gas containing SO_2 becomes a gas clean enough that its valuable heat can easily be extracted and the SO_2 can be converted to acid without additional filtration.

HISTORY OF INSTALLATION

The metal fiber filter has a successful history in chemical process use in hot gas fluidized bed catalyst reactions. Details of the processes cannot be revealed because of confidentiality agreements. In the event of a process upset, or as a chemical process "turnaround", the elements can be cleaned by backwashing with water or a cleaning solution and returned to service. Importantly, not one filter has yet been replaced. The following systems have been built.

Stainless Steel

1	456 elements	63mm Dia. × 1550mm Lg.	3.6 cm/sec*	Start up July 1988
2	840 elements	89mm Dia. × 3660mm Lg.	5.0 cm/sec*	Start up June 1989
3	220 elements	63mm Dia. × 1520mm Lg.	6.0 cm/sec*	Start up September 1991
4	300 elements	89mm Dia. × 3660mm Lg.	5.5 cm/sec*	Start up December 1992
5	336 element	89mm Dia. × 3660mm Lg.	6.3 cm/sec*	Start up scheduled March 1993
6	300 elements	89mm Dia. × 3660mm Lg.	4.9 cm/sec*	Start up scheduled June 1993

*Face velocity

The above operating temperatures range from 200°C–385°C. There has been no apparent wear or deterioration in structure or performance. Typical solids loading are 30 gm/m³ with bulk density of 640–9 60 kg/cubic meter. Installation number 2 is reverse air and the rest are pulse jet.

Inconel® (530°C–590°C in a Fume Silica Fluidized Bed — with 1% HCl by-product)

1	30 elements	50mm Dia. × 914mm Lg.	2.5 cm/sec	Start up mid 1988
2	300 elements	50mm Dia. × 1220mm Lg.	1.5 cm/sec	Start up early 1990

More recently, we have worked on applications that are more typical of a dust collector with low pressure, less severe service. Installations include a bin-vent filter and a product receiver filter.

Bin-Vent Filter

In a used catalyst disposal operation, catalyst is pneumatically conveyed to a hopper where it is cooled with fluidization air and discharged. A twenty-one (21) element filter was supplied to operate at 1869 AM³/hr at 400°C during the conveying cycle. The face velocity is 4.6 cm/sec with very high solids loading. Average particle sizes are 24 micron. During cooling the flow is 1000 AM³/hr (2.5 cm/sec) with temperatures to 425°C. The solids loading is lower but the particle sizes much smaller, averaging 2.2 micron. Start up was November 1992 and no problems have been seen.

Receiver Filter

A small Mikropul baghouse was retrofit to collect a product of metal fume. Twenty-five (25) metallic filters, 2.4 meters long, operate with 2040 AM^3/hr (3.4 cm/sec) at a temperature that varies in the range of 475°C – 500°C. The operation is continuous around the clock. Start up was in November 1992. The pressure drop is steady at 220mm w.g. with a material that is largely submicron.

Several other installations have been run as short-term pilot tests with similar success. Generally, with similar extremely fine fume material.

CONCLUSION

Metallic fiber filters have repeatedly been proven in high temperature high pressure chemical process applications. The use of these filters in more traditional dust collection equipment is proving viable both technically and economically where temperature and durability requirements warrant. Engineers and process designers have the opportunity to improve high temperature processes where the filter media has been the limiting factor. New clean air regulations can easily be met and the compliance can be maintained without constant worry about bag wear and breakage.

AN INTEGRATED CONCEPT FOR THE APPLICATION OF HIGH-TEMPERATURE GRANULAR BED FILTERS IN THE ALUMINUM INDUSTRY

WOLFGANG PEUKERT
Hosokawa MicroPul
Welserstrasse 9-11, 5000 Köln, Germany
FRIEDRICH LÖFFLER
Institut fuer mechanische Verfahrenstechnik und Mechanik
Universität Karlsruhe
Kaiserstrasse 12, 7500 Karlsruhe, Germany

ABSTRACT

Granular bed filters can be used for the simultaneous collection of fine particles and gaseous pollutants in a wide temperature range up to 1300 K. A concept for flue-gas cleaning in the secondary aluminum industry has been developed based on the results of theoretical and experimental investigations of granular bed filters in the laboratory, as well as on the characterization of the flue-gas and under consideration of the industrial process. The cleaning concept with granular bed filters provides two independently regenerated granular beds, separating the solids and acid gaseous components, respectively. The dust particles are collected in the first fixed bed at a temperature of around 973 K using quartz collectors with diameters smaller than 0.5 mm. In the second stage, the sorption of the gaseous pollutants can be realized with either CaO- or $NaHCO_3$-pellets in a counter-current bed. The supply of $NaHCO_3$ enables the recovery of the reaction products, thus minimizing the residues. The proposed concept was verified with the aid of a test-rig which was installed in the secondary aluminum plant. The emission standards were easily satisfied. With this concept it is possible to use the heat content of the flue-gases for the operation of a crystallization unit.

INTRODUCTION

Collection of particles and gaseous components from process- and flue-gases is gaining steadily in significance. In big plants standard technology for particle collection provides electrostatic precipitators or bag filters. Gaseous pollutants such as SO_2/SO_3, HCl or HF are mainly collected in wet scrubbers. For small and average-sized plants with often specific boundary conditions for which standard processes are not practical or too expensive, however, market demands refined and cost-effective cleaning systems. Combined collection of fine disperse dust particles and gaseous components in granular-bed filters offers a promising alternative. Granular bed filters may be used - in contrast to conventional cleaning methods - at high temperatures up to 1300 K, so that exploitation of the thermal energy of the flue-gases becomes possible.

This paper describes a concept for flue-gas cleaning with granular bed filters in the aluminum industry. The concept is based on the results of theoretical and experimental investigations of granular bed filters in the laboratory, as well as on the characterization of the industrial process. It was not intended to develop an end-of-pipe solution but to integrate the flue-gas cleaning in the overall process so that the input of energy and materials is minimized. The important question of the residues of the cleaning process has also been considered.

In order to transfer the experience with combined collection of particles and gases in the laboratory [Peukert 1990] to industrial application and in order to verify the proposed concept a test-bench was built in a secondary aluminum plant. The tests in the aluminum plant are therefore a consequent continuation and extension of the measurements in the laboratory. Particle collection at high temperatures enables the exploitation of the thermal energy of the flue-gases which can be used for energy supply to a crystallization process. Sorption of the gaseous pollutants SO_2 and HCl can be made with CaO or $NaHCO_3$-pellets in a counter-current bed. The supply of $NaHCO_3$ enables the recovery of the reaction products thus minimizing the residues.

DESCRIPTION OF THE SECONDARY ALUMINUM PLANT

The considered secondary aluminum plant produces about 50 000 tons/year raw aluminum from aluminum scraps with an aluminum content of at least 40%. The scrap which contains plastics (e.g. PVC), oil or paints depending on its origin is melted in 5 rotary kilns under a liquid mixture of $NaCl$ and KCl. The salt is used to prevent oxidation of the melted aluminum and absorbs impurities from liquid aluminum.

Figure 1 shows a simplified flow-chart of the process. First, a mixture of $NaCl/KCl$ is melted in the rotary kilns which are heated by heavy fuel burners. Sulphur content of the heavy oil is up to 2%. Feed of aluminum scrap is provided discontinuously so that process conditions, flue gas composition and temperature are nonstationary. The period of one batch process is about 8 hours. Finally, melted aluminum is discharged and casted into bars. Slag is also discharged and will be processed in the dissolving unit. Brine is fed into the crystallization unit which partly recovers KCl and $NaCl$ salts.

The flue-gas has temperatures between 700 and 1100 K with a flow rate of 4200 m^3/h (standard conditions (s.t.p.), dry state). Heat content of the flue-gas (about 900 kW per kiln) is completely lost because the gas is quenched before it enters the venturi wet scrubber. Performance of the wet scrubber is not satisfactory since particle emissions are too high and because the waste produced, which contains mainly $CaCl_2$ and $CaSO_4$, is contaminated with heavy metals and possibly dioxins. Deposition of solid waste faces growing difficulties due both to lacking space for landfill and hazards of soil contamination.

Fig.1 gives the composition of the flue-gas. Gas and particle concentrations are mean values measured over a period of one week, maximum values are given in brackets. SO_2 is mainly produced by the burning of heavy oil with high sulphur content. High HCl-concentrations are due to the melted salts NaCl and KCl and due to burning of chlorine-containing plastics. HF emissions have been found to be low and can therefore be easily absorbed with flue-gas cleaning.

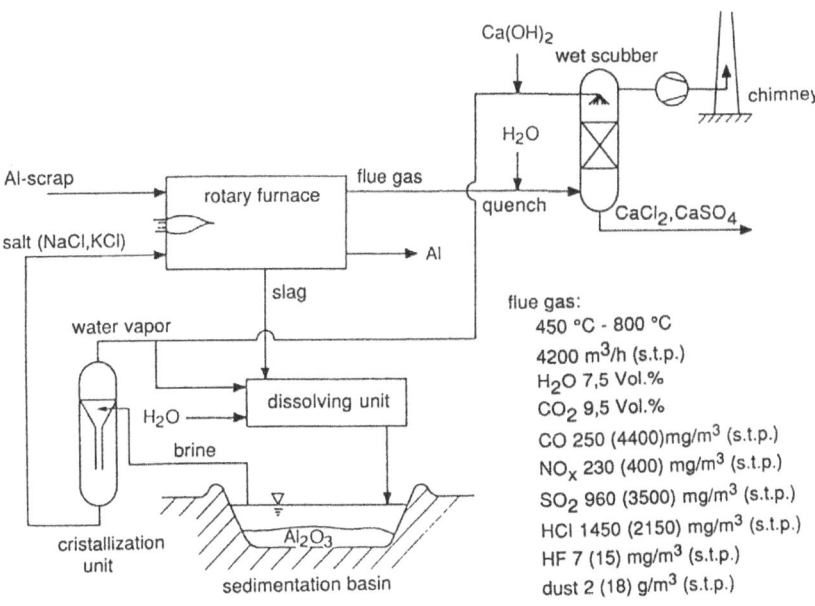

flue gas:
450 °C - 800 °C
4200 m^3/h (s.t.p.)
H_2O 7,5 Vol.%
CO_2 9,5 Vol.%
CO 250 (4400)mg/m^3 (s.t.p.)
NO_x 230 (400) mg/m^3 (s.t.p.)
SO_2 960 (3500) mg/m^3 (s.t.p.)
HCl 1450 (2150) mg/m^3 (s.t.p.)
HF 7 (15) mg/m^3 (s.t.p.)
dust 2 (18) g/m^3 (s.t.p.)

Figure 1   Simplified flow chart of the secondary aluminum plant

Flue-gas contains organic components, possibly also dioxins because of the high chlorine concentration. Measurement and collection of organic components, are beyond the scope of this investigation and are therefore not covered in this paper. It should be noted, however, that gas cleaning at high temperatures may suppress the build-up of dioxins which is catalyzed by heavy metal components.

Particles in the flue-gas are mainly fine condensation aerosols produced by evaporation of liquid NaCl and KCl. These are strongly adhesive so that a recovery of the heat content was so far not possible. Exploitation of the thermal energy is therefore only possible if particles are collected at high temperatures. Mean dust concentrations measured by isokinetic sampling were found to be 2 g/m^3 (s.t.p.) with maximum values of 18 g/m^3 (s.t.p.). Particle size distribution have been measured with an Andersen cascade impactor. Mean mass particle diameter was 0.5 μm with all particles smaller than 2 μm.

Emissions standards for secondary aluminum plants in Germany are defined by TA-Luft (Davids & Lange, 1986):

dust: 20 mg/m^3 (s.t.p., dry state)
HCl: 30 mg/m^3 (s.t.p., dry state)
HF: 5 mg/ m^3 (s.t.p., dry state)

Mean separation efficiencies for dust particles must be higher than 99% and for HCl higher than 98% on the basis of raw gas concentrations shown in Fig.1. For secondary aluminum plants, no emission standards exist so far for SO_2 while HF is of minor importance because of low raw gas concentrations.

DESCRIPTION OF THE FLUE-GAS CLEANING CONCEPT

The developed concept is based on the theoretical and experimental investigations of dust and gaseous components (SO_2 and HCl) in granular bed filters performed at the University of Karlsruhe. Laboratory experiments have been made up to temperatures of 1300 K [Peukert & Löffler, 1991, Peukert 1990]. It was found that collection of solid and gas phase pollutants can be made under identical conditions in one granular bed filter. However, life-time of a filter loaded with particles is in most cases considerable shorter than the life-time due to absorption of SO_2 or HCl. In case of particle collection filter life-time is restricted by the increasing pressure drop due to collected particles inside the filter or, from point of view of collection efficiency, preferably at the filter surface. (For high collection efficiencies, measurements in the laboratory revealed also that collection efficiency as well as pressure drop are more favorable if particles are collected at the filter surface.) After reaching the highest tolerable pressure drop the filter must be regenerated. In case of the considered aluminum plant maximum pressure drop is about 100-150 mbar if the pressure drop of the wet scrubber is used as a reference. Filter life-times have been found to be usually smaller than 1 hour for the given conditions, with filtration velocities smaller than 0.25 m/s.

Time scales for the sorption of SO_2 and HCl are generally in the range of some hours (even in beds of only 0.05 m height) and therefore considerably longer than that of particle collection (based on raw gas concentrations of 1500 vol.-ppm., i.e. 2.5 g/m^3 HCl and 4.5 g/m^3 SO_2, respectively). Therefore a two-stage concept is proposed with particle collection in the first stage at a temperature of about 1000 K and sorption of gaseous pollutants in the second stage so that both filters can be regenerated independently. The proposed concept is shown in Fig.2.

PARTICLE COLLECTION

Investigations into the optimization of particle collection have revealed that filtration conditions are more favorable for small collectors ($d_K < 0.5$ mm) at low filtration velocities ($u_0 < 0.25$ m/s) than under conventional conditions ($u_0 > 0.25$ m/s, $d_K > 1$ mm). This is true from the point of view of separation efficiency as well as from energy consumption (pressure drop). Collection of very fine particles ($x < 2$ μm) with separation efficiencies higher than 99% is only possible if collection is enhanced by the build-up of a dust cake at the filter surface. This, however, requires the utilization of small granules.

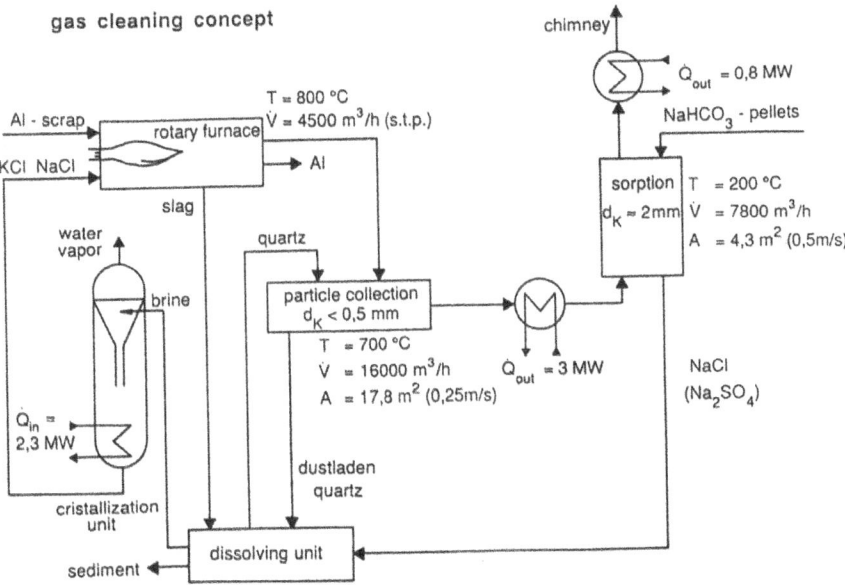

Figure 2   Schematic illustration of the proposed cleaning concept

389

Grade efficiency measurements at temperatures up to 1300 K have shown that collection within the diffusion regime for particles smaller than 1 µm becomes more efficient with increasing temperature. For larger particles a drop in separation efficiency was measured due to reduction in the inertial collection efficiency at higher temperatures. This is true for short filtration times when the filters dust load is still small. Investigations into filtration behavior with time (and thus increasing dust load) revealed that improved adhesion of dust at high temperatures favored cake formation and improved collection efficiencies [Peukert & Löffler 1991]. Because most are fine particles in the flue-gas of the aluminum plant ($x < 2$ µm, $x_{50} = 0.5$ µm), separation efficiency can be expected to increase with temperature. Separation should therefore be made at the highest possible temperature. Thermal energy of the flue-gases ($4.5 \cdot 10^6$ W for 5 kilns) can be used for operation of the crystallization unit (energy consumption $2.3 \cdot 10^6$ W). If sorption is made at 473 K with $NaHCO_3$ sorbents, thermal energy of at least $3 \cdot 10^6$ W can be exploited.

An other advantage of particle separation at high temperatures may be seen in the suppression of dioxin formation. Dioxins are formed at temperatures between 800 and 500 K (de novo synthesis) by reactions which are catalyzed by heavy metals. Since particles and thus possible catalysts are separated in the filter dioxin formation is expected to be much lower.

Beyond that, use of an inert granular bed filter may inhibit penetration of inflammable substances (particles and droplets). This caused problems in the operation of the plant, especially when large amounts of combustible substances have been fed into the kiln. If burning substances reach the scrubber the liner may be damaged with subsequent corrosion problems. Additional damage may result from explosive evaporation of scrubbing water.

Regeneration of the filter media can be made in combination with the dissolving unit where soluble substances are removed. The supply of inert quartz collectors allows regeneration of filter media so that the amount of residues remains small. Regeneration itself, however, was beyond the scope of this study since favorable conditions for collection of both particles and gaseous components had first to be found. Optimal regeneration and handling of solids at high temperatures have thus still to be investigated.

SORPTION OF GASEOUS POLLUTANTS

Sorption of gaseous components is made in the second stage. One possibility is to use calcium based sorbents ($Ca(OH)_2$, $CaCO_3$, CaO). Experiments in the laboratory revealed that collection of both SO_2 and HCl gases is efficient when the sorbents are supplied in form of pellets (diameter d_K between 0.5 and 2 mm) which have been agglomerated from fine limestone powders with particle diameters smaller than 60 µm. Decisive factor to reach high solid conversions was found to be the primary particle diameter of the limestone powder [Peukert & Löffler, 1993].

Temperature should not be higher than 950 K. At higher temperatures, $CaCl_2$ which is formed by the reaction of limestone or lime with HCl becomes plastic thus increasing the filter pressure drop. This was found by measurements of breakthrough curves of HCl and SO_2 through beds of lime and limestone between 700 and 1100 K. Additionally, it has to be mentioned that temperatures below 850 K may enhance recombination of CO_2 and CaO which may impair the sorption of SO_2. There exists therefore a temperature "window" in which collection of both SO_2 and HCl is favorable.

Main reaction products in the aluminum plant are $CaSO_4$ and $CaCl_2$ which have to be treated as solid waste. A considerable reduction of residues can be made by the substitution of heavy oil by fuel with low sulphur content. A far-reaching reduction of residues is also possible if $NaHCO_3$ sorbents are supplied. $NaHCO_3$ calcinates above 380 K to form more reactive Na_2CO_3 which exhibits excellent sorption efficiencies at temperatures up to 750 K. This is demonstrated in Fig.3 which shows conversion-time curves of Na_2CO_3 pellets at 523 K for the reaction with both 1500 vol-ppm SO_2 and HCl. Reaction is very fast. Overall conversion of Na_2CO_3 reaches 100% after 30 minutes, while 70% of the solid has been converted to NaCl and 30% to Na_2SO_4. With CaO-pellets at a higher temperature of 773 K under otherwise similar conditions a conversion of only 72% has been achieved. This shows that Na_2CO_3 is more reactive than CaO at lower temperatures.

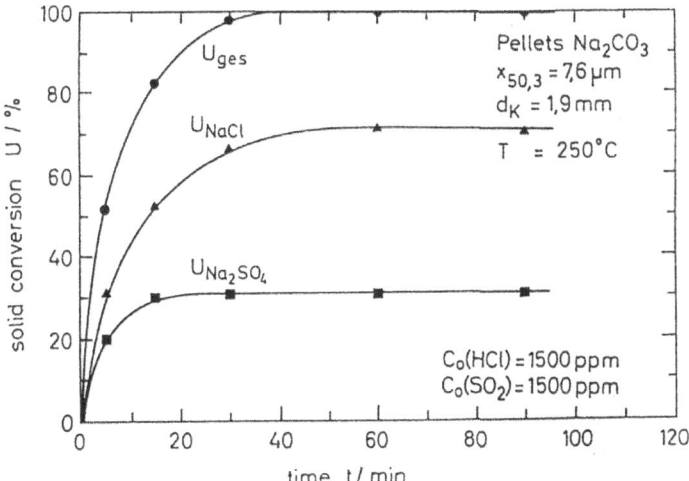

Figure 3   Measured conversion-time behavior of Na_2CO_3 -pellets in the reaction with SO_2 and HCl (carrier gas air); U_{ges} denotes the overall conversion

NaCl which is formed in the reaction with HCl can be recycled to the dissolving unit. Burning of low sulphur fuel in the kiln would further reduce the residues since formation of Na_2SO_4 would be almost suppressed. Additionally, Na_2CO_3 has a certain potential to reduce NO_X - components [Neal & Haslbeck, 1985].

Sorption should be made in a counter-current bed with independent regeneration. Counter-current operation allows an almost complete exploitation of the sorbents in combination with low emission values. Fig.4 shows calculated profiles of solid conversion U for the reaction of HCl with Na_2CO_3 at 623 K. Parameter is the reaction time. Filter life-time exceeds 24 hours even for thin beds of 0.2 m. Raw gas concentration was 1500 vol-ppm at a filtration velocity of 0.5 m/s.

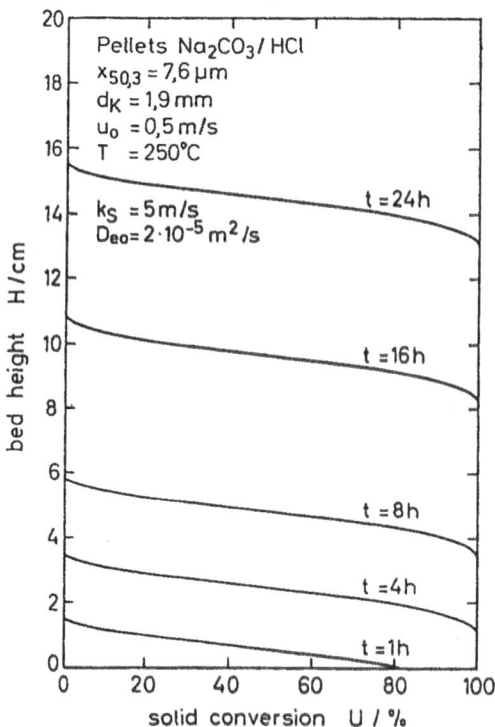

Figure 4   Calculated profiles of local solid conversions of Na_2CO_3-pellets in the reaction with HCl at 523 K (raw gas feed from below)

Within the scope of laboratory investigations, the measured conversion-time behavior under differential gas-conditions as well as breakthrough curves through the integral reactor of HCl could be described theoretically with a relatively simple shrinking-core model (Ishida & Wen, 1968). Theoretical description of sulphation is only possible with a more sophisticated pore model which takes the pellet's internal pore size distribution into account (Christman & Edgar, 1983). Calculated conversion profiles are relatively steep and therefore limited to a small range of bed. According to the calculations conversion of the pellets reaches 100% after 8 hours at a bed height of 0.03 m. More than 0.14 m at that specific time are completely unused. Similar results have been obtained for sulphation both for calcium and sodium based sorbents.

After sorption of acidic gaseous components another cleaning stage becomes necessary in order to bind organic components such as dioxins or furans which are supposed to be in the flue gas. A promising possibility which utilizes sediment particles (mainly Al_2O_3 with high inner surface) has been tested after completion of the measurements described in this paper.

In order to check the proposed concept a test bench was installed in the aluminum plant. Experimental set-up and results of the measurements are described below.

EXPERIMENTAL SET-UP

Basic arrangement of the high-temperature apparatus for determination of particle collection efficiencies and for measurement of breakthrough curves of SO_2 and HCl is shown schematically in Fig.5. The test-bench has been designed for temperatures up to 1300 K and flow rates up to 100 m³/h. A specified flow-rate is extracted from the flue-gas duct by isokinetic sampling. Gas temperature is maintained by heated pipes. Problems were encountered due to clogging of isokinetic probe by adhesive dust particles. This was strongly dependent on operational conditions of the kiln. The situation improved considerably after exchanging the fuel from heavy oil to light oil.

The fixed bed granular bed filter is situated at the end of the heating section. Filter bed is supported by an appropriate grid. Filter height could be changed between 1 and 20 cm. Temperature and pressure drop of the filter have been recorded continuously. Gas concentrations after the test-filter and in the flue-gas duct have been measured continuously after passing an absolute filter which separates the particles. For the determination of HCl a process photometer SPECTRAN 677 IR was used, while SO_2 concentrations have been determined with a BINOS photometer. At the end of the heated section particle concentration could be determined by using a particle filter or a cascade impactor for measuring particle size distributions. A wet scrubber was installed after the particle filter to avoid contamination of the flow control system and blower.

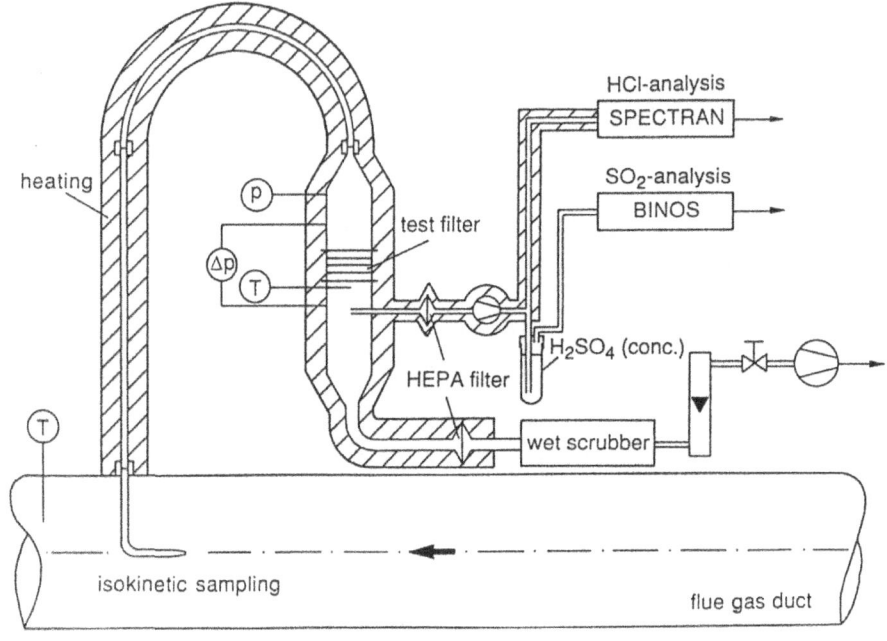

Figure 5   Schematic representation of the experimental set up in the aluminum plant

RESULTS

PRIMARY MEASURES

Emissions from the secondary aluminum plant could be considerably reduced with primary measures by improvement of the burner and by substitution of heavy oil with high sulphur content. After installing light oil burners mean SO_2 emissions dropped by a factor of 6 from 900 mg/m³ to below 150 mg/m³. With gas burners SO_2 concentrations could be even lowered by a factor of 100, i.e. to concentrations below 10 mg/m³. A gas burner would additionally allow a better control of the operation of the kiln. Mean particle concentrations were reduced from 2 g/m³ to 0.8 g/m³. Soot content of the flue-gas was also considerably reduced, albeit the particle size distribution remained unchanged. It can be expected that the concentration of unburned organic components is also lower although no experiments have been conducted in order to verify this assumption.

PARTICLE COLLECTION

Particle collection experiments have been performed by using inert quartz granules with mean diameters of d_K = 0.1, 0.29, 0.66 and 1.1 mm. Filtration velocity was varied between u_0 = 0.05 and 2 m/s. Filtration temperature was near 973 K. Table 1 shows some results of the measurements. Mass specific total separation efficiency E_3 (subscript 3 denotes for mass in contrast to number related properties with subscript 0) is given together with mean raw gas and clean gas dust concentrations c_e and c_f, respectively. Experiments were run until the filter pressure drop reached 10^4 Pa [0.1 bar].This value has been chosen since the wet scrubber of the conventional flue-gas cleaning operates between 0.1 and 0.15 bar. Filtration time up to the maximum pressure drop was reached varied between 30 and 90 min.

Table 1

Experimental results of separation efficiency measurements

(bed height 50 mm, temperature 973 K)

d_K / mm	u_0 / m/s	c_e / mg/m^3	c_f / mg/m^3	E_3 / %
1.1	2.0	1820	1320	27.5
1.1	1.0	1680	780	53.9
0.6	1.0	620	390	36.6
0.6	0.25	1900	9.0	99.5
0.6	0.1	740	33.0	95.6
0.6	0.05	370	1.8	99.5
0.3	0.25	5800	13.8	99.8
0.3	0.1	640	1.8	99.7
0.3	0.05	410	1.5	99.6
0.1	0.1	2050	9.1	99.6
0.1	0.05	1860	6.1	99.7

Table 1 shows that operation of the filter with conventional parameters, i.e. filter media diameter d_K > 1 mm and filtration velocities u_0 > 0.25 m/s exhibits only poor separation efficiencies. Even at high filtration velocities above 1 m/s inertial separation of the fine dust particles under consideration (x < 2 μm) is not sufficient to meet the emission standards. One possibility to use the advantage of coarse filter media (low pressure drop, high flow rates) and to improve collection efficiency could be by precoating the filter surface with a thin layer, e.g. with particles in the size range between 100 and 50 μm. This would greatly enhance the build-up of a dust cake which is otherwise impossible. A precoated filter is supposed to have excellent separation characteristics. Further investigations are needed in this field.

Experiments with the smaller granule diameters showed the expected high separation efficiencies. Reason for this are twofold: on the one hand separation efficiency is higher at low filter loadings (here emission is greatest), on the other hand build-up of a dust cake at the filter surface becomes possible with small granules and low filtration velocities ($u_0 < 0.25$ m/s). The measured results prove that emission standards of 20 mg/m^3 can be satisfied with granule diameters of 0.3 mm. In some cases even mean clean gas concentration below 2 mg/m^3 have been measured although filter height was only 50 mm. With higher layers still smaller values can be achieved since particle collection is improved especially at low filter loadings where emission is highest.

SORPTION OF SO$_2$ AND HCl

Sorption experiments have been performed with CaO and NaHCO$_3$ pellets. Pellets have been produced by agglomeration of fine limestone ($x_{50} = 1.8$ μm) and NaHCO$_3$ powders ($x_{50} = 7.6$ μm) in a rotary plate agglomerator. After agglomeration pellets have been sieved in narrow size distributions between 1.7 and 2.0 mm. Limestone pellets (CaCO$_3$) have been additionally calcined at 1073 K in order to produce more reactive CaO-pellets. Bed height was 50 mm in all experiments. Temperature was near 800 K in the case of CaO-pellets and 473 K in the experiments with NaHCO$_3$. NaHCO$_3$ calcinates in situ at temperatures above 373 K to form reactive Na$_2$CO$_3$.

Figures 6 and 7 show clean gas concentrations of SO$_2$ and HCl for CaO-pellets measured at 793 K and for Na$_2$CO$_3$-pellets measured at 503 K. Clean gas concentrations are below the emission standards in both cases. During these experiments no pre-filter for particles has been used, only the isokinetic probe was turned in flow direction. This has no influence on gas concentrations, but reduces particle concentration at the test filter. Otherwise pressure drop of the test filter will increase too fast so that gas concentration measurements would only be possible over a considerable shorter time period. Influence of dust particles on breakthrough curves of SO$_2$ has been studied experimentally in the laboratory. It was found that collection of particles in the test-filter has only a minor influence on sorption of gaseous components since collected particles will only affect the transport of the gases to the outer

Pellets - CaO / SO2 + HCl
F1 x50,3 = 1,8 µm
dK = 1,9 mm
H = 5 cm
u0 = 0,25 m/s
T = 520 °C
C0(SO2) = 960 mg/m3
C0(HCl) = 1450 mg/m3
SO2
HCl
concentration C / mg/m3
time t / min
0
20
40
60
80
100
30
90
120
150
180

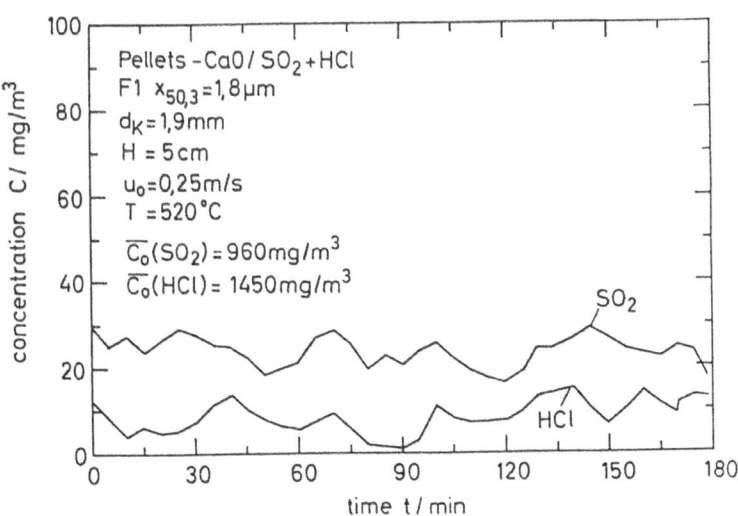

Figure 6 Measured clean gas concentrations of SO_2 and HCl after a bed of CaO-

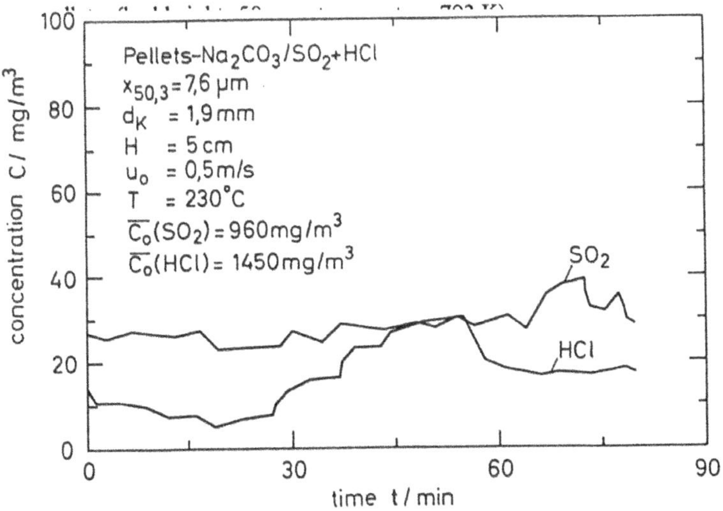

Pellets-Na2CO3/SO2+HCl
x50,3 = 7,6 µm
dK = 1,9 mm
H = 5 cm
u0 = 0,5 m/s
T = 230 °C
C0(SO2) = 960 mg/m3
C0(HCl) = 1450 mg/m3
SO2
HCl
concentration C / mg/m3
time t / min
0
20
40
60
80
100
30
90

Figure 7 Measured clean gas concentrations after a bed of Na_2CO_3-pellets (bed height: 50 mm, temperature 503 K)

surface of the sorbent grain but not the diffusion inside the grain which is the rate determining resistance.

With CaO pellets a clean gas concentration of around 10 mg/m^3 (s.t.p., dry state) for HCl and values between 20 and 30 mg/m^3 (s.t.p., dry state) for SO$_2$ were measured. It has to be mentioned that the detection limit (5% of measuring range) of the photometers employed in this work is 10 mg/m^3 for HCl and 30 mg/m^3 for SO$_2$. Similar results have been found for NaHCO$_3$ at 503 K. In the case of NaHCO$_3$ a peak in raw gas concentration was observed with HCl values up to 2500 mg/m^3. The somewhat higher HCl clean gas concentrations may also be attributed to the fact that NaHCO$_3$ showed some shrinkage due to calcination. This may have influenced the structure of the fixed bed. However, this problem will not occur in a continuous operated countercurrent bed adsorber.

Measured data were below the emission standards. These results demonstrate exemplary that emission standards can easily be met with granular bed filters although bed height was only 50 mm. With higher filters emission values can be further reduced. For comparison, the height of active carbon filters are often a factor 10 to 30 higher.

Solid conversions could not be determined in these experiments because NaCl and KCl particles have been collected in the filter, too. Determination of absorbed chlorine is therefore not possible. Experiments under similar conditions in the laboratory revealed, however, that solid conversions higher than 90% are possible, i.e. the stoichiometric factor is between 1.1 and 1.2. As already mentioned above counter-current operation will allow high extent of sorbent exploitation which are almost comparable with those achieved only in wet processes.

Summary

This study describes a flue-gas cleaning concept in a secondary aluminum smelting plant with granular bed filters. In a first inert bed of quartz granules particles are separated at a temperature of around 1000 K. In the second stage sorption of gaseous pollutants SO$_2$ and HCl is made with CaO- or NaHCO$_3$-sorbents. Supply of NaHCO$_3$ enables a significant reduction of residues since the main reaction product NaCl can be recycled and used in the smelting process.

In order to verify the proposed concept tests have been performed in the aluminum plant. Experimental results confirmed the feasibility of the proposed concept. Although particle size distributions of the sticky dust under consideration was very fine (x < 2 μm, x$_{50}$ = 0.5 μm), mean particle outlet concentrations below 2 mg/m^3 could be achieved. Also the emission standards for gaseous components have been met. The proposed concept enables an exploitation of the thermal energy of the flue-gases which had been completely lost with the conventional cleaning method. The saved energy is sufficient to run a crystallization process.

It has been shown that granular bed filters can be operated under various conditions. Appropriate choice of the filter media for particle collection and sorption in coordination with the whole process often allows for a minimization of residues. Since granular bed filters can be operated in a wide range of temperatures recovery of thermal energy becomes possible. Suitable solutions have to be found individually for each respective process.

REFERENCES

Christman E., Edgar G. (1983). Distributed Pore Size-Model for Sulfation of Limestone. AIChE Journal 29, 388-395

Davids P., Lange M. (1986). Technische Anleitung zur Reinhaltung der Luft, Technischer Kommentar. VDI-Verlag, Düsseldorf

Neal L.G., Haslbeck J.L. (1985). A Dry Simultaneous SO_2/NO_x Control Technology. Modern Power Systems, 45-49

Ishida M, Wen, C.Y (1968). Comparison of Kinetic and Diffusional Models for Gas-Solid-Reactions. AlChE Journal 14, 311-317

Peukert W. (1990). Die kombinierte Abscheidung von Partikeln und Gasen in Schüttschichtfiltern, PhD-thesis, Universität Karlsruhe, Germany

Peukert W., Löffler F. (1991). Influence of Temperature on Particle Separation in Granular Bed Filters. Powder Technology 68, 263-27

Peukert W., Löffler F. (1993). The Sorption of SO_2 and HCl in Granular Bed Filters. Submitted to 2. Int. Symposium on Gas Cleaning at High Temperatures, Surrey

HIGH TEMPERATURE GAS CLEANING FOR PFBC USING A MOVING GRANULAR BED FILTER

C.A.P. ZEVENHOVEN[2] [*], J. ANDRIES[2], K.R.G. HEIN[2] & B. SCARLETT[1]

Delft University of Technology
[1] Lab. for Chemical Process Technology, P.O. Box 5045, 2600 GA Delft, Holland
[2] Lab. for Thermal Power Engineering, P.O. Box 5037, 2600 GA Delft, Holland

[*] Currently at Åbo Akademi University, Combustion Chemistry Research Group
Lemminkäisenkatu 14-18B, FIN-20520 Åbo / Turku, Finland

ABSTRACT

The use of a moving granular bed filter has great potential for high temperature gas cleaning. The major advantages are the low cost of the filter material and the possibility to operate the filter at a constant pressure drop without the need for periodic shut-down and cleaning.

At Delft University of Technology a combination of a primary cylone and a granular bed filter has been tested in the exhaust of a pilot scale (1.6 MW$_{th}$) Pressurised Fluidised Bed Combustor (PFBC). After some tests with a fixed bed filter, a continuously moving granular bed filter was tested.

The various aspects related to the continuous operation of the filter at PFBC conditions and the filtration performance are presented in this paper. This includes a description of the method for high temperature, high pressure dust concentration measurement and the results of filter efficiency measurements. In addition, the granular material recirculation and regeneration system is discussed.

INTRODUCTION

The operation and optimisation of pressurised fluidised bed combustion (PFBC) of coal and gas cleaning at high temperature (HTG) has been investigated at the Delft Univ, of Technology since 1976. From 1986 to 1990 the Univ. carried out an EC funded project called "Closed loop controlled integrated hot gas clean-up". The goal of the project was to "develop, construct and operate a closed loop controlled hot gas clean-up system designed to operate at 850 °C and 10 bar" (Andries et al., 1991). Because particles in a wide range are emitted from the combustor, the research on hot gas clean-up concentrated on the use of cyclones as a first cleaning step followed by a granular bed filter. The PFBC/HTG test facility is shown schematically in Figure 1.

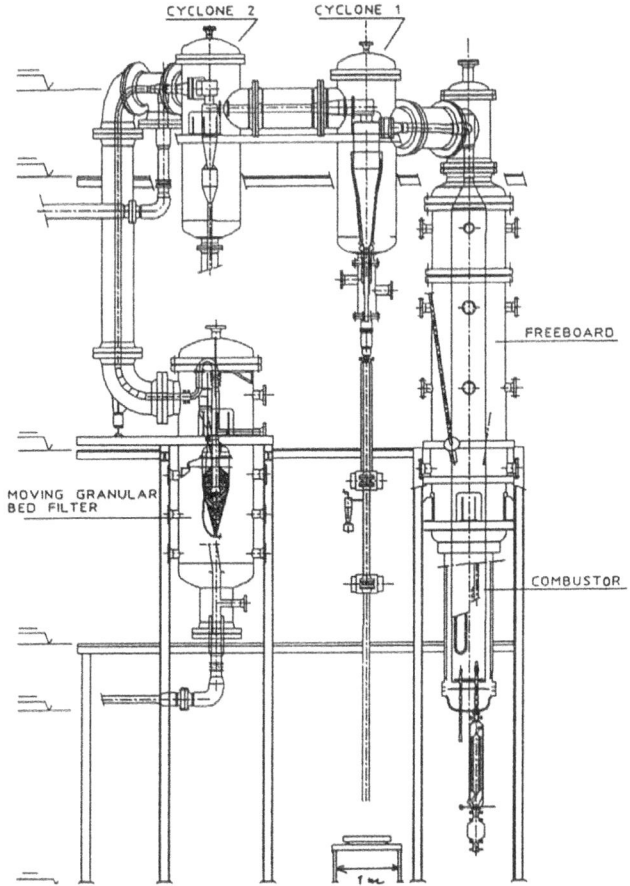

Figure 1   *The Delft University PFBC / Hot Gas Clean up test facility*

FIXED GRANULAR BED FILTRATION

Introduction

The first stage in the research on granular bed filtration was to study the operation of a high temperature, high pressure (HTHP) fixed granular bed filter (GBF). By including electrostatic enhancement as a special feature at a later date, it was intended to operate a filter with a high, controllable efficiency at varying operating conditions of the combustor. Initially, it was planned to proceed from an electrostatically enhanced fixed GBF to a moving bed filter with electrostatic enhancement. In this way, mechanical and electrical particle collection mechanisms are combined with a method for continuously

401

removing the collected dust. The results of tests with the fixed bed filter (without electrostatic enhancement) presented in this section, made it necessary to change this approach.

(It is noted that the work on electrostatic enhancement of granular bed filter is presented elsewhere, e.g. Zevenhoven 1992; Zevenhoven et al. 1993, 1991).

It was decided to first replace the filter by a moving bed filter which allowed continuous operation, and to equip this filter later with the facilities for electrostatic enhancement (Andries et al., 1991). The development and implementation of the equipment for HTHP dust sampling for the cyclone section (and in the filter inlet during the moving granular bed filter experiments) proceeded in parallel with the fixed bed filter operation tests. Dust sampling and analysis of the samples taken from the inlet and outlet of the filter will be discussed in a later section.

The test facility

The dimensions of the (cylindrical) fixed GBF which was installed inside a pressure vessel (see Figure 6) are shown in Figure 2. The filter support was composed of two layers of a steel plate perforated with 6.5 mm diameter holes giving a free surface of approximately 30%, and a 0.87 mm square mesh 'sandwiched' between these plates.

The direction of the gas flow through the filter was vertical and downwards. The flow through the filter was controlled using a vortex flow meter and a control valve.

The following process variables were controlled and/or recorded every 10 seconds:

•	system pressure	0 - 10 bar
•	filter pressure drop	0 - 500 mbar
•	filter temperature	0 - 1000 °C
•	flow through the filter	0 - 600 m^3_{STP}/h
•	superficial filter velocity	0 - 2 m/s

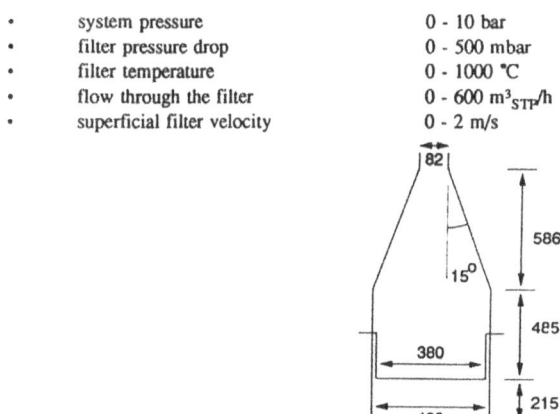

Figure 2   *The dimensions of the HTHP fixed granular bed filter (mm)*

402

Test results

After introductory tests in order to check the mechanical integrity (Andries et al., 1991) a series of filtration tests was carried out. The results of these tests are given in Table 1. The height of the layer of filter material was always 0.10 m. Three different granular filter materials were used:

S	quartz sand as also applied in the combustor	$d_{50} \approx 0.9$ mm
A	aluminum oxide	$d_{50} \approx 2.5$ mm
Q	quartz sand	$d_{50} \approx 1.7$ mm

A Polish coal was burnt in the PFBC during all the experiments. During most of the experiments sorbent material for sulphur dioxide capture was added.

TABLE 1

The results of the high temperature, high pressure fixed granular bed filtration tests.

Test Day / Exp.	Filter Material	Pressure (bar)	Temperature (°C)	Gas Flow (m^3_{STP}/h)	Filter Velocity (m/s)	Pressure Drop Increase (mbar/h)
1.1a	S	8	440 - 445	58	0.05	104
1.1b	S	8	425 - 430	52	0.045	61
1.2	S	6	420 - 460	46	0.05	73
1.3	S	6	360 - 450	54 - 45	0.05	104
1.4	A	6	500 - 610	85 - 77	0.10	14
1.5a	Q	3	300 - 350	30 - 28	0.05	10
1.5b	Q	3	350 - 420	47	0.10	50
1.6*	Q	6	380 - 570	85	0.10	60
1.7	Q	4	420 - 500	62 - 58	0.10	94
1.8a*	Q	6	350 - 480	90 - 85	0.10	232
1.8b	Q	6	400 - 380	45 - 50	0.05	59
1.9a	Q	5	480 - 560	75 - 70	0.10	116
1.9b	Q	5	500 - 480	37	0.05	103
1.10	Q	6	200 - 430	110	0.05	20
1.11**	Q	7.5	420 - 500	108	0.05	93

*   Continued with the same filter material
**   Second cyclone removed. Cyclone outlet measurements gave a dust concentration varying between 250 and 750 mg/m^3_{STP}.

Discussion of fixed bed filter performance

Due to the small size of the fly ash particles in the outlet gas stream of the (second) cyclone, a rapidly increasing filter pressure drop was found as shown in Table 1. As a result of this, a small filter velocity must be used in order to avoid excessive mechanical forces in the filter. This gave a 300 °C temperature drop between the cyclone section and the filter, which was unacceptable (Andries et al., 1991).

It was therefore decided to remove the second cyclone and thus to increase the average size of the particulates entering the filter. This decision was supported by the relatively low collection efficiency of the second cyclone (20% - 85% at pressures above 4 bar) and by problems in removing the bottom catch of the cyclone from the pressurised system due to valve blocking. The beneficial effect of a wider size distribution on the rate of pressure drop increase is moderated by the effect of the increased inlet concentration.

DUST SAMPLING

Introduction

To enable off-line monitoring of the inlet and outlet dust concentration and particle size distributions in the cyclone section (and the inlet of the filter), a HTHP dust sampling system was developed at Delft Univ. by Bernard and Pitchumani (Bernard et al., 1990).

A sampling tube, extended with an S-pitot tube, designed according to DIN and NEN standards, was used (see NPR-standard, 1985, VDI-standard, 1975).

Usually a sampling probe for taking dust samples from a gas stream is a tube located at the centre line of the flow, mounted parallel and opposite to the direction of the flow. To obtain a representative sample the following conditions must be carefully controlled:

- The samples must be taken iso-kinetically (Belayev & Levin, 1974; Cooper, 1986; Fissan & Schweintek, 1987)
- Effects of the presence of the sampling tube and external effects (e.g. gravity, turbulence) must be minimised (Stern, 1968; Wiener et al., 1988)

An extra effect in sampling from a HTHP combustion gas is the influence of pressure and temperature reductions, and the problems arising from condensation of components in the gas phase (Cooper, 1986). In the current application this means particularly the condensation of sulphuric acid (pH = 1.5 - 3).

In the sampling tubes used for the cyclones and the filter inlet, a dry sample is obtained for off-line analysis which is composed of two parts. These are obtained at temperatures slightly above the condensation temperature, with pressure reduction and further cooling downstream of the dust collection. Both parts of the sample were analysed with the Coulter Multisizer II (Bernard, 1992).

The sample taken from the outlet of the granular bed filter also consisted of two parts: a wet sample, which is a mixture of condensate and coarse particulates, while the fine fraction of the sample was analysed (dry) on-line in a Climet (Ci-226M + Ci-1000) optical particle counter. The wet sample was analysed using a Coulter Multisizer II.

The HTHP filter inlet dust sampling system

The HTHP dust sampling tube used for the cyclone efficiency measurements and for monitoring the inlet of the HTHP moving granular bed filter is shown schematically in Figure 3. The inner diameter of the main tube, d_{tube}, was 82 mm.

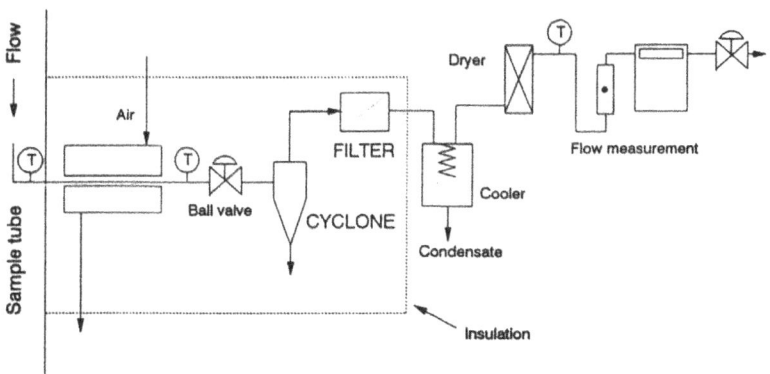

Figure 3   *The HTHP dust sampling system for cyclone section and filter inlet*

The sampling tube, which was directed against the flow was equipped with a thermocouple and an S-pitot tube in order to measure velocity at the sampling tube inlet. The inner diameter of the sampling tube, $d_{sampling\ tube}$, was 8 mm, with a sharpness of 6° at the inlet. The calibration factor of the S-pitot tube has been determined experimentally (Bernard, 1992). In practice, however, the volumetric flow for iso-kinetic sampling was determined using the following relationship, which was within the 10% error of the S-pitot velocity measurement (Andries et al., 1991).

$$\frac{Sample\ flow}{Main\ flow} = \frac{Surface\ area\ of\ sampling\ tube}{Surface\ area\ of\ main\ flow} = \left(\frac{d_{sampling\ tube}}{d_{tube}}\right)^2 = 0.00952 \qquad (1)$$

The flow entering the sampling tube was cooled by a concentric tube cooler to a temperature between 150 °C and 200 °C, using air or water. This is the maximum allowable temperature for the ball valve which separated the sampling system from the pressurised PFBC/Hot Gas Clean-up facility. The dust sample was collected directly after this ball valve, partly in a small cyclone and partly in a glass fibre filter element. Downstream of this gas/solid separation the gas flow was further cooled with water to ambient temperature and dried, after which the flow was measured in a flow meter and a cumulative dry gas flow meter. The sampling system upstream of the second cooler is insulated and equipped with trace heating. During the experiment the sample flow had to be controlled due to increasing pressure drop in the fibre filter. The samples are analysed at ambient, atmospheric conditions using the Coulter Multisizer II (Bernard, 1992).

The HTHP filter outlet dust sampling system

The HTHP dust sampling system used for monitoring the filter outlet was derived partly from the sampling system discussed in the previous sub-section. Since extending the sampling tube with an S-pitot tube was considered not to be essential for velocity information with less than 10% error, only an 8 mm sampling tube with a 6° sharpened inlet was installed on the centre line of the 82 mm outlet tube of the filter, opposing the direction of the flow. Since it could be expected that the particles leaving the filter were smaller than those entering the filter, the error due to an-isokinetic sampling would be small compared to the error in the filter inlet (cyclone section outlet) sample.

The objective was to obtain on-line measurement of the particle concentration at the filter outlet using the Climet optical particle counter (OPC) Ci-226M, with electronic module Ci-1000 (Zevenhoven, 1992). Because the OPC could not operate at temperatures above 50 °C and pressures above 1 bar, it was necessary to cool the sample and to reduce the pressure before it entered the OPC. This led to the design of the filter outlet dust monitoring system shown in Figure 4 (Zevenhoven et al. 1992a, 1992b, 1991).

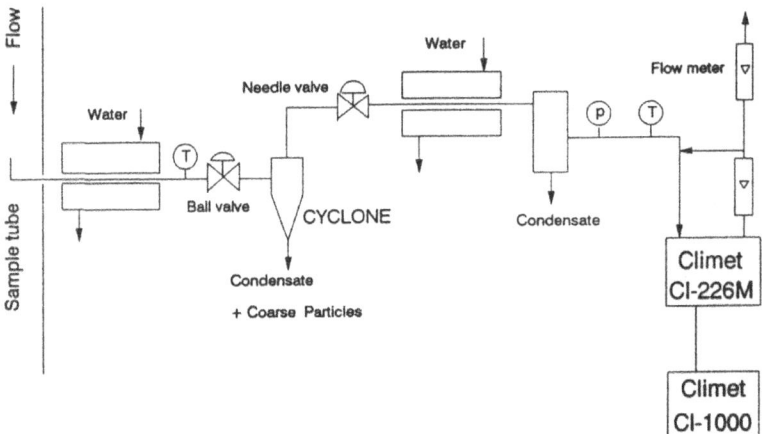

Figure 4   The HTHP sampling system at the outlet of the filter

The sampling line was opened or closed with a ball valve. In the first stage the aerosol sample was cooled with water to a temperature of 100 °C or less in a concentric tube cooler. A mixture of condensate and 'coarse' particles was collected using a cyclone identical to the cyclone in the filter inlet sampling system. Downstream of a needle valve, which was used to control the flow, additional water cooling was used to reduce the temperature to a maximum of 50 °C in order to protect the OPC for attack by acidic condensate droplets. If still more condensate was produced in this second cooling stage, the cooling capacity of the first stage is increased. Evidence for fouling and blocking of the needle valve was not found, whilst the sampled gas stream didn't decrease in time. Measurements using the OPC were carried out only when no 'secondary' condensate was collected before the inlet of the OPC. The sample flow and that through the OPC (0.0283 m$^3_{STP}$/min) was checked and controlled using flow meters. The collected condensate and the coarse dust sample were analysed using the Coulter Multisizer II. Since this instrument analyses liquid dispersions the condensate did not introduce problems.

The total weight of solids was determined from the mass of the total wet sample minus the mass of the condensate which was removed by washing and drying. A disadvantage of this procedure is that the result from the Coulter Multisizer must be combined to the result from the optical particle counter. The error of this procedure is reduced by the fact that the mass of the pre-collected dust is much higher than that of the dust analysed in the OPC (mass ratio approx. 10:1). Thus, the filter outlet concentration is approximately 10 times the concentration measured by the Climet optical particle counter.

MOVING GRANULAR BED FILTRATION

Introduction

Operating a granular bed filter in combination with a system for removing the contaminated filter material from the filter and recycling after cleaning allows continuous operation of the filter. It was decided to adopt the design of the Screenless Moving Granular Bed Filter by the Combustion Power Company for further development and testing. The design of this filter is shown in Figure 5.

The objective of the filtration and performance tests was to verify whether this filter would operate reliably at constant pressure drop and to obtain insight into the effect of the motion of the filter material on the efficiency. In addition it was intended to study the problems of this filter, which had been reported in the literature (gas leaking along the gas inlet pipe, bed transport problems) (Zakkay et al. 1989; Zakkay & Gbordzoe, 1989).

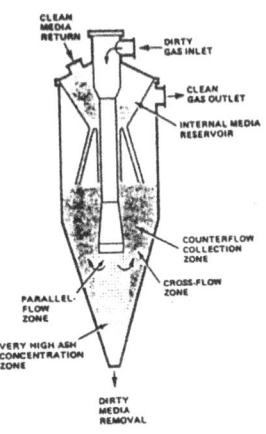

Figure 5 The CPC Screenless Moving Granular Bed Filter

The test facility

The moving GBF was mounted inside the pressure vessel (diameter 0.95 m height 3.2 m) as shown in Figure 6. The total filter length was 1.5 m, whilst the diameter of the cylindrical section was 0.4 m at a length of 0.5 m (maximum filling height 0.4 m). The total angle of the conical section was 30°. Filter material was recirculated at a rate of approximately 1 kg/min. The material was withdrawn from the filter and transported using an injector (see Figure 6) and a pneumatic conveying line. The captured fly ash particles were stripped from the filter granules during the high velocity pneumatic transport. The granules were collected in a vessel from which they flowed back into the filter vessel.

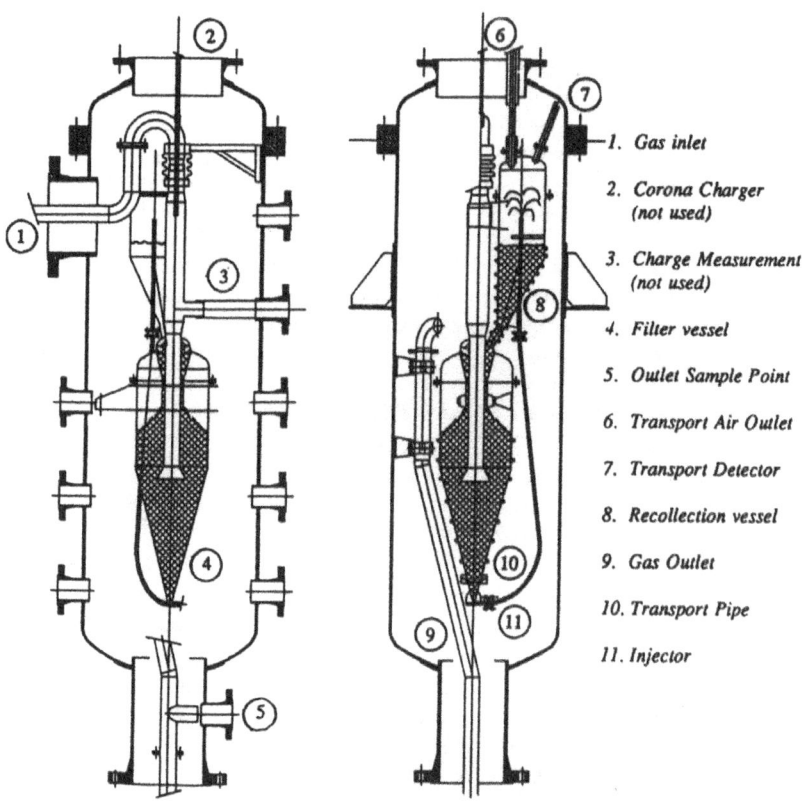

	1. Gas inlet
	2. Corona Charger (not used)
	3. Charge Measurement (not used)
	4. Filter vessel
	5. Outlet Sample Point
	6. Transport Air Outlet
	7. Transport Detector
	8. Recollection vessel
	9. Gas Outlet
	10. Transport Pipe
	11. Injector

Figure 6   The high temperature high pressure moving granular bed filter

The gas stream containing the fly ash particles was cooled after leaving the filter pressure vessel and is cleaned using a small inertial separator before expanding to atmospheric pressure. The following variables were controlled and/or recorded every 10 seconds:

• system pressure	0 - 10 bar
• filter pressure drop	0 - 300 mbar
• filter temperature	0 - 1000 °C
• flow through the filter	0 - 600 m^3_{STP}/h
• superficial filter velocity	0 - 2 m/s
• temperature of the lift pipe	0 - 1000 °C
• temperature of the recollector vessel	0 - 1000 °C
• temperature at the transport pipe outlet	0 - 1000 °C
• pressure drop in the conveying line (added later)	0 - 400 mbar

In all the HTHP moving granular bed filtration experiments the 1.7 mm quartz granules were used.

408

Start-up experiments

In the first experiments the bed material transport was driven by a small portion (2% maximum) of the hot gas stream entering the filter, at system pressures of 3 bar, 4 bar and 5 bar, using injector design I. This injector was later replaced by the improved injector design II, as shown in Figure 7. The transport rate of the bed material could not be measured nor could hard proof of the bed recirculation be obtained. On theoretical grounds a bed transport rate of approximately 1 kg/min was expected at a gas velocity in the lift pipe varying from 8 to 12 m/s, depending on the system pressure.

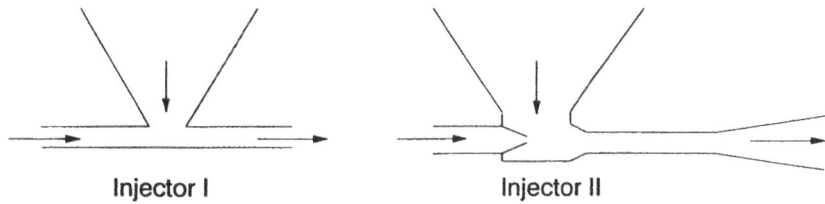

Injector I   Injector II

Figure 7   The design of injector I (←) and injector II (→)

Outlet dust concentrations between 5 and 200 ppmw were measured, with a volume mean diameter, $d_{p\,4,3}$ of 1.5 - 2.5 µm. In a second series of performance tests the injector (injector I) was modified to allow the bed transport to be achieved with externally supplied cold, dry, pressurised air, partly supplementing the hot gas from the filter inlet. A detector of the bed material transport, based on a piezoelectric element was developed and installed in the recollector vessel.

A procedure for starting up to bed material transport was determined in a cold pressurised test. It was found that bed transport could be started by opening the outlet valve to release the amount of gas corresponding to the pre-determined transport velocity and directly afterwards providing the corresponding amount of pressurised gas to the injector. A maximum overpressure of 2 bar relative to the pressurised test facility appeared to be sufficient.

It was decided to drive the injector entirely by cold pressurised air, dropping the alternative of using a small portion of the hot inlet gas stream.

A small inertial separator to remove the dust from the transport gas stream leaving the filter system was installed in order to collect the fly ash and thus protect the control valve for this flow. At this point, the operation of the filter was sufficient for efficiency tests based on inlet and outlet samples.

Filtration performance

An overview of the filtration tests with the HTHP moving granular bed filter is given in Table 2. In experiments 2.1-3 a Polish coal was burnt in the PFBC, while in experiments 2.4-7 an English coal (Kiveton Park) was used. In all experiments except experiment 2.7

a calcium-based sorbent for SO_2 removal was added as bed material in the combustor. Filter efficiency measurements were made during steady state operation. Samples were taken simultaneously from the inlet and outlet of the filter for periods of 30 min.

The performance of the filter as a function of time during experiments 2.1, 2.2 and 2.2, respectively, is presented in Figures 8, 9 and 11.

Particle size distributions are given in Figures 10 and 12 for experiments 2.2 and 2.7.

A particle size distribution for a fly ash fraction extracted from the transport air outlet, i.e. the transport air flow which removes the collected ash from the system after the recirculation / regeneration of the filter material, is given in Figure 13 (experiment 2.7c).

Experiments 2.1-3 showed that the major problems of operating a moving granular bed filter are related to the transport system for the granular bed material. It was decided to improve the design of the injector in order to give better reliability and a higher material recirculation rate, at a higher pressure drop as compared to injector I. This led to the implementation and testing of injector II. In experiments 2.4-7 this injector was used, driven by cold, dry, pressurised air only. In addition, a pressure drop meter for the material transport system was installed, with the option of including or excluding the pressure drop over the injector.

TABLE 2

The results of the high temperature, high pressure moving granular bed filtration tests

Test Day / Exp.	Pressure (bar)	Filter Velocity (m/s)	Transport without problems	Number of Inlet Samples	Number of Outlet Samples	Inlet Conc. (g/m^3_{STP})	Total Mass Efficiency (%)
2.1a	5	0.1	+ 1		1 (28) *		
2.1b†	5	0.2	+ 1		- (3)		
2.2a**	5	0.2	+ 1	2	3 (9)	1.04 - 1.05	89 - 97
2.2b	8	0.4	+ / - 1	2	2 (20)	1.33 - 1.48	93 - 98
2.3	6	0.2	- 1				
2.4	4	0.1	+ 1				
2.5†	8	0.1	+ 2				
2.6†	5	0.05-0.2	+ 2				
2.7a	5	0.1	* + 2	2	3 (18)	1.79 - 1.32	82 -97
2.7b	5	0.2	+ 2	1	1 (6)	1.07	76
2.7c†	8	0.2	+ 2	2	2 (12)	0.75 - 0.68	86 - 93

*   Number between brackets is the number of measurements with optical particle counter
**   Continued with the same filter material
†   Experiment had to be stopped due to malfunction of combustor or control valves
1   Injector I / 2   Injector II

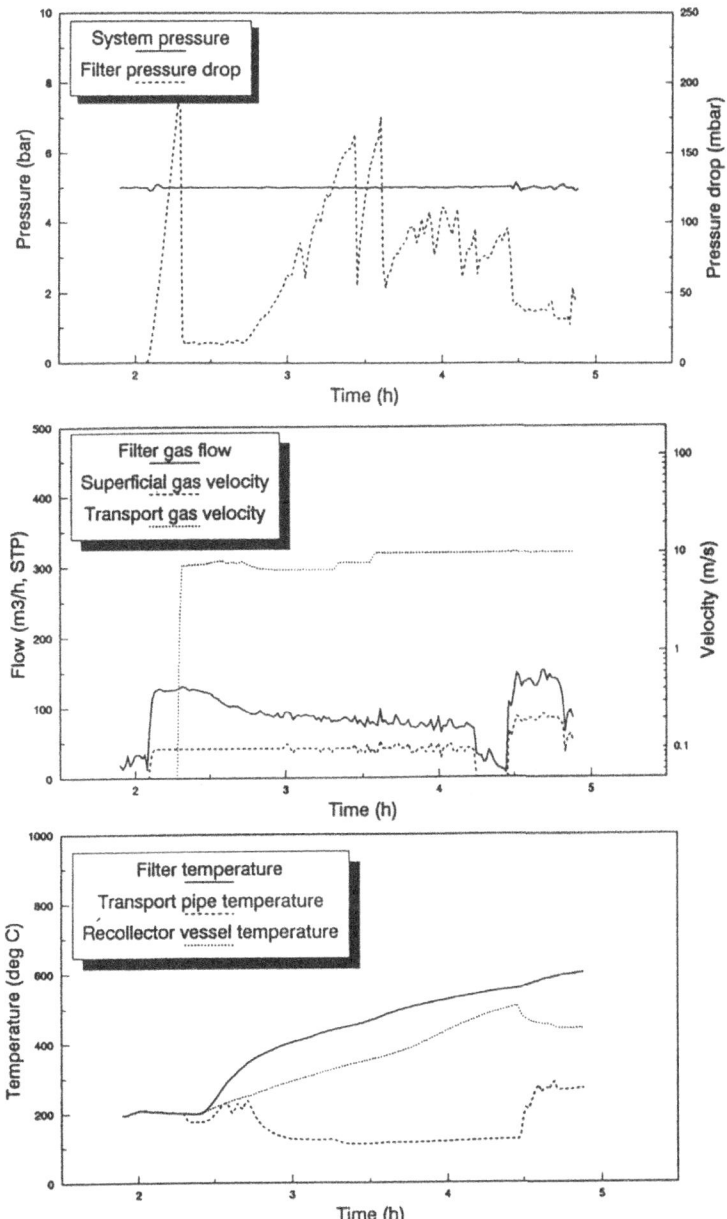

Figure 8   *HTHP moving granular bed filter test 2.1*

411

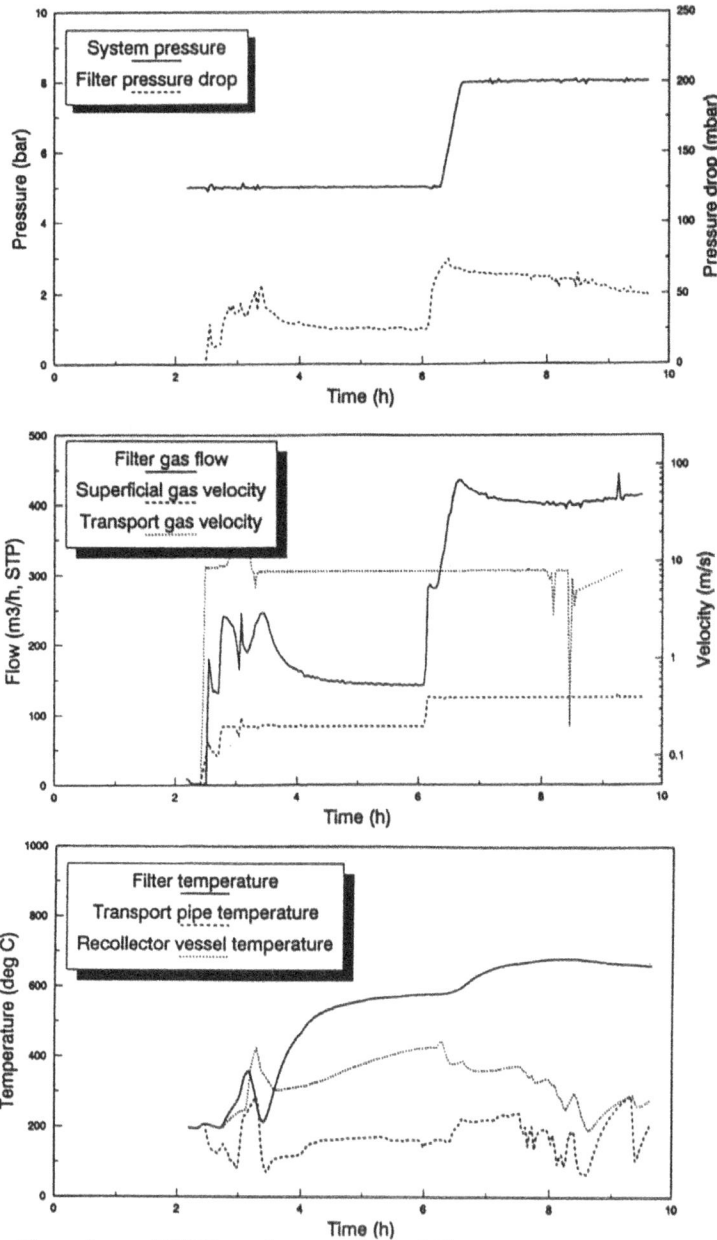

Figure 9   *HTHP moving granular bed filter test 2.2*

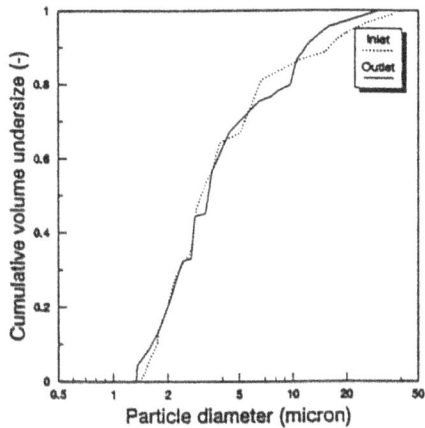

Figure 10a   Inlet and outlet particle size distributions for the HTHP moving granular bed filter. Experiment 2.2a-I.

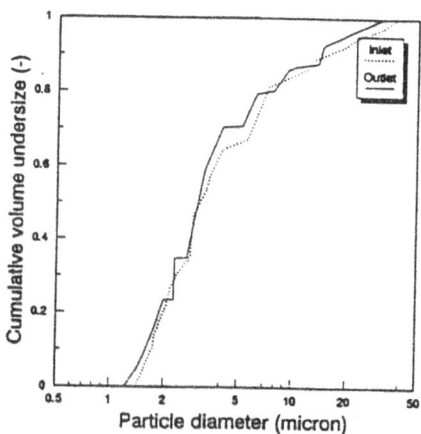

Figure 10b   Inlet and outlet particle size distributions for the HTHP moving granular bed filter. Experiment 2.2a-II.

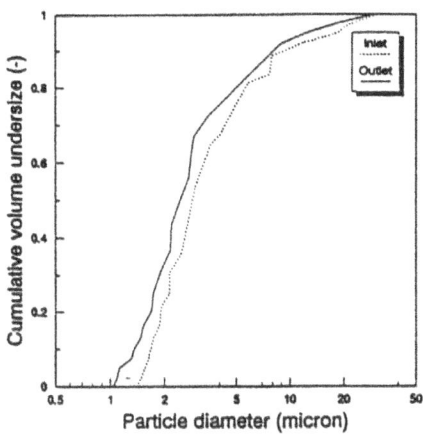

Figure 10c   Inlet and outlet particle size distributions for the HTHP moving granular bed filter. Experiment 2.2b-I.

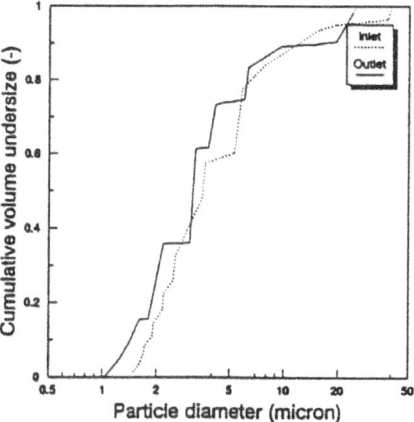

Figure 10d   Inlet and outlet particle size distributions for the HTHP moving granular bed filter. Experiment 2.2b-II.

413

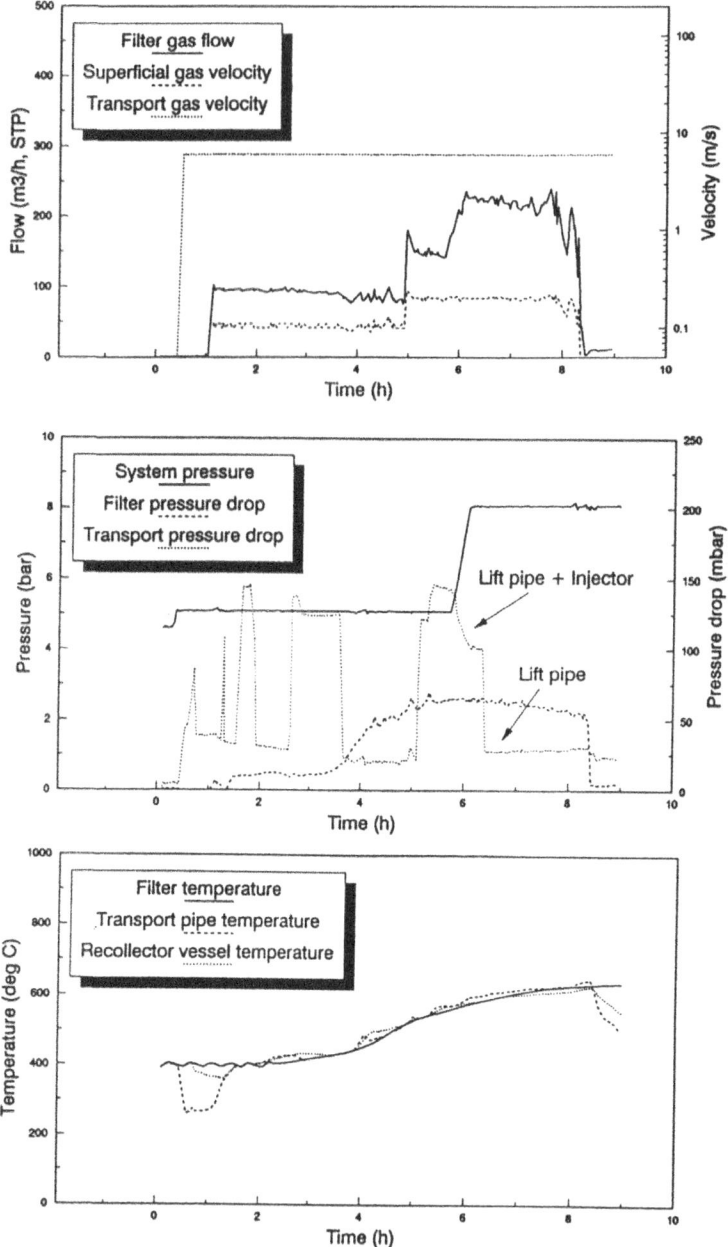

Figure 11   HTHP moving granular bed filter test 2.7

414

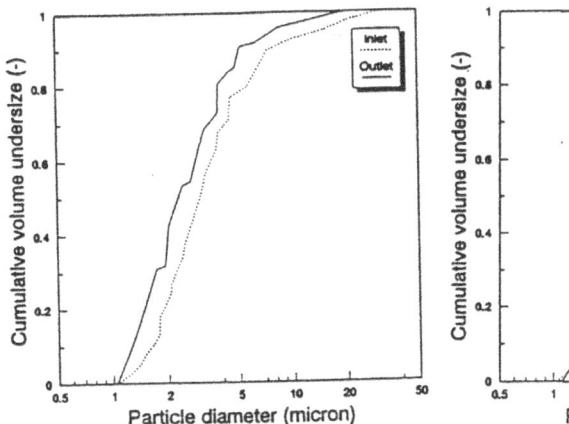

Figure 12a   Inlet and outlet particle size distributions for the HTHP moving granular bed filter. Experiment 2.7a-I.

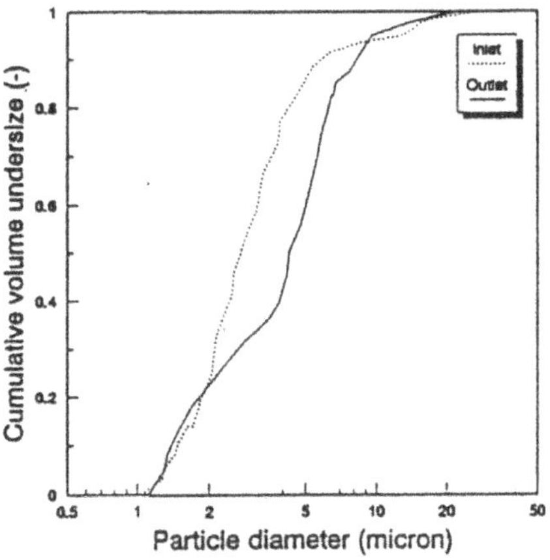

Figure 12b   Inlet and outlet particle size distributions for the HTHP moving granular bed filter. Experiment 2.7a-II.

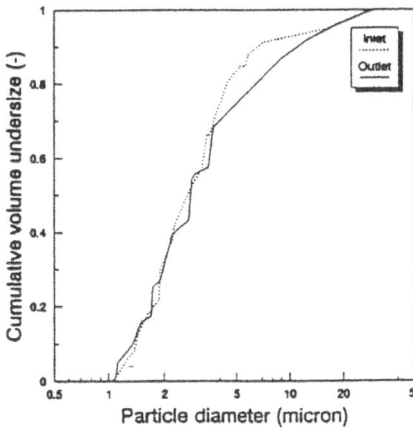

Figure 12c   Inlet and outlet particle size distributions for the HTHP moving granular bed filter. Experiment 2.7b.

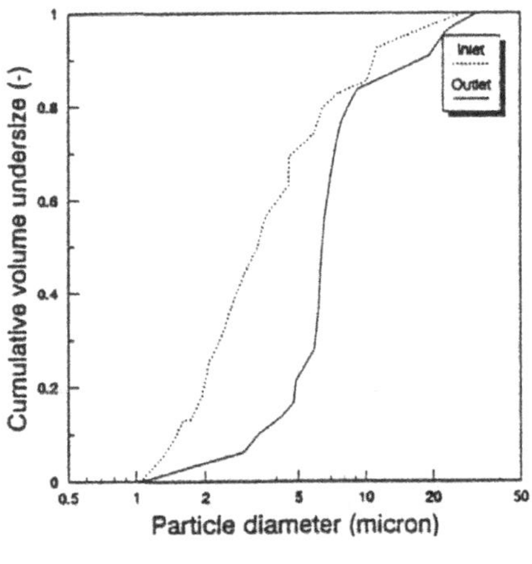

Figure 12d   Inlet and outlet particle size distributions for the HTHP moving granular bed filter. Experiment 2.7c-I.

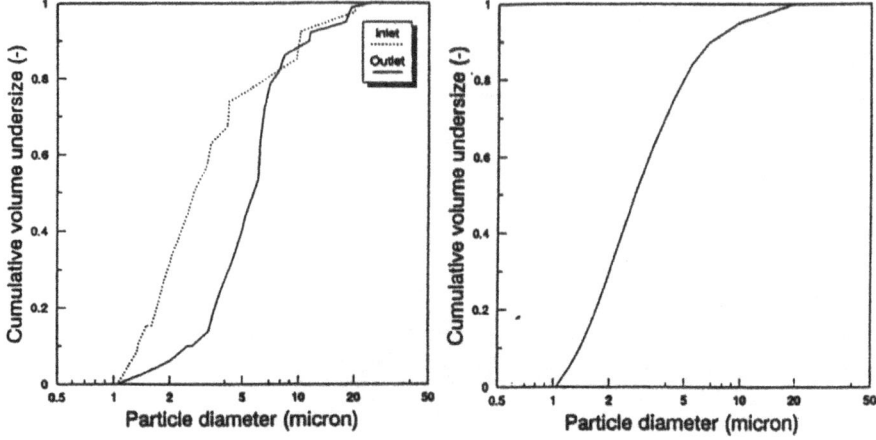

Figure 12e   Inlet and outlet particle size
distributions for the HTHP
moving granular bed filter.
Experiment 2.7c-II.

Figure 13   Particle size distribution for
the filter outlet: fly ash
collected in transport air
outlet. Experiment 2.7c.

The regeneration efficiency of the material recirculation and regeneration system could be estimated from a mass balance for the fly ash. This is based on the filter inlet sample and the amount of fly ash collected in the transport air outlet, which contains the fly ash removed from the granules during transport.

Alternatively, as was done after experiment 2.7, the mass fraction of fly ash in the contaminated filter material was analysed, comparing the dirtiest granules in the bottom of the filter, close to the injector, to the cleanest granules which were taken from the recollector vessel.

It was found that the contaminated filter granules contained 3.2 - 2.5 %wt fly ash, whilst the fly ash load on the granules after transport / regeneration was approximately 1.5 %wt. This indicates that the efficiency of the bed regeneration is between 30% and 40%. During the filtration process no effects of fly ash accumulation were noticed.

A mass balance for the fly ash is as follows:

$$Filter\,inlet\, -\, Filter\,outlet\, -\, Transport\,air\,outlet\, =\, Accumulation \qquad (2)$$

The efficiency of the regeneration system varied between 43% and 95% (experiment 2.2) and between 61% and 67% (experiment 2.7). It is noted that in both experiments at 8 bar the fly ash catch is contaminated with approximately 5 %wt filter granule material. This is accounted for in the calculations.

Discussion of moving bed filter performance

It was found that the HTHP dust sampling method is very successful and reproducible although the application of the optical particle counter includes a potential risk of damaging the optical system by sulphuric acid droplets.
The outlet measurements are disturbed by dynamic effects of the filter: due to high residence time of granules in the filter collected material can be reentrained at a later stage under altered operation conditions. Regarding the size distribution in the outlet of the filter it must be concluded from the size distributions that a major part of the particles in the outlet gas stream of the filter are reentrained agglomerates.
The moving granular bed filter can be operated at a pressure drop that shows no variation with time. The efficiency of the moving granular bed filter varies between
76 % and 98%. The bed transport system operation was much improved by improving the injector design, although it needs further optimisation, for example using an absolute method for measuring the bed transport rate. Another problem is the reeentrainment of filter granules with the outlet gas stream which may result in damage to the control valve for the flow.

CONCLUSIONS

Tests with a fixed granular bed filter showed a rapid increase in pressure drop. For this reason it was decided to change the route of the development towards an electrostatically enhanced moving granular bed filter, giving first priority to continuous filter operation. It was found that the HTHP dust sampling method is very successful and reproducible, although the application of the optical particle counter is not without problems. There is a potential risk of damaging the optical system by sulphuric acid droplets. It is advised to use the sampling system for filter inlet measurement also for the outlet measurement. It will, however, be necessary to find an alternative for the Coulter Multisizer II in order to cover the size range which is covered by the optical particle counter.
The outlet measurements on the moving granular bed filter are disturbed by dynamic effects: due to the high residence time of granules in the filter (1 - 2 h) collected material can be reentrained at a later stage under altered operation conditions. For this reason no grade efficiency results are given. It must be concluded from the size distributions that a major part of the particles in the outlet gas stream of the filter are reentrained agglomerates. The efficiency of the moving granular bed filter is of the order of 80-98%, which is too low for large scale application. In addition, it must be concluded that the bed transport system is not reliable enough, despite the various improvements that were made.
NOTE: Further plans for the work on the HTHP moving granular bed filter at Delft University of Technology are given in (Nykänen et al., 1993).

ACKNOWLEDGEMENTS

The presented work was funded by the Commission of the European Communities Non-Nuclear Energy R&D Programme. It was also supported by the Dutch Ministry of Economic Affairs (PEO contracts 20.36-0110.10 and 20.36-0110.11), TNO/KRI and Stork Boilers.

REFERENCES

Andries, J., Scarlett, B., Bernard, J.G., Zevenhoven, C.A.P., van de Leur, R.H.M., Ennis, B., de Haan, P.H., Hogervorst, A.C.R., Nikolić, M. (1987) "Closed loop controlled integrated hot gas clean up" Final report EC Contract EN3F-0028-NL(GDF) Delft Univ. of Technology

Belyaev, S.P., Levin, L.M. (1974) "Techniques for collection of representative aerosol samples" J. Aerosol Sci. $\underline{5}$(4) pp. 325-338

Bernard, J.G. (1992) "Experimental investigation and numerical modelling of cyclones for application at high temperatures" Ph.D. Thesis Delft Univ. of Technology

Bernard, J.G., Andries, J., Scarlett, B., Pitchumani, B. (1990) "Cyclone performance at high temperatures and pressures" Proc. 5th World Filtration Congress, Nice (France) 1990 Vol. 2 pp. 510-515

Cooper, D.W. (1986) "Minimising bias in sampling contaminant particles" Solid State Technology $\underline{29}$(1) pp. 73-79

Fissan, H., Schweintek, G. (1987) "Sampling and transport of aerosols" TSI J. Particle Instr. $\underline{2}$(2) pp. 3-10

NPR-Standard 2788 (1985) "Air quality - Exhaust and process gases - Gravimetric determination of concentration and flow of particulates" (in Dutch) NNI Nederlandse Praktijkrichtlijn

Nykänen, J., Zevenhoven, C.A.P., Andries, J., Hein, K.R.G. (1993) "Moving granular bed filter research at Delft University of Technology" accepted for presentation at the 5th Finnish National Aerosol Symposium, June 1-3, 1993, Finland, 6 pp.

Stern, A.C. (1968) "Air pollution" Vol. II, Academic Press

VDI-Standard 2066 (1975) "Measurement of particles - Dust measurements in flowing gases, gravimetric determination of dustconcentration - overview" (in German) VDI-Handbuch Reinhaltung der Luft $\underline{8}$ / VDI-Handbuch Staub $\underline{4}$

Wiener, R.W., Okazaki, K., Willeke, K. (1988) "Influence if turbulence on aerosol sampling efficiency" Atmos. Environm. $\underline{22}$(8) pp. 917-928

Zakkay, V., Gbordzoe, E.A.M., Radhakrishnan, R., Sellakumar, K.M., Patel, J., Kasinathan, R., Haas, W.J., Eckels, D.E. (1989) "Particulate and alkali capture from PFBC flue gas utilising granular bed filter (GBF) Comb. Sci. and Technol. $\underline{6}$ pp. 113-130

Zakkay, V., Gbordzoe, E.A.M. (1989) "A review for hot-gas cleanup for PFBC" Conf. Proc. "Combustion en lechos fluidizados" (EC-Comett) May-June 1989, Zaragoza, Spain, pp. 216-275

Zevenhoven, C.A.P., Hein, K.R.G., Scarlett, B. (1993) "Moving granular bed filtration with electrostatic enhancement for high temperature gas clean-up" paper presented at the 6th World Filtration Congress, May 1993, Nagoya (Japan)

Zevenhoven, C.A.P. (1992) "Particle charging and granular bed filtration for high temperature application" Ph.D. Thesis Delft Univ. of Technology

Zevenhoven, C.A.P., Andries, J., Hein, K.R.G., Scarlett, B. (1992a) "The electrostatic enhancement of a moving granular bed filter for PFBC combustion gas" Preprints PARTEC / 2nd European Symp. on the Separation of Particles from Gases" March 1992, Nürnberg (BRD) pp. 261-273

Zevenhoven, C.A.P., Bernard, J.G., Andries, J., Scarlett, B., Hein, K.R.G. (1992b) "The development of a high temperature gas clean-up system for pressurised fluidised bed combustion (PFBC)" paper presented at the 10th Members Conf. of the Int. Flame Res. Found. (IFRF) May 1992, Noordwijkerhout (NL)

Zevenhoven, C.A.P., Andries, J., Scarlett, B. (1991) "The filtration of a PFBC combustion gas in a granular bed filter" Conf. Proc. Filtech 1991, Oct. 1991 Karlsruhe (BRD) Vol. II pp. 415-423, also published in Filtration & Separation $\underline{29}$(3) (1992) pp. 239-244

DEVELOPMENT OF A SIMULTANEOUS SULFUR AND DUST REMOVAL PROCESS FOR IGCC POWER GENERATION SYSTEM

KATSUYA ISHIKAWA*, NAOYUKI KAWAMATA**, KENJI KAMEI**
Kawasaki Heavy Industries, LTD.
* Technology Group, Akashi Technical Institute, 1-1, Kawasaki-cho, Akashi 673 Japan
** Research and Development Office, 2-4-1, Hamamatsu-cho, Minato-ku, Tokyo 105 Japan

ABSTRACT

Kawasaki Heavy Industries, LTD.(KHI) has been developing a moving bed gas cleanup process for the Integrated Coal Gasification Combined Cycle Power Plant(IGCC). Its main features are cross flow type reactor removing sulfur and particulate from gas simultaneously and the granular material which contains iron oxide as an absorbent of H_2S and COS. Results of laboratory studies of sulfidation and regeneration characteristics of iron oxide sorbent, bench scale moving bed experiments, and reaction model analysis are discussed. Moreover, direct sulfur recovery from the sorbent regenerator offgas is studied and a suitable catalytic reduction process of SO_2 with coal gas directly to elemental sulfur is obtained. An advanced test treating 1,000m³N/h of product gas from an entrained flow coal gasifier is being prepared.

INTRODUCTION

IGCC needs gas cleanup at high temperature and high pressure for ensuring its expected high energy efficiency. KHI has been concentrating its effort to develop a moving bed type gas cleanup process. R/D project started at 1976, as a moving bed filter. Through a series of pilot tests treating generated gas from a fluidized bed gasifier with a capacity of 40t-coal/d at Yubari, a larger scale test facility of the moving bed granular filter with 200t-coal/d entrained flow gasifier has been constructed at Nakoso and its test operation is now undertaken.

For further simplification of gas cleanup process, KHI started the development of simultaneous sulfur and particulate removal process in 1988 as shown in Table 1. As the

TABLE 1

The schedule of research and development of the moving bed gas cleanup system by KHI

Experiments		Test Scale / Conditions	1976	1977	1978	1979	1980	1981	1982	1983	1984	1985	1986	1987	1988	1989	1990	1991	1992	1993
Basic Researches	Bench Scale Test	Temperature : 723K Pressure : atmospheric Gas Volume : 500 m³/h Filtration Area : 0.25m²	Heavy oil ash	Converter dust	Blast furnace dust															
	Cold Visible Test	Temperature : 293K Pressure : atmospheric Gas Volume : 500 m³/h Filtration Area : 0.25m²	Granule flow test																	
Granular Bed Filter	Blast Furnace	Temperature : 423K Pressure : 4.5 atm. Gas Volume : 6000 m³$_w$/H Filtration Area : 0.25m²		Test operations																
	FBC Boiler	Temperature : 423 to 633K Pressure : atmospheric Gas Volume : 1000 m³$_w$/H Filtration Area : 1.0 m²				Tube & bench tests			20T/h FBC test											
	Fly Ash Treating Plant	Temperature : 623 to 773K Pressure : atmospheric Gas Volume : 1600 m³$_w$/H Filtration Area : 1.0 m²					Test operations													
	5t/d Coal Gasifier (Yubari)	Temperature : 573 to 673K Pressure : 8 atm. Gas Volume : 150 m³$_w$/H Filtration Area : 0.25m²				Tube & bench tests	Test operations													
	40t/d Coal Gasifier (Yubari)	Temperature : 733K Pressure : 20 atm. Gas Volume : 4500 m³$_w$/H Filtration Area : 1.2 m²					Design&Construction			Test operations										
	200t/d Coal Gasifier (Nakoso)	Temperature : 698K Pressure : 25.7 atm. Gas Volume : 48100 m³$_w$/H Filtration Area : 14m²											Design	200t/d Support researches	Construction		Test operations			
Moving Bed Sulfur and Dust Removal	Bench Scale Test	Temperature : 673K Pressure : atmospheric Gas Volume : 102 m³$_w$/H Filtration Area : 0.3m²													Tube test		Bench test			
	4t/d Coal Gasifier (Nakoso)	Temperature : 698K Pressure : 25.7 atm. Gas Volume : 1000 m³$_w$/H Filtration Area : 0.3m²															Design&Construction	Test operations		

performance and durability of main components of the moving bed facility have been already established through the test operations as a granular filter, the granular sulfur sorbent was the key of the development. We have obtained a ceramic based sorbent containing iron oxide as the sulfur absorbing material. The sorbent has enough desulfurization performance, regeneration ability, and strength for moving bed use. In this report, results of sorbent screening, sulfidation and regeneration characteristics of the sorbent, and the result of bench scale moving bed tests will be discussed. Additionally, several processes for captured sulfur recovery have been considered and a direct catalytic reduction process of SO_2 to elemental sulfur was selected. The result of SO_2 reduction experiments will be also reported.

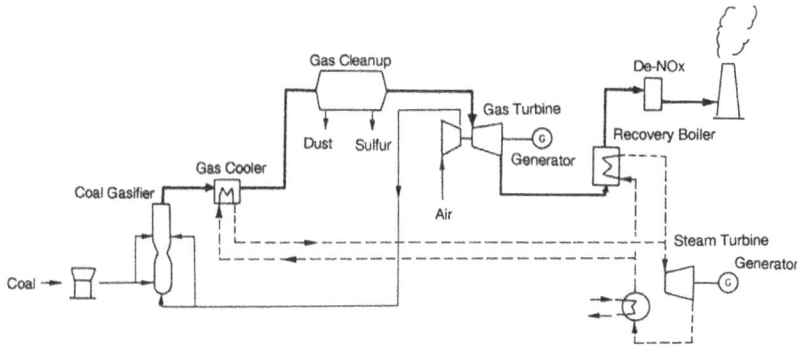

Figure 1. Process flowsheet of an IGCC with a moving bed gas cleanup

PROCESS PRINCIPLES

Figure 2 shows the process flow of the moving bed gas cleanup system. The reactor contains two or three moving beds according to the dust and sulfur loading. The cross sectional area of gas permeating zone of each bed changes according to the treating gas capacity. All beds are filled with the same granular material moving downward, while the treating gas permeates horizontally. These beds are so controlled that dust particulate is mostly removed at the first bed and desulfurization reaction occurs mainly in the second and the third beds, that is, the moving speed of the sorbent at the first bed is several times faster than the second bed for discharging the captured dust particulate and the re-activated sorbent is fed to the second and third beds directly from the regenerator. The regenerator, a counter flow cylindrical reactor, is placed just above the reactor and connected directly to the second and the third beds, and the sorbent moving downward is oxidized by the regeneration gas flowing upward. Discharged sorbent from each bed in the reactor is carried and separated from the dust particulate by a pneumatic conveyer and returns to the top of the facility. Table 2 shows the operation condition and target performance of the moving bed gas cleanup system.

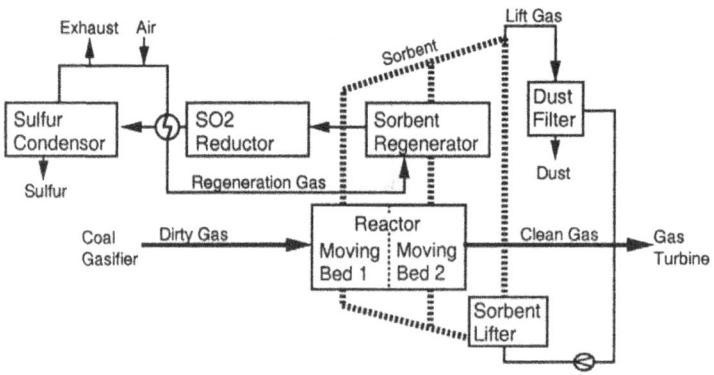

Figure 2. Process flowsheet of the moving bed gas cleanup system

TABLE 2
The operation condition and target performance of the moving bed gas cleanup system

Operating Conditions	
Temperature	673 to 773 K
Gas Pressure	2.1 to 3.5 MPa (20 to 35 kg/cm^{2}G)
Inlet Sulfur Content	1000 to 3000 ppm (as H2S + COS)
Inlet Dust Content	1000 to 3000 mg/m^{3}N
Target Performance	
Outlet Sulfur Content	< 100 ppm (> 90% desulfurization)
Outlet Dust Content	< 10 mg/m^{3}N (> 99% dust collection)

SORBENTS

Granular sorbent materials for moving bed gas cleanup system are required to have the following physical and chemical properties,

Granular shape	: Spherical is desirable for smooth flow in moving beds,
Strength	: Squeezing strength should be higher than 98N(10kgf) with no reduction after sulfidation and regeneration,
Thermal Resistance	: Durable in static exposure at 1073K and the thermal shock in the regenerator,
Desulfurization Performance	: Equilibrium concentration of H2S should be lower than 100ppm in the IGCC gas cleanup condition,
Regeneration Ability	: Reaction activity should be maintained after more than 2000 times of regeneration.

Jalan evaluated twelve metal oxides as sulfur sorbent for desulfurization of coal gas and selected four effective candidates, i.e. Zn, V, W, and Cu(Jalan 1981). Steinfeld found zinc oxide and zinc-ferrite to be effective and regenerative sulfur sorbent(Steinfeld et al. 1982). Flytzani-Stephanopoulos reported the desulfurization performance and regenerability of several mixed oxides of Zn, Fe, Cu, and Al(Flytzani-Stephanopoulos et al. 1985). The high desulfurizing ability and regeneration efficiency of iron oxide are also reported(Hasatani, Wen, et al. 1980 & Hasatani et al. 1982). Table 3 shows the reaction schemes and characteristics of several metal oxides with H2S(Ishikawa et al. 1991).

TABLE 3

The reaction schemes and characteristics of metal oxides with H2S

Metal	Reaction Scheme	Characteristics
Ca	$CaO + H_2S \rightarrow CaS + H_2O$ $CaCO_3 + H_2S \rightarrow CaS + CO_2 + H_2O$	Appropriate temperature range is over 1100K. Produced CaS is toxic.
Cu	$2CuO + H_2S + H_2 \rightarrow Cu_2S + 2H_2O$	Equilibrium H2S conc. is relatively low. Reduced Cu may be produced.
Fe	$Fe_3O_4 + 3H_2S + H_2 \rightarrow 3FeS + 4H_2O$	Equilibrium H2S conc. is relatively high. Easily regenerated by oxidation.
Mn	$MnO + H_2S \rightarrow MnS + H_2O$	Equilibrium H2S conc. is relatively high.
Pb	$PbO + H_2S \rightarrow PbS + H_2O$	Unstable reduced Pb may be produced. Possibility of toxicity and corrosion trouble.
Zn	$ZnO + H_2S \rightarrow ZnS + H_2O$	Equilibrium H2S conc. is relatively low. Regeneration without Zn loss is not easy.

Among these candidate metals, we have selected iron oxide for IGCC gas cleanup because of its high desulfurization performance within the target temperature range, regeneration ability, and easiness of handling and disposing. The chemical equilibrium concentration of H_2S with iron oxide is higher than that with zinc oxide, but within the target temperature range of 673 to 723K, it is low enough for the required desulfurization level for IGCC. For the use at higher temperature, we also considered other metal oxides such as zinc oxide and zinc-ferrite, but they still have some problems in sulfur absorption capacity and regeneration stability.

The reaction scheme with iron oxide is,

(Reduction)	$3Fe_2O_3 + H_2(CO)$	$\rightarrow 2Fe_3O_4 + H_2O(CO_2)$
(Sulfidation)	$Fe_3O_4 + 3H_2S + H_2(CO)$	$\rightarrow 3FeS + 4H_2O(3H_2O + CO_2)$
(Regeneration)	$2FeS + 3.5O_2$	$\rightarrow Fe_2O_3 + 2SO_2$

In reduction and sulfidation, products changes according to the reduction agent, H_2 or CO. The reaction scheme with CO is shown in the brackets. Besides H_2S, coal gasified gas also contains COS, and we have experimentally confirmed that COS can be converted rapidly to H_2S with the presence of iron oxide as,

(COS conversion) $COS + H_2 \quad \rightarrow \quad H_2S + CO$

As shown above, the form of iron compound undergoes cyclic changes, and it is important for sorbent to keep its strength under such structural changes of containing iron. To obtain the stability in such cyclic and long term usage, the sorbent, spherical granule about 1mm in diameter, has a porous ceramic base with enough strength for the use in the moving bed facility and enough pore volume for containing required amount of iron oxide for the sulfur absorbing capacity.

EXPERIMENTAL

Desulfurization and Regeneration

Evaluation and Selection of Sulfur Sorbents : We have evaluated more than a hundred kinds of sorbents containing iron using a fixed bed laboratory scale test facility shown in Figure 3. The facility has a tubular reactor with 22.1mm I.D., simulated gas feed system controlled by thermal mass flow controllers, a pressure regulator, and electrical heaters. Sulfur content in the inlet and outlet gas is measured by FPD and TCD gas chromatography. Evaluation of sorbents were made by the measurement of desulfurization efficiency and strength after at least five times of repeated sulfidation and regeneration. With the sorbents screening, it is shown that for the required strength and physical stability, the sorbent should have inert carrier supporting reacting iron oxide. We have obtained a sorbent with an alumina-silica based porous carrier which has sufficient pore volume to hold enough iron oxide.

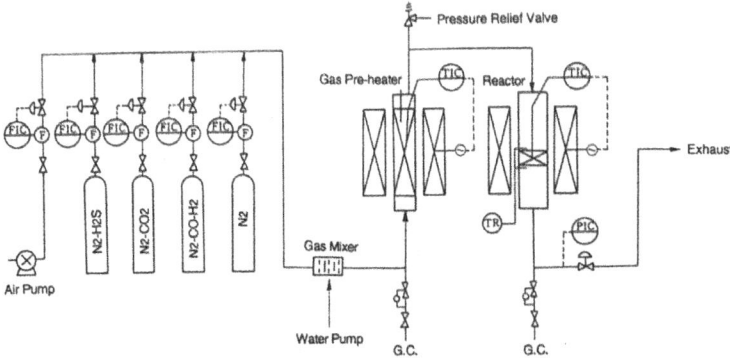

Figure 3. Fixed bed laboratory scale test facility

Desulfurization Reaction Characteristics : Figure 4 shows the breakthrough curve of Fe and Zn-Fe sorbents on the same alumina-silica carrier. Fe sorbent has higher absorption capacity and no less desulfurization efficiency than Zn-Fe sorbent at 773K. In our tests, desulfurization efficiency of Zn-Fe sorbent becomes superior to Fe sorbent over about 800K.

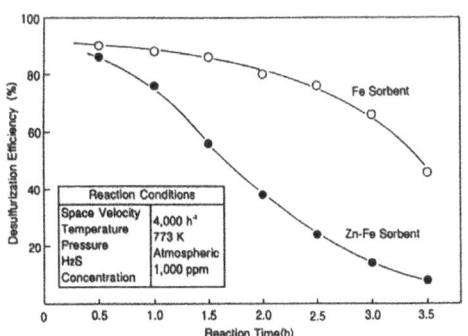

Figure 4. Breakthrough curves of Fe and Zn-Fe sorbents

Figure 5 shows the effect of gas pressure on the breakthrough curve of Fe sorbent(Ishikawa et al., 1991). Breakthrough time becomes longer with pressure because high

425

pressure increases the desulfurization reaction rate and the absorption capacity of the iron sorbent is sufficient for the absorption time in the moving beds. The durability of sorbent was also tested using the fixed bed reactor, and as shown in Figure 6, no drop of breakthrough time and strength of sorbent granule were observed with repeated sulfidation and regeneration up to 2000 times (Ishikawa et al., 1992).

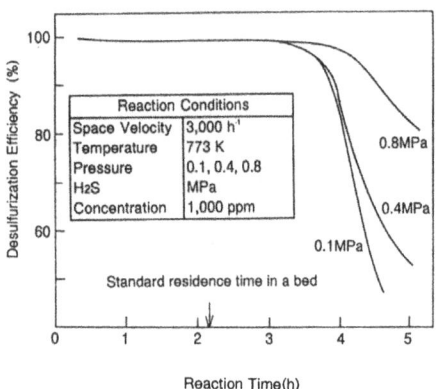

Figure 5. Effect of gas pressure on desulfurization reaction

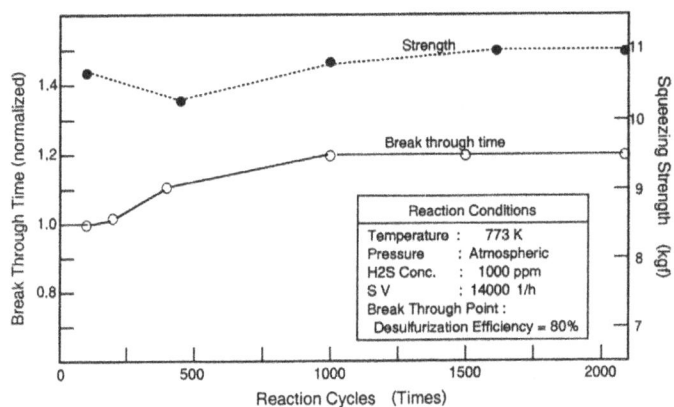

Figure 6. Change of breakthrough time and squeezing strength of sorbent granule with repeated sulfidation and regeneration

Regeneration Reaction Characteristics : Hasatani reported that FeS could be converted to Fe_2O_3 completely over 873K, but the regeneration became incomplete under 873K because of the formation of sulfates(Hasatani et al. 1982). It was also reported that the O_2 partial pressure

426

in the regeneration gas had strong influence on the reaction temperature(Hasatani et al. 1989). The regeneration reaction characteristics under the reaction condition of our process were tested with the same fixed bed test facility used for the sorbent screening and the thermo-gravimetric analyzer. Figure 7 and 8 show the effect of temperature and O2 partial pressure on the regeneration efficiency(Ishikawa et al., 1991). At temperature lower than 673K, the regeneration efficiency drops considerably, and even over 673K, high O2 partial pressure decreases the regeneration efficiency by promoting formation of FeSO4, which is stable up to 840K and is not completely decomposed even at 973K in our thermogravimetric analysis(TGA) tests.

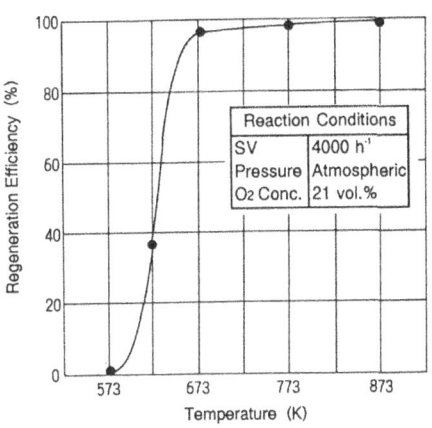

Figure 7. Effect of temperature on the regeneration efficiency

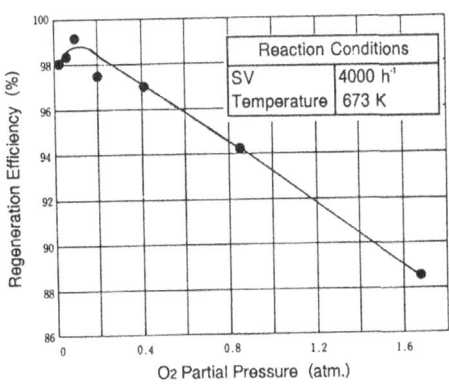

Figure 8. Effect of O2 partial pressure on the regeneration efficiency

Regeneration efficiency higher than 90% is required for the moving bed gas cleanup system, and because of it, temperature in the reacting zone in the regenerator should be kept over 673K. and O_2 partial pressure in the regeneration gas should be no higher than 0.14MPa. In a steady state of operation with enough treating gas volume and sulfur load, this temperature may be easily obtained by the heat from the oxidation of FeS and Fe_3O_4, but for the startup and partial load operations, regenerating gas heater should be considered.

Bench Scale Moving Bed Experiments : Figure 9 and Table 4 show the bench scale moving bed test facility.

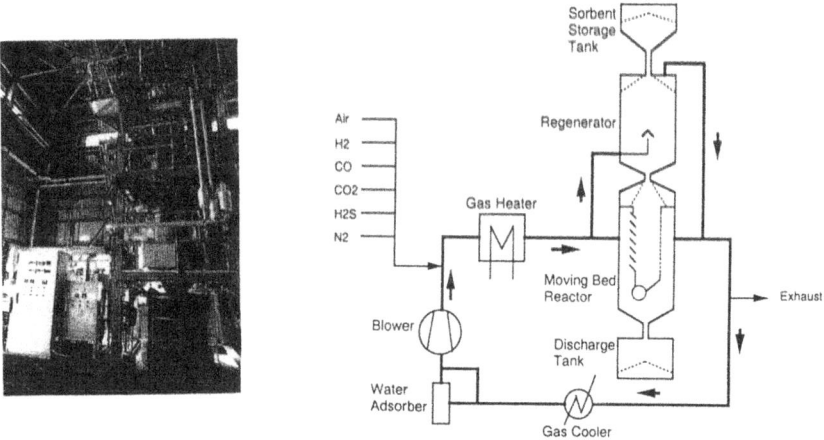

Figure 9. Bench scale moving bed test facility

TABLE 4
Specifications of bench scale moving bed test facility

Reactor	Single cross flow moving bed	^{H}2000, ^{W}150, ^{T}300 mm
Regenerator	Cylindrical counter flow moving bed	$^{I.D.}$300, ^{H}500 mm
Gas Feed	From mixed gas bombs controlled by four mass flow controllers	
Gas Flow	max. 102 m^3N/h	
Temperature	max. 750 K	
Pressure	Atmospheric	

The simulated gas circulates in the facility and is heated before entering the reactor. The

428

components consumed in the reactor, i.e. H_2 and H_2S, are fed by mass flow controllers to keep the content in the inlet gas stable, and H_2O produced in the reactor is removed by silica-gel after the gas cooler. Figure 10 shows the sulfur content in the outlet gas from the reactor (Ishikawa et al., 1992). At the beginning of a desulfurization test operation, sorbent in the reactor contains sulfur as FeS and $FeSO_4$ which remains at the shut-down of the preceding regeneration operation. Except the exhaustion of this residual sulfur as both H_2S and SO_2, H_2S concentration in outlet gas is kept around its equilibrium concentration at given condition.

Influence of dust masking on the desulfurization performance was also observed by feeding dust particulate to the inlet gas. Dust particulate of 40T/D fluidized bed coal gasifier captured by the moving bed granular filter at Yubari IGCC Pilot Plant was used and the dust load was around $10g//m^3N$, which matches the real IGCC dust load in the actual gas volume. In this test, there was almost no decrease in the desulfurization efficiency. Figure 11 shows temperature and content of outlet gas of the regenerator through a test operation. Regeneration reaction with the temperature rise and caused SO_2 emission are shown to be quite stable until the supply of sorbent stops in Figure 11.

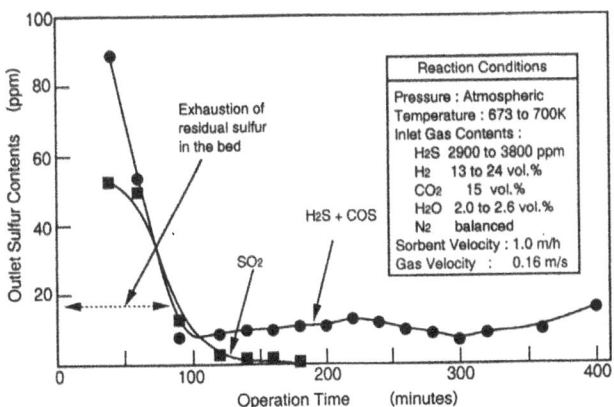

Figure 10. Sulfur content in the outlet gas from the moving bed reactor

429

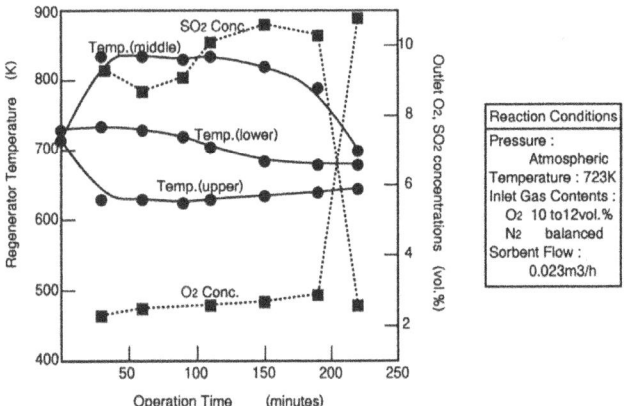

Figure 11. Temperature and content of outlet gas of the moving bed regenerator

<u>Modeling of Desulfurization Reaction</u> : From the thermo-gravimetric analysis of single sorbent granule sulfidation, the reaction rate is shown to match the unreacted core model and the rate-controlling step is the surface chemical reaction while the conversion of Fe in the sorbent to FeS is less than 70%. The reaction rate is

$$r_p = 4 \pi R^2 k_s C_{H2S} (1-x_p)^{2/3}$$

where r_p reaction rate (mol./s)
       R   granule radius (m)
       k_s rate constant per surface area (m/s)
       C_{H2S} H2S concentration (mol./m^3)
       x_p conversion of sorbent (-)

The value of k_s is obtained by TGA tests as 0.00320 at 673K and 0.00412 at 773K. From this relation of reaction rate, H2S concentration, and sorbent conversion, concentration distribution of H2S in the sorbent moving bed(counter flow) is calculated under IGCC conditions. The result is given in Figure 12. The breakthrough curves of laboratory scale fixed bed reactor and bench scale moving bed reactor are also calculated using the same model and quite good conformity with experimentally observed data is obtained as shown in Figure 13 and 14. Provided that the influence of the variety of gas and sorbent speed in the real moving bed is not significant, less than 50mm of depth of moving bed with active sorbent will be sufficient to reduce H2S concentration from 1000ppm to the chemical equilibrium concentration, which is lower enough than the required level for the IGCC gas cleanup, and there will be little difference in the required bed depth between 673K and 773K.

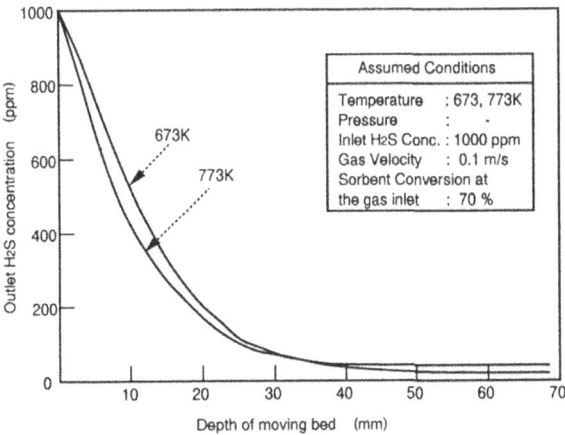

Figure 12. Calculated concentration distribution of H2S in the counter flow moving bed

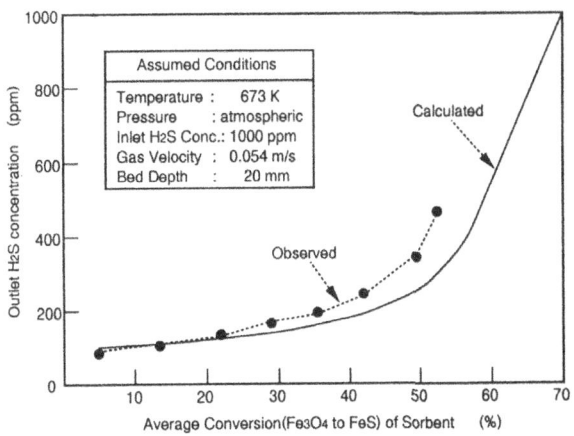

Figure 13. Calculated and observed desulfurization breakthrough curves of the
laboratory scale fixed bed reactor

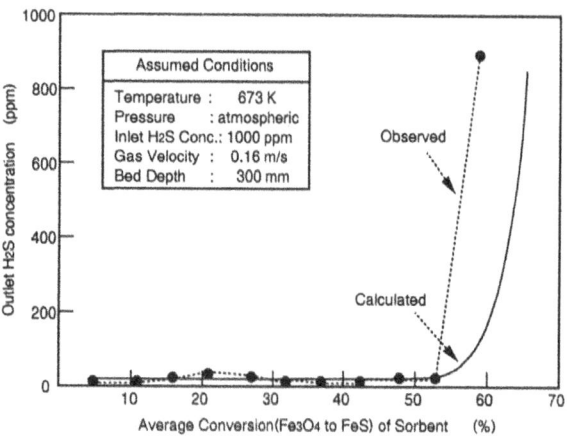

Figure 14. Calculated and observed desulfurization breakthrough curves of the
bench scale moving bed reactor

Sulfur Recovery

Iron oxide sorbent regeneration off-gas contains less than 2 vol.% of SO_2. Among several
processes of sulfur recovery, we have selected direct reduction of SO_2 with coal gasified gas to
elemental sulfur because of the economic feasibility and simplicity of this process. Catalysts
which have activity at the regenerator off-gas temperature of 573 to 673K have been screened
using the laboratory scale fixed bed reactor. Figure 15 shows the effect of SV on sulfur
recovery efficiency with an Alumina catalyst and a Co-Mo catalyst, and Figure 16 shows the
effect of pressure on sulfur recovery efficiency with the Co-Mo catalyst. As shown in Figure
15, Co-Mo catalyst has superior activity to Alumina catalyst and there is an optimum SV to the
sulfur recovery efficiency at 573K. When SV is lower(contact time is longer) than the optimum
point, formation of H_2S decreases the sulfur recovery efficiency. In Figure 16, only a little
influence of pressure is observed, but under further higher pressures, sulfur recovery efficiency
in a direct reduction process of SO_2 to elemental sulfur is reported to increase significantly
(Gangwal et al. 1991). Sulfur recovery efficiency of our system with the real coal gasified gas
and under the real IGCC conditions should be tested in the future, but it can be expected to be
no less than 80% in the real operating pressure of 2.1 to 3.5 MPa.

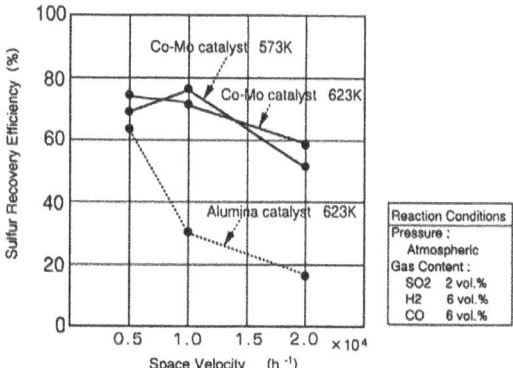

Figure 15. Effect of SV on sulfur recovery efficiency

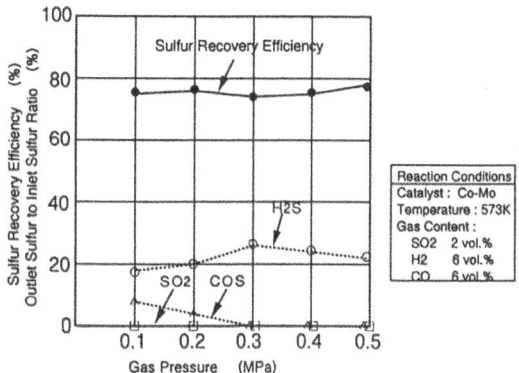

Figure 16. Effect of pressure on sulfur recovery efficiency

Figure 16. Effect of pressure on sulfur recovery efficiency

FUTURE PLANS

As the advanced phase of R/D, test operations of the moving bed gas cleanup system treating 1,000m3N/h of coal gasified gas are scheduled to start at the beginning of 1993 as a joint research of Engineering Research Association for Integrated Coal Gasification Combined Cycle

Power Systems(IGC) and KHI. Coal gasified gas will be supplied from the entrained flow coal gasifier of 200t-coal/d capacity of IGC Nakoso pilot plant. Table 5 shows the specifications of this test facility. This test facility is equipped with all elemental components of the moving bed gas cleanup system except sulfur recovery, so the result of these test operations will confirm the applicability of this system to the commercial IGCC power plant.

TABLE 5

Specifications of coal gasified gas treating moving bed test facility

Reactor	3 cross flow moving beds
Regenerator	Cylindrical counter flow moving bed
Gas Feed	From the entrained flow coal gasifier of 200t-coal/d capacity
Gas Flow	max. 1000 m^3N/h (correspond to 4t-coal/d)
Temperature	693 K
Pressure	2.6 MPa

CONCLUSIONS

A moving bed gas cleanup process with simultaneous sulfur and dust removal functions for IGCC has been developed on the base of the moving bed granular filter. Appropriate granular sulfur sorbent which contains iron oxide as the sulfur absorbing component and enough strength for the use in the moving bed facility has been developed, and its desulfurization performance and regeneration ability are confirmed by atmospheric simulated gas experiments. The characteristics of SO_2 direct reduction with coal gasified gas to elemental sulfur are also tested and suitable catalysts and reaction conditions have been obtained. An advanced test treating product gas from an entrained flow coal gasifier is planned to start near in the future and the application of the moving bed gas cleanup system to IGCC power plant is expected to be confirmed.

REFERENCES

Flytzani-Stephanopoulos, M., Tamhankar, S. S., Gavalas, G. R., Bagajewicz, M. J., & Sharma, P. K.(1985). High-Temperature Regenerative Removal of H2S by Porous Mixed Oxide Sorbents. Preprints of Papers, American Chemical Society, Division of Fuel Energy, 30, 4, 16-25

Gangwal, S. K., McMichael, W. J., & Dorchak, T. P. (1991). The Direct Sulfur Recovery Process. Environmental Progress, 10, 3, 186-191

Jalan, V. (1981). High Temperature Desulfurization of Coal Gases by Regenerative Sorption. Proceedings of 1981 International Gas Research Conference, 291-303

Hasatani, M., Yuzawa, M., Sugiyama, S., & Wen, C. Y. (1980). Reactivity of Fe_2O_3 with H_2S Contained in Low Calorie Syngas. Kagaku Kogaku Ronbunshu, 6, 5, 515

Hasatani, M., Yuzawa, M., Ogura, K., & Matsuda, H. (1982). Regeneration of Sulfated Iron with Oxygen for High-Temperature Desulfurization Cycle of Low Calorie Syngas. Kagaku Kogaku Ronbunshu, 8, 5, 611

Hasatani, M., Matsuda, H., & Kumazawa, K. (1989). Regeneration of Iron Sulfide Particle Packed Bed in High-Temperature Desulfurization Cycle of Low Calorie Coal Gasified Gas. Kagaku Kogaku Ronbunshu, 15, 2, 372

Ishikawa, K., Itoh, H., Kubo, Y., Hirao, M., & Yoshida, K. (1991). Development of a Simultaneous Sulfur and Dust Removal Process for IGCC Power Generation System. K.H.I. Technical Review, 109, 30-38

Ishikawa, K., Itoh, H., Kawamata, N., & Kamei, K. (1992). Development of a Moving Bed Gas Cleanup System for IGCC. Chemical Engineering, 37, 699-706

Steinfeld, G. (1982). Hot-Gas Desulfurization. DOE/MC/16545-T1

Section 3
Chemical Separations

HIGH TEMPERATURE DESULFURIZATION OF COAL DERIVED SYNGAS: BENCH SCALE AND PILOT PLANT SORBENT EVALUATIONS

A. M. ROBIN, J. S. KASSMAN, T. F. LEININGER,
J. K. WOLFENBARGER, AND P. P. YANG
Texaco, Inc.
P.O. Box 400, Montebello, CA 91733, USA

ABSTRACT

Texaco Inc. and the U. S. Department of Energy have conducted a joint 5-year cooperative research program to demonstrate the integration of hot syngas clean-up with the Texaco Coal Gasification Process. As part of this program, theoretical studies and bench scale experiments were performed at the Massachusetts Institute of Technology, Research Triangle Institute, and Texaco's Beacon and Montebello Research Laboratories to identify effective in-situ and external desulfurization sorbents. Pilot scale demonstrations with the most promising sorbents were performed in the pilot scale gasifiers at Texaco's Montebello Research Laboratory. The results of this high temperature desulfurization research are summarized, and a comparison of the bench scale tests and pilot scale demonstrations is provided.

INTRODUCTION

IGCC is gaining momentum as a commercially viable source of clean energy. Evidence for this is the numerous, recent announcements of commercial power generation projects. For instance, four projects using Texaco Power Generation Systems (TGPS) were recently announced (McLaughlin & Heuer, 1992): a 260 MW coal fueled Texaco Gasification Power System (TGPS) in Polk County, Florida; a 250 MW coke fueled TGPS in Delaware City Delaware; a 500 MW heavy oil fueled TGPS in Sicily, Italy; and a 250 MW heavy oil fueled TGPS in Venice, Italy. The driving force behind this trend is that IGCC based power generation is both cleaner and more efficient than conventional coal-fired boilers with scrubbers. Also, as in the case of the Texaco Gasification Process, a wide range of feedstocks can be used such as coal, heavy oil, petroleum coke and sewage sludge.

Sulfur and particulates are removed from the hydrocarbon derived synthesis gas (syngas) before it is delivered to the combustion turbine. Sulfur components, which are composed primarily of H_2S and COS, historically have been removed by wet chemical absorption in a solvent at low temperatures (40 to 120°F). Although IGCC with cold gas clean-up is more efficient than traditional coal fired plants, further gains in efficiency are possible if the syngas can be cleaned at higher temperatures (>1000°F). Recent economic studies (Robin, et. al. 1992), with a fully integrated Texaco based IGCC plant show that a 3% reduction in the heat rate, a 6% decrease in capital cost, and a 6% decrease in the levelized cost of electricity are possible if hot gas cleanup with high temperature desulfurization is used instead of cold gas cleanup.

From a practical standpoint, two methods of desulfurizing the syngas at high temperature are 1) in-situ and external desulfurization with once-through or disposable sorbents, and 2) external desulfurization with regenerable sorbents. Once-through sorbents can be added to the slurry with desulfurization taking place in the gasifier (in-situ desulfurization). Also, the sorbent can be injected downstream of the gasifier into the hot syngas, and subsequently removed with a high temperature cyclone or filter. Desulfurization with a once-through sorbent is attractive because little additional equipment is needed. Regenerable sorbents contact the raw syngas in a reactor, e.g., a fixed bed or fluid bed, that is downstream of the gasifier. The sulfided sorbent subsequently is regenerated and recycled for additional desulfurization. This method is attractive because solid waste is minimized, and very low sulfur emissions can be obtained.

A technical challenge to the success of high temperature desulfurization with regenerable sorbents is the development of a durable sorbent. Economic studies suggest that these desulfurization sorbents must withstand at least 200 cycles of high temperature sulfidation and regeneration to satisfy economic constraints. The sorbent and the desulfurization process should resist or avoid the following likely degradation mechanisms: 1) metal loss due to the reduction and vaporization of the metal when exposed to the reducing syngas; 2) pellet cracking and pore blockage from chemical swelling and contraction that occurs when the metal oxide is sulfided or regenerated. Other sorbent degradation mechanisms are possible.

Texaco Inc. and the U. S. Department of Energy have conducted a joint 5-year cooperative research program to investigate high temperature desulfurization and demonstrate hot gas clean-up. As part of this program, theoretical studies and bench scale experiments were performed at the Massachusetts Institute of Technology (MIT), Research Triangle Institute (RTI), Texaco's Beacon Laboratory (R&D-B) and Texaco's Montebello Research Laboratory (MRL) to identify effective once-through and regenerable desulfurization sorbents. Pilot scale demonstrations with the most promising sorbents and process configurations were performed in the pilot scale gasifiers at MRL. A summary of this high temperature desulfurization research program is provided in this paper.

THEORETICAL AND BENCH SCALE IN-SITU DESULFURIZATION STUDIES

Theoretical studies and bench scale tests were used to select in-situ (once-through) desulfurization sorbents for evaluation in the PDU. Preferred operating conditions were identified. Practical considerations of sorbent cost and availability eliminated all but the most common elements, such as iron, calcium and sodium from the study.

The studies were performed by Texaco at MRL and R&D-B (Najaar, 1989; Robin, *et. al.*, 1989) and included the following two key activities: 1) bench scale studies of sulfur capture with in-situ sorbents; and 2) theoretical studies of in-situ desulfurization and slag sorbent interactions. Additionally, high temperature bench scale viscosity measurements of in-situ desulfurization slag and coal slurry rheology with in-situ sorbents were studied.

Two bench scale units were used in the slag sulfur capture studies. Each consisted of a 32" long by 1-1/4" I.D. alumina tube that was surrounded by four KERAMAX heating elements. Coal ash and sorbent samples (50-100mg) were placed in a crucible and suspended inside the furnace by a thin platinum wire. After at least 18 hours of continuous exposure to simulated syngas at atmospheric pressure and fixed temperature, the wire was broken by an electrical current and the sample rapidly quenched in water. The reacted sample was then analyzed to determine the phases and amounts of each phase present.

Multi-phase thermodynamic calculation were performed with the CSIRO Thermochemistry System from the CSIRO Division of Mineral Products, an in-house version of SOLGASMIX (Eriksson, 1971), and F*A*C*T which is operated by McGill University.

Important results were obtained from experiments and thermodynamic calculations on the use of calcium based sorbents for in-situ desulfurization. At the temperatures and the reducing conditions that are typical of the Texaco gasifier, calcium is soluble in the silicate phase in the coal slag. This is shown in the phase diagram in Figure 1 for typical Pittsburgh #8 slag at 2200°F. Calcium that is absorbed by the silicate phase has little ability to capture

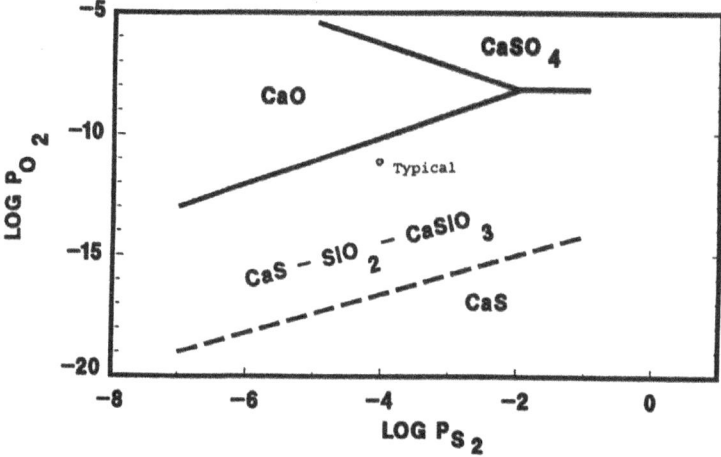

Figure 1. Calcium sulfide - oxide - silicate equilibrium at 2200°F

sulfur. Also, as shown by thermodynamic calculations in Figure 2, the silicate phase has

440

a large capacity to absorb calcium. In this case, no calcium is available to absorb sulfur at Ca:S molar ratios below 1.2:1. The results of these thermodynamic calculations were confirmed by the bench scale slag capture experiments.

However, the tests did indicate that calcium is an attractive desulfurization sorbent if it is injected when the syngas is approximately 1750°F; at these temperatures, the slag has solidified and no calcium is absorbed by the silicate phase. Sulfur removal efficiencies from 70 to 90% can be expected depending on gasifier operation and sulfur content of the feed.

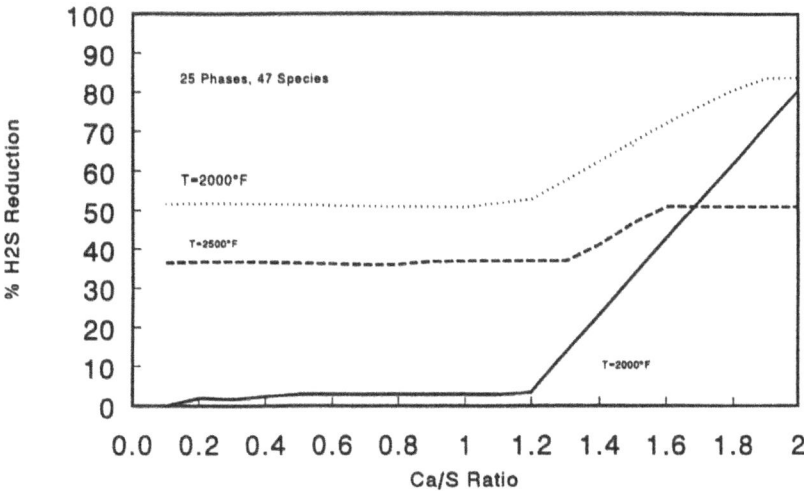

Figure 2. Pitts. #8 - calcium - iron desulfurization thermodynamics.

Figure 2 also shows that iron oxide has some potential for in-situ desulfurization. Over 50% desulfurization can be achieved with oxygen-blown Texaco syngas, but the level is sensitive to the gasifier temperature. As one would expect, the calculations also show that higher levels of desulfurization are obtained with higher initial H_2S concentrations in the gas. As in the case of calcium, iron will react with the silicate phase where little desulfurization takes place.

Both thermodynamic calculations and bench scale tests showed that sodium reduces the solubility of iron in the silicate phase, and as a result, improves iron utilization.

PILOT SCALE DESULFURIZATION WITH ONCE-THROUGH SORBENTS

To test desulfurization with once-through sorbents on the pilot scale and obtain data for program economic studies, 14 runs were performed from 1988 to 1991 (Robin, *et. al.*, 1989, Robin, *et. al.*, 1991). A total of 492 hours of in-situ desulfurization were accumulated with iron, calcium and sodium in-situ sorbents that were added to the coal slurry. Also, 166 hours of desulfurization with calcium based sorbents downstream of the gasifier at 1750°F were completed. Approximately 500 tons of coal were gasified during these experiments. The conditions and the extent of desulfurization for the key experiments are summarized in Table 1.

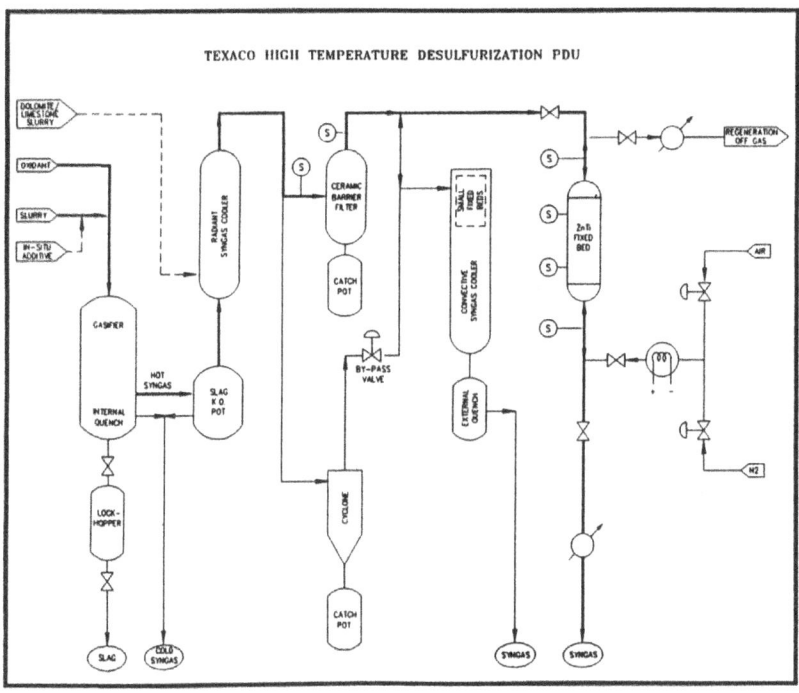

Figure 3. Pilot scale gasifier for desulfurization experiments.

    The pilot scale gasification runs primarily were performed in MRL's Low Pressure Solids Gasification Unit (LPSGU), which is shown schematically in Figure 3. The LPSGU normally operates at 350 psig at a throughput of 15-20 tons/day of coal. Both oxygen (99.75 vol. % purity) and air were used as the oxidant. The slurry, which is prepared in the MRL coal grinding and slurry preparation unit, is pumped to the gasifier burner and mixed with oxidant in the reaction chamber. As shown, the LPSGU is equipped with a Radiant Vessel and Convection Vessel for high temperature syngas cooling. The 1750°F calcium desulfurization experiments were performed by injecting dolomite upward from the bottom (inlet) of the Radiant Vessel and with fixed beds of lime or dolomite in the Radiant Vessel.

The highest level of desulfurization reported in Table 1 was obtained with calcium at 1750°F in the Radiant Vessel. Because of geometry and size limitations in the LPSGU and the Radiant Vessel, dolomite slurry injection was less successful than operation with the fixed bed of dolomite.

The highest steady-state in-situ sulfur removal with iron (64%) occurred during oxygen gasification at the lowest operating temperature. In order to increase this level, an iron-calcium mixture, which saturated the silicate phase, was used. Non-steady state sulfur removal of 66% was obtained with this iron-calcium mixture even at gasifier temperatures that were 600°F higher than the best iron only test; however, slag removal difficulties developed.

The pilot scale results are in general agreement with the bench scale and thermodynamic predictions except that higher levels of desulfurization were observed in the PDU. In general, the PDU experiments confirm that sulfur removal increases as the gasifier temperature is lowered. Higher levels of desulfurization are achieved during oxygen gasification than during air gasification; a maximum of 27% desulfurization was achieved with Pittsburgh #8 coal and Fine-Ox during air gasification, which is not sufficient for commercial application.

BENCH SCALE DESULFURIZATION WITH REGENERABLE SORBENTS

Bench scale screening and parametric studies were used to select external high temperature desulfurization sorbents for evaluation in the PDU. Preferred operating conditions also were identified. The bench scale studies were performed for Texaco at Research Triangle Institute (Harkins, et. al., 1989, Gangwal, et. al., 1991) and Massachusetts Institute of Technology (Flytzani-Stephanopoulos, et. al., 1989). High temperature sulfidations and regenerations were performed with simulated oxygen and air-blown syngas. The following investigations were performed: 1) preparation and fixed bed, multi-cycle screening of copper and zinc based sorbent grains and extrudates at MIT; 2) multi-cycle screening of zinc titanate extrudates in a pressurized fixed bed reactor at RTI; 3) multi-cycle studies on the impact of chloride on zinc titanate in an atmospheric thermogravimetric reactor (TGR) at RTI.

Sorbent Preparation and Screening Studies at MIT

MIT synthesized four promising zinc and copper based mixed metal oxide desulfurization sorbents for the screening studies. Additionally, a commercial iron oxide sorbent, and a zinc ferrite sorbent which was synthesized by United Catalysts, Inc. (UCI, L-3201), were obtained for screening. Sorbent grains and extrudates were sulfided and regenerated in isothermal, packed quartz tubes at atmospheric pressure over the temperature range of 1000-1400°F and space velocities of 2000 hr⁻¹ STP. The quartz tubes were 15 mm in diameter and 60 mm in length. H_2S breakthrough curves were monitored during the experiments, and the physical and chemical properties of the regenerated sorbents were examined after the experiments.

Several key findings and recommendations were made as part of the MIT fixed bed screening study. Zinc oxide and zinc ferrite were found to be unsuitable for highly reducing syngas (such as Texaco syngas) at temperatures above 600°C. Severe structural changes, solid phase segregation, and formation of occluded iron-rich surfaces were identified in the

TABLE 1. Summary of key PDU runs with once-through sorbent.

Run*/ Test Period	Length of Run (hrs)	Length of Test Period (hrs)	Oxidant/ Air Inlet Temp (°F)	Gasifier Temp (°F)	Fine-Ox Dosage**	Soda Ash Dosage**	% Sulfur Removal
1/	26.4		O_2	T_o			-
2/	69.4		O_2				
A		14.4	N/A	T_o	3.5	0	44.7
B		16.7	N/A	T_o-200	3.5	0	51.1
C		9.3	N/A	T_o-300	3.5	0	63.8
D		15.5	N/A	T_o-100	3.5	0	44.7
E		11.0	N/A	T_o-100	4.5	0	44.7
3/	59.3		O_2				
A		11.6	N/A	T_o	3.5	1.5	-
B		13.0	N/A	T_o-200	3.5	1.5	-
C		23.0	N/A	T_o	3.5	2.0	57.4
D		9.0	N/A	T_o-100	3.5	2.0	57.4
4/	84.0		O_2				
A		15.3	N/A	T_o	0	0	-
B		12.0	N/A	T_o	1.67	0	40.0
C		12.0	N/A	T_o-150	1.67	0	44.7
D		12.0	N/A	T_o	2.56	0	49.3
E		12.0	N/A	T_o-150	2.56	0	52.7
F		12.0	N/A	T_o-125	2.56	0.25	54.0
G		6.2	N/A	T_o-175	2.56	0.25	52.7
5/	67.5		O_2				
A		8.5	N/A	T_o	0	0	-
B		7.0	N/A	T_o	0	0	-
C		10.0	N/A	T_o	2.5	0	47
D		11.0	N/A	T_o	2.5	0	42
E		10.0	N/A	T_o-100	2.5	0	52
G		7.7	N/A	T_o-200	2.5	0	51
H		11.7	N/A	T_o-200	3.5	0	44
6/	66		Air				
A		11.5	1050	T_o	3.5	0	11
B		11.5	1050	T_o-100	3.5	0	19
C		12.1	1050	T_o-180	3.5	0	19
D		11.5	1050	T_o-200	4.5	0	23
E		8.2	950	T_o-250	4.5	0	27
7***	62		Air				
A		17	1050	T_o+50	0	0	-
B		9	1050	T_o+50	0	0	72
8****/	11		O_2				
A		7	N/A	T_o+50	0	0	77-82

*   All runs were performed with Pittsburgh #8 coal (2.14% S) except Run 4, where Powhatan coal (5.1% S) was used.

**   Sorbent dosage is in lbs. active sorbent (i.e. Fe or Ca) /lb S in coal feed.

***   Desulfurization during Run 7 used dolomite injection in the Radiant Vessel.

****   Desulfurization during Run 8 used a fixed bed of dolomite in the Radiant Vessel.

zinc ferrite extrudates after only six cycles of sulfidation and regeneration. Iron and zinc oxide reduction, and subsequent vaporization appear to play a role in this.

Zinc titanates were identified as the most promising sorbents with the potential to achieve effluent H_2S levels below 10 ppmv. The TiO_2 stabilizes the zinc oxide and slows the rate of zinc reduction. Other studies have shown that the rate of zinc reduction with zinc titanate can be 9 times slower than with pure zinc oxide (Lew, et. al., 1992). Also, zinc titanates sorbents can be regenerated to their original chemical phases (M. C. Woods, et. al., 1989), which may be important for multi-cycle durability.

Multi-cycle Fixed Bed and TGR Sorbent Screening at RTI

Two formulations of zinc titanate extrudates were evaluated at RTI in a larger pressurized fixed bed reactor (3" diameter x 15" length) during five cycles of sulfidation and regeneration in both simulated air-blown and oxygen-blown syngas. Zinc titanates with a 0.8 Zn/Ti molar ratio (ZT-0.8) and a 1.5 Zn/Ti molar ratio from UCI were tested. The sulfidations were performed at 350 psig and 1300°F. Regenerations were performed at 135 psig and temperatures between 1247°F and 1436°F using air diluted with nitrogen.

As expected, ZT-1.5 achieved higher sulfur loadings than sorbent ZT-0.8 due to its higher zinc content. The sulfur loadings were 18.7% higher in simulated oxygen-blown syngas and 28.1% higher in simulated air-blown syngas. Up to 60 percent of theoretical sulfur capacity was achieved by both sorbents with a bed height to diameter ratio of 5 and a wet gas space velocity of approximately 2000 hr^{-1} STP. The sorbents did not show a statistically significant decline in sulfur capacity over five cycles. All of the samples retained their original shape.

Analyses of the fresh and regenerated sorbents are provided in Table 2. The data show that some differences exist between the fresh and regenerated sorbents. The phase $ZnO \cdot TiO_2$, which was not noted in any of the fresh samples, was detected in significant amounts in the regenerated samples. Also, there appears to be a slight and increasing reduction in the crush strength, pore volume and surface area of the regenerated sorbents from the syngas inlet to the syngas outlet.

A separate study was performed at RTI to evaluate the effect of chloride in the syngas on zinc titanates. A series of three ten-cycle sulfidation-regeneration tests at 1300°F were run in a thermogravimetric reactor (TGR) at RTI. Simulated oxygen-blown syngas was used in all tests, and 500 ppmv HCl was added to the simulated syngas during two of the tests.

The study showed that a chloride level of 500 ppmv in Texaco oxygen-blown syngas did not have a deleterious effect on the sulfur capacity of zinc titanate sorbent over 10 cycles at 700°C when nitrogen and air were used during regeneration. Sorbent regeneration appeared to remove the residual chloride.

PILOT SCALE DESULFURIZATION WITH REGENERABLE SORBENTS

Based on the bench scale studies, zinc titanates were chosen for multi-cycle evaluations in the pilot unit. Three pilot plant runs were performed with a 1 ft^3 regenerable bed that was downstream of the high temperature ceramic candle filter assembly as shown in Figure 3. A schematic diagram of the regenerable bed assembly, which was designed and constructed at MRL, is shown in Figure 4. The sulfidations were performed with Pittsburgh #8 oxygen-

Table 2. New and regenerated Zn/Ti sorbents from RTI testing.

Sorbent: ZT-0.8 (Zn/Ti=0.8)

	Unused	Simulated Air-Blown Syngas			Simulated Oxygen-Blown Syngas	
Sample Bed Section	-	1	3	4	1	5
Rel. Phase Conc.,						
$2ZnO \cdot 3TiO_2$	3	3	1	1	3	2
$2ZnO \cdot TiO_2$	1	0	0	0	0	0
$ZnO \cdot TiO_2$	0	1	2	2	0	2
TiO_2	1	1	1	1	0	1
Crush Strength, lbs.	36.8	26.1	27.3	15.2	25.5	17.5
Pore Volume, cc/gm	0.396	0.386	0.363	0.325	0.393	0.363

Sorbent: ZT-1.5 (Zn/Ti=1.5)

	Unused	Simulated Air-Blown Syngas			Simulated Oxygen-Blown Syngas		
Sample Bed Section	-	2	3	5	1	3	5
Rel. Phase Conc.,							
$2ZnO \cdot 3TiO_2$	1	1	0	0	1	1	0
$2ZnO \cdot TiO_2$	3	3	2	2	3	2	2
$ZnO \cdot TiO_2$	0	0	1	2	0	1	2
TiO_2	0	0	0	0	0	0	0
Crush Strength, lbs.	32.5	32.8	18.9	10.3	39.1	27.1	22.3
Pore Volume, cc/gm	0.385	0.337	0.374	0.342	0.367	0.346	0.333

Note: Crush strength is the average of 6 tests. A plus sign after the number indicates there was at least one full scale (50 lbs) reading used in the average.

blown syngas from the PDU, and regenerations were performed with hot air and nitrogen.

Two- and Five-Cycle Zinc Titanate Test

The first two runs, a two-cycle (cycles A-1 and A-2) and a five-cycle test (cycles B-1 through B-5), were performed with ZT-1.5 from UCI (Zn/Ti = 1.5). Table 3 summarizes the average operating conditions for both tests.

Following each test, the extrudates were removed in eight sections for sampling and inspection. Pellet cracking and spalling was apparent throughout both beds, with the exception of the sulfidation outlet, where little sulfidation and regeneration occurred. Typical examples of the sorbent degradation that was observed is shown in the photograph

TABLE 3. Zinc titanate regenerable bed operating averages.

Cycle Number [1]	A-1	A-2	B-1	B-2	B-3	B-4	B-5
Average Sulfidation Temp., °F	1080	1040	1090	970	1015	1030	925
Maximum Sulfidation Temp., °F	1180	1110	1170	1040	1050	1080	1008
Sulfidation S.V. [2], dry, NTP, hr^{-1}	2880	3010	3840	5520	5510	5900	5900
Sulfidation Pressure, psig	331	341	335	336	324	321	330
Total Sulfur Capture, g-moles	151	109	128	138	128	120	114
Sorbent Util., % (mole S/mole Zn)	73	53	67	72	66	62	64
Average Regeneration Temp.[3], °F	1010	1100	1140	1260	1280	1260	1270
Maximum Regeneration Temp., °F	1390	1290	1270	1390	1400	1400	1320
Regeneration S.V.[3], dry, NTP, hr^{-1}	2730	3020	3580	3390	2980	3060	2910
Regeneration O$_2$ Partial Press., psia	3.45	4.55	2.75	1.75	1.44	1.19	0.59
Regeneration O$_2$ Conc., mol%	1.57	1.37	1.07	1.08	0.98	1.04	0.45
Total Regeneration O$_2$ Fed, g-mole	406	474	392	216	246	204	84
Total SO$_2$ Produced During Regeneration[4], g-mole (calculated from energy balance)	103	155	140	126	135	110	50

[1] During cycles A-1 through B-4, a 19-element ceramic candle filter was used to dedust the syngas upstream of the fixed bed. During cycle B-5, a cyclone was used.

[2] The sulfidation space velocity was increased during test B in order to shorten cycle length so that more cycles could be obtained during the test.

[3] The average regeneration temperature was increased during test B in order to increase the regeneration kinetics and reduce the O$_2$ requirement. Also, the O$_2$ partial pressure was decreased in order to reduce the likelihood of sulfate formation.

[4] SO$_2$ produced during regeneration was measured using an on-line mass spectrometer and a UV process gas analyzer. However, the data obtained by these instruments do not appear to be as reliable as the calculated SO$_2$ production obtained from an energy balance around the bed. Cycle B-5 regeneration was incomplete. When the sorbent was analyzed after shutdown, 56 g-mole sulfur was found remaining in the bed.

showed no physical signs of degradation in rapid atmospheric TGR cycling (Gupta, et. al., 1993). The two RTI formulations and the UCI ZT-2.0 were spherical pellets with nominal diameters of five millimeters.

The sorbents were placed in the fixed bed reactor with the UCI ZT-1.5 in thin alternating layers for comparison and were exposed to 5-1/2 cycles of sulfidation and regeneration. The conditions were similar to those in the two- and five-cycle tests except the space velocity was doubled. After the sixth sulfidation, the fixed bed reactor was cooled down and the sorbents were inspected in the sulfided state.

Severe cracking and spalling were observed in all four formulations in the following order of increasing degradation ZT-2.0 < ZT-1.5 < XZT-0.8 < XZT-1.1. Approximately 10 percent of the UCI ZT-2.0 suffered no visible signs of degradation while 100% of the experimental RTI sorbents cracked and spalled into multiple pieces. Sulfur analysis of the sulfided sorbent at the sulfidation inlet showed that the UCI ZT-2.0 contained only 2.0

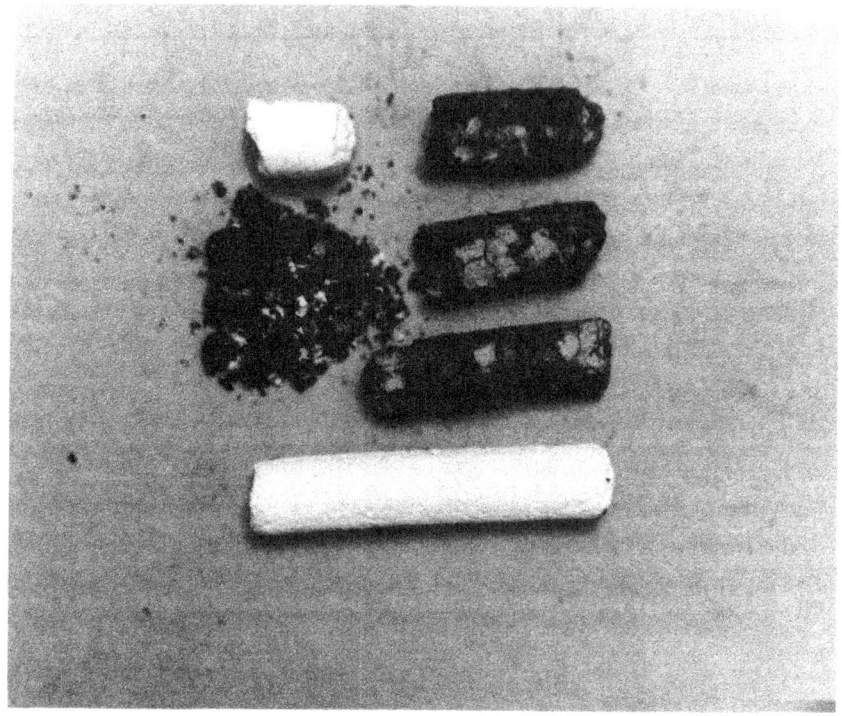

Figure 5.   Photograph of typical zinc titanate degradation in fixed bed pilot unit test.

percent sulfur while the other formulations contained between eight and eleven percent sulfur.   The UCI ZT-2.0 may have suffered less degradation during the six cycles because the sorbent was less reactive.

XRD, mercury intrusion porosimetry and nitrogen adsorption were performed on the fresh and the fully regenerated sorbents.   Little change in the chemical phases was detected; however, ZnO was detected in the regenerated RTI formulations.   After six cycles, 10-50% pore volume loss occurred for pore sizes below 1000 Å, but little pore volume loss occurred in the 1000-2000 Å regime.   An increase in pore volume occurred for the larger pores (2000-10,000 Å), which may be related to the sorbent cracking.

DISCUSSION

In-situ Desulfurization
The pilot scale in-situ desulfurization results generally are consistent and agree with the bench scale and thermodynamic predictions, except that slightly higher levels of

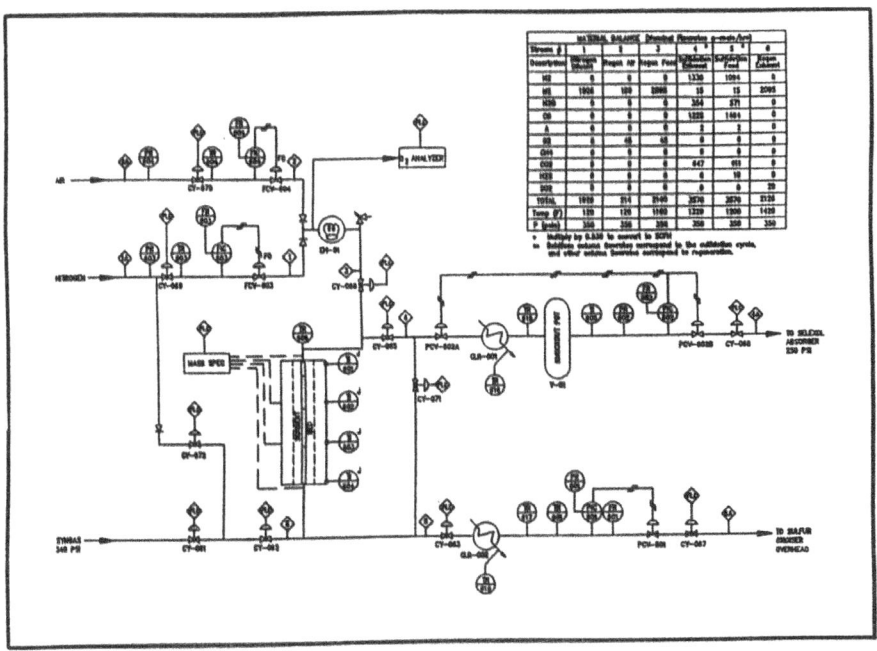

Figure 4. Regenerable sorbent bed assembly.

in Figure 5. The degradation was more severe in the bed exposed to five cycles, where 3 wt% of the sorbent passed through a 14-mesh screen (1.4 mm openings). Samples from each section were analyzed for residual sulfur, crystal phases, crush strength, pore volume, surface area and zinc and titanium content. Sulfate was also analyzed for, but none was found in any of the samples. Sorbent data for the five-cycle tests are shown in Table 4.

From X-Ray Florescence (XRF), Direct Current Plasma spectroscopy (DCP), and Inductively Coupled Plasma spectroscopy (ICP), it is not clear if the Zn/Ti molar ratio changed significantly during the two-cycle test. However, Table 3 shows that in addition to the physical degradation, the sorbent exposed to five cycles suffered an average decrease of 7% in the Zn/Ti molar ratio, as detected by both XRF and DCP. A general loss of pore volume and surface area also were observed.

Six-Cycle Multi-Sorbent Test
After high levels of decrepitation were observed in the pilot plant runs, emphasis was placed on finding a more durable sorbent. Three promising alternatives to ZT-1.5 were identified: ZT-2.0 (Zn/Ti=2) from UCI, and two experimental sorbents from Research Triangle Institute, XZT-0.8 (Zn/Ti=0.8) and XZT-1.1 (Zn/Ti=1.1). Both formulations from RTI

TABLE 4. ZT-1.5 sorbent analyses after cycles B-1 through B-5.

Sorbent Bed Position[1]	Unused	Top							Bottom
Sample No.	0	1	2	3	4	5	6	7	8
wt% Sulfur	0.09	14.3	14.3	9.41	6.78	3.37	0.49	0.17	0.14
Phases[2]									
ZnS	0	4	4	2	2	2	0	0	0
$2ZnO \cdot 3TiO_2$	1	2	2	2	1	1	1	1	1
$2ZnO \cdot TiO_2$	3	1	1	2	3	3	3	3	3
ZnO	0	0	0	0	0	0	0	0	0
TiO_2	0	2	2	1	1	0	0	0	0
Zinc Sulfide [3] Zinc Titanate	0	1.3	1.3	0.5	0.5	0.5	0	0	0
Crush Strength[4], lb	25.0	27.0	----	15.0	17.0	22.0	20.0	32.0	20.0
Pore Volume, cc/gm	0.40	0.32	0.31	0.32	0.33	0.34	0.36	0.37	0.38
Surface Area, m²/gm	3.1	6.2	5.8	3.1	4.0	1.9	1.7	2.5	2.1
Zn/Ti molar ratio, DCP	1.51	1.39	1.38	1.41	1.36	1.39	1.37	1.39	1.36
Zn/Ti molar ratio, XRF	1.47	1.42	1.38	1.40	1.37	1.36	1.34	1.36	1.42

[1] UNUSED = sample in "as received" condition; TOP = sulfidation inlet, regeneration outlet; BOTTOM = sulfidation outlet, regeneration inlet
[2] Digits represent relative presence of each phase
[3] Zinc Sulfide/Zinc Titanate = $ZnS/(2ZnO \cdot 3TiO_2 + 2ZnO \cdot TiO_2)$
[4] Average of six tests
[5] DCP = Direct Current Plasma Spectroscopy; XRF = X-Ray Fluorescence Spectroscopy

desulfurization were observed in the PDU. In general, the PDU experiments confirmed that sulfur removal increased as the gasifier temperature was lowered, and as expected, higher levels of desulfurization were achieved during oxygen gasification (higher H_2S partial pressure in raw syngas) than during air gasification.

Sulfur removal levels during in-situ desulfurization alone are not sufficient at the conditions tested for new plants, and an additional external desulfurization step will be required. This, combined with the additional solid waste that is produced as a byproduct, reduces the economic viability of in-situ desulfurization in most cases. Economic potential for in-situ desulfurization may exist in some retrofit applications.

External Desulfurization with Regenerable Sorbents
Although fixed bed bench scale tests have identified zinc titanate as a durable sorbent for high temperature desulfurization, the pilot scale fixed bed tests were less encouraging. Significantly, more degradation was observed in the pilot scale tests in terms of loss of zinc, pore volume, surface area and crush strength.

As a result of the sorbent degradation that was observed in the pilot scale tests, a closer inspection of the bench scale sorbents and additional bench scale tests have been performed. Fine cracks were observed in later inspections of the zinc titanates from the five-cycle test in the bench scale pressurized fixed bed at RTI. Some cracking of the UCI sorbents in atmospheric TGR rapid cycling was recently reported to Texaco by RTI (Gupta, et. al., 1993); however, the most severe cracking occurred when the cycles were performed at high pressure.

The apparent correlation between sorbent cracking and (1) operating pressure and (2) axial bed position may be related to the formation of zinc sulfate during regeneration. Higher regeneration pressures lead to higher O_2 and SO_2 partial pressures which favors the thermodynamics and accelerates the kinetics of sulfate formation. The sorbent at the sulfidation inlet (regeneration outlet) is bathed in higher concentrations of SO_2 and O_2, and as a result more sulfate formation is expected.

Experimental evidence to support sulfate formation as a mechanism for sorbent degradation in zinc titanate has been provided by RTI (Gupta, 1993), who reported that sulfate formation, sorbent swelling and sorbent cracking were observed in TGR experiments during the regeneration of fresh, un-sulfided sorbent when SO_2 was added to the regeneration gas.

The identification of zinc sulfate as a likely mechanism for sorbent degradation is encouraging because sulfate formation can be minimized and prevented. For instance, the use of a fluidized bed desulfurizer and regenerator, which will minimize the contact time between the sorbent and the regeneration gas, should reduce the potential for sorbent degradation. Additionally, alternate metals, such as copper which was identified in the MIT subcontract, may eventually be useful because sulfate formation is thermodynamically less favorable.

CONCLUSIONS

High temperature desulfurization and hot gas clean-up of synthesis gas can offer capital savings and efficiency gains over traditional cold gas clean-up. Development of this technology, however, continues to offer technical challenges.

Based on the thermodynamic calculations, bench scale tests, and pilot scale demonstrations, in-situ desulfurization is limited to applications where 50% desulfurization is sufficient, e.g., some retrofit application. For many coals, calcium based sorbents are not attractive for in-situ desulfurization because the calcium reacts with the silicate phase; the calcium silicate phase is viscous and will not readily flow from the slagging gasifier.

Greater than 99% desulfurization was demonstrated in pilot plant tests for multiple cycles of sulfidation and regeneration with external desulfurization sorbents; however, the tests indicate that this approach continues to pose technical challenges in sorbent and process development. Severe physical and chemical degradation of zinc ferrite was observed in fixed bed sulfidations (half-cycle tests), which was traced to iron and zinc vaporization in the highly reducing syngas.

Severe physical degradation was observed in zinc titanates in the fixed bed pilot plant tests and in some of the bench scale tests. Preliminary information suggests that sulfate formation may be responsible for this degradation.

Zinc titanate continues to offer promise for high temperature desulfurization because process modification, such as the use of fluid beds instead of fixed beds, can reduce the potential for sorbent degradation.

ACKNOWLEDGEMENTS

The authors thank the dedicated staff at the Research Triangle Institute, the Massachusetts Institute of Technology and Texaco's Research Laboratory in Beacon, New York, who made critical contributions to this project. Also, we acknowledge the support and advise of the U. S. Department of Energy.

REFERENCES

Eriksson, G. (1971). Acta Chem. Scand. Vol. 25, p. 2651.

Flytzani-Stephanopoulos, M., Yu, T. & Lew, S. (1989). Bench Scale Research at MIT, Preliminary Desulfurization Research, Integration and Testing of Hot Desulfurization and Entrained Flow Gasification for Power Generation Systems. Phase I Topical Report to U.S. Department of Energy, Vol. III, Contract No. DE-FC21-87MC23277.

Gangwal, S.K., Paar, T.M. & McMichael, W.J. (1991). Effect/Fate of Chlorides in the Zinc-Titanate Hot Gas Desulfurization Process, Process Optimization, Integration and Testing of Hot Desulfurization and Entrained Flow Gasification for Power Generation Systems. Phase II Topical Report to U.S. Department of Energy, Vol. III, Contract No. DE-FC21-87MC23277.

Gupta, R.P. (1993). Personal Communication.

Gupta, R.P., Gangwal, S.K. & Johnson, E.W. (1993). Evaluation of Zinc Loss From Zinc Titanate Sorbents During Hot-Gas Desulfurization, Integration and Testing of Hot Desulfurization and Entrained Flow Gasification for Power Generation Systems. Final Report to U.S. Department of Energy, Vol. II, Contract No. DE-FC21-87MC23277.

Harkins, S.M., Folsom, G.G. & Gangwal, S.K. (1991). Study of Zinc Titanate Desulfurization Sorbents, Process Optimization, Integration and Testing of Hot Desulfurization and Entrained Flow Gasification for Power Generation Systems. Topical Report to U.S. Department of Energy, Vol. II, Contract No. DE-FC21-87MC23277.

Lew, S., Sarofim, A.F. & Flytzani-Stephanopoulos, M. (1992). Sulfidation of Zinc Titanate and Zinc Oxide Solids. Ind. Eng. Chem. Res. Vol. 31. pp. 1890-1899.

McLaughlin, C.M. & Heuer, M.C. (1992). Texaco's Gasification Initiative - Meeting Power Industry Needs. Paper presented at 11th Annual EPRI Meeting, San Fransico, CA, USA, 22 October.

Najjar, M.S. (1989). Theoretical and Bench Scale Research and TCRB, Preliminary

Desulfurization Research, Integration and Testing of Hot Desulfurization and Entrained Flow Gasification for Power Generation Systems. Phase I Topical Report to U.S. Department of Energy, Vol. II, Contract No. DE-FC21-87MC23277.

Robin, A.M., Wu, C.M. & Kassman, J.S. (1989). Preliminary Desulfurization Research, Program Summary and PDU Operations, Integration and Testing of Hot Desulfurization and Entrained Flow Gasification for Power Generation Systems. Phase I Topical Report to U.S. Department of Energy, Vol. I, Contract No. DE-FC21-87MC23277.

Robin, A.M., Kassman, J.S., Leininger, T.F., Wolfenbarger, J.K., Wu, C.M. & Yang, P.P. (1991). Process Optimization, Program Summary and PDU Operations, Integration and Testing of Hot Desulfurization and Entrained Flow Gasification for Power Generation Systems. Phase II Topical Report to U.S. Department of Energy, Vol. I, Contract No. DE-FC21-87MC23277.

Robin, A.M., Jung, D.Y., Kassman, J.S., Leininger, T.F., Wolfenbarger, J.K. & Yang, P.P. (1992). Integration and Testing of Hot Desulfurization and Entrained Flow Gasification for Power Generation Systems. In Proceedings of the 12th Annual Gasification and Gas Cleanup Systems Contractors Review Meeting, Vol. I. ed. R.A. Johnson & S.C. Jain. pp. 95-104.

Woods, M.C., Leese, K.E., Gangwal, S.K., Harrison, D.P. & Jothimurugesan K. (1989). Reaction Kinetics and Simulation Models for Novel High-Temperature Desulfurization Sorbents. Final Report to U.S. Department of Energy, Contract No. DE-AC21-87MC24160.

DEVELOPMENT AND CHARACTERIZATION OF STEAM REGENERABLE SORBENTS FOR HOT GAS DESULPHURIZATION IN COAL GASIFICATION BASED COMBINED CYCLE PLANT

P.E. HØJLUND NIELSEN & INGA DÓRA SIGURÐARDÓTTIR
Research and Development Department
Haldor Topsøe A/S, Nymøllevej 55, DK-2800 Lyngby, Denmark

ABSTRACT

A process and sorbents have been developed for the desulphurization of coal gases at temperatures above 350°C. Characteristic features of these sorbents are the utilization of the reversible gas solid reaction

$$H_2 + H_2S + SnO_2 = 2H_2O + SnS$$

which has an equilibrium constant close to 400 at 400°C. A low sulphur emission can be achieved when using tin-sorbents for the cleaning of coal gases. As the reaction is reversible, steam can be used to oxidize SnS back to SnO_2 and this is typically done at 450°C to 500°C.

The sorbents have been examined for other possible reactions in the coal gas. Whereas the sorbents have high catalytic activity for carbonyl sulphide conversion, they have a vanishing catalytic activity for the water gas shift reaction and the hydrocarbon synthesis reaction.

With respect to mechanical properties, repeated sulphidation and regeneration in steam have not resulted in deterioration of strength or mechanical breakdown.

Application of sorbent based upon tin dioxide in IGCC (Integrated Gasification Combined Cycle) plants along with other units processes needed in a hot gas cleaning train result in an improvement of the thermal efficiency by several percent compared to systems based upon wet scrubbers.

INTRODUCTION

The hot gas cleaning concept is thought to be applied in connection with IGCC power plants based on coal. The possible use of such plants, instead of the conventional coal-fired plants, is being vigorously debated world-wide. The hot gas cleaning system may provide the desired improvement of the IGCC plant, i.e. the 2-4% (absolute) in thermal efficiency (Højlund Nielsen *et al.*, 1991). For most feedstocks and for most locations, this would be an advantage for the IGCC plant. However, the Danish situation is more balanced. The conventional power plant benefits by our access to cold water for cooling and the expected thermal efficiencies for IGCC and conventional coal-fired plants differ only slightly. Continental based plants using low grade coal would, on the other hand, be able to obtain a significantly higher efficiency than coal dust fired plants with advanced steam systems.

Another advantage of the IGCC technology using hot gas cleaning is that it is expected to have very low emissions of harmful gases such as nitrogen oxides and sulphur dioxide.

At Topsøe we have been engaged in the work on coal gases for many years. We have commercialized catalysts for sulphur resistant water gas shift conversion and for carbonyl sulphide conversion.

The possibility of using SnO_2 as a steam regenerable S-removing agent was initially identified through a desk study at Topsøe. Subsequent experiments showed that this was also possible in practice. Patents and patent applications covering this use of SnO_2 have been obtained or are pending in a number of countries.

THEORY AND METHODS

Thermodynamics

What can be concluded from a thermodynamic study is that the equilibrium constant for the absorption:

$$H_2 + H_2S + SnO_2 \rightarrow SnS + 2H_2O \tag{1}$$

at 400°C is:

$$K_p = \frac{P_{H_2O}^2}{P_{H_2} \cdot P_{H_2S}} \cong 400 \tag{2}$$

where Kp is the thermodynamic equilibrium constant and P_{H_2O}, P_{H_2} and P_{H_2S} are the partial pressures of steam, hydrogen and hydrogen sulphide, respectively.

By dividing the partial pressure by the total pressure, we obtain the molar fractions, X_{H_2}, X_{H_2O} and X_{H_2S}, respectively, and equation (3)

$$K_x = K_p = \frac{X_{H_2O}^2}{X_{H_2} \cdot X_{H_2S}} \tag{3}$$

since equation 1 is independent of pressure $K_p = K_x$, (see Figure 1), which means that the reverse reaction, the regeneration,

$$2H_2O + SnS \rightarrow SnO_2 + H_2 + H_2S \tag{4}$$

has the equilibrium constant

$$K_x^{-1} = \frac{X_{H_2} \cdot X_{H_2S}}{X_{H_2O}^2} \cong \frac{1}{400} \quad (at\ 400°C). \tag{5}$$

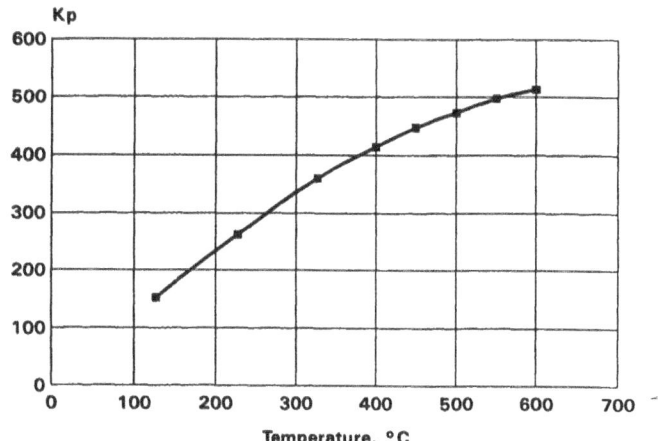

Figure 1. SnO$_2$/SnS equilibrium constant for reaction (1)

If the regeneration reaction (4) is carried out in 100% steam at 400°C, a consequence of the latter is

$$X_{H_2} = X_{H_2S} = X, \ X_{H_2O} = 1-2X$$

and

$$\frac{X^2}{(1 - 2X)^2} \cong \frac{1}{400}$$

leading to

$$\frac{X}{1 - 2X} = \frac{1}{20} \qquad \text{i.e. } X_{H_2S} = X \cong 0.045$$

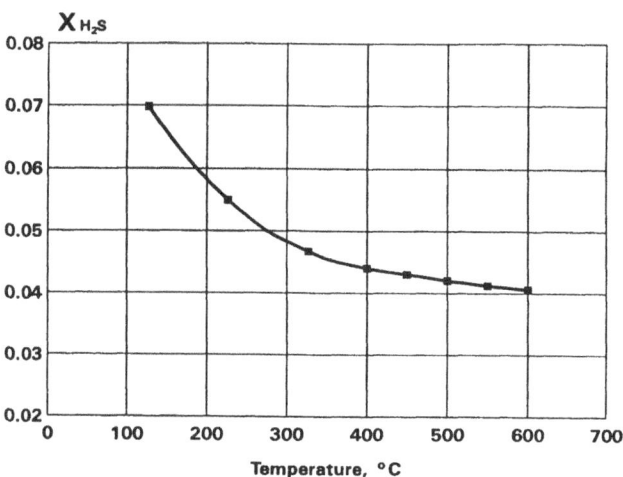

Figure 2: H_2S equilibrium concentration in steam in the presence of SnO_2 and SnS.

On the other hand, in a typical gas from a dry coal gasifier like the Shell, Prenflo or GSP (Gaskombinat Schwarze Pumpe) one has :

$$CO = 60 \text{ vol}\%, \quad CO_2 = 6 \text{ vol}\%, \quad H_2 = 25 \text{ vol}\%, \quad N_2 = 5.5 \text{ vol}\%$$
$$H_2O = 3 \text{ vol}\%, \quad H_2S = 0.5 \text{ vol}\%,$$

one has

$$X_{H_2S} \cong \frac{0.04^2}{0.245 \cdot 400} \approx 1.6 \cdot 10^{-5}$$

which will yield an exit concentration of H_2S at 16 ppm, if equilibrium is reached at 400°C.

The application of tin oxide has one drawback, i.e. that SnO_2 may be reduced to metal:

$$SnO_2 + 2H_2 \rightarrow Sn + 2H_2O \qquad (6)$$

Above 231.9°C, Sn is a liquid. To avoid the reduction to metal, the H_2O/H_2 must be controlled see the phase diagram in Figure 3. Gases stemming from a wet gasifiers, like the Texaco, Dow and HTW, are easily handled below 550°C without any risk of SnO_2 reduction, whereas gas from dry gasifiers, like the Shell, Prenflo and GSP, may require the addition of a small amount of steam. CO does not react with SnO_2 at temperatures below 500°C (Oates & Todd, 1962) so tin oxide is not reduced even if there is a thermodynamic potential for it (Westmoreland & Harrison, 1976).

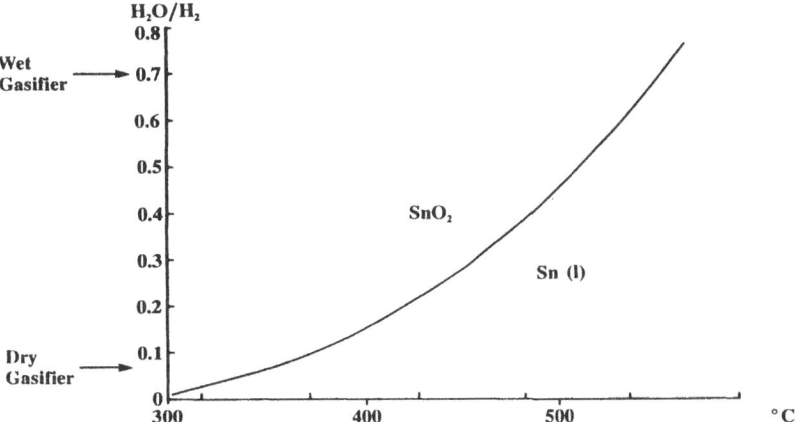

Figure 3: Phase Diagram water/hydrogen molar ratio versus the temperature in centigrade.

Another potential drawback may be the volatility of tin halogenides, see Table 1.

TABLE 1

Melting and boiling points of some tin compounds

Compound	Density	Melting Point °C	Boiling Point °C
Sn	7.3	231.9	2623
SnO	6.45	1050	1425
SnO$_2$	6.95	2000	2500
SnS	5.08	881	1210
SnSe	6.18	880	
SnTl	6.45	806	
SnF$_2$	4.57	213	853
SnF$_4$	4.78	442 ?	708
SnCl$_2$	3.90	245	650
SnCl$_4$	2.22	-34	115
SnBr$_2$	4.92	230	640
SnBr$_4$	3.35	30	208
SnI$_2$	5.29	320	717
SnI$_4$	4.47	144	364

The only halogen present in substantial amounts is chlorine. A typical amount in a coal gas would be 300 ppm. Gaseous tin chlorides may be formed by:

$$SnO_2 + 2HCl + H_2 \rightarrow SnCl_2(g) + 2H_2O \tag{7}$$

$$SnO_2 + 4HCl \rightarrow SnCl_4 + 2H_2O \tag{8}$$

In other words, water suppresses the formation of tin chlorides. Potentially, their formation is more likely in a dry than in a wet gasifier.

For typical coal gas from a dry gasifier (25 vol% H$_2$, 3 vol% H$_2$O and 300 ppm HCl) there could be up to 15 ppm SnCl$_2$ in the gas at 400°C and 25 bar.

In the worst possible cases, the HCl concentration is about 2000 ppm and then several hundred ppm tin chloride may be present in the "cleaned" gas which shows that there are a number of good reasons for having a halogen removal upstream the sulphur removal. Based on the above, it may be concluded that a modest leakage of hydrogen chloride, i.e. ~ 10 ppm, will leave the sulphur absorption mass untouched.

The trace components in the coal gas, such as Sb, Cd, and Hg, should not react with SnO_2 at 400°C due to their extremely low concentrations.

Experimental

All experiments were carried out in a stainless steel (Sanicro 28) reactor. The gas could flow either down or up through the reactor. A down-flow was used in the absorption, but in most regenerations a reversed flow was used.

Absorption was carried out at pressures between 2-3 MPa and at temperatures between 350-500°C using gases derived from catalytic decomposition of methanol. A typical dry gas composition was: H_2 ~ 70 vol%, CO ~ 10-20 vol%, CO_2 ~ 10-20 vol% and H_2S between 500 and 7000 ppm.

Regeneration was carried out using steam at pressures between 1-3 MPa, temperatures between 350-550°C. After the reactor, a known quantity of nitrogen was added and the steam condensed out. The gases were analysed by means of gas chromatography and in the first experiments, as regards the sulphur, by means of Dräger tubes.

The absorption masses tested were either prepared as pellets or as extrudates.

Calculation

An existing mathematical model developed by Topsøe was used to calculate the H_2S absorption break-through curves. This model is based on the following assumptions:

1. The absorption H_2S is described by a core-shell model restricted by gas diffusion through the pores and gas film transport (Yagi & Kunii, 1955). Reaction rate and solid state diffusion are not taken into account.
2. The structure of the particles does not change during absorption.

3. The absorption is isothermal and the linear velocity in the catalyst bed is constant.
4. Axial back-mixing is negligible.

Mass balances of the gas and the phases give two coupled partial differential equations. They are solved by a trapezoidal method, where the absorption time and bed length are divided by the number of integration points, thereby forming a rectangular grid. Integration is then preformed through the grid for a given time. This reduces the differential equations to two nonlinear equations, which are then solved by the Newton-Raphson method.

The effective diffusion coefficient of H_2S and the gas film transport coefficient are calculated by the methods described by J. Kjær (Kjær, 1972).

RESULTS

We found that neither the water gas shift reaction

$$CO + H_2O = CO_2 + H_2 \tag{9}$$

nor

$$SnO_2 + 2\ CO = Sn + 2\ CO_2 \tag{10}$$

took place on the SnO_2 absorption mass within the investigated temperature range (350-500°C). However, a COS conversion reaction takes place, thus removing the COS along with H_2S, most likely through the hydrolysing reaction:

$$COS + H_2O = CO_2 + H_2S \tag{11}$$

and then the H_2S is captured as SnS, according to reaction (1).

The sulphur capacity as well as the sulphur leakage were determined for a number of absorption masses. Up to 30 sulphidation/regeneration cycles were made with a single absorption mass. Samples were taken both as sulphide and oxide. By examining the crushing strength, no loss of mechanical strength was seen, see Table 2.

TABLE 2

Properties of H₂S absorption mass.

Density:	1.4-2.5 kg/m³
Bulk Density:	1.0-2.0 kg/m³
Crushing Strength **(Pellets)**	
Before Use:	170 kg/cm²
As Oxide (after 6 cycles):	200 kg/cm²

The calculated H₂S breakthrough curves are in good agreement with the curves measured above 400°C (see figure 4).

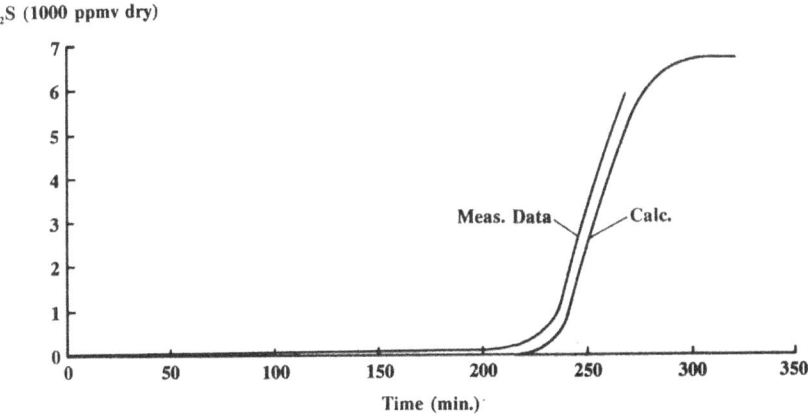

Figure 4. Measured and calculated absorption at 405°, 25 bar.

In the calculation, only the gas diffusion in the pores and the gas film transport resistance are taken into account. Figure 5 shows how other resistances, such as the chemical reaction rate, become rate controlling for adsorption below 400°C. The model can, therefore, only be used to predict the H_2S profile above 400°C.

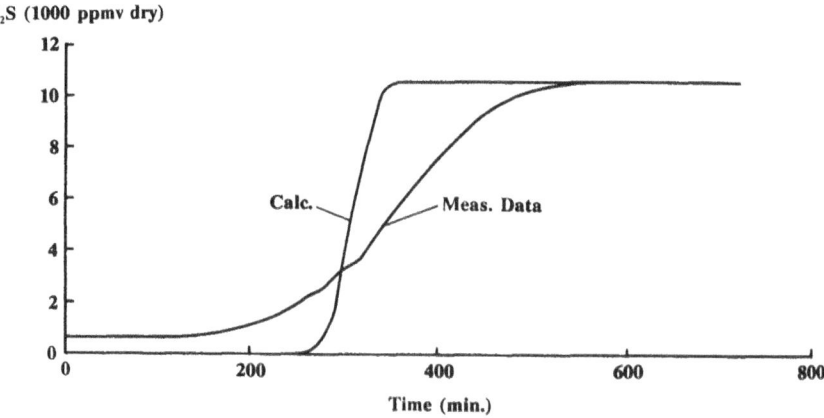

Figure 5. Measured and calculated absorption at 364°C, 24 bar.

The absorption mass retains its absorption capacity, as can be seen in Figure 6 where the H_2S breakthrough starts after approximately 190 minutes on stream in both the 2nd and the 7th absorption.

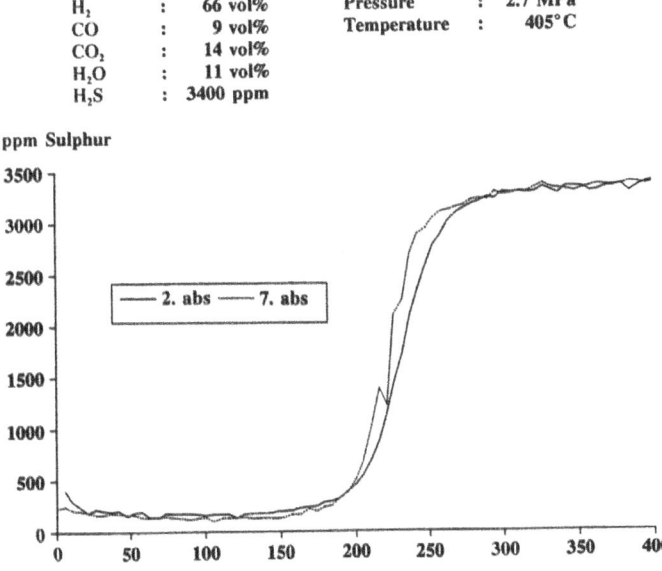

H$_2$	:	66 vol%	Pressure	:	2.7 MPa
CO	:	9 vol%	Temperature	:	405°C
CO$_2$	:	14 vol%			
H$_2$O	:	11 vol%			
H$_2$S	:	3400 ppm			

Figure 6. H$_2$S absorption

As regards absorptions below 400°C, the slow chemical reaction rate makes it difficult to reach equilibrium, unless the space velocity is low. However, during the regeneration the equilibrium was approached at temperatures around 450°C. A typical regeneration is shown in Figure 7.

465

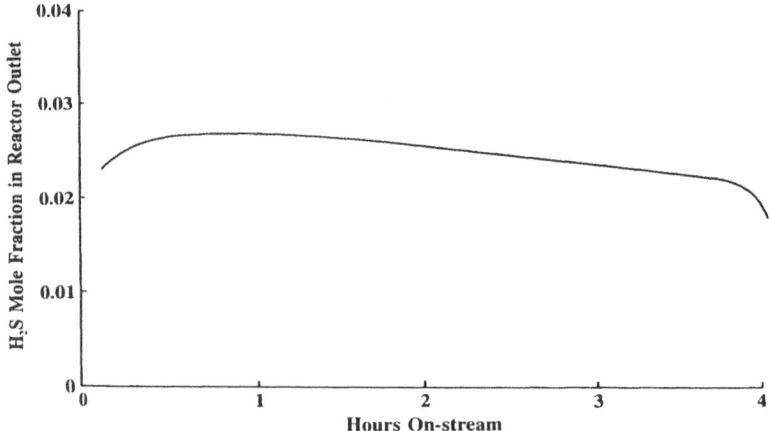

Figure 7. Regeneration of a sulphided sorbent in 100% steam.

Regeneration of a sulphided sorbent with steam has a number of advantages compared to e.g. regeneration with air/oxygen.

First of all, it is almost thermo-neutral compared to the strongly exothermic regeneration with air/oxygen. Secondly, the sulphur is recovered as pure H_2S, see below, instead of a dilute SO_2. Besides, steam and coal gas do not form explosive mixtures whereas air/oxygen certainly does.

Below we briefly outline how we expect to use such a system.

DISCUSSION

We have demonstrated in the above and in previous publications (Højlund Nielsen *et al.*, 1991 and 1993) that the interaction between solid tin oxide and tin sulphide and gases like steam, hydrogen, and hydrogen sulphide can approach thermodynamic equilibrium at temperatures above 400°C and at space velocities of industrial relevance. Despite repeated cycling between the oxide and sulphide system, the tin sorbent maintained its mechanical strength. Thus a cleaning mass based upon tin oxide can form the basis for a hot gas cleaning train like the one shown in Figure 8.

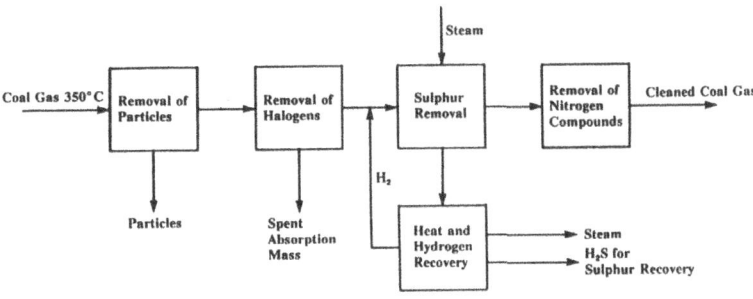

Figure 8. The Topsoe Hot Gas Cleaning Process.

Besides sulphur removal, the following unit operations are used:

Particle removal;	here ceramic candles are already in application today.
Halogen removal;	commercial guards are available

| Optional Shift converter; | commercial catalysts are available. |
| Ammonia removal; | no definite solutions, but a number of possibilities exist. |

Catalytic ammonia decomposition may be carried out as a part of the combustion in the gas turbine or by using a modified combustion system (Staged Combustion).

Figure 9 shows, in more detail, how a hot, dry sulphur recovery system may work. By carrying out the regeneration using steam at a slightly higher pressure than the coal gas, one may return the hydrogen formed in the regeneration process to the coal gas. The product hydrogen sulphide gas may be processed to elemental sulphur in a Claus unit.

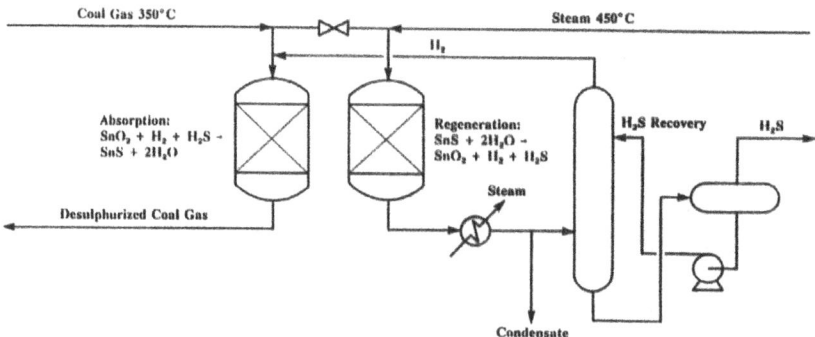

Figure 9. Sulphur recovery system.

A more advanced regeneration system has been described recently by Højlund Nielsen and Rudbeck (Højlund Nielsen & Rudbeck 1993). In that system the operation of the Claus plant is integrated with the regeneration of the sulphur absorber.

Sulphur removal carried out as shown in Figure 9 would yield a gas containing from 10 to 50 ppm of sulphur. In general, this would represent more than 99% sulphur removal and as such it will in most cases be sufficient. As regards special systems, e.g. fuel cell based plants, a fine sulphur removal system will be needed.

CONCLUSIONS

Tin oxide reacts very easily with hydrogen sulphide and we have shown that tin sulphide may be easily regenerated with steam at temperatures above 450°C. This knowledge can be used in a hot gas cleaning process in which the sulphur is recovered as elemental sulphur.

REFERENCES

Højlund Nielsen, P.E., Rudbeck, P. & Christiansen, H. (1991). Steam Regenerable Sulfur Absorption Masses and Their Application in IGCC Plants. Paper presented at The Tenth EPRI Conference of Coal Gasification Power Plants, San Francisco, California, 16-18th October.

Højlund Nielsen, P.E. and Rudbeck,P. Hot Gas Cleaning in IGCC Power Plants, in Power Generation Technology 1993. Sterling Publication London 1992.

European Patent Publication No. 426833.

Kjær, J. (1972). Computer Methods in Gas Phase Thermodynamics. Haldor Topsøe, Denmark.

Oates, W.A. & Todd, D.D. (1962). J. Australian Inst. Metals. 7. pp. 109.

South African Patent No. 9013943. US Patent 5.169.612.

Westmoreland, P.R. & Harrison, D.P. (1976). Environ.Sci.Technol. 10. pp. 659.

Yagi, S. & Kunii, D. (1955). 5th Symposium (International) on Combustion. Reinhold, New York, pp. 231.

DEPENDENCE OF SULPHUR CAPTURE PERFORMANCE ON AIR STAGING IN A 12 MW CIRCULATING FLUIDISED BED BOILER

ANDERS LYNGFELT, KLAS BERGQVIST, FILIP JOHNSSON,
LARS-ERIK ÅMAND AND BO LECKNER
Dept. of Energy Conversion
Chalmers University of Technology
Göteborg, 412 96 SWEDEN

ABSTRACT

Three cases of air staging were examined in a 12 MW circulating fluidised bed boiler: *i)* no staging, *ii)* normal staging and *iii)* intensified staging. The conditions inside the combustion chamber were investigated by zirconia cell measurements of the oxygen partial pressure, 0.35, 0.65 and 8 m above the bottom air distributor plate. A significant effect of the degree of staging was seen in the two lower locations: At 0.65 m height the fraction of time under substoichiometric conditions was low in the no-staging case (2—35%), at normal staging it was 70—90%, whereas at intensified staging it was 100%. At 0.35 m height, *i.e.* in the dense bed, a similar effect was seen, although the fraction of time under reducing conditions was lower. The fraction of time under reducing conditions was low in the top of the combustion chamber in all three cases.

The increase in the fraction of time under reducing conditions with a higher degree of staging is associated with a decrease in sulphur capture. It is assumed that a release of SO_2 from $CaSO_4$ takes place during the transitions between oxidising and reducing conditions. Thus, the rapid alternations between oxidising and reducing conditions, as seen with the zirconia cell, offer an explanation of the reductive decomposition and, accordingly, of the dependence of sulphur capture on temperature and on the extent of staging.

INTRODUCTION

Fluidised bed combustion provides a possibility of abatement of sulphur dioxide emissions from solid fuels. The temperature in the combustion chamber of a fluidised bed boiler (FBB), 800—900°C, is sufficient to allow for calcination of a sorbent, *e.g.* limestone, $CaCO_3$. Yet the temperature is low enough to prevent sintering of the resulting porous calcine, CaO, which

470

reacts with SO_2 in the presence of oxygen according to

$$CaO + SO_2 + \tfrac{1}{2}O_2 \rightarrow CaSO_4 \qquad\qquad (1)$$

The average sorbent residence time in a commercial FBB is in the order of 10 h (Lyngfelt & Leckner, 1992) and typical sorbent conversions, *i.e.* molar ratio of $CaSO_4$ to total Ca, are 40–50% (Mjörnell *et al.*, 1991).

Despite extensive laboratory research on limestone behaviour, as well as a large number of studies showing global measurements of sulphur retention in commercial scale FBBs, there is still no satisfactory understanding of the parameters which limit the sulphur capture performance in FBBs. The limiting factors for efficient sulphur capture include:

1) Sorbent residence time.

2) Sorbent reactivity.

3) Reducing conditions.

The present investigation is related to the last of these three items, *i.e.* the limitation of conversion owing to the reductive decomposition of the reaction product. The net sulphur retention is the result of competition between sulphur capture, reaction (1), and reductive decomposition, *i.e.* the release of sulphur previously retained by calcium, represented by the overall reaction (2).

$$CaSO_4 + CO \rightarrow CaO + CO_2 + SO_2 \qquad\qquad (2)$$

The adverse effect of reductive decomposition on sulphur capture has been studied in laboratory scale combustors (Khan & Gibbs, 1991), pilot scale combustors (Bramer *et al.*, 1988) and in a 16 MW stationary (bubbling) FBB (Lyngfelt & Leckner, 1989*a*). In the 16 MW FBB, conclusive evidence of reaction (2) was obtained by raising the bed temperature above 890°C, where reaction (2) was faster than reaction (1), resulting in a net release of sulphur from the partially sulphated sorbent accumulated in the boiler. At the highest temperature, 940°C, the emission of sulphur was 2.5 times greater than the amount of sulphur introduced into the boiler with the fuel. Zirconia cell measurements showed that reducing conditions prevailed in the bed of the boiler (Cooper & Ljungström, 1987). These measurements were made in 14 locations at approximately half bed height and showed highly reducing conditions, *i.e.* a partial pressure of O_2, $P_{O_2} < 10^{-10}$ bar, for 80–90% of the time, although no air staging was used and the excess air ratio was high, 1.4.

Under constant reducing conditions the fuel sulphur can be expected to react with CaO to CaS. Under conditions changing between oxidising and reducing, however, an intermediate region is passed, see Fig. 1, where neither CaS nor CaSO$_4$, but CaO is stable. Thus, the captured sulphur is released during the shifts between oxidising and reducing conditions. A comprehensive study of the release of sulphur under such transitions between oxidising and reducing conditions was made by Hansen (1991; Hansen *et al.*, 1993). Hansen's study was performed in a fixed-bed reactor containing the sorbent, through which a gas flow containing 1500 ppm SO$_2$ and 10% CO$_2$ was led. The gas flow contained alternatingly 4% CO or 4% O$_2$. A typical result for a rather low conversion (about 10%) is shown in Fig. 2. The peaks in SO$_2$ during the shifts from reducing to oxidising conditions, caused by the oxidation of CaS, exceed the inlet concentration of SO$_2$ thus resulting in a net release in

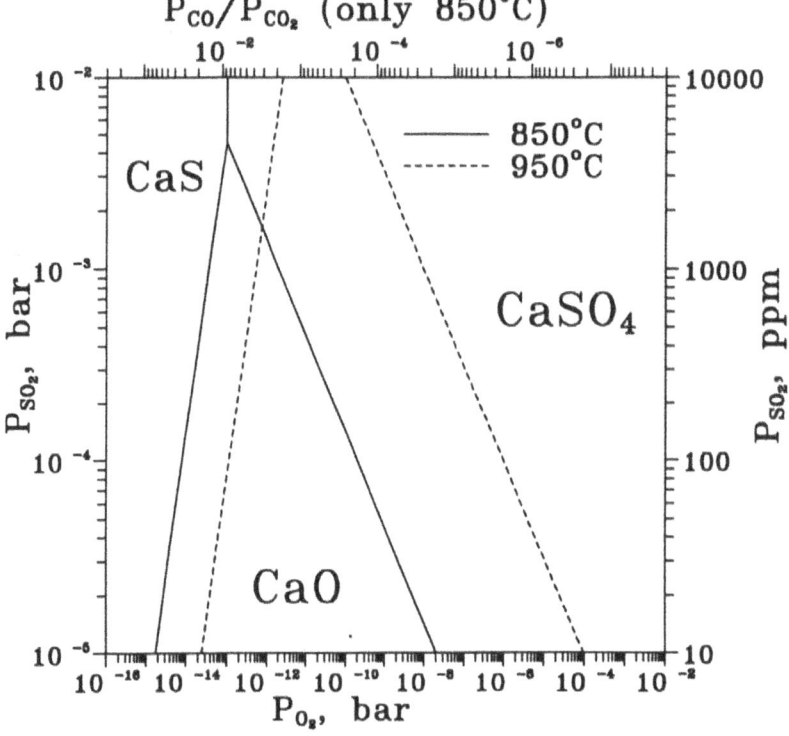

FIGURE 1. Phase equilibrium diagram for the system CaO, CaS, CaSO$_4$, SO$_2$ and O$_2$. Total pressure 1 bar.

SO$_2$. In the second shift from oxidising to reducing conditions, the SO$_2$ peak supersedes the inlet SO$_2$ concentration indicating a net release due to the reductive decomposition of CaSO$_4$. In general the SO$_2$ peaks caused by oxidation of CaS decrease with increased conversion, while the SO$_2$ peaks caused by reductive decomposition increase with increased conversion but shift to decrease at high conversions. There were, however, important differences between the various sorbents investigated and also depending on experimental conditions, for instance particle size.

The reducing conditions noted in the stationary FBB is an effect of the gas flow pattern in the bed and the consequence is a net sulphur release at increased temperature, but also a less efficient sulphur capture at normal operating conditions (Lyngfelt & Leckner, 1989b). A model of the release and capture of SO$_2$ suggests that the adverse effect of reducing conditions on sulphur capture is more or less present at all temperatures (Lyngfelt & Leckner, 1993b).

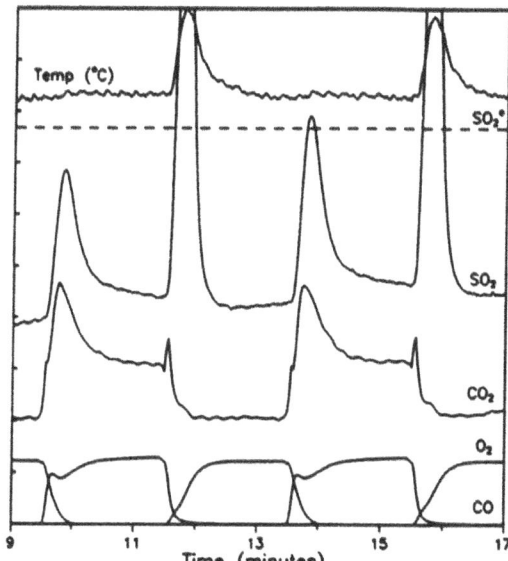

Fig. 2. Example showing temperature, SO$_2$, CO$_2$, O$_2$ and CO versus time under conditions alternating between oxidising and reducing (Hansen, 1991). SO$_2^*$ indicates inlet SO$_2$ concentration. Range of ordinate axis: temperature 780→880°C; SO$_2$ −150→2000 ppm; CO$_2$ 9→14%; O$_2$ 0→30%; CO 0→35%.

The information regarding the distribution and effect of reducing conditions in *circulating* FBB's is, however, incomplete and therefore tests were performed in the 12 MW circulating FBB at Chalmers University of Technology. The purpose was to examine the presence of reducing conditions in the combustion chamber for three cases of air staging, and to relate these to the sulphur capture efficiency. The object is to provide information about the conditions to which the sorbent is exposed, including where and how decomposition takes place. Increased understanding of how operating parameters, such as air staging, affect these conditions is valuable for minimising sorbent costs and SO_2 emissions. This is a' the more important in view of the desire to reduce NO and N_2O emissions, since a measure implemented to reduce one emission often increases others.

Additional results from these tests, regarding the effect of air staging and temperature on SO_2 as well as N_2O and NO emissions, are previously published (Lyngfelt & Leckner, 1993a).

EXPERIMENTAL CONDITIONS

The tests involved variation of the degrees of air staging. Air staging is defined as the process by which a part of the combustion air, called secondary air, is added at a later stage in the combustion chamber. Thus, a primary combustion zone with a reduced air supply is formed. The extent of staging can be varied either by altering the ratio of secondary air to total air or by altering the position where secondary air is introduced.

Three cases of air staging were studied with limestone addition at a bed temperature of 850°C, see Table 1. Test conditions are shown in Table 2. The overall excess air ratio was held constant at about 1.2 in all three cases.

In order to obtain supplementary data from the lower part of the combustion chamber, a second test series was made with the same coal and under otherwise similar conditions. Operational parameters such as load, excess air, primary/secondary air ratio, and bed heights were close to identical, see Table 2. The only difference was that a dolomite was used instead of a limestone. The same Ca/S ratio, 3, was used which means a higher mass flow of dolomite.

474

TABLE 1.
The three cases of staging.

Normal staging	Reference case with approximately 55% primary air and 35% secondary air, which was introduced at level 1/2 about 2 m above the air-distributor plate. (The remaining 10% air was introduced elsewhere, *e.g.* in the particle cooler.)
No staging	No secondary air.
Intensified staging	Approximately 40% primary air and 50% secondary air which was introduced at a high level (level 4, at 5.5 m height).

TABLE 2.
Average values of operation parameters and gas measurements.

Degree of staging	No	Normal	Intensified
Bed temperature, °C	851	850	850
Top temperature, °C	858	857	867
Pressure drop, bed, kPa	3.7	4.9	5.8
Pressure drop, total, kPa	5.8	6.2	6.9
Total air flow, kg/s	3.5	3.4	3.5
Primary air flow, kg/s	3.1	1.9	1.3
Secondary air flow, kg/s	0	1.3	1.8
O_2, %, flue gas	3.7	3.8	3.9
CO (6% O_2), ppm, flue gas	114	117	151
SO_2 (6% O_2), ppm, flue gas	35	45	231
O_2, %, top of furnace, H11	3.3	3.0	7.4
CO, ppm, top of furnace, H11	660	866	1038
Second test series:			
Bed temperature, °C	848	853	851
Top temperature, °C	847	851	859
Pressure drop, bed, kPa	3.0	4.8	5.8
Pressure drop, total, kPa	5.4	6.3	6.7
Total air flow, kg/s	3.6	3.5	3.5
Primary air flow, kg/s	3.0	1.8	1.3
Secondary air flow, kg/s	0	1.3	1.7
O_2, %, flue gas	3.3	3.5	3.9
CO (6% O_2), ppm, flue gas	196	127	158
SO_2 (6% O_2), ppm, flue gas	28	40	184
O_2, %, below sec. air, H4	10.7	2.1	1.1
CO, ppm, below sec. air, H4	5400	5800	>10,000

<u>The boiler</u>

The 12 MW circulating FBB used for the experiments has the features of a
commercial boiler, but was built for the purpose of research. The boiler is
equipped for special measurements and has facilities that make it possible
to vary parameters independently and in a wider range than is possible in a
commercial boiler. It is also possible to run the boiler under extreme
conditions inappropriate to commercial boilers.

The boiler is shown in Fig. 3. The height of the combustion chamber is
13.5 m and the square cross-section is about 2.5 m². Fuel is fed to the bottom
of the combustion chamber through a fuel chute (3). Primary air is
introduced through nozzles in the bottom plate (2) and secondary air can be
injected through several nozzle registers located horizontally at both sides
of the combustion chamber, as indicated by the arrows in Fig. 3. Entrained

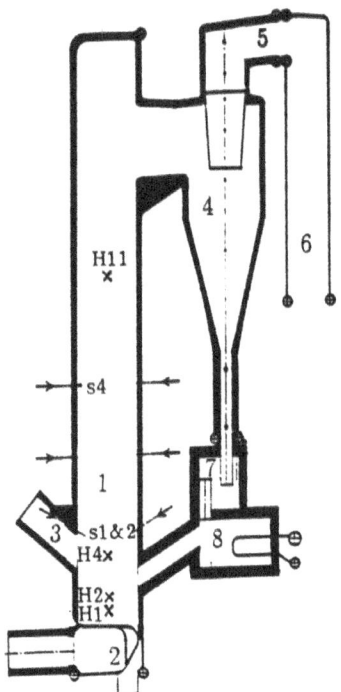

FIGURE 3. The Chalmers boiler. 1 combustion chamber, 2 air plenum and
start-up combustion chamber, 3 fuel feed chute, 4 cyclone, 5 exit duct, 6
convection cooling section, 7 particle seal, 8 particle cooler; → secondary
air nozzle registers (s1&2 and s4); x holes (H1, H2, H4 and H11).

bed material is captured in the hot, refractory-lined cyclone (4) and returned to the combustion chamber through the return leg and particle seal (7).

Figure 3 does not show the flue-gas recycling system, which supplies flue gas to the combustion chamber. This system is used for fine tuning of the bed temperature. Large, intentional changes in bed temperature can be made using the external, adjustable particle cooler (8).

The fuel and the sorbent

The fuel was a bituminous coal with low sulphur content and the sorbents were Ignaberga limestone (main test series) and a dolomite called Myanit (second test series). Sorbent and fuel data are shown in Table 3.

TABLE 3.
Fuel and limestone (daf=dry, ash—free).

	Mass fraction	Particle size
Fuel, bitum. coal		<16 mm, 50% < 3 mm
moisture	0.15	
ash	0.05	
volatiles, daf	0.39	
carbon, daf	0.88	
hydrogen, daf	0.06	
nitrogen, daf	0.01	
oxygen, daf	0.04, calculated	
sulphur, daf	0.008	
Sorbent, Ignaberga		0.2–2 mm, 50% < 0.6 mm
$CaCO_3$	0.90	
Sorbent, Myanit		0.1–1 mm, 50% < 0.4 mm
$CaCO_3$	0.525	
$MgCO_3$	0.439	

Measurements

Regularly calibrated on-line gas analysers were used for continuous monitoring of O_2, SO_2 and CO after the bag-house filter. In addition, gases were sampled from the centre of the combustion chamber with a suction probe in hole H11, at 8 m height. In the second test series the suction probe instead sampled in hole H4, at 1.6 m height, also in the centre.

Two zirconia cell oxygen probes were inserted into the combustion chamber through the side walls in hole H2, 0.65 m above the bottom, and hole H11, 8 m above the bottom, see Fig. 3. Hole H2 was in the splash zone of the bed and well below the secondary air inlet. On each level three measurements were made: one in the centre, *i.e.* halfway between the side walls, one halfway between the centre position and the side wall and finally the centre position was repeated. In the second test series the two probes were inserted at two levels: at 0.35 m height, H1, and at 0.65 m height, H2, and two measurements were made on each level: one in the centre position and one halfway between the centre and the side wall. Hole H1 is in the dense bottom bed.

In each measurement data from the oxygen probes were sampled every 0.05 s (20 Hz) during 30 minutes. The choice of sampling frequency was based on previous experimental experience, which shows that higher frequencies do not significantly improve the resolution. This is an effect of the response time of the probes.

One of the probes was water-cooled and the other was air-cooled. The water-cooled probe was inserted in hole H2 in the first test series and in hole H1 in the second. Thus, measurement data from hole H2 were obtained from both probes.

Both probes have Bosch zirconia cells, protected from larger particles by a shield with slits. The water-cooled probe is described by Ljungström (1985). The air-cooled probe cell is mounted on a lance, which is water—cooled except for the 10 cm nearest to the cell. Inside the lance, which has a rectangular cross section, and fitted to two sides is a cylindrical tube. Cooling water is introduced into one of the two channels surrounding this tube and returned, via a tube deformation, in the other. The cylindrical tube contains the electric circuits and two tubes through which cooling air is introduced.

The voltage signal from the zirconia cell measurements was somewhat off-set. The off-set was small for the air-cooled probe and larger for the water-cooled probe. The values have been corrected for this effect. The reason for the off-set seems to be related to the cooling flow, *i.e.* to the temperature gradients over the cell. Except for this off-set, the probes are reliable and their function has been checked in a number of ways, including comparitive tests which show that the probes react likewise to similar conditions. The amplitude and the character of the signal affirm that the probes worked properly.

The use of zirconia cell oxygen probes for measurements inside fluidised bed combustors is established (*e.g.* Cooke et al., 1972; Saari & Davini, 1982; Rocazella 1983; Stubington & Chan 1990).

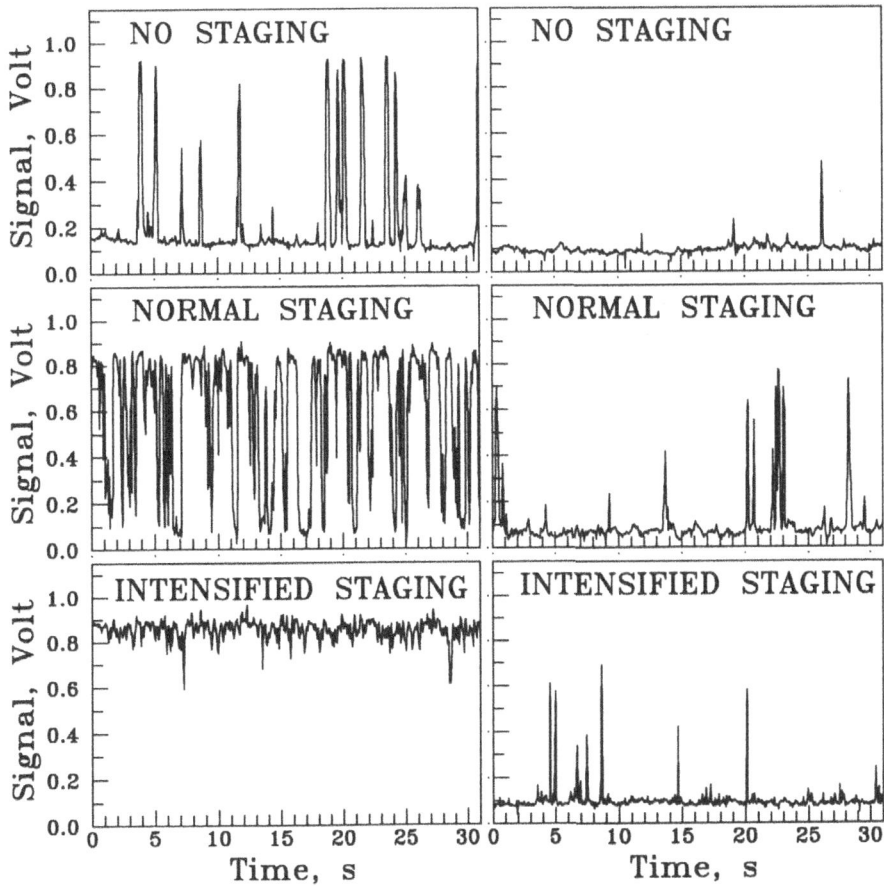

FIGURE 4. Examples of oxygen probe measurements.
To the left, at 0.65 m height (H2); to the right at 8 m height (H11).
From top to bottom: no staging; normal staging; intensified staging.

RESULTS

Zirconia cell measurements

The three diagrams to the left in Fig. 4 show typical examples of the signal
measured with the zirconia cell in hole H2, in the splash zone 0.65 m above
the distributor plate, for the three cases of staging. The low voltage level

corresponds to oxidising conditions and the higher level to reducing (substoichiometric) conditions. In the no-staging case, oxidising conditions dominate with detours to reducing conditions. Under normal staging the pattern is reversed with reducing conditions dominating, but with frequent detours to oxidising conditions. In the case of intensified staging, reducing conditions predominate and occasional detours towards oxidising conditions are not successful.

The corresponding diagrams to the right in Fig. 4 show the zirconia cell signal measured in hole H11, 8 m above the distributor plate. In all three cases of staging oxidising conditions prevail with occasional detours towards reducing conditions.

In the following, a level of 0.3 V is chosen to differentiate between oxidising and reducing conditions. This is the point of stoichiometry, *cf.* Appendix A, which corresponds to an equilibrium oxygen partial pressure of $3 \cdot 10^{-7}$ bar and a P_{CO}/P_{CO_2} ratio of $4 \cdot 10^{-6}$. With this choice of voltage level the fraction of time under reducing conditions, f_{red}, can be calculated and the results are shown in Fig. 5.

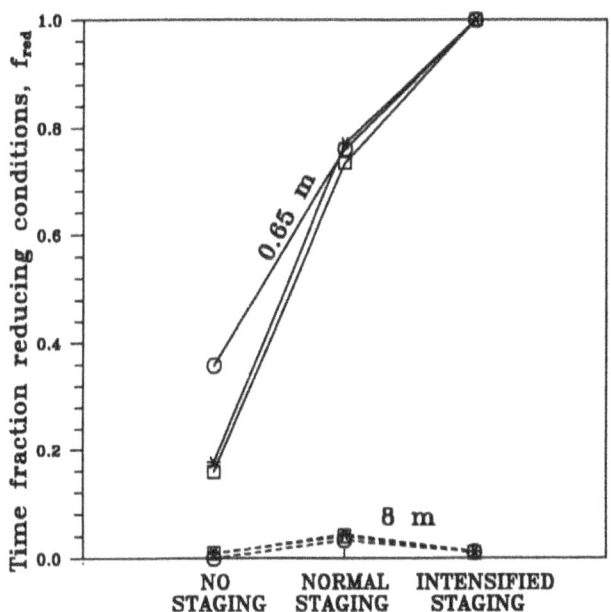

FIGURE 5. Fraction of time under reducing conditions at 0.65 m height (solid lines) and 8 m height (dashed lines): □, × centre; ○ halfway to wall

The signal character from the second test series was similar to that of the first test series. The time fraction under reducing conditions, f_{red}, is shown in Fig. 6. The data obtained in hole H2, at 0.65 m, are similar to the first test series, with the exception that the fraction of time under reducing conditions is lower in the no-staging case. The reason for this is not known. The fraction of time under reducing conditions measured in hole H1, at 0.35 m and in the dense bed, shows the same dependence on air staging, but the values of f_{red} are lower than in the measurements in H2 in the splash zone. The rather low fraction of time under reducing conditions in the dense bed for normal staging, about 30%, indicates that the dense particle phase is not predominantly reducing, which suggests a good gas-solids mixing.

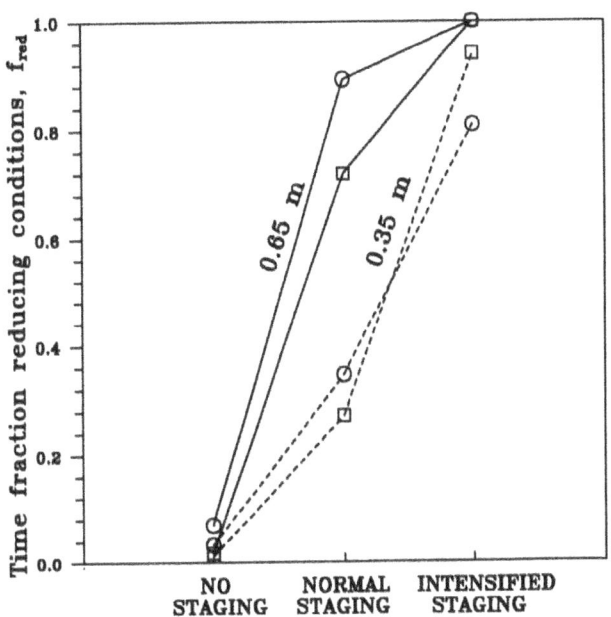

FIGURE 6. Fraction of time under reducing conditions in second test series
at 0.65 m, H2, (full lines) and 0.35 m, H1, (dashed lines):
□ centre; o halfway to wall.

Bed ash analysis

The results of the oxygen probe measurements can be compared with the amount of fuel in the bottom part of the furnace. This is done in Table 4 as follows: the amount of combustible solids is estimated from analysis of bed material samples, and the bed mass is determined from the pressure drop in the bottom part of the furnace. The fraction of reducing time is assumed to correspond to the probability of the probe of being in the fuel-rich plumes from burning or devolatilising particles in accordance with the simplified model derived in Appendix B:

$$f_{red} = 1-\exp(-c_1 m_{comb}/\dot{m}_{prim.\,air}) \qquad (3)$$

With the values for the normal staging case, $f_{red} = 0.75$ and $m_{comb}/\dot{m}_{prim \cdot air} = 32.6$, c_1 can be computed and with this value of c_1, f_{red} for the two other cases of staging can be estimated, see Table 4. The good agreement with the measured values in Fig. 5, is probably partly co-incidental: the mass fractions of combustibles in the bed samples are somewhat uncertain, nor does the simple model account for differences in fluidising conditions caused by different bed heights and fluidising velocities. Also, the relationship between the amount of combustibles and volatiles release, char combustion and gasification, is not as simple as assumed in eqn. (3). Still, it appears that eqn. (3) can be used to relate f_{red} to changes in the $m_{comb}/\dot{m}_{prim.\,air}$ ratio caused by variation in the degree of air staging.

TABLE 4.
Calculation of the ratio of mass of combustibles to mass flow of air in the bottom part of the furnace and f_{red} according to eqn. (3).

Degree of staging	No	Normal	Intensified
Fraction combustibles, (-)	0.022	0.045	0.099
Pressure drop, kPa	3.80	4.50	5.85
Bed mass, kg	1120	1330	1730
Mass combustibles, m_{comb}, kg	24.8	60.4	170
Primary air flow, $\dot{m}_{prim.air}$, kg/s	3.10	1.85	1.30
$m_{comb}/\dot{m}_{prim.\,air}$	8.01	32.6	131
f_{red} eqn(3) with $c_1=0.0425$	0.29	0.75	0.996

Combustion chamber suction probe measurements

Average values from suction probe gas measurements made at 8 m height in the centre of the combustion chamber are shown in Table 2 for the three cases of staging. The O_2 concentration varied with an amplitude of approximately 1%, and the measured CO peaks did not exceed 4000 ppm. The reducing periods were not detected in these measurements because of the longer response time.

The most important observation in relation to the oxygen probe measurements is the much higher oxygen level in the case of intensified staging. This may explain the lower fraction of time under reducing conditions, f_{red}, in the top, see Fig. 5, as compared with normal staging. Since the load and the excess air ratio were constant, the difference between the three cases of staging should, in principle, only relate to the lower half of the combustion chamber. In reality, the conditions in the top were affected in several ways. The extent of secondary air penetration affects the O_2 concentration profile over the cross-section. In addition, an increased degree of air staging results in a higher concentration of solid combustibles which may involve an increased oxygen consumption in the cyclone, *i.e.* a higher O_2 concentration in the upper part of the furnace since the exit O_2 is held constant. A possible explanation for the measured oxygen levels is:

- Shift from no staging to normal staging: The secondary air does not penetrate in full and results in lower O_2 concentration in the centre.

- Shift from normal to intensified staging: The high degree of staging increases the concentration of combustible solids and thus the cyclone oxygen consumption. (The increased combustion in the cyclone can be seen from the smaller temperature fall over the cyclone for the case of intensified staging.)

In the second test series, the suction probe samples gas from hole H4, at 1.6 m height, well below the secondary air inlets. As expected, these measurements show an increase in CO and a significant decrease in O_2 with higher degree of staging, *cf.* Table 2.

The effect of staging on sulphur capture

The effect of air staging on sulphur capture performance is shown in Fig. 7. The effect of staging in a high temperature case, 930°C, is also included in the figure. The horizontal dashed line shows the "zero level", 414 ppm, *i.e.* the sulphur emission with a sand bed prior to the start of limestone addition. In the case of no staging, there was a decrease in sulphur capture efficiency from 90 to 60% when the temperature was increased to 930°C. This is attributed to the fact that reducing conditions are present, even in the case of no staging. For normal staging the sulphur capture efficiency was somewhat lower than in the case of no staging at 850°C (Lyngfelt & Leckner, 1993a). A large increase in SO_2 emission is seen at the high temperature where the sulphur capture has decreased from 90 to −20%. The latter, negative value indicates a net sulphur release from sulphur previously captured in the form of $CaSO_4$. (Of course, negative sulphur capture is not possible at steady state: the negative values are averages for time periods of 2–3 hours following a temperature increase.) A very pronounced effect on sulphur capture is seen in the case of intensified staging: here the sulphur capture efficiency already is reduced to 40% at 850°C and at the high temperature the net release of sulphur is large, with a sulphur capture efficiency of less than −100%.

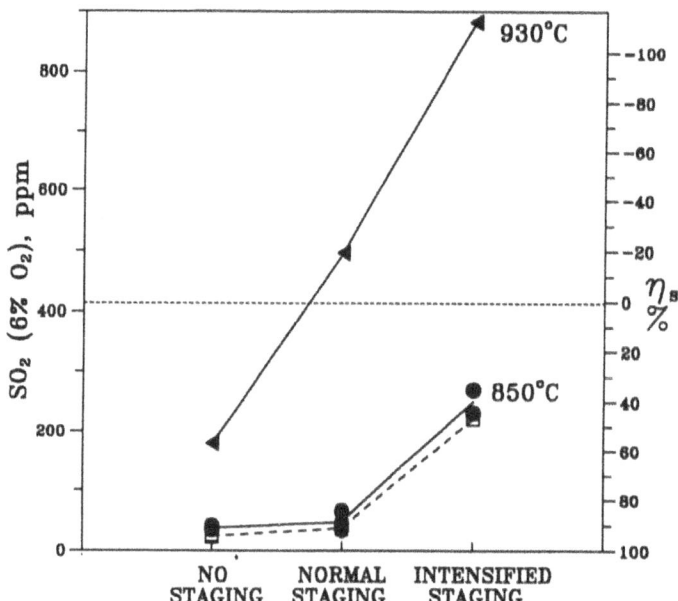

FIGURE 7. SO_2 (6% O_2) and sulphur capture efficiency, η_s, for three cases of staging: • 850°C; ▲ 930°C. Ca/S = 3. Second test series: □, dashed line.

DISCUSSION

The oxygen probe measurements verify that the effect of a higher degree of staging is to increase the extent of substoichiometric conditions in the bottom part of the boiler. The fraction of time under reducing conditions in the top of the boiler was only a few per cent.

The bed height varied in the different cases of staging. For the measurements in H2, at 0.65 m height, the probe was in the splash zone, but the distance from the probe to the dense bottom bed decreased with increasing staging. The measurements in H1, at 0.35 m were, at least for the normal and intensified staging cases, made in the dense bottom bed. The strong effect of staging on the fraction of reducing conditions was seen at both levels, *i.e.* both in and above the dense bottom bed.

The combustion rate (mass of combustibles burnt per time unit divided by total mass of combustibles present) is inversely proportional to the mass of combustibles since the load, *i.e.* the mass of combustibles burnt per time unit, is constant. Increased staging results in a decreased combustion rate which is evident from the increase in the mass of combustibles, *cf.* Table 4. This decrease in combustion rate with increased staging is, of course, explained by the oxygen deficiency and is thus consistent with both the effects on sulphur capture and the oxygen probe data.

The average frequency of the shifts between oxidising and reducing conditions was in the order of 1 Hz. An important fraction of the number of shifts had a period length not much longer than the response time of the probe. It is not improbable that the resolution prevents detection of shifts with shorter duration.

It should be pointed out that the conditions experienced by small freely-moving sorbent particles may differ significantly from measurements with a 20 mm probe made in a few fixed positions.

Comparison of laboratory and boiler data

Although Hansen (1991; Hansen *et al.*, 1993) noted a net release of SO_2 during the transitions between oxidising and reducing conditions, he found no significant effect of the shifts on the degree of sulphation, except at high temperatures, 950°C. In the present measurements a significant effect of staging can already be seen at 850°C. This is probably explained by some important differences between the laboratory tests and the boiler measurements:

485

i) Time period length. The periods of sorbent exposure to oxidising or reducing conditions were 30 s each in Hansen's study, whereas they may be in the order of 1 s in the fluidised bed. Hansen showed that SO_2 is released during the transitions, and the expected effect of the much higher frequency of the shifts in the boiler is a higher rate of reductive decomposition.

ii) Fraction of time under reducing conditions. The sulphur capture is good at entirely reducing or oxidising conditions. In between these extremes, the sulphur capture should be at a minimum for a certain value of f_{red}. This "worst case" value of f_{red} may be significantly higher than that used in Hansen's study, 0.5. Experiments by Hansen (1991) indicate that the conversion of the sorbent is decreased when f_{red} is increased. No experimental data to determine a "worst case" value of f_{red} are available, however.

iii) Reducing agent. Hansen also studied the influence of H_2 and found a more pronounced effect than with CO. H_2 is found in volatile plumes in the combustion chamber.

Thus, the strong effect of reducing conditions in the case of intensified staging at 850°C as opposed to the small overall effect seen in laboratory tests may be explained by 1) higher frequency of shifts, 2) higher fraction of reducing conditions and, 3) presence of a more reactive reducing agent, H_2. At present it is not possible to safely conclude which of these is most important.

The effect of alternations on SO_2 release

It is not evident where and how SO_2 is released from the sorbent. The following discussion will point to some possibilities, based on the intensified staging case. In this case f_{red} was 1.0 in the splash zone at 0.65 m (below called reducing bottom zone), 0.8—0.9 at lower levels in the bed, at 0.35 m (lowest bottom zone), and about 0.01 at 8 m height. This indicates that sorbent particles in the top of the boiler as well as in the cyclone will experience predominantly oxidising conditions, while particles in the lower part of the combustion chamber will experience either conditions changing between reducing and oxidising (lowest bottom zone) or constant reducing conditions (reducing bottom zone). Thus, some general mechanisms for SO_2 release can be outlined: movement of sorbent particles to a reducing zone involving SO_2 release from $CaSO_4$ (items 1—3 and 8, below), movement to an oxidising zone involving SO_2 release from CaS (items 4—6), and

486

local shifts (items 7 and 9):

1) Particles move into the reducing bottom zone after falling down from the oxidising zone higher up in the combustion chamber.

2) Particles enter the reducing bottom zone after being recycled via the cyclone.

3) Particles enter the reducing bottom zone after having experienced some degree of oxidising conditions in the lowest bottom zone.

4) Particles are thrown up from the reducing bottom zone into the oxidising zone above the secondary air inlets.

5) Particles are entrained from the reducing bottom zone through the combustion chamber to the cyclone.

6) Particles move from the reducing bottom zone to the lowest bottom zone where there are partly oxidising conditions.

7) Particles in the lowest bottom zone are subject to shifts without moving. (In fact, this may also be the case in the reducing bottom zone if the measured value of $f_{red} = 1.0$ is partly an effect of response time and probe size. The size of the probe, 20 mm, prevents the detection of smaller oxidising zones and the probe response time prevents detection of oxidising periods shorter than appr. 0.05 s.)

8) In addition, it cannot be excluded that recycled material is subjected to reducing conditions in the return leg.

9) Sorbent particles may also be exposed to reducing conditions in the upper part of the combustion chamber. This is probably not an important mechanism. If Figs. 5 and 7 are compared it is seen that the poor sulphur capture in the case of intensified staging is not associated with a high f_{red} at 8 m, instead it is significantly lower than in the normal staging case.

Several mechanisms are available to explain the release of sulphur involving movement of sorbent particles in and out of reducing zones as well as shifting conditions in one specific location. It is not possible at present to determine to what degree the various mechanisms contribute to the loss in sulphur capture efficiency.

Undesired parameter variations

A circulating FBB is a complex system which is difficult to run under otherwise constant conditions when individual parameters are intentionally varied. Unintentional variations in parameters are inevitable, but most parameters in the present study have remained reasonably constant, and the quantitave effect of undesired parameter variations is judged to be minor, see Lyngfelt & Leckner (1993a).

CONCLUSIONS

The fraction of time under reducing conditions was measured in the dense bed, in the splash zone and in the top of the combustion chamber. Increasing the degree of air staging from no staging to intensified staging significantly increased the fraction of time under reducing conditions in the lower part of the combustion chamber. The effect in the top of the combustion chamber was minor, which is also expected since the variation in air staging mainly affected the lower part of the furnace, while load, overall excess—air ratio etc., were constant. The fraction of time under reducing conditions as measured by the oxygen probe in the lower part of the combustion chamber appears to be related to the ratio of mass of combustibles and primary air flow.

The more substoichiometric conditions in the bottom zone following increased staging are also related to a reduction in sulphur capture. This can be explained by the reductive decomposition of $CaSO_4$ during the frequent transitions between oxidising and reducing conditions.

ACKNOWLEDGEMENTS

This work has received financial support from the Swedish National Energy Administration, the Swedish Energy Development Corporation (SEU) and Kvaerner Generator. The tests were performed with the assistance of Abbas Zarrinpour, Hans Schmidt, Per Nilsson and the staff of the Chalmers boiler. Maria Karlsson, Henrik Brodén and Håkan Kassman were also helpful. Analyses of solids were performed by Kazimiera Puromäki and Tobias Mattisson.

REFERENCES

Bramer, E., de Jong, P. and Tossaint, H. (1988). Sulphur capture under reducing conditions at AFB combustion, *Proc. Inst. Energy 4th Int Fluidised Combustion Conference*, London, Vol I, pp. I/11/1—11.

Cooper, D., and Ljungström, E. (1987). The influence of bed temperature on the in—bed O_2 partial pressures in a 16 MW AFBC fired with petroleum coke. Internal Rep. OOK A87 002, Department of Inorganic Chemistry, Chalmers University of Technology, Göteborg.

Cooke, M.J., Cutler, A.J.B. and Raask, E. (1972). Oxygen measurements in flue gases with a solid electrolyte probe, *J. Inst. Fuel*, **45**, 153—156.

Hansen, P.F.B. (1991). Sulphur capture in fluidized bed coal combustors, Ph. D. Thesis, Dept. of Chemical Engingeering, Technical University of Denmark.

Hansen, P.F.B., Dam—Johansen, K., and Østergaard, K. (1993). High temperature reaction between sulphur dioxide and limestone. V. The effect of periodically changing oxidizing and reducing conditions, *Chem. Eng. Sci.*, **48**, 1325—1341.

Khan, W.U.Z., and Gibbs, B.M. (1991). The effect of staged combustion on in—situ desulphurization by limestone in a fluidised bed combustor, *Inst. Chem. Eng. Symp. Ser.*, No 123, 193—203.

Ljungström, E. (1985). In—bed oxygen measurements in a commercial size AFBC, *Proc. Int. Conf. Fluid. Bed Combustion*, **8**, 853—864.

Lyngfelt, A., and Leckner, B. (1989a). SO_2 capture in fluidised bed boilers: re—emission of SO_2 due to reduction of $CaSO_4$, *Chem. Eng. Sci.*, **44**, 207—213.

Lyngfelt, A., and Leckner, B. (1989b). Sulphur capture in fluidised bed combustors — temperature dependence and lime conversion, *J. Inst. Energy*, **62**, No. 450, 62—72.

Lyngfelt, A. and Leckner, B. (1992). Residence time distribution of sorbent particles in a circulating fluidised bed boiler, *Powder Technol.*, **70**, 285—292.

Lyngfelt, A. and Leckner, B. (1993a). SO_2 capture and N_2O reduction in a circulating fluidised bed boiler: influence of temperature and air staging. Accepted for publication in *Fuel*. (Expected in September 1993.)

Lyngfelt, A. and Leckner, B. (1993b). Model of sulphur capture in fluidised bed boilers under conditions changing between oxidising and reducing. *Chem. Engng Sci.*, **48**, 1131—1141.

Minchener, A.J. and Stringer (1984). The use of electrochemical probes for measuring oxygen partial pressures within a fluidised bed, *J. Inst. Energy*, **57**, 240—251.

Mjörnell, M, Leckner, B., Karlsson, M., and Lyngfelt, A. (1991). Emission Control With Additives in CFB Coal Combustion, *Proc. Int. Conf. Fluid. Bed Combustion*, **11**, 655—664.

Rocazella, M.A. (1983). The use and limitations of stabilized zirconia oxygen sensors in fluidized—bed coal combustors, *Electrochem. Soc. Proc.*, **83**(5), 85—100.

Saari, D.P. and Davini, R.J. (1982). Evaluation of instruments for in—bed oxygen measurements in a fluidized bed combustor, *Proc. Int. Conf. Fluid. Bed Combustion*, **7**, 995—1009.

Stubington, J., and Chan, S. (1990). The interpretation of oxygen—probe measurements in fluidised—bed combustors, *J. Inst. Energy*, **63**, 136—142.

APPENDIX A. POINT OF STOICHIMETRY.

The gas mixture is considered to be stoichiometric, *i.e.* excess oxygen = 0, when the partial pressure of O_2 is balanced by CO (for simplicity assumed to be the only oxygen consuming gas present), *i.e.*

$$P_{O_2} = \tfrac{1}{2}P_{CO} \tag{A1}$$

At equilibrium the concentrations of CO and O_2 are given by

$$K_{eq} = P_{CO_2}/P_{CO}P_{O_2}^{\frac{1}{2}} \tag{A2}$$

where K_{eq} is the equilibrium constant, $4 \cdot 10^8$ bar$^{-1/2}$. Assuming $P_{CO_2} = 0.15$ bar, and combining eqns A1 and A2 yields an equilibrium O_2 partial pressure of $3 \cdot 10^{-7}$ bar for a stoichiometric mixture at 850°C. This value may be inserted in the Nernst equation, which relates oxygen partial pressure to voltage signal, U, from a zirconia cell probe

$$U = \frac{RT}{4F}\ln(P_{O_2,\,ref}/P_{O_2}) \tag{A3}$$

where R is the universal gas constant, 8.31 J/mol,K, T is the absolute temperature, K, and F is the Faraday constant, 96487 As/mol. With a reference oxygen partial pressure, $P_{O_2,\,ref}$, of 0.21 this yields a voltage signal of 0.3 V for a stoichiometric mixture at 850°C.

APPENDIX B. CONNECTION BETWEEN f_{red} AND MASS OF COMBUSTIBLES

Assume that the gas has a uniform flow rate and that a burning/devolatilising particle causes an understoichiometric gas plume. The probability that an oxygen measuring probe of infinitesimal size located at a higher level will be in this plume is A_p/A_{cs}, where A_p and A_{cs} are the cross–section areas of the plume and of the combustor. Assuming that the bed contains n such particles with plumes of cross–section areas of A_p yields the following probability for not being in a plume:

$$P_{ox} = (1 - A_p/A_{cs})^n \qquad (B1)$$

and the probability for being in a plume

$$P_{red} = 1 - P_{ox} = 1 - (1 - A_p/A_{cs})^n = 1 - \exp(n \cdot \ln(1 - A_p/A_{cs})) \qquad (B2)$$

which, since $(1 - A_p/A_{cs}) \simeq 1$, can be reduced to

$$P_{red} = 1 - \exp(-n \cdot A_p/A_{cs}) \qquad (B3)$$

Here n is proportional to the mass of combustible particles, m_{comb}, and, assuming that the volume flow of reducing gas from a particle is independent of the gas flow rate, A_p is inversely proportional to the gas flow rate, $\dot{m}_{prim.air}$, which yields

$$P_{red} = 1 - \exp(-c_1 m_{comb}/\dot{m}_{prim.air}) \qquad (B4)$$

where c_1 is a constant.

It should be pointed out that the model is a simplification and does not take into account, for example, the complicated gas flow pattern in a fluidised bed.

HYDROGEN SULFIDE RETENTION ON LIMESTONE AT HIGH TEMPERATURE AND HIGH PRESSURE

J.B. Illerup, K. Dam-Johansen and J.E. Johnsson
Department of Chemical Engineering Technical University of Denmark
Building 229, DK-2800 Lyngby, Denmark

ABSTRACT

The suitability of limestone as a sorbent for hydrogen sulfide was tested at coal gasification conditions. Experiments were carried out in a fixed bed reactor at pressures and temperatures up to 10 bars and 1223 K respectively. The influence of pressure, temperature, particle size and gas composition was tested. Especially the partial pressure of CO_2 relative to the calcination pressure is important for the sulfur capacity of the limestone. This partial pressure is in practical systems determined by the gasification process employed. At low partial pressure of CO_2 the limestone calcines and $CaCO_3$ is converted into CaO. At these conditions an almost quantitative conversion of CaO to CaS was observed. The rate of reaction between CaO and H_2S increased with decreasing particle diameter. At high partial pressure of CO_2, $CaCO_3$ is stable and the limestone remains uncalcined. In this case the ultimate degree of conversion into CaS was below 0.25 even for very porous limestones. The conclusion is that for gasification processes with a partial pressure of CO_2 below the equilibrium pressure over $CaCO_3$, e.g. 0.2 bar at 1073 K and 2.0 bar at 1173 K, limestone is a suitable sorbent for hydrogen sulfide.

INTRODUCTION

Increasing requirements to lower the environmental impact from power plants has resulted in an intensive research in order to develop technologies with high thermal efficiency and with low emission of air polluting species. Advanced electricity generation systems in which coal gasification is coupled with gas turbines or fuel cells are of interest due to potentially high electrical efficiency and low environmental impact. Coal gasification is an established technology in the chemical industry but the application of coal gas directly for power generation is still unproven. In order to avoid irreversible damage of the hardware and environmental problems highly efficient sulfur removal from concentrations of several thousand ppm H_2S in the gas down to about 1 ppm for molten carbonate fuel cells and about 100 ppm for gas turbines is needed. Among the possible sorbents for H_2S retention at high temperature, limestone has the advantage of being a cheap and commonly used bulk chemical. The objective of this work is to evaluate the suitability of limestone as a sorbent for hydrogen sulfide at coal gasification conditions.

The overall reactions involved in the sulfidation of limestone may take place in two different ways. At low CO_2 partial pressures and/or high temperatures $CaCO_3$ calcines rapidly to CaO and CO_2:

$$CaCO_3(s) \rightarrow CaO(s) + CO_2(g) \tag{1}$$

This reaction is followed by the sulfidation of the CaO:

$$CaO(s) + H_2S(g) \rightarrow CaS(s) + H_2O(g) \tag{2}$$

At high CO_2 partial pressures and/or low temperatures reaction (1) is strongly thermodynamically disfavoured. Under these conditions removal of H_2S may, however, still be possible but now by direct sulfidation of $CaCO_3$:

$$CaCO_3(s) + H_2S(g) \rightarrow CaS(s) + H_2O(g) + CO_2(g) \tag{3}$$

The equilibrium CO_2 partial pressure depends on the temperature. The partial pressure of CO_2 in the coal gas varies with the gasification process employed and both uncalcined and calcined conditions may be encountered when using limestone as a sorbent.

Naturally occurring $CaCO_3$, including limestone and dolomite ($CaMg(CO_3)_2$) have previously been tested as H_2S sorbents. The $CaCO_3/H_2S$-reaction and the CaO/H_2S-reaction was studied by Borgwardt and Roache (1984) and Borgwardt et al. (1984), respectively. In both cases H_2 inhibited the reaction between $CaCO_3$ or CaO and H_2S. The maximum conversion of $CaCO_3$ to CaS was limited to about 25% for H_2-concentrations above 15%. The rate of reaction between CaO and H_2S decreased with increasing H_2-concentration. However, for large surface areas no effect of the H_2-concentration was seen. The negative influence of H_2 was explained by an entry of H_2 molecules into the CaS lattice and thereby blocking the mobility of O^{2-} and S^{2-} (Borgwardt et al., 1984). Borgwardt and Roache (1984) showed that $CaCO_3$ particles with a diameter larger than 15 μm showed a high initial reaction rate rapidly decreasing above 10% of solid conversion. The reduced rate is suggested to be due to filling of the available pore volume with reaction product, and the reaction becomes limited to the particle surface. The smallest particles (1.6 μm) maintained high reaction rate for conversions exceeding 60%. The BET surface area was increased with decreasing particle diameter indicating that new surface was generated by fracture of the grains. The activation energy for 28 μm particles was estimated to be 176 kJ/mol for temperatures ranging from 823 K to 1023 K, but for temperatures above 1023 K no increase in the rate was observed. Most of the loss in reactivity was explained by sintering of the $CaCO_3$ particles at temperatures above 1023 K. The overall reaction rate for reaction (2) increased with increasing temperature (873 to 1173 K) and the apparent activation energy was found to be 130 kJ/mol (Borgwardt et al., 1984). Also large CaO particles with a diameter of 1.5 mm showed an increased H_2S retention capacity with increasing temperature (873 to 1073 K), (Yumura and Furimsky, 1985).

Ruth et al. (1972) studied the reaction between H_2S and half calcined dolomite and showed that the rate of reaction was much higher than for fully calcined solid. At low temperatures, 773 K and 873 K, some surprising results were obtained: A significant increase in the rate of reaction with increasing concentration of either CO_2 or H_2S, and pulses of oxygen (0.3%) could increase the reactivity. They suggested that the gas environment may affect the way in which CaS grows within the solid microstructure. The role of an oxygen pulse or the presence of CO_2 or H_2O while CaS is forming may be to create an opening in the CaS layer, so the diffusion of H_2S molecules to the adsorption sites can take place.

An in-situ desulfurization method was tested in circulating fluidized bed gasification under atmospheric pressure (Pintsch and Gudenau, 1990). The coal feed rate was 27 kg coal/h and the gasifying agent was either air or a mixture of air and steam. It was found that $Ca(OH)_2$ was an ineffective desulfurization agent both with air and with oxygen/steam gasification. The ineffectiveness of $Ca(OH)_2$ was explained with the reaction kinetics under reducing conditions. Also the high steam content (10 vol% and higher) prevented the total decomposition of $Ca(OH)_2$ to CaO, and with a CO_2 concentration of 15 vol% and higher, CaO may be converted to $CaCO_3$ at temperatures up to 1073 K.

In the present work three different limestones are tested with respect to their capacity for reaction with H_2S at atmospheric and pressurized conditions.

THE EXPERIMENTAL APPARATUS

The main components of the experimental apparatus previously used by Illerup et al. (1993) are a fixed-bed reactor, a gas mixing system, gas analyzers and a data collecting/control system.

The feed gas is supplied from gas cylinders and its flow rate controlled by mass flow controllers. Depending upon the position of a three-way valve, the reaction gas can either enter the top of the reactor or by-pass the reactor. The pressure in the reactor is maintained by a back-pressure valve through which the gas is expanded to atmospheric pressure. The product gas passes to the gas analyzers through a water condenser.

The reactor is made of quartz and enclosed in a pressure vessel, Fig. 1. It is able to stand simultaneously pressures and temperatures of up to at least 10 bar and 1273 K. The quartz reactor consists of three parts: an injection tube (B) at the top of the reactor, a removable sample holder containing a sintered quartz basket at the middle of the reactor (C), and an exhaust tube at the bottom of the reactor (I). The quartz basket can easily be removed and replaced with another one containing a fresh sample. The inner diameter of the sample holder is 18 mm.

The reaction gas enters into the reactor through the injection tube. In order to avoid contact between the corrosive gas and the walls of the pressure vessel, nitrogen or CO_2 is injected to the bottom of the vessel. The inert gas passes upward between the steel wall and the exhaust tube. Above the sintered basket the inert gas and the reaction gas injected through the injection tube are mixed in cross-flow. This design makes it possible

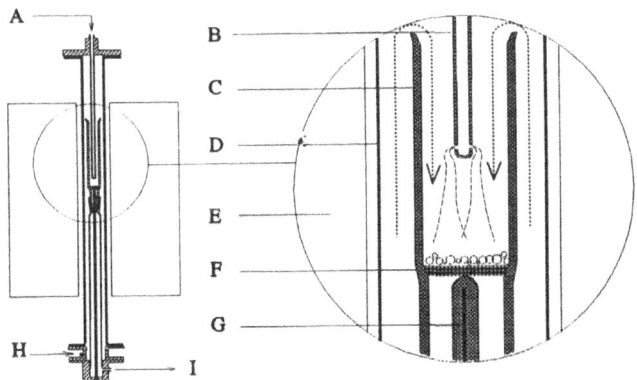

FIGURE 1. Reactor: (A) reaction gas inlet. (B) injection tube. (C) sample holder. (D) pressure vessel. (E) electric furnace. (F) sintered quartz basket. (G) thermocouple. (H) inert gas inlet. (I) product gas outlet.

to carry out experiments in a quartz reactor at elevated pressures, and at the same time to avoid contact between the corrosive gases and the pressure cap. Problems resulting from different expansion coefficients for quartz and steel are also avoided by this design. Pressurization of the reactor to 10 bar takes only a few minutes due to the relatively small volume of the pressure cap.

The pressure in the reactor is measured by an electronic pressure monitor and the temperature is measured by a thermocouple (Ni-CrNi) located just below the sintered quartz basket and protected by a quartz tube. The outlet concentration of H_2S, CO and CO_2 were continuously measured by IR-analyzers, calibrated and checked for cross sensitivities.

EXPERIMENTS

Experimental Conditions

The investigation of the reaction between H_2S and three types of limestone was carried out under the experimental conditions summarized in Table 1. The limestones have previously been tested for sulfur dioxide capture capacity at atmospheric pressure by Dam-Johansen and Østergaard (1991 a,b), Hansen et al. (1993) and at elevated pressure by Illerup et al. (1993). The most reactive limestone towards SO_2, Stevns Chalk, has a very open and porous texture composed of small grains of about 1 μm in diameter. Faxe Bryozo and more pronounced Gotland limestone has a more compact texture of larger grains and are classified as limestones of intermediate reactivity towards SO_2 by Dam-Johansen and Østergaard (1991 a). The chemical compositions and the BET surface areas of the three limestones prior to and after calcination at 1123 K and 1 bar are listed in Table 2.

TABLE 1
Experimental conditions

Type of limestone		Stevns Chalk, Faxe Bryozo, Gotland				
Particle size	(mm)	0.36	0.65	0.93	1.5	
Temperature	(K)	973	1073	1123	1173	1223
Pressure	(bar)	1	6	10		
Gas composition	(vol%)					
H₂S		0.25				
CO₂		2	20			
CO		2	20			
H₂		2	20			
N₂		Balance				
Mass of limestone						
(uncalcined)	(mg)	300				
Bed height	(mm)	1.5				
Total gas flow rate	(m³ (STP) s⁻¹)	5·10⁻⁵				

The limestone in the reactor will calcine if the CO_2 partial pressure is below the equilibrium pressure at the given temperature. Experiments are carried out at both uncalcined and calcined conditions. Table 3 shows the equilibrium pressure and the partial pressures of CO_2 at the reaction temperatures. Experiments to the left and below the solid line (bolded) are performed at uncalcined conditions while to the right and above the line the limestone can calcine. In order to study both the effect of total pressure and the influence of the CO_2 partial pressure on the H_2S retention on limestone, the concentrations of CO_2 and the total pressures were chosen to give the same partial pressure of CO_2 at different total pressures in the experiments with 2 vol% CO_2 for 10 bar and 20 vol% CO_2 for 1 bar.

Experimental Procedure

The course of an experiment is illustrated in Fig. 2. Each time period is designated with a letter used in the explanation below. The reactor is heated to the desired temperature and the pressure and gas flows are adjusted to preset values while the gas is by-passing the reactor (period A). Then the gas is diverted to the empty reactor by a three-way valve. A drop in the outlet H_2S concentration is observed (period B). The reactor is flushed with nitrogen to 'avoid ignition of CO or H_2, the bottom part of the reactor is removed and put back with the limestone sample on the quartz plate and the reactor is flushed with nitrogen again (period C). In this period the limestone calcines if calcination is possible, and the temperature becomes steady. The reaction gas mixture is turned on again and the reaction between H_2S and limestone proceeds until a steady

TABLE 2
Chemical compositions and BET areas of the limestones

| Limestone | Weight percent | | | | | | BET area (m^2/g) | |
	CO_2	SiO_2	Al_2O_3	Fe_2O_3	CaO	MgO	Uncalcined	Calcined
Faxe Bryozo	43.6	0.45	0.10	0.08	55.1	0.43	0.69	16.4
Stevns Chalk	43.6	0.36	0.07	0.04	55.2	0.29	1.2-1.8	9.9-11.8
Gotland	41.5	3.33	1.27	0.65	50.7	1.37	2.5	12.9

concentration level of H_2S is obtained (period D). The gas mixture is diverted to the by-pass to measure any drift in mass flow controllers or analyzers (period E). Drift in the inlet concentration of H_2S is corrected for by linear interpolation. A slight conversion of H_2S in the empty reactor (period B) and in the loaded reactor when the sulfidation is finished (end of period D) is observed. Thermodynamic calculations indicate that conversion of H_2S to COS is most probably responsible for this slightly lower (5%) H_2S concentration.

The fractional conversion of the limestone to CaS was calculated by three different methods. The two first mentioned were used previously by Dam-Johansen and Østergaard (1991 a,b), Hansen et al (1993) and Illerup et al (1993) in the study of SO_2 retention on limestone.

Integration of Concentration Profile. The degree of sulfidation was calculated as a function of time by integrating the difference between the inlet and outlet H_2S concentration in period D. The concentration was corrected for the conversion of H_2S to COS and for the conversion of H_2 to H_2O (shift reaction).

Gravimetric Method. The final conversion is calculated from the weight change of the particles, assuming that the reacted particles consist of CaO and CaS for calcined conditions and $CaCO_3$ and CaS for uncalcined conditions.

Titrimetric Method. The sulfur content of the reacted limestone is measured by dissolution of the sample followed by precipitation of S^{2-} by Cu^{2+}. The excess Cu^{2+} was then reduced with a KI solution and the I_2 formed was titrated using the thiosulfate method.

No significant difference between the methods was observed. The uncertainty limits for the conversions shown later in Figs. 5a and b are based on integration of concentration profiles for duplicate tests.

RESULTS

The initial reaction between H_2S and fresh limestone is very fast. In Fig. 2, period D the outlet H_2S concentration is near to zero for the first minutes and then increases as the sulfidation proceeds and the reaction rate decreases.

TABLE 3
Partial pressures of CO_2 compared with equilibrium pressures.
Experiments with calcining conditions are bolded.

	Temperature (K)				Total pressure (bar)
	973	1073	1123	1173	
Equilibrium pressure of CO_2 (bar)	0.03	0.21	0.48	1.05	-
Partial pressure of CO_2 in gas with 2 vol% CO_2 (bar)	0.02 **0.12** **0.20**	0.02 0.12 0.20	0.02 0.12 0.20	0.02 0.12 0.20	1 6 10
Partial pressure of CO_2 in gas with 20 vol% CO_2 (bar)	**0.20** **1.2** **2.0**	0.20 **1.2** **1.2**	0.20 **1.2** **2.0**	0.20 **1.2** **2.0**	1 6 10

Temperature and Pressure Effects

The conversion of Stevns Chalk to CaS versus time at 973, 1073 and 1173 K and 1 and 10 bar is shown in Fig. 3. The limestone is uncalcined at the low temperature but calcined at 1073 and 1173 K, Fig. 3a. At the high pressure, the limestone is uncalcined at all temperatures, Fig. 3b. If the limestone calcines, the ultimate conversion to CaS becomes almost 100% within 25 minutes, but if it is uncalcined, the final conversion is only 15 to 20%. It is seen from Fig. 3 that the conversion to CaS is almost independent of the temperature for both uncalcined and calcined limestone. The same independence is found for the other two types of limestone at both calcined and uncalcined conditions. The influence of total pressure at calcined conditions is dependent on the type of limestone. For Stevns Chalk and Faxe Bryozo, the conversion to CaS seems to be independent of the total pressure, but for Gotland limestone the conversion to CaS seems to increase with total pressure, Fig. 4. For uncalcined conditions no influence of total pressure was observed for any of the three limestones tested.

Figure 5 summarizes the results for the porous Stevns Chalk showing the conversion to CaS versus the temperature after 10 minutes of sulfidation with the pressure as a parameter. The estimated experimental uncertainty at 1073 K is shown as error bars. The variation of conversion to CaS with temperature and pressure is almost within the

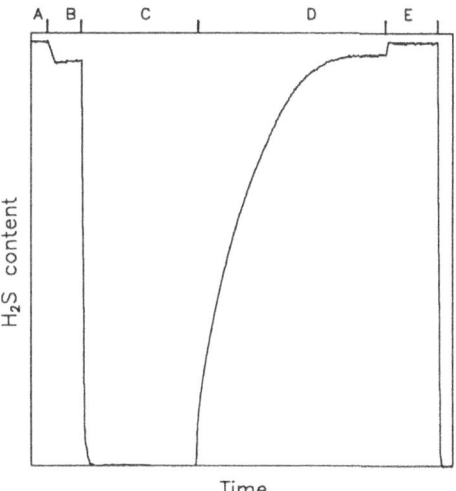

Figure 2. Outlet H_2S concentration versus time for a typical experiments (Stevns Chalk, 973 K, 1 bar, CO_2: 2 vol %)

experimental error, except for changes from uncalcined to calcined limestone, resulting in a change from about 15% to over 70% of solids conversion. To have uncalcined conditions the partial pressure of CO_2 was kept above the equilibrium value in two different ways, either by a large mole fraction of CO_2 at a low total pressure or by a small mole fraction of CO_2 at a high total pressure. The observed sulfidation did not depend on how the limestone was kept uncalcined.

Gas Composition Effect

The fractional conversion to CaS was influenced by the CO_2 partial pressure for calcined conditions. An increase from 2 to 20 vol% CO_2 at 1 bar and 1073 and 1173 K caused a decrease in the reaction rate, but the final conversion was still 100%. At uncalcined conditions, 973 K and 1, 6 and 10 bar no significant influence of the CO_2 partial pressure was observed. The CO content has no influence on the rate of sulfidation or the sulfur capacity in the range from 2 to 20 vol%. A slight increase in the final sulfur capacity was observed at the highest temperature and pressure for uncalcined conditions, when the H_2 content was lowered from 20 to 2 vol%.

The Effect of Limestone Type and Particle Size

The porous limestones (Stevns Chalk and Faxe Bryozo) have a higher rate of reaction and final degree of conversion than the more compact Gotland limestone. This is most pronounced for large particles, where a difference between Stevns Chalk and Faxe Bryozo is also observed, and the final conversion for Gotland limestone is as low

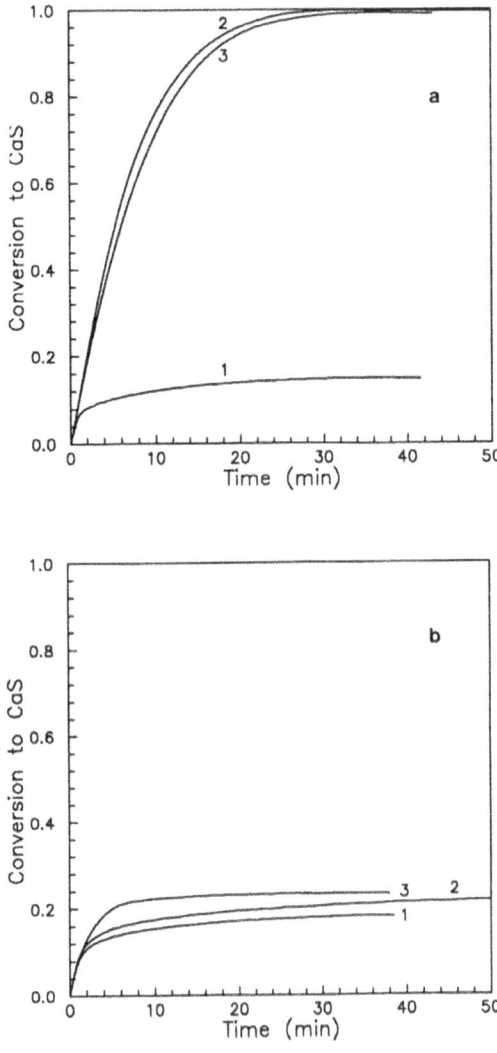

FIGURE 3. The conversion of Stevns Chalk to CaS versus time. Particle size: 0.30-0.42 mm; CO_2: 20 vol%; 1: 973 K; 2: 1073 K; 3: 1173 K. a) 1 bar, b) 10 bar.

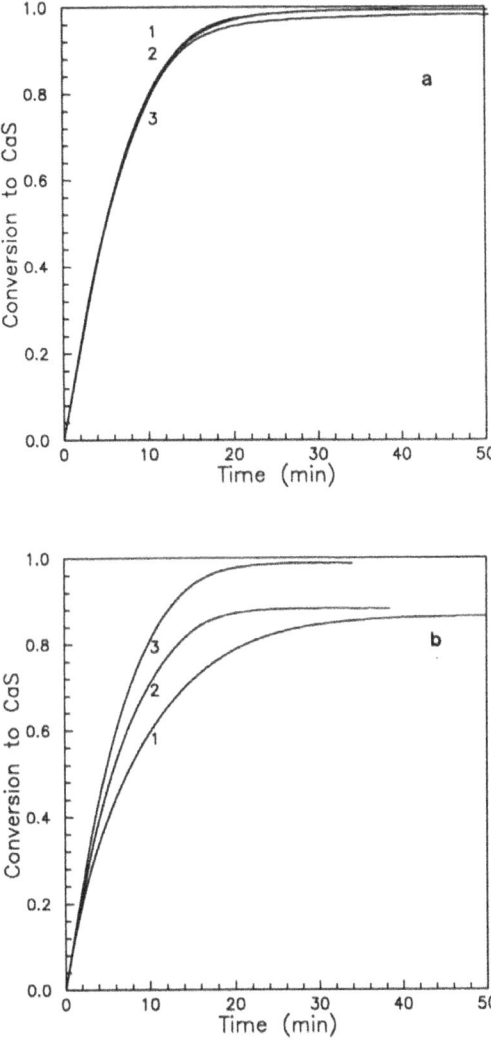

FIGURE 4. The conversion of limestone to CaS versus time. Particle size: 0.30-0.42 mm; CO_2: 2 vol%; 1123 K; 1: 1 bar; 2: 6 bar; 3: 10 bar. a) Faxe Bryozo, b) Gotland.

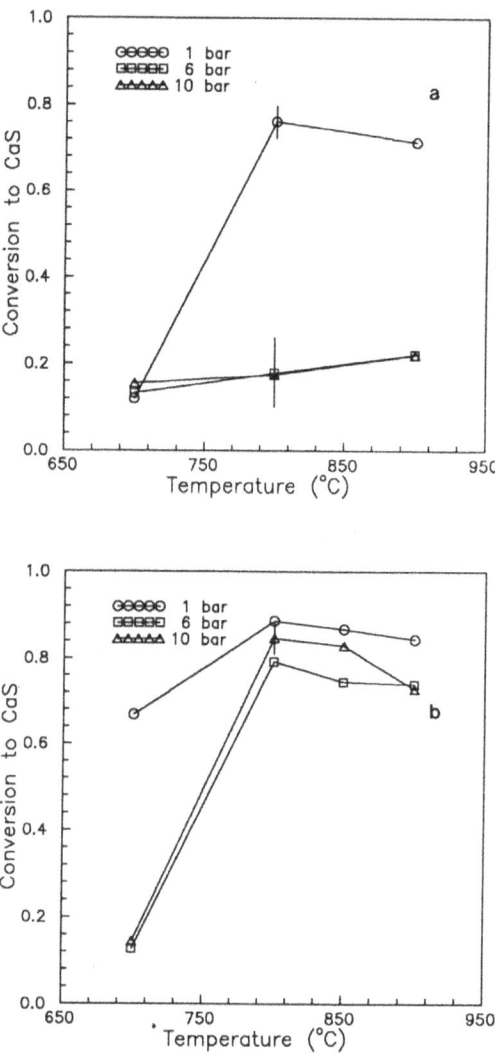

FIGURE 5. The conversion of Stevns Chalk after 10 minutes of sulfidation versus the temperature at different pressures. Particle size: 0.30-0.42 mm. a) CO_2: 2 vol%,b) CO_2: 20 vol%.

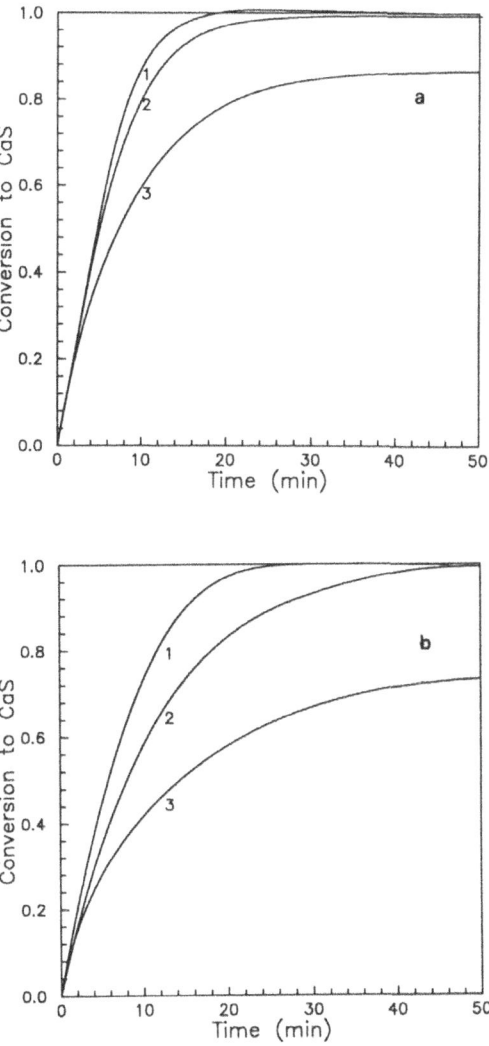

FIGURE 6. Conversion of limestone to CaS versus time at calcined conditions. 1123 K; 1 bar; CO_2: 2 vol%; 1: Stevns Chalk; 2: Faxe Bryozo; 3: Gotland. a) Particle size 0.30-0.42 mm, b) Particle size 0.85-1 mm.

as 72%, Fig. 6. However, at calcined conditions and for a high pressure of 10 bar the final fractional conversion of all three limestones reaches 100%. At uncalcined conditions no influence of type of limestone or particle size is observed.

DISCUSSION

At calcining conditions the reactivity of the three limestones was found to be closely related to the physical texture of the material. The most porous limestone, Stevns Chalk, was found to be the most reactive, and the most compact limestone, Gotland limestone, was found to be the least reactive with Faxe Bryozo in between and close to Stevns Chalk. This order of reactivity is in agreement with earlier investigations of the same types of limestone in the reaction with SO_2 at atmospheric pressure under constant oxidizing conditions (Dam-Johansen & Østergaard, 1991 a), alternating oxidizing and reducing conditions (Hansen et al., 1993), and at elevated pressure (Illerup et al., 1993). The difference between sulfation and sulfidation is probably related to the structure of natural limestones and the molar volumes of $CaCO_3$: 36.9, CaO: 16.9, $CaSO_4$: 46.0 and CaS: 28.9 cm³/mol. A particle of Stevns Chalk and many other limestones consists of grains and each calcined grain can be viewed as consisting of micrograins (Dam-Johansen et al. 1991). The grains become porous upon calcination because the remaining CaO has a lower molar volume than $CaCO_3$. During reaction the CaO is converted to CaS, but because the molar volume of CaS is smaller than the molar volume of the initial $CaCO_3$, some of the porosity is retained and no pore plugging is expected to occur, even for the more compact limestone.

The most important factor for the reactivity and the final conversion to CaS is whether the limestone calcines or not, contrary to the reaction between SO_2 and limestone at oxidizing conditions, where the fractional conversion of the porous Stevns Chalk is high for both calcined and uncalcined conditions at 1023 K, while it is low for the more compact Gotland limestone in both cases (Illerup et al. 1993). The observed distinct difference between calcined and uncalcined limestone in the reaction with H_2S indicates that the reaction mechanism is different in the two cases. This conclusion is supported by measurement of the local solids conversion in partly and completely reacted limestone particles. Stevns Chalk particles were withdrawn from the reactor after varying reaction times, imbedded in a PVC resin, and ground halfway through. The intraparticle degree of sulfidation was determined qualitatively by energy-dispersive X-ray analysis (EDAX) in a scanning electron microscope. The results are shown in Figs. 7 and 8, where the local sulfur concentration is shown by the intensity of the white dots.

At calcined conditions, the reaction takes place in an outer shell moving towards the center as the reaction progresses, Fig. 7. Quantitative measurement of the sulfur concentration in the shell indicates that the CaO is completely converted to CaS, and so the unreacted shrinking core model (Levenspiel, 1972) seems to be valid.

At uncalcined conditions the sulfur is distributed evenly in the whole limestone particle independent of particle size, and the concentration of sulfur in the particle increases as the reaction proceeds, Fig. 8. In this case the progressive conversion model (Levenspiel, 1972) seems to be valid, but only 15 to 25% conversion to CaS is obtained. In the uncalcined limestone there is no porosity in the grains, and the even distribution of CaS shown in Fig. 8 indicates a fast diffusion of H_2S in the pore system between the

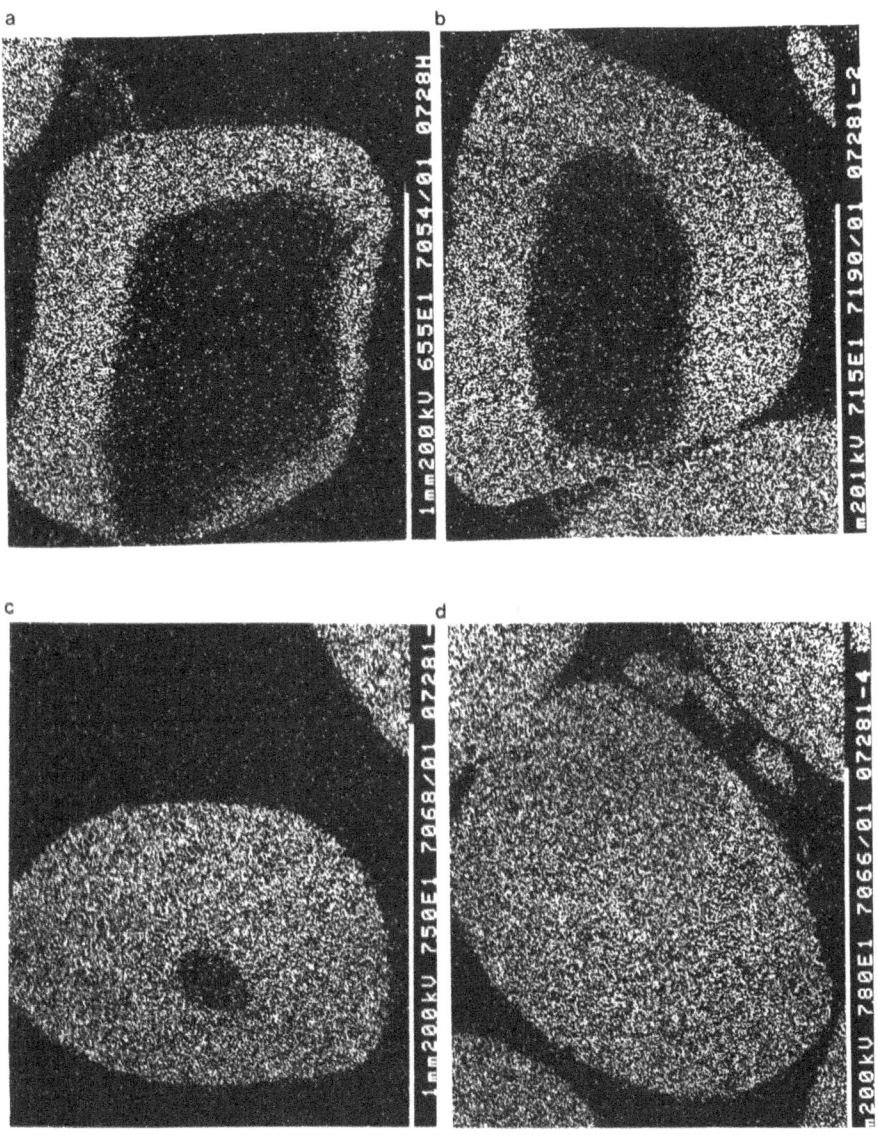

FIGURE 7. Dot picture (EDAX-analysis) of sulfur distribution in Stevns Chalk at calcined conditions. 1123 K; 1 bar; particle size: 0.85-1.0 mm. Reaction times in minutes: a) 5, b) 10, c) 15, d) 25.

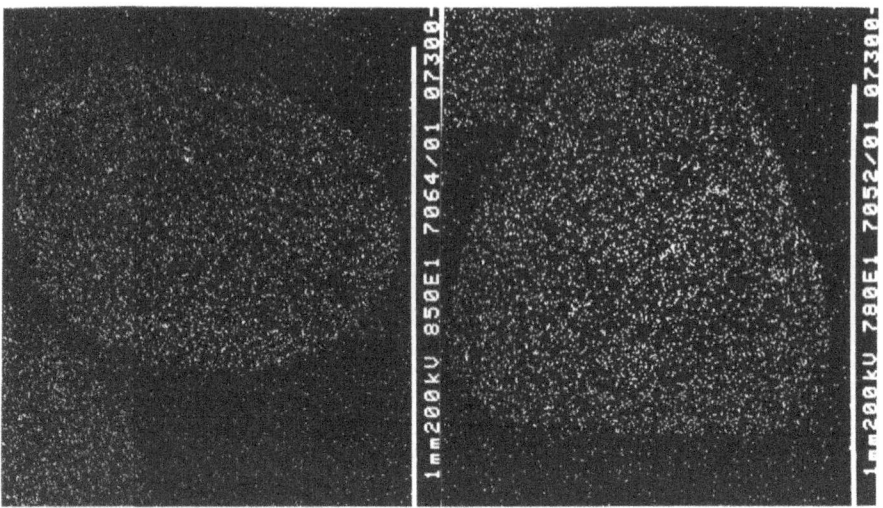

FIGURE 8. Dot picture (EDAX-analysis) of sulfur distribution in Stevns Chalk at uncalcined conditions. 1123 K, 6 bar, CO_2: 20 vol%, particle size: 0.85 to 1.0 mm. Reaction times in minutes: a) 4, b) 15.

$CaCO_3$ grains compared to the sulfidation rate. This conclusion is also supported by experiments with different particle sizes, Fig. 9. The progress of sulfidation and the final conversion degree is independent of particle size for uncalcined conditions and diffusion between grains is not limiting. The rate
limiting step can be either surface reaction on the grains or solid state diffusion in the CaS product layer. The initial reaction rate between H_2S and $CaCO_3$ is very fast and comparable in size to the rate of reaction between H_2S and CaO, indicating that either product layer diffusion in the grains limits the reaction rate for fractional conversions larger than about 20% or deactivation of the $CaCO_3$ surface is enhanced by H_2S, H_2, CO or the CaS formation. Borgwardt and Roache (1984) found that product layer diffusion was not a major resistance for particle sizes comparable to grain size (1.6 to 10 μm), in favour of the hypothesis of deactivation by sintering. CaO is more stable than $CaCO_3$ and so the sintering effect is negligible at calcined conditions.

Industrial Application

The possible use of limestone as a regenerative sorbent for H_2S in gasification processes depends on the gasification conditions, because the limestone must be calcined at the reaction conditions in order to obtain a high conversion to CaS. Typical gasification conditions are total pressures of 10 to 40 bar, temperatures of 1100 to 1300 K and the CO_2 partial pressure varies typically between 0.2 and 6 bar. The calcination temperature for 0.2 and 6 bar CO_2 partial pressure is 1070 and 1300 K respectively and

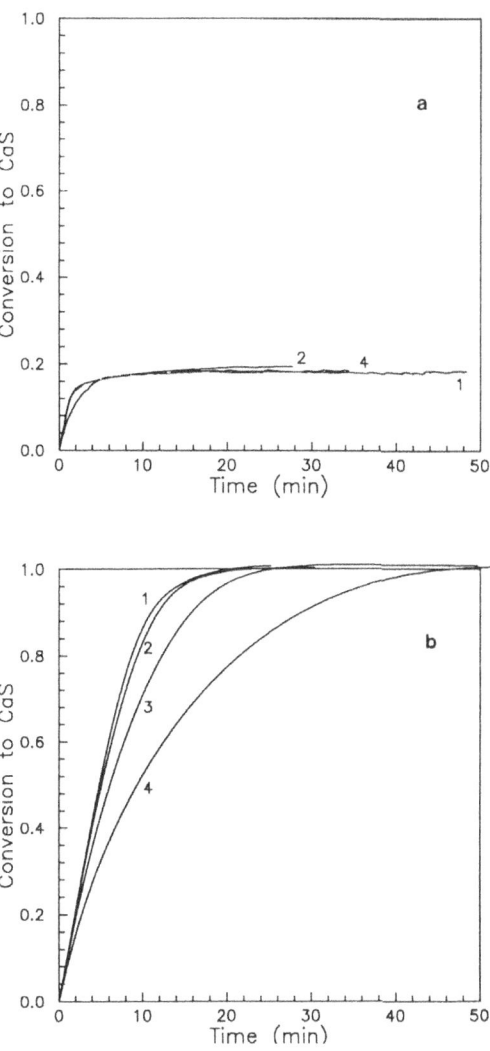

FIGURE 9. The conversion of Stevns Chalk to CaS versus time for different particle sizes, at 1123 K, 1: 0.30-0.42 mm, 2: 0.60-0.71 mm, 3: 0.85-1.0 mm, 4: 1.4-1.7 mm. a) Uncalcined conditions (6 bar, CO_2: 20 vol%), b) calcined conditions (1 bar, CO_2: 2 vol%).

507

there is a wide range of operating conditions where the limestone calcines and becomes a possible sorbent for H_2S. However, for low temperature and high CO_2 content, Ca-based sorbents are ineffective (Pintsch & Gudenau, 1990).

Regeneration of Limestone

Regeneration of sulfided limestone was tested in preliminary experiments at three different temperatures, 1123, 1173 and 1223 K. The regeneration gas was a mixture of N_2 and CO_2 at 1 bar and the products were CaO, CO and SO_2:

$$CaS(s) + 3CO_2(g) \rightarrow CaO(s) + 3CO(g) + SO_2(g) \qquad (4)$$

Sulfidation and regeneration was carried out at the same temperature and the optimum temperature was found to be around 1173 K, where the ultimate degree of sulfidation was about 80% after 10 sulfidation-regeneration cycles. At 1223 K the degree of sulfidation was only 30% after 10 cycles probably due to sintering of the particles.

CONCLUSIONS

Calcined limestone was found to be a promising sorbent for H_2S in industrial gasification processes. Complete conversion of CaO to CaS was obtained for 0.36-1.6 mm particle sizes, 973-1173 K temperatures, 1-10 bar pressures and 0.25% H_2S content. The porous limestones (Stevns Chalk and Faxe Bryozo) were fully converted at all pressures, but for the more compact Gotland limestone the final degree of sulfidation increased from about 72 to 100% when the pressure was increased from 1 to 10 bar. The most important conclusion is, that 100% conversion of CaO to CaS can be obtained at gasification conditions even for compact limestones in contrast to the sulfation reaction at atmospheric and elevated pressures, where compact limestones have a low conversion of about 25% because of pore plugging.

Uncalcined limestone is a poor sorbent for H_2S. The final fractional $CaCO_3$ conversion to CaS is 15 to 25% almost independent of temperature, pressure and particle size, but may increase a little at low H_2 concentrations. The poor performance of uncalcined limestone is probably due to sintering of $CaCO_3$ in the presence of H_2S or CaS.

ACKNOWLEDGEMENTS

This work is a part of the research program CHEC (Combustion and Harmful Emission Control) funded by the Danish Technical Research Council, Elsam (the Jutland-Funen Electricity Consortium), Elkraft (the Zealand Electricity Consortium) and the Danish Ministry of Energy.

REFERENCES

Borgwardt, R.H. & Roache, N.F. (1984). Reaction of H_2S and sulfur with Limestone Particles. Ind.Eng.Chem. Process Des. Dev., 23, pp. 742-748.

Borgwardt, R.H., Roache, N.F. & Bruce, K.R. (1984). Surface Area of Calcium Oxide and Kinetics of Calcium Sulfide Formation. Environmental Progress 3, pp. 129-135.

Dam-Johansen, K. & Østergaard, K. (1991a). High Temperature Reaction between Sulphur Dioxide and Limestone - I. Comparison of Limestone in two Laboratory Reactors and a Pilot Plant. Chem.Eng.Sci. 46, pp. 827-837.

Dam-Johansen, K. & Østergaard K. (1991b). High-temperature Reaction between Sulphur Dioxide and Limestone - II. An Improved Experimental Basis for a Mathematical Model. Chem.Eng.Sci. 46, pp. 839-845.

Dam-Johansen, K.; Hansen, P.F.B. & Østergaard, K. (1991). High-temperature Reaction between Sulphur Dioxide and Limestone - III. A Grain-micrograin Model and its Verification. Chem.Eng.Sci. 46, pp. 847-853.

Hansen, P.F.B., Dam-Johansen, K. & Østergaard, K. (1993). High-temperature Reaction between Sulphur Dioxide and Limestone - V. The Effect of Periodically Changing Oxidizing and Reducing Conditions. Chem.Eng.Sci., 48, pp. 1325-1341.

Illerup, J.B.; Dam-Johansen, K. & Lundén, K. (1993). High Temperature Reaction between Sulfur Dioxide and Limestone - VI. The Influence of High Pressure. Chem.Eng.Sci., 48, pp. 2151-2157.

Levenspiel, O. (1972). Chemical Reaction Engineering, John Wiley & Sons, Inc.

Pintsch, S. & Gudenau, H.W. (1990). In-situ Desulfurization in an HTW-Coal-Gasifier Development Unit. Coal Sci.Technol., 16 (Process. Util. High-sulfur Coal 3), pp. 697-707.

Ruth, L. A., Squires, A.M. & Graff, R.A. (1972). Desulfurization of Fuels with Half-Calcined Dolomite: First Kinetic Data. Envir.Sci.Techn. 6, No. 12, pp. 1009-1014.

Yumura, M. & Furimsky, E. (1985). Comparison of CaO, ZnO, and Fe_2O_3 as H_2S Adsorbents at High Temperatures. Ind.Eng.Chem. Process Des. Dev., 24, pp. 1165-1168.

SO$_2$ REMOVAL FROM STACK GASES BY HIGH TEMPERATURE DRY PROCESS WITH HIGH REACTIVE Ca-Mg BASED SORBENTS

TOMOHIDE WATANABE, MOTOHITO HAYASHI, AKIRA TAKAHASHI, HITOKI MATSUDA and MASANOBU HASATANI
Department of Chemical Engineering, Nagoya University, Furo-cho, Chikusa-ku, Nagoya 464-01, JAPAN

ABSTRACT

As a first step of developing an effective and economical desulfurization technique for coal combustion, the sulfurization reactivity of calcined calcite, dolomite and magnesite were compared for various experimental conditions by changing the calcination and the sulfurization temperature ranged from 973K.to 1473K by a TGA(thermogravimetric analyzer). The effect of water vapor on the calcination and sulfurization characteristics of desulfurizing sorbents were studied. The structural change of desulfurizing sorbent during calcination and sulfurization was studied by the measurements of pore volume and surface area and observation of SEM photographs. It was recognized that calcined sorbent treated with hydration/dehydration showed a higher reactivity for sulfurization by restoring pores and by modifying the crystal structure of the sorbent.

The desulfurization experiments conducted with a laboratory-scale fixed bed-type gas flow reactor under pressurized atmosphere showed that the desulfurization capacity of MgO particle bed was increased with an increase in the reactant gas pressure.

INTRODUCTION

Acid rain derived from SOx and NOx has become an important global environmental problem with an increasing use of coal. Particularly, in developing countries, SO2 has been almost released from coal combustors without SO2 abatement process. It is, therefore, urgently required to develop the SO2 removal process with a high efficiency and economy in order to be applied to these countries. As is well known, the major desulfurization technique for power plant boiler usually consists of wet limestone-gypsum process. However, from an economical point of view, the dry desulfurization process is considered much more attractive in the terms of cost performance and energy utilization efficency. One of the important factors to decide the performance of the dry desulfurization process is the reactivity and cost of the sorbents. Different kinds of desulfurizing sorbents like calcium oxide[1,2,12,14] magnesium oxide[11,22], dolomite[7,19], zinc oxide[16,17], manganese oxide[15], and copper oxide[5], etc.[23], so far have been studied for applying to the dry desulfurization of coal combustion gas. Especially, Ca-Mg based carbonate like calcite, magnesite, and dolomite, of which the oxide of CaO,MgO and [CaO · MgO] can thermodynamically react with SO2 under desulfurization condition for sorbent injection process in coal combustor as FBC. It is important to make clear the fundamental reaction characteristics of these Ca-Mg based sorbents with SO2. A lot of reserches concerning the reaction cheracteristics of CaO derived from limestone with SO2 have been reported to evaluate the reaction kinetics and overall reaction rate[7,10,12], but that of MgO and calcined dolomite are not sufficiently made clear because of lack of data in wide experimental condition.

In this study, the sulfurization reactivity of calcined calcite, dolomite and magnesite were compared by means of TG under various calcination and sulfurization temperatures. The effect of water vapor involved in flue gas on the calcination and sulfurization characteristics was also studied. Taking into account the difference in the pore structure for each sorbent, we compared the sulfurization characteristics of calcium carbonate, magnesium carbonate and dolomite with each other. Then, an attempt was made to make clear the sulfurization mechanism of these sorbents.

Considering the previous data, one of the main problem of dry SO_2 removal process using limestone arises from a low utilization of sorbent owing to an increasing intraparticle diffusion resistance of reactant gas during sulfurization with pore plugging. It is nenecessary to enhance the reactivity of sorbents to attain more effective dry desulfurization process. To overcome such essential defects of calcium oxide caused by the structural change accompanied by a chemical reaction with molar volume expansion, we tried to change the pore structure of calcium oxide by treating it with hydration/ dehydration procedure.

PFBCs(pressureized coal combustors), recently, are becoming much more promising, which enables coal combustion with a higher efficiency under pressurized conditions with SO_2 removal in the combustor. Few data, however, are available for the desulfurization characteristics of sorbent under pressurized conditions except those by Iisa and Hupa[13] who investigated the effect of reactant gas pressure on the reaction rate and the capacity of $CaCO_3$ for desulfurization by means of a pressurized TG. In the present study, the desulfurization experiments were carried out in a laboratory-scale fixed bed type reactor under pressurized conditions, and the effect of reactant gas pressure on the surfurization characteristics was examined.

EXPERIMENTAL

Samples

The samples used for TG experiments were magnesite, calcite and dolomite. The chemical composition of the sample is shown in Table 1. Prior to the TG experiments, all the specimen was pulverized and sieved with micro sieve(5 μm mesh) by wet screening to prevent the sulfurization reaction from proceed under diffusion control step. The mean particle diameter of these samples determined by a laser scattering method was about 2.3 μm.

The specimen of about 1 mm in the diameter was used for the desulfurization experiment with a tubular reactor under pressurized conditions.

To evaluate the characteristics of sample, the pore size, pore volume and the surface area were measured by N_2 adsorption isotherm using a Micrometritics ASAP 2000. Surface area was calculated by a BET method and the pore size distributions were obtained from N_2 adsorption isotherm. The particle structure was observed with SEM photographs.

TABLE 1

Chemical components of samples

Sample	Source	CaO	MgO	SiO_2	Fe_2O_3	Ig.Loss
Magnesite	China	0.66	46.2	2.51	0.24	49.82
Calcite	Japan(Hiroshima)	55.74	-	-	-	43.74
Dolomite(1)	China	29.15	20.15	4.00	1.2	44.80
Dolomite(2)	China	32.23	21.16	-	0.55	46.61

Thermogravimetric experiment

Figure 1 shows a TG apparatus employed for the experiment. The unit was connected with a humidifier to investigate the sulfurization reactivity of calcined samples: The influence of calcination temperature and water vapor pressure on the reactivity of calcines was examined. The weight change of sample was continuously measured during both the calcination and the sulfurization reactions. The temperature was measured by a thermocouple(Pt-PtRh) attached to the sample holder. About 3mg of the carbonate specimen were placed in a platinum sample holder of 5 mm height and 6mm diameter. Total gas flow rate through the reaction tube was constant about $6.6 \times 10^{-3} m^3 \cdot hr^{-1}$ at 293K, which was found to be free from gas film diffusion resistance for both the calcination and the sulfurization reactions.

i)Calcination. The sample was first calcined under N_2 atmosphere at various temperatures. The heating rate of the furnace was $5K \cdot min^{-1}$ for calcination process. After the sample had been completely decomposed, the furnace was cooled down to the prescribed sulfurization temperature with a cooling rate of $20K \cdot min^{-1}$, and then the sulfurization experiment was started. The calcination condition was as follows;
calcination temperature : 973-1473K, atmosphere : N_2 or N_2+H_2O(0-27.2vol%)

ii)Sulfurization. After the calcination of specimen had been completed and a steady state conditions had been confirmed by monitoring the temperature and weight change, the sulfurization reaction was initiated by quick switching to the reaction atmosphere. The sulfurization experiment was carried out under the following conditions;
sulfurization temperature : 973-1473K
simulated flue gas: 0.32-0.35vol% SO_2, 4.8-4.9vol% O_2, 0-24.4vol% H_2O, N_2 balance

iii)Hydration/dehydration treatment. For the purpose of improving the capacity of desulfurizing sorbent, the calcined sample was once hydrated with water or water vapor. The hydrate was then decomposed to calcium oxide and subjected to the desulfurization experiment. To evaluate the reactivity of the sample treated by the following procedure, the sulfurization experiment was conducted at 973K under SO_2, O_2 and N_2 gas mixture. The hydration was achieved by the following two methods:

a) hydration with water. The calcinated sample was cooled down from the calcination temperature to the room temperature under N_2 atmosphere and then hydrated with distilled water with the stoichiometric ratio of 100 to the calcines. The produced hydrate of which the conversion of hydration was nearly 100% was then heated to 973K under N_2 atmosphere and was completely dehydrated.

b) hydration with water vapor. The calcined sample was cooled down to 433K under N_2 atmosphere and then the hydration was made complete by introducing the water vapor of 5.46-14.6vol%. As the same case of hydration with water, the produced hydrate was again heated to 973K to obtain the calcine.

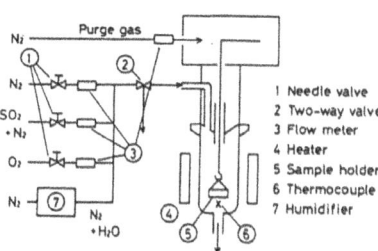

Figure 1 Schematic diagram of Thermogravimetric analyzer

1 Needle valve
2 Two-way valve
3 Flow meter
4 Heater
5 Sample holder
6 Thermocouple
7 Humidifier

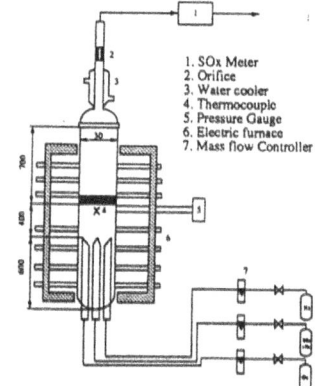

1. SOx Meter
2. Orifice
3. Water cooler
4. Thermocouple
5. Pressure Gauge
6. Electric furnace
7. Mass flow Controller

Figure 2 Schematic diagram of fixed bed rector

Experiment with fixed bed reactor under pressurized conditions

A schematic diagram of the fixed bed-type reactor is shown in Fig.2. The apparatus is made of quartz glass of 30mm inner diameter and 1700mm height. The pressurized conditions were achieved by making flow the reactant gas through an orifice located at the exit of the reactor. The gas preheating section was 600mm length and the gas mixing section was 400mm length.

Each gas of SO₂, O₂ and N₂ was supplied separately from the bottom of the reactor to prevent SO₂ and O₂ from reacting before reaching the reaction part. The calcined sample was set to form a particle bed in the middle part of the reaction tube. Gas leaving from the reactor was cooled down with a water cooler to prevent from the reaction between SO₂ and O₂ after passing through the sample particle bed. The specimen particles were put on SiO₂ gauze over a SiO₂-distributor with 35 mesh holes. The reaction temperature was measured with a type-K thermocouple sheathed by quartz glass at the position beneath the distributor.

Before starting the desulfurization experiment, the carbonate sample of 3.6-12g which dispersed uniformly over SiO₂ gauze was decomposed to the oxide sample at a constant reaction temperature under N₂ flowing condition. The sulfurization experiment was then initiated by switching N₂ to SO₂, O₂ and N₂ gas mixture.The SO₂ concentration in the exhaust gas from the reactor was measured continuously with an infrared analyzer. The composition of the gas employed was 0.34vol% SO₂, 5.0vol% O₂ and balance gas of N₂. The total absolute gas pressure was regulated at a constant value ranged from about 101.3 to 303.9kPa. The weight change of the sample was measured after the completion of reaction and then the mass balance with respect to SO₂ mole was checked.

RESULTS AND DISCUSSION

Thermodynamic equilibrium of $CaCO_3/CaO/CaSO_4$ and $MgCO_3/MgO/MgSO_4$

Figure 3(a) shows the equiribrium pressure of CO₂ for Ca-Mg based oxide/carbonates systems obtained by Haul et al[9]. Assuming that the concentration of CO₂ in real flue gas is about 15 vol% under atmospheric condition, the temperature at the equiribrium pressure of CO₂ for decomposition of calcium and magnesium carbonates are about 615K and 1060K, respectivity.

Figure 3(b) shows the equilibrium pressure of SO₂ for MgO/MgSO₄ and CaO/CaSO₄ systems, which was calculated on the basis of the change of standard free energy. Assuming that the partial pressure of SO₂ is about 0.35 kPa in real flue gas containing 5 kPa of O₂, the decomposition temperatures of CaSO₄ and MgSO₄ at the equilibrium pressure of SO₂ are about 1630K and 1200K, respectivity.

Thus, it is considered that from thermodynamic point of views, calcined oxides like CaO, MgO and [CaO · MgO] can react with SO₂ under operation conditions of convetional FBC boilers.

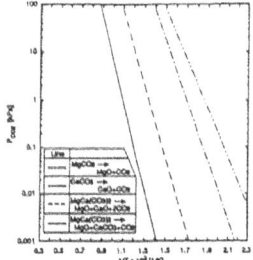

Figure 3(a) Equiribrium of Ca-Mg based
carbonate/oxide reaction system

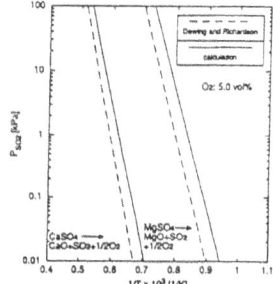

Figure3(b) Equiribrium of CaO/CaSO₄ and
MgO/MgSO₄ reactions system

TG experiment under atmospheric pressure

Calcination

Table 2 shows the measured values of BET surface area, the pore volume and the average pore radius of calcines produced at various calcination temperatures, together with those of initial carbonates. It can be seen from the table that the values of specific surface area and pore volume decreased significantly with an increase in the calcination temperature. This was caused by the sintering which took place in a higher temperature region and progressed with the exposure time.

Figure 4 shows the pore size distributions of initial carbonates and calcines produced at various temperatures. The large pore volume was observed at the mean pore size of about 4, 30 and 50nm for calcined magnesite, dolomite and calcite, respectivity. The pore radius of the calcined magnesite was the smallest among three kinds of calcines; about ten times smaller than calcined calcite. Thibault et al.[21] reported that there was a difference of the pore size distribution between calcined magnesite and calcite. Our data are consistent with their results in terms of qualitative trend of pore size distribution. The value of pore sizewhere the maximum value of dV/dlog r was obtained, however, was different from their data.

The pore size where the maximum value of pore size distribution curve was obtained, was similar for each calcines produced in the present work, despite the difference in calcination temperature. However, the pore volume tended to decrease with an increase in the calcination temperature.

TABLE 2
Characteristics of Calcined Samples employed for TG experiment

Sample	Temp.[K]	Calcination Atmosphere	$S[m^2 \cdot kg^{-1}]$	$V[m^3 \cdot kg^{-1}]$	Av.pore radius[nm]
MgO	973K(5K $\cdot$ min^{-1})	N$_2$	1.17×10^5	3.00×10^{-4}	4.86
MgO	1273K(5K $\cdot$ min^{-1})	N$_2$	7.33×10^4	2.06×10^{-4}	5.16
CaO	1073K(5K $\cdot$ min^{-1})	N$_2$	2.41×10^4	1.79×10^{-4}	14.9
CaO	1473K(5K $\cdot$ min^{-1})	N$_2$	5.56×10^3	2.26×10^{-5}	8.14
CaO $\cdot$ MgO(1)	1073K(5K $\cdot$ min^{-1})	N$_2$	2.59×10^4	1.25×10^{-4}	9.62
CaO $\cdot$ MgO(1)	1373K(5K $\cdot$ min^{-1})	N$_2$	6.16×10^3	1.10×10^{-5}	3.47
CaO $\cdot$ MgO(2)	1073K(5K $\cdot$ min^{-1})	N$_2$	4.07×10^4	2.11×10^{-4}	10.3
CaO $\cdot$ MgO(2)	1373K(5K $\cdot$ min^{-1})	N$_2$	2.76×10^4	1.39×10^{-4}	10.1

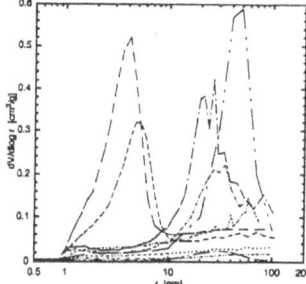

Line	Sample	Calcination temp. [K]
·············		Raw
— — —	Magnesite	973
– – – –		1273
·············		Raw
—·—·—	Dolomite (1)	1073
—··—··—		1373
·—··—··—		Raw
—— · ·—	Calcite	1073
—··—··—		1473

Figure 4   Pore size distribution of samples

Sulfurization

The overall reactions between the calcines and SO$_2$ with excess O$_2$ are described as the followings:

$$CaO + SO_2 + 1/2O_2 \rightarrow CaSO_4 \tag{1}$$
$$MgO + SO_2 + 1/2O_2 \rightarrow MgSO_4 \tag{2}$$
$$[CaO \cdot MgO] + 2SO_2 + O_2 \rightarrow CaSO_4 + MgSO_4 \tag{3}$$

The degree of calcination and sulfurization was expressed by the conversion based on the weight change of sample by means of TG analysis. Thus, the surfurization conversion of the calcines can be expressed as:

$$X = \frac{1-W/W_0}{Y(1-M_{sulfate}/M_{oxide})} \qquad (4)$$

where W is the weight of the calcined sample during sulfurization and W_0 is the initial weight of the calcined sample, Y is the weight fraction of the solid reactant in the calcine and M is the molecular weight.

Effect of calcination temperature.   Figures 5(a), 5(b), 5(c) show the comparison of the time-change of sulfurization of magnesite, dolomite and calcite calcined at various calcination temperatures. The reactivity of sulfurizaiton of calcined calcite and dolomite decreased with increasing in the calcination temperature. As shown in Table 2 and Fig.4, this decrease in the reactivity was related to a decrease in the surface area and pore volume of the calcined particle at higher calcination temperatures. Concerning the relation between the sulfurization reactivity and the surface area of the calcine particles, Borgwardt et al.[3] noted that the reactivity increased with the square of the BET surface area of calcined limestone over the range of 2 to 63$m^2 \cdot g^{-1}$. Larger pore volume and surface area of calcines particles result in decreasing the intraparticle diffusion resistance of the reactant gas. The calcined magnesite, however, had the highest reactivity at a calcination temperature of 1273K, while its pore volume and surface area decreased with an increase in the calcination temperature. At the present stage, it is supposed that this result may be caused by the difference of the active center on the surface of calcined particle. The active center on the surface of the calcined magnesite may increase in the number with an increase in the calcination temperature or have maximum value for a calcination temperature about 1273K. This effect may be larger than that of an increase in the intraparticle diffusion resistance.

Effect of water vapor.   Figure 6 shows the effect of water vapor on the conversion of sulfurization after the time lapse of 10 hours from the beginning of the reaction. The calcined magnesite did not show any more reactivity for sulfurization after the elapsed time of 10 hours. In the case that the water vapor was involved in the sulfurization atmosphere, the sulfurization capacity of calcined magnesite was slightly increased with an increase in the concentration of water vapor. To the contrary, when the water vapor coexists in both the calcination and the sulfurization atmosphere, the sulfurization capacity of calcined magnesite remarkably decreased. This suggests that the water vapor may have a greater influence on the calcination than the sulfurization of calcines. Murthi et al.[19] observed by the experiment with the fixed bed reactor that the reactivity of CaO particle for sulfurization was remarkably decreased under the coexistence of 10vol% moisture in flue gases. As a possible explanation for the decreased reactivity of CaO, it was supposed that the active center on the surface of the calcines was decreased owing to the adsorption of H_2O molecules on the active site during calcination.

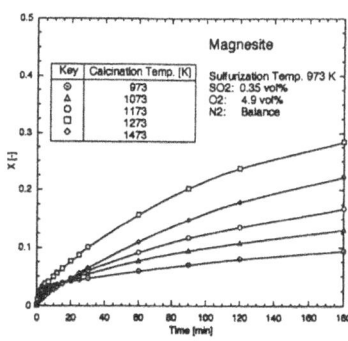

Figure 5(a)   Comparison of time-cource of sulfurization of magnesite calcined at various temperature

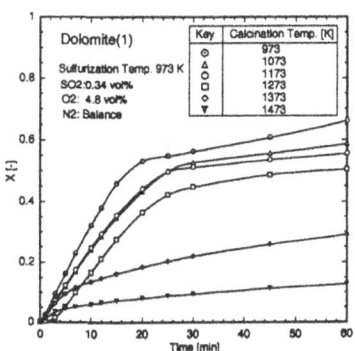

Figure5(b)   Comparison of time-cource of sulfurization of dolomite calcined at various temperature

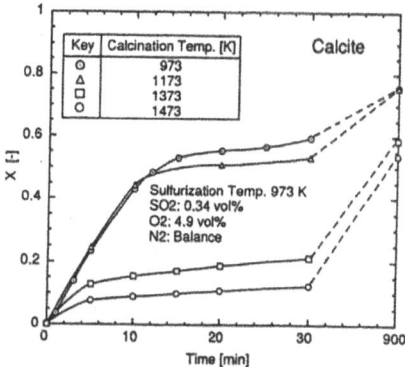

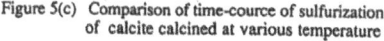

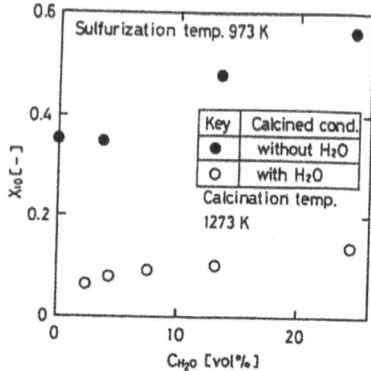

Figure 5(c) Comparison of time-cource of sulfurization
of calcite calcined at various temperature

Figure 6   Effect of water vapor on sulfurization and
calcination of MgO

Reactivity of Ca-Mg based sorbents.   Figures 7 and 8 show the comparison of sulfurization of Ca-Mg based sorbents at 1173K and 1373K, respectively. It is seen in Fig. 7 that the calcined calcite shows a higher reactivity with SO_2 than the calcined magnesite at a sulfurization temperature of 1173K. The same tendency was observed for the sulfurization temperature below 1173K. The broken lines in the figures show the calculated conversion value on the assumption that CaO involved in dolomite only reacts with SO_2 up to 100%: MgO is assumed not to take part in sulfurization. In fact, the saturated conversion values for dolomite (1) and (2) was situated, more or less, above the calculated line (broken line). This means that MgO as well as CaO reacted with SO_2. Calcined dolomite(1), especially, had a higher reactivity and a higher reaction capacity than calcined calcite in spite of a relatively large content (39.6wt%) of the involved MgO. Such a high reactivity and reaction capacity of dolomite(1) may be caused by the presence of impurities like iron oxide. The crystal structure of dolomite which differs for the producing district, may also affect these reaction characteristics. Borgwardt[1] and Ishihara[14] reported that iron oxide impurities in calcines participated in the sulfurization. Marier et al.[18] and Murthi et al.[19] suggested that iron oxide in calcines might play a role of catalyst for the oxidation of SO_2 to SO_3. Furthermore, Thibault et al. [21] reported that the reactivity of MgO with SO_3 was higher than that of CaO with SO_3.

In the present experimental condition (the partial pressure of SO_2 is at most 0.35kPa), MgO did not react with SO_2 above about 1200K from the point of view of the thermodynamic equilibrium of $MgO/MgSO_4$. As can be seen in Fig.8, the reaction capacity of calcined dolomite was lower than that of calcined calcite, since in this reaction temperature of 1373K, MgO did not react with SO_2. As shown in Fig.8, the reaction between CaO and SO_2 did not attain the conversion of 100% because of the plugging of pores within the solid reactant during sulfurization. Under the present experimental condition, the conversions of 72% of sulfurization by calcined calcite was obtained after the time lapse of 300 minutes. The conversion of nearly 100% was obtained by the component of CaO in dolomite (1) and (2) in relatively short time lapse of 20 minutes. Such a high reaction capacity of CaO in calcined dolomite may be explained by a smaller intraparticle diffusion resistence of SO_2 and O_2 through the pores: MgO is considered to play a sort of buffer role for preventing from pore plugging during the sulfurization of CaO. Fan et al.[7] indicated in their theoretical analysis that the diffusion coefficient of reactant gas into calcined dolomite was 40-500 times higher compared to calcined calcite.

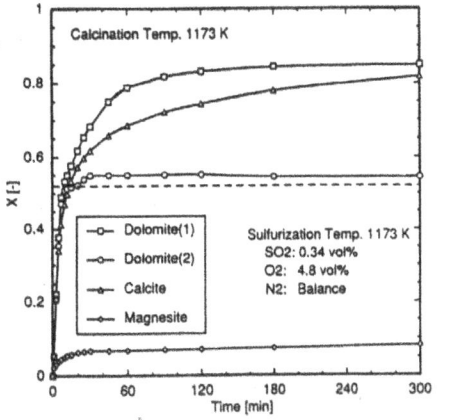

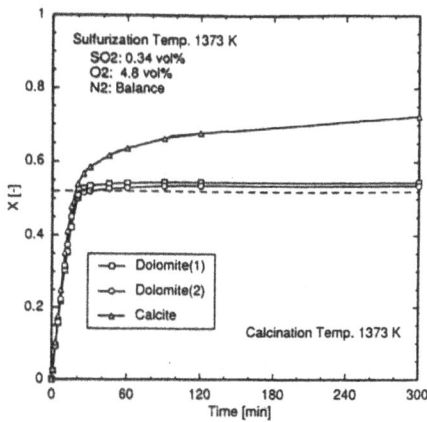

Figure 7   Comparison of the reactiviy of sulfurization at 1173K with calcined samples

Figure 8   Comparison of the reactiviy of sulfurization at 1373K with calcined samples

Figure 9 shows F1 and F2-plots for the sulfurization of calcined calcite, magnesite and dolomite at 1173K which corresponds mostly to the actual operating temperature in a FBC boiler. As seen in the figure, the sulfurization of calcined calcite and dolomite proceeded under the chemical reaction controlling step up to the conversion of about 50%. On the other hand, the diffusion controlling steps became dominant after the conversion had exceeded over 50%. In the case of sulfurization of calcined magnesite, however, the chemical reaction control regime is kept only up to 4% conversion. It is, therefore, concluded that the sulfurization of calcined magnesite proceeded practically under the diffusion control. This is consistent with the pore size distribution shown in Fig.4: the pore size of calcined magnesite was much finer than those of calcined calcite and dolomite. Chemical reaction rate constants of calcined magnesite, dolomite and calcite obtained by grain model[20] under the present experimental condition were 2.44×10^3, 1.25×10^4 and $1.21 \times 10^4 m^3 \cdot mol^{-1} \cdot sec^{-1}$, respectively.

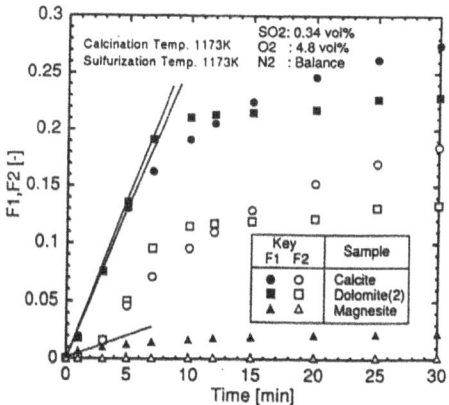

Figure 9   F1 and F2 plots

517

Revival effect of sulfurization reactivity by hydration/dehydration treatment.
Figure 10 shows the comparison of reactivity of calcined calcites treated with a hydration/dehydration procedure. As it appeared in Figs. 5(a), 5(b) and 5(c), the calcium oxide calcined at a higher temperature shows a lower reactivity for sulfation reaction. This decrease in the reactivity was explained by the decrease in the surface area and the pore volume. It can be seen in Fig. 10 that the calcium oxide treated with the hydration/dehydration procedure recovered its reactivity for sulfurization: a remarkable revival of the reactivity was recognized after the hydration/dehydration treatment had been conducted, although the surface area and the pore volume of the calcined sample were diminished in such a high temperature region by sintering. The extent of recovery of the reactivity was proportional to the water vapor pressure. The liquid phase-hydration showed the highest revival effect: a high water vapor concentration brings about a high fragmentation of CaO particles by the incerased hydration rate. Similar results are reported by Gullett et al.[7] and Bruce et al.[4] that the calcium oxide derived from Ca(OH)2 showed a higher reactivity for sulfurization than that obtained from the calcination of CaCO3.

Figure 11 shows the comparison of pore size distribution between calcine treated with hydration/dehydration and non-treated calcine. It is obvious in the figure that the pore volume of calcium oxide treated with hydration/dehydration, on the whole, was larger than that of calcined calcite. Moreover, it was found that the degree of decrease in pore volume caused by the plugging of pores during sulfurization was lower for calcine treated with hydration/dehydration than for that non-treated.

As the summary of the obtained experimental results, it is considered that the hydration/dehydration treatement may be one of the effective method for the restoration of pores by reforming the crystal structure of calcine.

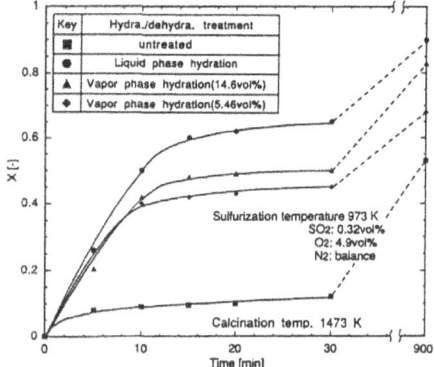
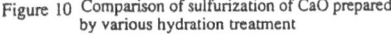
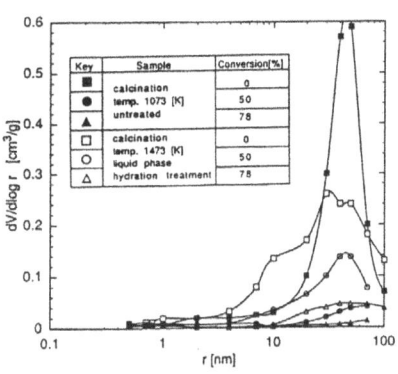

Figure 10 Comparison of sulfurization of CaO prepared by various hydration treatment

Figure 11 Pore size distribution change with the progress of sulfurization reaction

SO2-absorption capacity of MgO under pressurized condition
Table 3 shows the characteristics of magnesite sample used for sulfurization experiment with a fixed bed reactor. Figure 12 shows the pore size distribution of calcined magnesite. The pore size distribution of MgO produced from MgCO3 of 850-1190µm in diameter was similar to that produced from MgCO3 of less than 5 µm in diameter.

Effect of sulfurization temperature.　　　Figure 13 shows the time-change of SO2 concentration at the exit of reactor "breakthrough curve of SO2" for various temperatures under pressurized atmosphere. The reactivity of magnesite calcined at 1073K was higher than that calcined at 1173K. In our previous study of TGA under atmospheric pressure, it was found that the reactivity of calcined magnesite was the highest at the sulfurization temperature of about 1073K[22]. Even under the pressurized conditions around 101.3-303.9kPa, the reactivity of MgO showed the same tendency for sulfurization temperature similarly to TG experimental results .

TABLE 3

Characteristics of Calcined Samples employed for the experiment with fixed bed reactor

Sample	Calcination Temp.[K]	Atmosphere	S[m² · kg⁻¹]	V[m³ · kg⁻¹]	Av.pore radius[nm]
Raw	-	N₂	4.30×10^2	3.11×10^{-7}	14.5
MgO(1)	1073	N₂	7.98×10^4	2.34×10^{-4}	5.68
MgO(2)	1173	N₂	6.10×10^4	2.11×10^{-4}	6.90

Effect of pressure. Figure 14 shows the effect of total pressure of SO₂ ,O₂ and N₂ in the reactor on the breakthrough curve of SO₂. As seen from the figure, the desulfurization characteristics of calcined magnesite particle bed were increased with an increase in the pressure. This experimental trend may be interpreted by the fact that the reactant gas of SO₂ and O₂ can easily diffuse into the inner surface of porous specimen through the pores with the aid of a higher reactant gas pressure. According to the sulfurization experiment by Iisa et al.[13] using pressurized TG apparatus, the rate of reaction between CaCO₃ and SO₂ is slightly enhanced by the elevated pressure in the range of 0.91-2.5MPa.

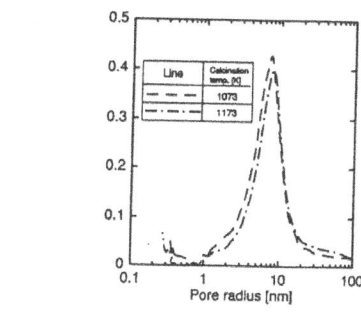

Figure 12 Pore size distribution of precalcined sample used fixed bed reactor

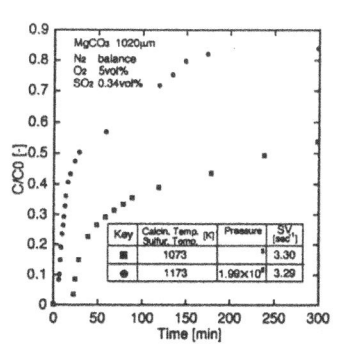

Figure 13 Break-through of SO₂ during sulfurization progress

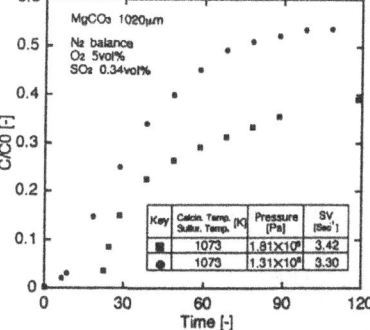

Figure 14 Effect of pressure on sulfurixzation

CONCLUSION

The sulfurization reactivity of calcined dolomite and calcite decreased with an increase in the calcination temperature. The order of magnitude of the reactivity of sulfurization was dolomite, calcite and magnesite. Magnesite calcined at 1273K had the highest reactivity despite the decrease in the surface area and pore volume with an increase in the calcination temperature. The sulfurization reactivity of calcined magnesite was decreased by the existence of water vapor involved in the reactant gas.

The difference in reaction characteristics of Ca-Mg based sorbents came not only from the intrinsic reaction rate but also from the reactant particle structure.

The hydration/dehydration treatement for calcium oxide showed a remarkable enhancement in sulfurization characteristics by reforming the crystal structure of CaO and by causing the fragmentation of particles.

Under the present experimental conditions, the desulfurization capacity of MgO particle bed was increased with an increase in the total pressure of reactant gas ranged from 101.3 to 303.9 kPa.

ACKNOWLEDGEMENT

This research was a part of the International Joint Research Program "INTERFLUID" supported by the New Energy and Industrial Technology Development Organization (NEDO, Japan).

NOMENCLATURE

C	: concentration of reactant gas	[vol%]
C_0	: initial concentration of reactant gas	[vol%]
F_1	: $(=1-(1-X))^{1/3})$	[-]
F_2	: $(=1-3(1-X)^{2/3}+2(1-X))$	[-]
k	: chemical reaction rate	$[m^4 \cdot mol^{-1} \cdot s^{-1}]$
M	: molecular weight	[kg]
P	: pressure	[Pa]
r	: pore radius	[nm]
r_g	: grain radius	[m]
S	: specific surface area	$[m^2 \cdot kg^{-1}]$
SV	: space velocity	$[s^{-1}]$
T	: temperature	[K]
V	: pore volume	$[m^2 \cdot kg^{-1}]$
W	: weight of sample	[kg]
W_0	: initial weight of sample	[kg]
X	: conversion	[-]
Y	: weight fraction of solid reactant in particle	[-]

REFERENCES

1) Borgwardt R.H. (1970): Kinetics of the reaction of SO2 with calcined limestone, Envi. Sci. & Tech., 4, 1, 59-63
2) Borgwardt R.H. and R.D.Harvey (1972): Properties of carbonate rocks related to SO2 reactivity, Envi. Sci. & Tech., 6, 4, 350-360
3) Borgwardt R.H. and K.R.Bruce (1986): Effect of specific surface area on the reactivity of CaO with SO2, AIChE J., 32, 2, 239-246
4) Bruce K.R., B.K.Gullett and L.O.Beach (1989): Comparative SO2 reactivity of CaO derived from CaCO3 and Ca(OH)2, AIChE J., 35, 1, 37-41
5) Centi G., N.Passarini, S.Perathoner and A.Riva (1992): Combined DeSOx/DeNOx reactions

on a copper on alumina sorbent-catalyst, Ind. Eng. Chem. Res., 31, 1947-1955

6) Dewing E.W. and F.D. Richardson (1959): Decomposition equiliblia for calcium and magnesium sulphates, Trans. Faraday Soc. , 55 , 611-615

7) Fan L.S., S. Satija, W.I.Wilson, D.C.Fee, K.M.Myles and I.Johnson (1984): Thermogravimetric analysis of sulfation kinetics of calcined limestones or dolomites, Chem. Eng. J., 28, 151-162

8) Gullett B.K. and K.R.Bruce (1987): Pore distribution changes of calcium-based sorbents reacting with sulfur dioxide, AIChE J., 33, 10, 1719-1726

9) Haul R.A.W. and J. Markus (1952): On the thermal decomposition of dolomite.IV. thermogravimetric investigation of the dolomite decomposition, J. Appl. Chem., 2, 298-306

10) Hartman M. and R.W.Coughlin (1976): Reaction of sulfur dioxide with limestone and the grain model, AIChE J., 22, 3, 490-498

11) Hartman M. (1978): Comparison of various carbonates as absorbents of sulfur dioxide from combustion gases, International Chemical Engineering, 18, 4, 712-717

12) Hasatani M., M.Yuzawa and N.Arai (1982): Reactivity of CaO produced by pyrolysis of micro-fine limestone particles with SO2, Kagaku Kogaku Ronbunshu, 8, 1, 45-50

13) Iisa K. and M.Hupa (1990): Sulfur absorption by limestone at pressurized fluidized bed conditions, The Combustion Institute, 943-948

14) Ishihara Y., C.Asagawa and H.Fukuzawa (1975): Studies on sulfur oxides removal from flue gas by dry limestone injection process (2), Journal of the fuel society of Japan, 54, 575, 175-183

15) Kiang K.D., K.Li and R.R. Rothfus (1976): Kinetics studies of sulfur dioxide adsorption by manganese dioxide, Envi. Sci. & Tech., 10, 9, 886-893

16) Kocaefe D., D.Karman and F.R.Steward (1985): Comparison of the sulfurization rates of calcium, magnesium and zinc oxides with SO2 and SO3, Can. J. of Chem. Eng., 63, 971-977

17) Kocaefe D.,D. Karman and F.R.Steward (1987): Interpretation of the sulfation rate of CaO, MgO and ZnO with SO2 and SO3, AIChE J., 33, 11, 1835-1843

18) Marier P. and H.P. Dibbs (1974): The catalytic conversion of SO2 to SO3 by fly ash and capture of SO2 and SO3 by CaO and MgO, Thermochem. Acta., 8, 155-165

19) Murthi K.S., D. Harrison and R.K.Chan (1971): Reaction of sulfur dioxide with calcined limestones and dolomites, Envi. Sci. & Tech., 5, 9, 776-781

20) Szekely J., J.W.Evans and H.Y.Sohn(1976): Gas-solid reactions, New York, Academic Press

21) Thibault J.D., F.R.Steward and D.M.Ruthven (1982): The kinetics of adsorption of SO3 in calcium and magnesium oxides, Can. J. of Chem. Eng., 60, 796-801

22) Watanabe T., H.Matsuda and M.Hasatani (1993): Reactivity of MgO calcined from magnesite particles with SO2, Kagaku Kogaku Ronbunshu, (in press)

23) Westmoreland P.R. and D.P.Harrison (1976): Evalution of candidate solids for high-temperature desulfurization of low-Btu gas, Envi. Sci. & Tech., 10, 7, 659-661

CLEANING OF HOT GASES FROM COAL GASIFIERS IN A DRY COMBINED PROCESS

Peter GÄNG and Friedrich LÖFFLER
Institut für Mechanische Verfahrenstechnik und Mechanik
Universität Karlsruhe (TH), Kaiserstr. 12, D-7500 Karlsruhe

ABSTRACT

The integrated gasification combined cycle (IGCC) is one of the most attractive options for electricity generation from coal with a high efficiency and relatively low environmental pollution. Thermodynamic calculations have shown that the application of a dry gas cleaning process at a high temperature level will improve the efficiency of such power plants by 2 to 3 % points compared to an installation with conventional wet gas cleaning |1|. The demands for a gas cleaning system are given by the required quality of the gasification gases, i. e. the fuel gas for the gas turbine. Thus, due to environmental regulations and for protection of the gas turbine and other components it is necessary to comprehensively reduce dust and gaseous pollutant, such as H_2S, COS, HCl, HF and especially alkali salt vapours, down to relatively low concentrations levels. An attractive concept for such a gas cleaning system is the use of cleanable filters for the simultaneous separation of dust and reactive gases in a combined dry-scrubbing process. In this process rigid cleanable filters will be used as high efficiency dust separators. Simultaneously the gases will react with dry, fine-grained sorbents being injected into the hot gas coming from the gasifier and also within the filter cake being formed on the filter's surface, i. e. the system represents a combination of an entrained flow reactor and a highly porous and efficient gas-solid fixed-bed reactor. The concept of this dry-scrubbing process and the scientific approach for the investigations being conducted in a feasibility study are discussed in this paper. In the filtration experiments a method is applied which is further developed from a test procedure and laboratory set-up recently proposed as a VDI standard for the testing and evaluation of cleanable filter media |2, 3|.

INTRODUCTION

The work described here is linked to the project "Pulse-jet Cleaning of Rigid Filter Elements at High Temperatures" also presented in this conference by S. Berbner and F. Löffler. The expected results are intended to be translated directly into the development of a dry cleaning process for hot gasifier gases. The save and reliable operation of cleanable filters at high temperature and high pressure conditions offers the opportunity to use their high dust collection efficiency and also the potential to remove gaseous pollutants by gas-solid reactions within the filter-cake and thus to clean coal derived fuel gas for gas turbines in a dry process.

Within the scope of this study theoretical and experimental investigations are done, as well as a comprehensive literature review, in order to clear up the feasibility of such a dry combined gas cleaning for an IGCC installation. IGCC power systems have been the subject of considerable research over the pas 15 years in the USA and in Europe, with the clear motivation of rising the efficiency of coal fired power plants to the 50 % level. In the IGCC process coal is gasified with air or oxygen and steam to produce a fuel gas at high temperature and high pressure conditions. The pressurized gas is fed to a gas turbine where it is combusted to produce electricity. The hot flue gases are subsequently used to generate steam for a conventional steam turbine. In conventional gasification processes the fuel gas is cooled down in a wet scrubbing process in order to remove dust, sulphur and other impurity, especially alkali-compounds.

The motivation for the use of a dry gas-cleaning system is the potential to further reduce the costs of electricity generation from coal and a significant lowering of CO_2 emission. Investigations on dry cleaning systems, especially for the removal of sulphur compounds and alkali vapours have aimed for a temperature range up to approx. 900 °C. A recent comprehensive study has shown, that the most benefit, i. e. a rise of 2 to 3 % in efficiency, will be gained by doing the step from wet-scrubbing to a dry gas cleaning process at approx. 350 °C |1|. Higher temperatures will result only in a slight rise in overall efficiency but will enhance the problems for the gas-cleaning considerably. Besides this the temperature of the fuel gas at the inlet of the burning chamber is also limited by the maximum inlet temperature for the gas turbine, for instance for low and medium calorific gas (3,000-25,000 kJ/kg) ABB limits the fuel gas temperature to 350 °C |4|. Taking into account these arguments, for IGCC the aim need not to be to clean the gases at temperatures as high as possible, but to optimize the filtration and cleaning conditions of the filter with respect to gas reduction and fuel gas properties. This could mean for example, to operate a rigid candle filter in the temperature range from 350 to 500 °C where adhesion and sintering behaviour of the dust are less critical and alkali vapour pressure is low enough to reach turbine limits without applying high temperature gettering-reactions. **Fig. 1** shows a basic schematic of an IGCC power plant installation with a conventional wet gas cleaning process, marked by the grey background. There is also included an air-separation process with the option to feed nitrogen to the fuel gas or into the burning chamber, in order to enhance the mass flow of the gas and to control the flue gas temperature.

The schematic of a basic proposal for a dry-scrubbing system, intended to substitute the grey marked area in Fig. 1, is drawn in **Fig. 2**. The temperatures for the filter

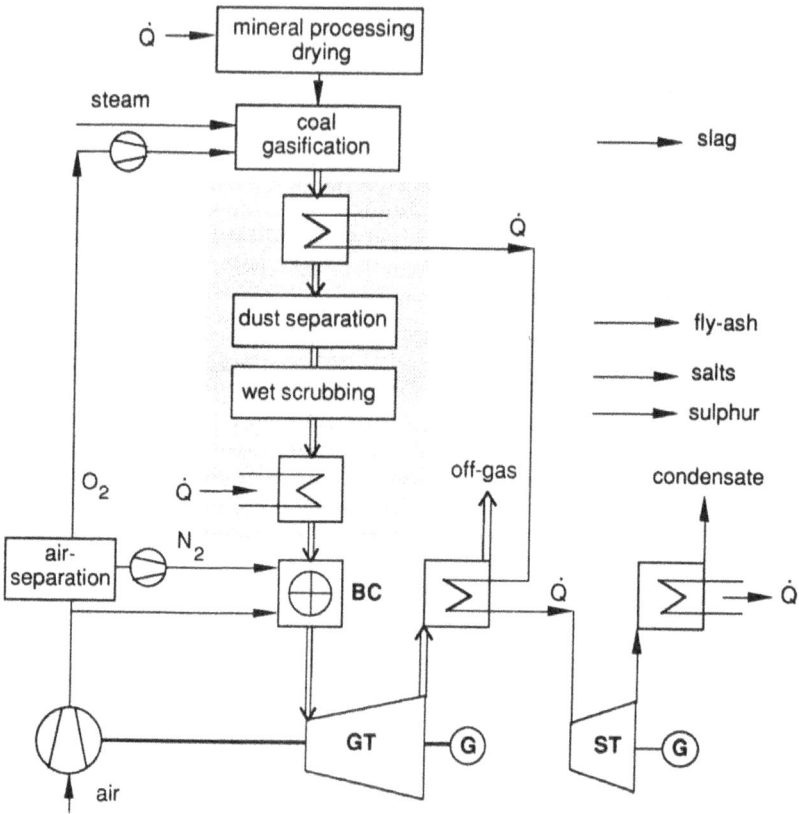

FIGURE 1 Basic schematic of an IGCC-installation; BC: burning-chamber;
GT: gas turbine; ST: steam-turbine; G: electricity-generator

unit, considered to be of any relevance for this study, range from 300 to a maximum
of 800 °C, the maximum temperatures applied for reactions in the entrained flow
can be higher, approx. from 800-1000 °C. The sorbents are intended to be injected in
the appropriate temperature windows for the specific gaseous components to be
scrubbed. The gas-solid reactions take place in the entrained flow as well as within
the filter cake. In the case of a one-stage filter-unit, the collected dust - a mixture of
gasifier dust, reacted and unreacted sorbents and condensation aerosols - could be
partly redispersed in the raw-gas, in order to enhance the conversion of the sorbent
and possibly to improve the filtration behaviour. The spent dust will probably have
to undergo special treatment before disposal.

The general goal of the currant investigation is a dry, one- or two-staged process, for
the combined separation of particulates and gases from coal derived fuel gases for
gas turbines. The aim of this specific study is the collection and generation of data
which give information about the feasibility of the concept and which can serve as a
sound basis for the further planning of the further leading project.

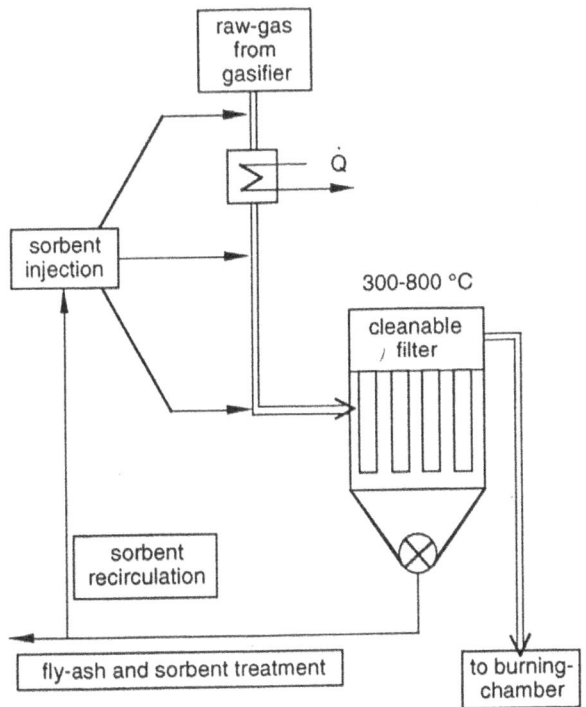

FIGURE 2 Basic schematic of a proposal for a dry-scrubbing system, using reactions in the entrained flow and a cleanable filter as dust collector and gas-solid reactor

SEPARATION TASKS AND PRINCIPLES

Starting points for this project are, on the one side, the gas compositions depending on the different main gasification processes currently applied, with very different gasification conditions (see **Fig. 3**) using a wide variety of coals. On the other side stands the fuel gas quality required from gas turbine manufacturers which defines the necessary reduction efficiency of the gas cleaning system.

In **Table 1** data on the composition of different gasifier gases are put together from different literature sources. A rather big difference is noticed between gasifiers operated with air and with oxygen. But also within one group the values for different pollutants vary considerably. The dust concentrations encountered depend from the pre-separator for particulates, in general a cyclone. The values given in literature range from some g/m^3 up to some hundred g/m^3.

In **Table 2** tolerable values for gas turbines given by different manufactures are shown. Although the values differ in a rather wide range there seems to be a tendency towards very low limits for trace metals. Regarding the high contents of H$_2$S and COS, sulphur removal is mandatory, - i.) in order to meet environmental legis-

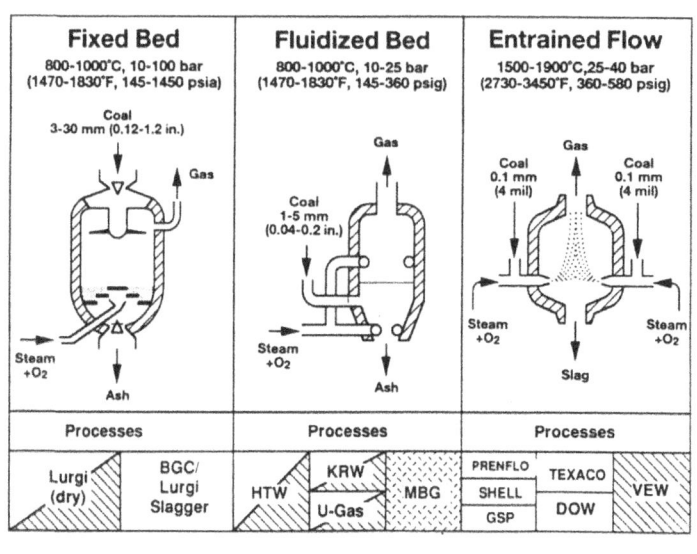

Fixed Bed	Fluidized Bed	Entrained Flow
800-1000°C, 10-100 bar (1470-1830°F, 145-1450 psia)	800-1000°C, 10-25 bar (1470-1830°F, 145-360 psig)	1500-1900°C, 25-40 bar (2730-3450°F, 360-580 psig)

Processes	Processes	Processes
Lurgi (dry) — BGC/ Lurgi Slagger	HTW — KRW / MBG / U-Gas	PRENFLO — TEXACO / SHELL / GSP — DOW — VEW

☐ Auto-Thermal Gasification with Oxygen ▨ Auto-Thermal Gasification with Air ▦ Allo-Thermal Steam Gasification

FIGURE 3   Basic principles of main coal gasification processes, taken from |6|

lative standards and - ii.) to protect the gas turbine blades from high temperature corrosion caused by molten layers of alkali sulphates |5|. Thus the limits for the sulphur contents are also linked to the alkali concentration.

Damage caused by HCl seems to be not clear but emission is limited by environmental legislation in various countries. NH_3 and HCN cause NO_x emission in the gas turbine flue gas and might be a a environmental problem.

Thus the main components to be separated from the gas stream in the discussed dry scrubbing process are:

solids:
- gasifier dust after cyclone
- particulate reaction products
- unreacted sorbents
- condensation aerosols (e. g. NaCl and KCl)

gaseous:
- H_2S and COS
- HCl and HF
- NaCl and KCl
- (NH_3 and HCN)

The basic principle for the reduction of a gas using solid sorbents is to choose a reaction system with a low equilibrium partial pressure of the specific gas component at

526

TABLE 1						
Composition of coal gases from different sources						
	NCB |7, 8|	U-Gas |9|	GSP |10|	BGL |11|	Prenflo |12|	Shell |13|
Component	Gasification with air		Gasification with oxygen			
CO/% vol	17.9-18.0	17.8-24.3	58.5-59.1	54.0-56.6	59.6-62.3	65
H_2/% vol	10.0-10.3	12.3-13.2	21.9-25.5	28.5-29.4	26.0-27.7	30
CO_2/% vol	8.9-9.0	5.2-8.3	4.0-4.2	3.9-5.4	3.4-3.7	1.6
H_2O/% vol	3.3-3.8	5.3-10.9			2	0.4
N_2/% vol	57.8	48.3-49.9	10.2-12.9	2.8-2.9	5.9-10.0	3.1
H_2S/% vol	0.02-0.1		0.3-0.78	0.6-0.7	0.31-1.3	0.3
COS/vpm			400-800	(H_2S+COS)		
HCl/vpm	300-1000			500		
NaCl/vpm	8.6 vpm					
KCl/vpm	1.7 vpm					
Dust/g/m^3	4.1	up to 30 g/m^3 (i. N.) ≈ 330 g/m^3 (350 °C, 25 bar)				

NCB: National Coal Board, UK; U-Gas: Tampella Power Inc.; GSP: Deutsche Babcock; BGL: British Gas/Lurgi; Prenflo: Krupp Koppers

TABLE 2							
Manufactures' tolerable values for gas turbine limits from different sources							
McLaughlin 1991 |7|						
Turbine manufacturer	Turbine inlet temperature °C	Na + K ppm	Ca ppm	Sulphur			
United Aircraft	1050-1110	0.2-0.6	0.1	0.18 mol%			
Pratt and Witney		0.1	0.01	0.18 mol% 0.25 H_2S			
Westinghouse	1080	0.07-0.15	1.4	0.28 % (wt)			
General Electric	1050-1110	0.1	0.23	1 ppm			
Company guidelines	Na+K	Va	Pb	Zn	Ca	Dust	H_2S
Siemens 1992 / ppb	10	10	20	40	200	< 5 µm 2000	30 ppm
ABB 1992 / ppb	500	1000	sum 1000		2000	20000	3000 vpm

the operational conditions given and to form a solid reaction-product with a low vapour pressure. If the reactions take place in the entrained flow with fine particulates, the reaction product has to be collected in a high efficient filter.

Considering the required fuel-gas quality for gas turbines, environmental regulations and the unalterable necessity to guarantee a save and continuous operation in the power plant, the following tasks have to be solved:

(i) find appropriate sorbents and reaction conditions to reduce the concentration of gaseous pollutants to values below the limits, given for gas turbines and by environmental legislation,

(ii) reduce particulate concentration to values below the limits for gas turbines,

(iii) guarantee the cleanability and a long service life of the filter elements and a high availability of the whole unit,

(iv) find appropriate filtration conditions in order to fulfill (iii) by taking into account the adhesion properties of the dust mixture and the possible sintering of the filter cake.

REDUCTION OF GASEOUS COMPONENTS

Sulphur Compounds

Although rather high concentrations of sulphur compounds seem to be tolerable to gas turbines (see Table 2) H_2S and COS need to be separated, in order to meet environmental legislation.

For dry scrubbing many different sorbents have been investigated by several working groups. Taking into account these results, sorbents containing the oxides of iron, zinc, titanium and calcium seem to be most promising.

In Japan iron oxide is mounted in titanium oxide honeycomb structures and operated in fixed beds between adsorption, regeneration and reduction |14, 15|. The reaction principles are as follows:

Adsorption: $Fe_3O_4(s) + 3 H_2S(g) + H_2(g)$ (CO) -----> $3 FeS(s) + 4 H_2O(g)$ (CO_2)

Regeneration: $4 FeS(s) + 7 O_2(g)$ ----> $2 Fe_2O_3(s) + 4 SO_2(g)$

Reduction: $Fe_2O_3(s) + H_2(g)$ (CO) ----> $Fe_3O_4(s) + H_2O(g)$ (CO_2)

The process works at 400 °C while regeneration is done with diluted air at 1.5 % oxygen. Iron oxide is limited in operating temperature to about 450 °C |16, 17, 18|. Higher temperature results in reduced desulphurization and the formation of carbon in high severity reducing gases with steam less than 15 %.

Higher sulphur removal and better structural stability was achieved with zinc ferrite sorbents. The sorbent preparation consists of high temperature calcination of stoichiometric amounts of oxides at 870 °C with small amounts of binder (e. g. bentonite). The suitable operating range for zinc ferrite was 500 to 650 °C in low-calorific gases with 15 % steam. The temperature range is limited because of potential reduction of Fe_2O_3 to FeO and Fe which results: -i.) in carbon formation and severe decrepitation of the extruded sorbent and -ii) in reduction of ZnO and evaporating of Zn. The overall reactions for zinc ferrite would write:

Adsorption: $ZnFe_2O_4(s) + 3 H_2S(g) + H_2$ -----> $ZnS(s) + 2 FeS(s) + 4 H_2O(g)$

Regeneration: $ZnS(s) + 2 FeS(s) + 5 O_2(g)$ (725-760 °C) -----> $ZnFe_2O_4(s) + 3 SO_2(g)$

For applications at higher temperatures up to 735 °C or in the presence of reducing gases containing low level of steam (5 %), zinc titanate exhibited good sulphur retention capacity and stability. The sorbent is made up of a mixture of ZnO and TiO_2 at a molar ratio of 0.8 to 2 with best results for the 1.5 ratio. Although the titanium dioxide does not take part in reaction, it seems to stabilize the zinc and prevents its vaporization.

Zinc titanate is also employed by Tampella Power Inc. [9, 19] as a post-bed desulphurization (polishing) of the fuel gas, which is precleaned by in-situ desulphurization using dolomite ($CaCO_3 \cdot MgCO_3$). In the precleaning about 90 % sulphur reduction can be reached. In this process the dolomite calcines to CaO and MgO but only CaO reacts with H_2S. Although MgO does not react it provides a more open pore structure compared to pure limestone. Thus the conversion of the calcium in the dolomite can reach 90 %. The remaining H_2S is reduced in fluidized beds of zinc titanate at approx. 550 °C. The regeneration is carried out with oxygen between 720 and 760 °C.

A variety of other metal oxides, for example CuO, $CuO-Fe_2O_3$, $CuO-Al_2O_3$, MnO, SnO were investigated, aiming for better stability, more capacity or more cost effective schemes for the sorption beds.

Attractive sorbents for high temperature application are Ca-compounds and dolomite. A comparison of CaO, ZnO and Fe_2O_3 [20] showed, that Fe_2O_3 had the largest capacity for sulphur per gram followed by CaO and ZnO. The trends with temperature show in the case of CaO a gradual increase in capacity from 600 to 800 °C, where CaO appears to be the best. During the early stages of reaction relatively large quantities of SO_2 were observed in the exiting gas, probably due to the reaction:

$$2 \, Fe_2O_3(s) + H_2S(g) \; \dashrightarrow \; 4 \, FeO(s) + H_2(g) + SO_2(g)$$

which indicates, that also in the case of Fe_2O_3 it is the FeO species which is effectively binding the H_2S at high temperatures. This SO_2 formation was not observed for ZnO and CaO.

Limestone and dolomite are the cheapest sorbents to be used at high temperature. If the partial pressure of CO_2 in the gas is low enough and for temperatures high enough, $CaCO_3$ decomposes to CaO and CO_2:

$$CaCO_3(s) \; \dashrightarrow \; CaO(s) + CO_2(g)$$

CaO reacts with H_2S according to:

$$CaO(s) + H_2S(g) \; \dashrightarrow \; CaS(s) + H_2O(g)$$

The direct sulfidation of $CaCO_3$, at temperatures where the equilibrium partial pressure of CO_2 due to the calcination reaction is smaller than the partial pressure of CO_2 in the gas, is thermodynamically still possible via reaction:

$$CaCO_3(s) + H_2S(g) \; \dashrightarrow \; CaS(s) + H_2O(g) + CO_2(g)$$

Borgwardt et al. [21, 22, 23] studied the sulfidation of CaO and $CaCO_3$ in the temperature range from 550 to 900 °C. The reactions of fresh calcined CaO with H_2S and COS at 700 °C are faster compared to the very intensively investigated sulfation with SO_2, the conversion of the solid is almost complete, probably due to less plugging of pores by the sulphide than by the sulphate. The reactive rate is very much dependant on temperature and the specific surface area of the sorbent, which is decisively influenced by the sintering of the particles.

The rate and conversion of $CaCO_3$ directly reacting with H_2S is very much dependent on temperature. At low temperatures (approx. < 600 °C) reaction rate is slow and conversion is low (approx. < 30 %). Above 700 °C the reaction is rapid and conversion is high, e. g. at 750 °C the conversion of 1.6 μm limestone particles with 5000 ppm H_2S was 70 % after 7 minutes.

Hydrogen Chloride Reduction

U. S. coals contain up to 0.5 % of chlorine, some coals in the U. K. as much as 1 %. Brown coal can have much higher chlorine concentration of up to several percent. Concentrations of HCl vapour in coal gas have been found in the range up to 1000 ppm (see Tab. 1). The removal of HCl from coal gas was investigated from 550 to 650 °C by Krishnan et al. |24| using alkali and alkaline earth compounds. It was found that the reaction is rapid and the HCl vapour concentration can be reduced to about 1 ppm. The presence of H_2S and trace metal impurities did not affect significantly the performance of the sorbent.

The thermodynamic stability and the volatility of the solid chloride products are usually the mayor criteria in selecting the sorbent for high temperature.

For the reaction: $Na_2CO_3(s) + 2 HCl(g)$ -----> $2 NaCl(s) + H_2O(g) + CO_2(g)$

the equilibrium partial pressure of HCl is 1.5×10^{-6} atm, the equilibrium vapour pressure of NaCl is 3.5×10^{-6} atm at 900 K.

For the reaction: $CaCO_3(s) + 2 HCl(g)$ -----> $CaCl_2(s) + H_2O(g) + CO_2(g)$

the corresponding values are 9.5×10^{-4} atm for HCl(g) ($\approx$ 950 ppm at 1 atm) and 2.8×10^{-9} atm for $CaCl_2(s)$ |24|. That means at 900 K limestone is not a suitable sorbent because of high HCl partial pressure. Sodium carbonate on the other hand exhibits high NaCl vapour pressure that might cause a problem for the gas turbines.

This problems could be avoided by operating the dust filter (see Fig. 1 and 2) at around 350 °C since partial pressures are low at this temperature and reaction with HCl will still takes place |25, 26|. The same also applies to the reduction of hydrogen fluoride (HF) due to its very high reactivity and the stability of the product.

Nitrogen Compounds

Fuel bound nitrogen in coal is partly released as NH_3 and HCN which might enhance the formation of nitrogen oxide in the gas turbine combustor. Although NO_x can be controlled during combustion, additional measures might be necessary to reduce high contents of NH_3 and HCN. Results by Leppalathi et al. |27| suggest that it should be possible to decompose NH_3 by contact with iron based minerals at temperatures > 900 °C or with dolomite at temperatures > 1000 °C. It was also found that HCN concentration could be reduced with dolomite and limestone at temperatures > 800 °C.

Alkali Salt Vapour

The vapour concentration of alkali chlorides increase with an increasing coal chlorine content. The total amount of alkali chloride in the coal gas, condensed and vapour phase, depends also on the gasification conditions (see Fig. 3), e. g. the gasification temperature. The alkali concentration found in the coal gas are around 10 ppm which is in the order of 100 to 1000 times higher than the limits for gas turbines. At temperatures around 900 °C alkali vapour can be reduced by high temperature reactions with alumino silicates, represented by the following equation, with x = 2 to 6 and M = Na or K:

$$2 \, MCl(g) + H_2O(g) + Al_2O_3 \cdot x \, SiO_2(s) \longrightarrow M_2O \cdot Al_2O_3 \cdot x \, SiO_2(s) + 2 \, HCl$$

A considerable number of investigators | 7, 28, 29, 30, 31 | have covered reactions with these and other "getter"-materials, mostly aiming for a cleaning process using fixed bed adsorbers. These studies suggest that, depending on the HCl content, the coal gas has to cooled to 800-900 °C when high temperature alkali reactions are applied. If not, the gas has to be cooled to approx. 600 °C in order to reach turbine limits by condensation of the salt vapours and collection of the particles. Thus if the gas filtration is taking place between 350 and 500 °C, as suggested above, alkali salt vapours should not be a problem, if a highly efficient cleanable filter is employed.

SCIENTIFIC APPROACH AND EXPERIMENTAL PROCEDURES

In the first phase of this feasibility study different methods are employed to get information about the best conditions for the **reactions of gases** using specific sorbents and also about the behaviour of the different dusts and dust mixtures during **dust collection** with rigid ceramic filter media and during **filter cleaning**.

In the field of gas separation a thermobalance and a differential tube reactor are used to investigate the reactions of HCl and H_2S with pre-selected sorbents (see **Figures 4** and **5**). Both set-ups are equipped to prepare appropriate gas mixtures using CO, CO_2, HCl, H_2S, N_2, O_2 and water vapour. HCl (and mixtures with H_2S) are only used in the quartz reactor to prevent corrosion problems in the thermobalance. With these methods data about the reaction rates, the conversion of the solids and the influences of parameters as temperature and gas concentration can be gained. These data are needed to be able to model the gas reduction in the filter cake an in the entrained flow. The samples are prepared by dispersing the pure sorbents and their mixtures with gasifier fly ashes using a dust feeder and collecting them on absolute filters made from quartz fibres. The samples can be cut from these filters. Thus the reacted samples can be extracted easily from the thermobalance or from the tube reactor and be examined in a heating microscope and in the SEM to get knowledge about the **sintering** behaviour of the dusts. This is an essential information in order to evaluate the dust properties and the influence of gas solid reactions (with the particles in the filter cake an in the entrained flow) with respect to the cleanability of filter media (e. g. rigid candle filters) to be used in a combined gas-solid separation process.

For filtration experiments including the **cleaning of the filters**, i. e. the detachment of the filter cake from the filters surface, a new experimental set-up and test-proce-

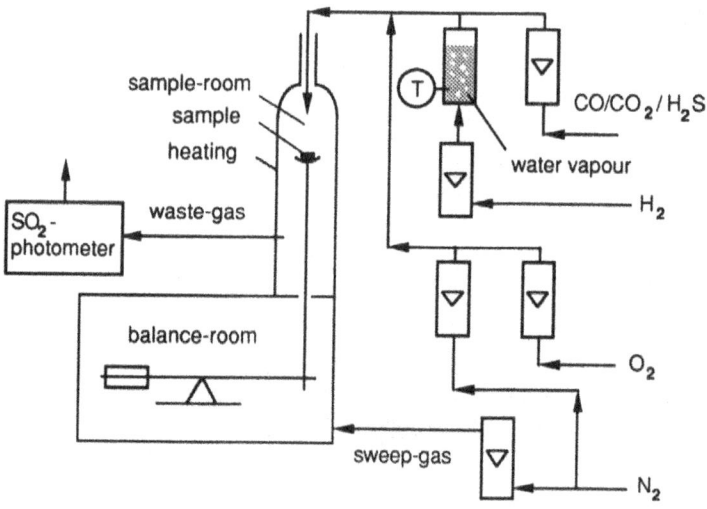

FIGURE 4   Thermobalance set-up with supply of reactive test-gas

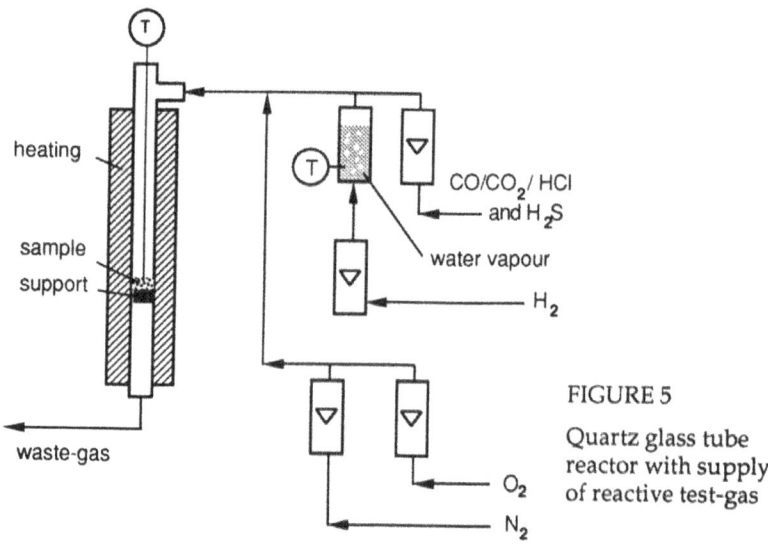

FIGURE 5

Quartz glass tube reactor with supply of reactive test-gas

dure is employed, which has been proposed as VDI standard 3926 |2, 3|. The apparatus offers the opportunity of realizing filter operation incorporating both filtration and pressure-pulse cleaning cycles and is designed to automatically execute any number of cleaning cycles in non-stop tests. In order to duplicate the conditions encountered around bags in industrial filters, a cross-flow arrangement for the filter sample was chosen.

Due to the design of the apparatus, of which a scale drawing can be seen in **Fig. 6**, various test-parameters such as filter-face velocity, raw-gas concentration, test-dust properties and cleaning conditions are both explicite and easy to change. This allows different filter media to be compared and assessed under definite conditions very similar to those encountered in a pulse-jet filter. For the measurement of fractional collection efficiencies and the particle penetration during the cleaning pulse the apparatus is additionally equipped with a sophisticated and fast measuring

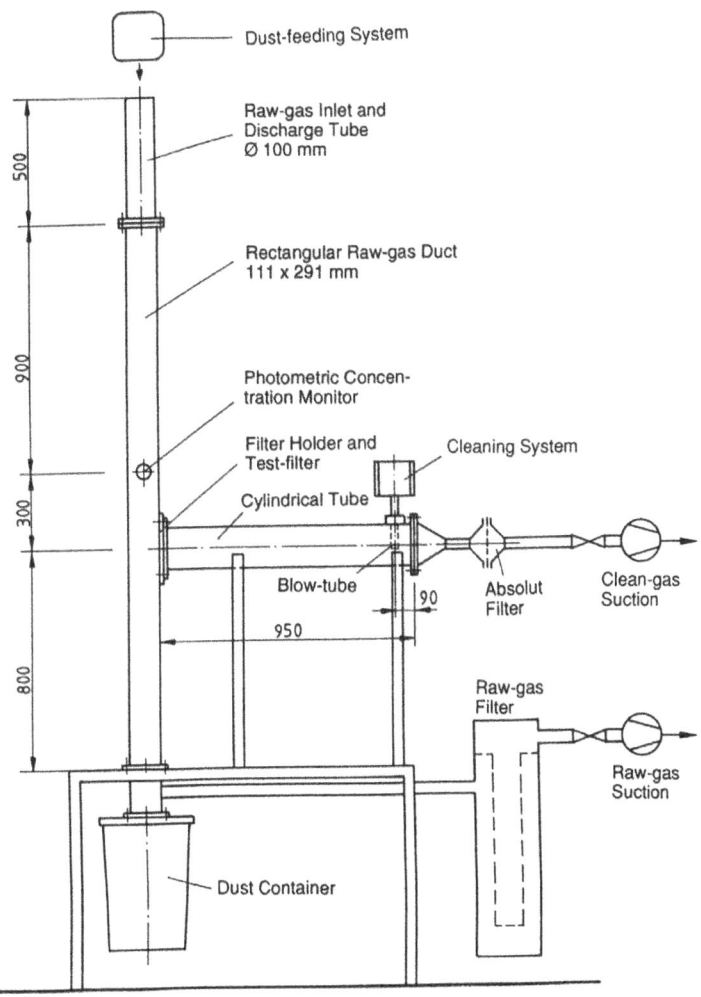

FIGURE 6   Laboratory filter test-rig proposed in VDI-standard 3926

scattered light particle size analyser. In addition to the measurement of the filter media properties, the application of rapid pressure transducers gives access to the determination of the separating forces required for filter-cake detachment and hence, the cake adhesion. Such data is vital for an efficient cleaning system design and operation.

The apparatus basically consists of a rectangular, vertical raw-gas duct, a cylindrical, horizontal suction unit incorporating the sample of the fabric to be tested (filtering area = 0.0154 m^2; Ø 140 mm). The sample is clamped in an aluminum filter holder which can be weighed without having to remove the sample. The cleaning-system consists of a 2.5 litre pressure tank, a quick acting diaphragm valve and a blow-tube. At the top of the raw-gas inlet a continuously working dust feeder provides a particle laden gas-stream (approx. 4.5 m^3/h) which is mixed in the cylindrical inlet-tube with additional atmospheric air, drawn into the apparatus. Inside the inlet-tube the raw-gas (5.5 m^3/h, dust concentration 5 g/m^3) passes a 3.7 MBq β–source, in order to discharge the particles, and is subsequently sucked through the rectangular duct, whereby a certain proportion is drawn through the test filter in a cross-flow filtration manner. The gas in the horizontal duct is subsequently led through a total filter in order to gravimetrically measure the particle concentration in the clean-gas. The rest of the uncleaned raw-gas leaves the apparatus at the bottom, whereby it is partially cleaned by the raw-gas exit, working as a inertial separator, to be finally cleaning by a back-up filter. Directly above the test filter the raw-gas dust concentration is continuously monitored by a photometric device which measures the extinction of the light of a laser diod.

In **Fig. 7** exemplary results of two filtration experiments using this apparatus are shown. The filtration and cleaning conditions were kept constant, the filter-face velocity was chosen at 5 cm/s, the cleaning was triggered at a pressure loss of 1000 Pa. Test dust was a very fine limestone fraction with a mass medium diameter of approx. 3.2 μm (see **Figs. 8** and **9**). In the plots differential pressure curves as functions of the cleaning cycle-number and the operating time may be seen, which includes the trends for the cycles 0 to 20, 50 to 60 and 90 to 100. The upper plot displays, what one would call a regular behaviour for a cleanable filter. Starting from a basic differential pressure for the clean filter, the residual pressure drop after cleaning rises slowly and should remain constant for an extensive period. Following a slight change during the first 10 to 20 cycles, the cycle duration remains constant with an almost linear pressure rise with each cycle, i. e. with the increase in the areal dust mass on the filter. This means, that after the filter cake is detached, the filter surface is rapidly coated and a new filter cake is built. Hardly any deep-bed filtration exists after cleaning. In the lower plot, especially during the first cycles, the sample reveals distinct deep-bed filter characteristics with the differential pressure rising progressively with a concave curve shape. In this case the filter surface was not thoroughly cleaned and retained a crust of dust which could not be removed, leading to a significant decrease in the filtration cycle-time from approx. 20 minutes in the beginning, to less than 3 minutes for the 100th cycle. As a result, the overall time for the experiment for the lower graph was less than half of that for the surface-treated medium. A close analysis reveals that the characteristics of the curve change during the course of the experiment, showing a linear differential pressure rise after the first 20 to

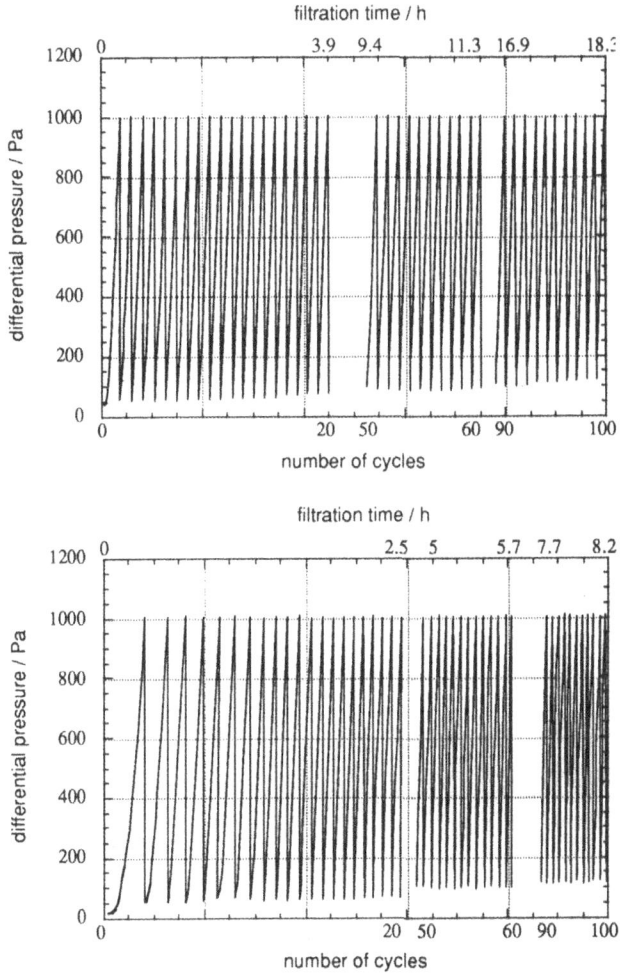

FIGURE 7   Characteristic pressure loss signal for filter samples with satisfactory
            filtration behaviour (top) and insufficient cleaning (bottom)

30 cycles and even developing a convex trend for higher cycle numbers. This is probably due to an incomplete cleaning of the filters surface and the formation of
cracks in the undetached filter cake. These cracks would indeed cause a low initial
residual pressure drop, to be closed immediately by fresh collected dust, leading to
a sharp initial differential pressure rise which becomes linear during progressive filter cake formation. The residual dust mass after 100 filtration cycles was 1.4 g for
the first filter and 8.9 g for the second one, mainly due to dust which could not be
removed from the filters surface. Besides the characteristic of the pressure loss curves, the residual pressure loss, the residual dust mass and the development of filtra-

tion cycle length, the clean gas dust concentration is additionally determined gravimetrically, in order to evaluate the filtration behaviour of a filter medium or a dust.

For **hot gas filtration experiments** a new test rig will be used, incorporating both, a single filter element (e. g. rigid candle) and the cross-flow arrangement employed in the VDI set-up. The apparatus which is introduced in this conference by S. Berbner and F. Löffler, is working in the temperature range up to 1000 °C. It is made up from SiC and will be equipped to handle gases as HCl, H_2S and SO_2. In this hot gas rig filtration tests will be done only with pre-selected test dusts, in order to determine suitable filtration conditions, taking into account the reaction of gases and the dust sintering.

PRELIMINARY RESULTS

Filtration

For the filtration experiments three different filter dusts from coal gasifiers ("GT", "BT" and "WT") and a well known fine limestone test-dust ("MC") were used. As can be seen in Figures 8 and 9 all of the gasifier dusts are coarser than the limestone test-dust, whereas two of them exhibit a bimodal particle size distribution.

The characteristic data extracted from filtration tests like those introduced in Fig. 7, are shown in **Figures 10** to **12**. The experiments were done with a rigid fibre ceramic at the same conditions as described above except of the dust concentration which was 2 g/m^3. The cleaning pulse was always triggered at a pressure drop of 1000 Pa across the filter. Looking at the development of cycle duration in Fig. 10 it is obvious that all three gasifier dusts are causing a very fast rise in pressure drop compared to the even finer limestone dust. The trends in Figures 11 and 12 suggest that the two bimodal dusts are behaving in a similar way. Filter dust "BT" develops the highest residual pressure drop although it is much coarser than the limestone. The trends for the rise of the residual areal dust mass of the two monomodal dusts seem to be very similar. A comparison of measured cleaning efficiencies show that in order to detach a filter ash dust cake, a considerable higher pressure pulse has to be exerted than for limestone, in order to reach the same efficiency.

Other experiments with dust mixtures are underway in order to find solutions to improve filtration behaviour. This might be possible by blending the dusts with potential sorbents, as will be the case in a combined separation of gases and dust. An other possible approach which is pursued is the pre-coating of the filter with an appropriate material.

Reactions of Gases and Sintering of the Dusts

As first results from work being done with the thermobalance and with the tube reactor the formation of considerable amounts of soot, as mentioned in |16|, can be confirmed for the reaction of H_2S with iron oxide and zinc ferrite, under reducing conditions at 500 °C and with 2 % water vapour. It was also observed that in the first reaction step at 500 °C Fe_2O_3 seems to be reduced to FeO before reacting with H_2S which causes an initial production of considerable amounts of SO_2, as already mentioned in |20|. For the reaction of H_2S with limestone it seems to be necessary

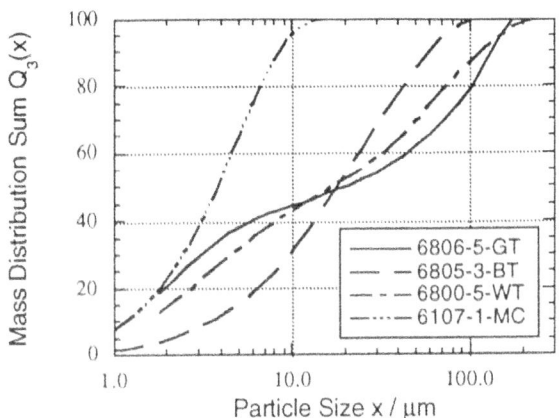

FIGURE 8   Mass distribution sum of three different filter dusts from coal gasifiers
GT, BT, WT and limestone test-dust MC

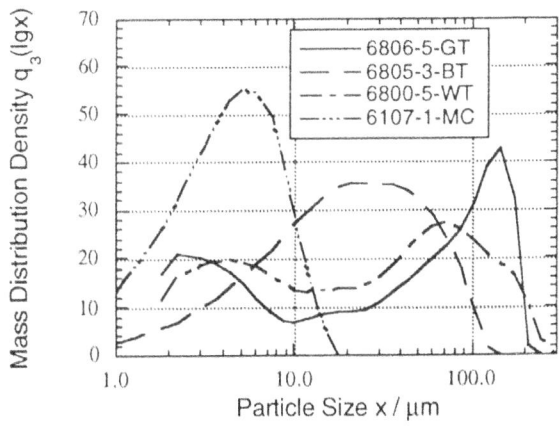

FIGURE 9   Mass distribution density of three different filter dusts from coal gasifiers GT, BT, WT and limestone test-dust MC

to reach calcination temperature, i. e. the CO_2 partial pressure due to calcination
must exceed the pressure in the gas, in order to realize a high reaction rate and conversion of the solid. Thermodynamic calculations confirm that approx. 600 °C have
to be reached at least, in order to get any appreciable reaction with limestone. CaO
will react much better than limestone but will preferably react to carbonate if temperature is too low and high CO_2-concentrations are present.

Filter cakes from mixtures of fine limestone with filter ashes from gasifiers (see Figs.
8 and 9) were also reacted with H_2S and HCl at different temperatures. The results

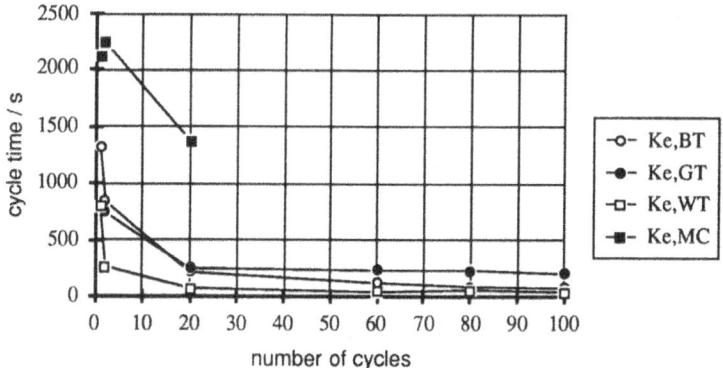

FIGURE 10   Development of cycle duration measured with filter Ke for four dusts

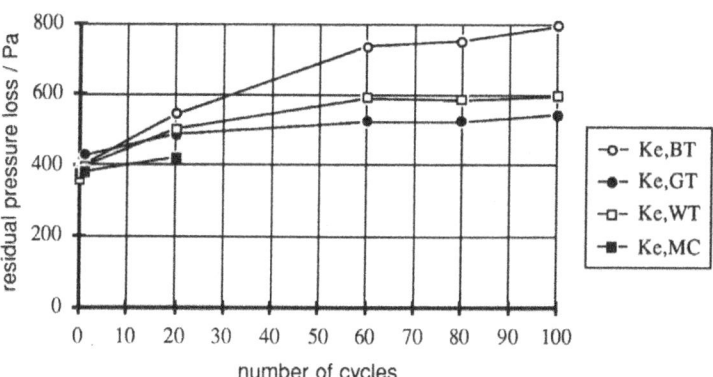

FIGURE 11   Residual pressure loss trends measured with filter Ke for four dusts

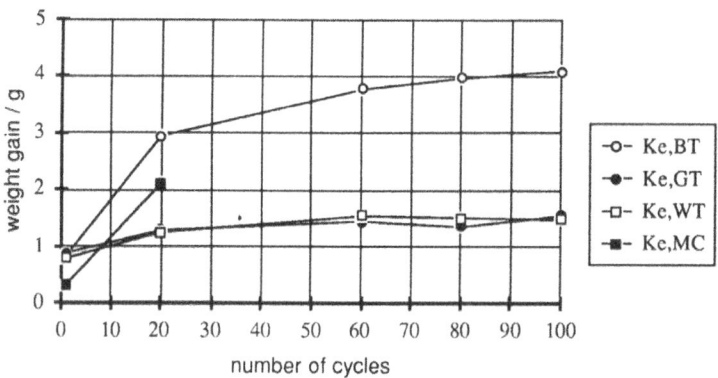

FIGURE 12   Development of weight gain measured with filter Ke for four dusts

derived at 500 and 750 °C show no noticeable sintering if only H_2S is present. With HCl in the gas a definite sintering of the filter cakes at both temperatures can be detected in the SEM.

This means it has to be taken into account, that enhanced adhesion of the filter cake will be encountered by using Ca-compounds as a sorbent, even at a temperature as low as 500 °C if HCl is present in the gas. If this is detrimental for the application of a cleanable filter is not quite clear and depends also on the amount of HCl and the composition and amount of the solid substances.

The ongoing investigations are aiming for further information about the optimal conditions and temperature windows for the injection of selected sorbents, as well as about filtration and cleaning behaviour of the dust mixtures, taking into account the influence of gas-solid reactions on the adhesion of the filter cakes.

ACKNOWLEDGEMENT

This project, running from June 1992 to May 1994, is being financed by "Technische Vereinigung der Großkraftwerksbetreiber e.V. (VGB)", Essen, Germany. The authors wish to thank this organisation for its support.

REFERENCES

| 1 | NOVEM, Netherlands agency for energy and the environment, High Temperature Gas Cleaning at IGCC Systems; System Study by ECN, KEMA, Stork Boilers and TNO as part of the Netherlands Clean Coal Programme, Novem, Swentiboldstraat 21, P.O. Box 17, NL - 6130 A A Sittard

| 2 | VDI Richtlinie 3926 (Entwurf); Prüfung von Filtermedien, Blatt 1, Teil 2: Prüfung von Filtermedien unter anwendungstechnischen Bedingungen, VDI Handbuch Reinhaltung der Luft, Band 6, Beuth Verlag, Berlin, 1992

| 3 | Gäng P., Löffler F.; New Procedure and Test-rig for the Characterization of Cleanable Filter Media, 9th World Clean Air Congress, Montreal, Canada, Aug. 30th-Sept. 4th, 1992

| 4 | ABB Fuel Specifications (VGT 11491 sales documentation); Fuels for ABB Gas Turbines with Standard Combustor Specification (Doc. No. HTCT71283)

| 5 | Schmitz F.; Hochtemperaturkorrosion von Schaufeln stationärer Gasturbinen durch Heißgasverunreinigungen (Siemens AG), GVC-Fachausschuß Energieverfahrenstechnik, Würzburg, Nov. 28th 1991

| 6 | Haupt G.; Status and Development of Integrated Coal Gasification Combined Cycle, COMETT Course "Pollutants Removal in Power Plants", Kernforschungszentrum Karlsruhe, June 22th-24th 1992

| 7 | McLaughlin J.; The Removal of Volatile Alkali Salt Vapours from Hot Coal-derived Gases, PhD thesis November 1990, Department of Chemical and Process Engineering, University of Surrey, Guildford

| 8 | Reed G. P.; The Economics of Hot Gas Cleaning for Coal Gasification Combined Cycle Power Generation, Filtech Conference, 1983

| 9 | Mojtahedi W., Horvath A., Salo K.; An Environmental Sound Scheme for Energy Production: Tampella´s IGCC Process, Intern. Symposium on Energy and Environment, August 25th-28th, 1991, Espoo, Finnland

| 10 | Meier H. J., Jelich W.; GSP-Verfahren - Stand und Pläne, Mitteilung Nr. 220, Deutsche Babcock Energie- und Umwelttechnik AG

| 11 | Gliddon B. J., Raymant A. P., Meadowcroft D. B., Gibbs G. B., Thompson B. H.; A Demonstration of the Gasification of Power Station Coal in the BGL Gasifier, VGB Konf. Kohlevergasung 16./17.5 1991 Dortmund, VGB-Technische Vereinigung der Großkraftwerksbetreiber e.V., Essen

| 12 | Schellberg W., Kuske E.; PRENFLO-Kohlevergasung im Flugstrom, VGB-Konferenz Kohlevergasung 16.-17. Mai 1991, Dortmund

| 13 | Shell Company Information

| 14 | Ishikawa H., Hamamatsu T., Moritsuka H., Toda H., Ishigami S., Furuya T.; Der Entwicklungsstand der integrierten Kohlevergasung in Japan, VGB Kraftwerkstechnik 72 (1992) 5, S. 405-412

| 15 | Nishizaki S., Konagai S.; Study of hot gas clean up system in IGCC, International Symposium on Energy and Environment, Espoo, Finland, August 25th-28th 1991

| 16 | Gangwal S. K.; Hot-Gas Desulpharisation Sorbent Development for IGCC Systems, Symposium "Technologies and Strategies for Reducing Sulphur Emissions", University of Sheffield, March 20-21 1991, IChemE Symposium Series No. 123

| 17 | Gangwal S. K., Stogner J. M., Harkins S. M., Bossart S. J.; Testing of Novel Sorbents for H_2S Removal from Coal Gas, Environmental Progress, Vol. 8(1989)1, p. 26-34

| 18 | Gangwal S. K., Harkins S. M., Woods U. C., Jain S. C., Bossart S. J.; Bench-Scale Testing of High-Temperature Desulfurization Sorbents, Environmental Progress, Vol. 8(1989)4, p. 265-269

| 19 | Mojtahedi W., Horvath A., Salo K., Gangwal S. K.; Development of Tampella IGCC Process, 10th EPRI Conf. of Coal Gasification Power Plants, October 16th-18th, 1991, San Francisco, USA

| 20 | Yumura M., Furimsky E.; Comparison of CaO, ZnO, Fe_2O_3 as H_2S Adsorbents at High Temoeratures, Ind. Eng. Chem. Process Des. Dev., 24 (1985), p. 1165-1168

| 21 | Borgwardt R. H., Bruce K. R.; EPA Experimental Studies of the Mechanisms of Sulfur Capture by Limestone, 1st Joint Symp. on Dry SO_2 and Simult. SO_2 and NOx Contr. Techn., Nov. 13th 1984, San Diego, USA

| 22 | Borgwardt R. H., Roache N. F.; Reaction of H2S and Sulfur with Limestone Particles, Ind. Eng. Chem. Process Des. Dev. 23 (1984), p. 742-748

| 23 | Borgwardt R. H., Roache N. F., Bruce K. R.; Surface Area of Calcium Oxide and Kinetics of Calcium Sulfide Formation, Environmental Progress 3(1984)2, p. 129-135

| 24 | Krishnan G. N., Wood B. J., Tong G. T., Kothari V. P.; Removal of Hydrogen Chloride from High Temperature Coal Gases, Prepr. of 195th ACS Nat.al Meeting and 3rd Chem. Congr. of North America, Am. Chem. Soc., Div. of Fuel Chem., Toronto, Canada, June 5-10 1988, Vol. 33, No. 1

| 25 | Balekdjian O.; Abscheidung von Chlorwasserstoff und Schwefeldioxid an calciumhaltigen Absorbentien, Dissertation, Universität Karlsruhe, 1987

| 26 | Peukert W.; Die kombinierte Abscheidung von Partikeln und Gasen in Schüttschichtfiltern, Dissertation, Universität Karlsruhe, 1990

| 27 | Leppalathi J., Kurkela E.; Behaviour of nitrogen compounds and tars in fluidized bed air gasification of peat, Fuel 70(1991)4, p. 491-497

| 28 | Singh A., Clift R., Reed G. P., Schofield P., Bower C. J.; Thermodynamic calculations of the effect of chlorine on alkali removal from hot coal-derived gases. IChemE Symp. University of Surrey, Guildford, UK, Sep. 16-18 1986, IChemE series number 99 (1986), p. 167-176

| 29 | Mulik P. A., Alvin M. A., Bachovchin D. M.; Simulaneous high-temperature removal of alkali and particulates in a pressurized gasification system, Progress Report, DOE MC-16372-8 1983, US Department of Energy, Morgantown Energy Technology Centre, WV, USA

| 30 | Scandrett L. A.; The removal of alkali compounds from gases at high temperature, PhD thesis 1983, University of Cambridge, Cambridge, UK,

| 31 | Scandrett L. A., Clift R.; The thermodynamics of alkali removal from coal-derived gases, Journal of the Institute of Energy, 57 (1984) p. 391-397

MEASUREMENT AND CONTROL OF ALKALI METAL VAPOURS IN COAL-DERIVED FUEL GAS

IAN ROBERT FANTOM BSc CChem MRSC
Coal Research Establishment, British Coal Corporation,
Stoke Orchard, Cheltenham, GL52 4RZ.

SYNOPSIS

The successful operation of a high performance gas turbine, fired on hot coal-derived fuel gas, is dependant in part on minimising the extent of hot corrosion of turbine blades. Hot corrosion is accelerated by deposition of alkali metal vapours onto the blade surfaces. The fuel gas produced from the British Coal Topping Cycle fluidised bed gasification process contains significant quantities of alkali metal vapours.

A series of tests has been completed at the British Coal Coal Research Establishment (CRE) on an atmospheric pressure spouted fluidised bed gasifier aimed at reducing the alkali metal content of the fuel gas to around 0.1 ppm wt Na + K . The work has been completed in three phases:-

1. Development of a suitable measurement system. Achievement of accurate and precise alkali vapour measurements is a prerequisite to development of alkali removal systems.
2. Reduction of alkali vapour content by cooling the fuel gas and condensing the alkalis on to particulates which are subsequently removed by filtration.
3. Removal of alkali vapours using sorbents.

This paper describes the rationale to alkali control work in the British Coal Topping Cycle, the experimental work, and results of each of the phases described above. Future programmes of work are also considered.

INTRODUCTION

British Coal is developing an advanced coal fired power generation system known as the British Coal Topping Cycle. The cycle involves partial gasification of coal at high pressure to drive a gas turbine with exhaust gas heat recovery. The residual char is burned in a Circulating Fluidised Bed Combustor (CFBC) to raise stream. During the gasification process alkali metal vapours are released into the gas phase. If these vapours pass into the gas turbine they have the potential to deposit onto the turbine blades resulting in enhanced corrosion of the blades and cementation of particulates leading to turbine fouling. In order to reduce the potential corrosion/fouling of the high performance gas turbine it is necessary to control the alkali vapour levels in the fuel gas entering the gas turbine combustor. A target concentration range of 0.07 to 0.1 ppm wt total sodium and potassium in the fuel gas has been set based on a target specification in the turbine expansion gas of 0.024 ppm wt (Na+K) (USDOE, 1986).

A prerequisite to the development of control technologies is the identification and development of a suitable measurement device.

This paper describes work carried out by and on behalf of British Coal to develop alkali control technologies over a range of fuel gas temperatures. The development of an alkali measurement system is also described together with details of the future work programme.

APPROACH TO ALKALI CONTROL

Development of alkali control systems funded by British Coal and the European Community started in the late 70's and was targeted at alkali vapour removal from flue gas streams at temperatures appropriate for Pressurised Fluidised Bed Combustor (PFBC) applications. In 1988 with the start of development of the British Coal Topping Cycle the direction of the work was changed to alkali control in fuel gas streams. A substantial programme of work was put in place to develop an alkali removal system. Studies were performed over a range of temperatures from the raw gas temperature of 950°C down to the lowest envisaged gas cleaning temperature of 400°C. Recent studies have shown the optimum gas cleaning temperature to be around 600°C. The technical risks of operating a high performance filter system are reduced further at by cleaning at 400°C with only a marginal loss in generating efficiency. At these temperatures much of the alkali vapour present in the fuel gas is expected to condense out on to the particulates and be removed in the filter. However, alkali control technology using sorbents continues to be developed as a backup process should cooling of the gas alone be insufficient to control alkali vapour levels.

DEVELOPMENT OF A MEASUREMENT TECHNIQUE

In order to assess the degree of alkali control required and to evaluate the performance of alkali control technology it was necessary to develop a suitable alkali measurement system. Alkali metal vapours are typically found at levels of 0.01 to 10 ppm wt total Na and K in hot coal-derived gases and are difficult to measure (Zarchy, 1980, Haas et al., 1981, Lee & Carls, 1990, Mojtahedi et al, 1990).

Methods of alkali vapour measurement fall into two broad categories; on-line measurement producing instantaneous alkali values; and batch sampling techniques requiring, typically several hours, to obtain an alkali value off-line.

542

On-line Methods

On-line methods provide the ideal method of measuring alkali vapour levels providing instantaneous results allowing time dependant changes and process perturbations to be quantified. A number of devices have been developed and tested. In the late 70's Zarchy developed a surface ionisation detector for the measurement of alkalis in PFBC streams. Initial laboratory scale results showed reasonable results. However, the instrument required frequent calibration and much maintenance. At the same time development of a flame atomic adsorption/emission spectrometer commenced. Over the last decade a number of variants including the Ames, METC (Morgantown Energy Technology Centre) INEL (Idaho National Engineering Laboratory), FOAM (Fibre Optic Alkali Meter) meter and the MAFOS (Morgantown Alkali Fibre Optic Spectrometer) have been constructed (Haas et al.,1981). The devices have been tested at many organisations including CRE (Coal Research Establishment), Grimethorpe PFBC Establishment, NYU (New York University), ANL (Argonne National Laboratory) and Westinghouse, under the auspices of the METC. Sodium and potassium vapour levels in the ppb to ppm range have been measured. Some questions remain over the accuracy and precision of the results; this has been attributed to the use of stainless steel feed lines reacting with the alkali vapours in the hot gases.

Over the last few years attention has turned to measuring specific alkali compounds such as NaCl, KOH etc. for the purpose of modelling the fate of alkalis in coal and oil fired systems. Laser Induced Photo-Fluorescence (LIPF) or Laser Induced Breakdown Spectroscopy (LIBS) provides quantitative measurements when used in purpose built laboratory rigs, predominantly flames. However, in real systems data are qualitative. This is due to the difficulty in calibrating the optical devices and in interpretation of the measurements. The device is not suitable yet for use on pressurised systems although a number of USDOE (Department of Energy) contractors are developing high pressure optical windows. The LIPF is being developed at organisations such as PSI (Physical Sciences Incorporated) Technology Company (Helble & Srinivasachar,1992).

Batch Sampling Techniques

Batch sampling techniques require a fixed volume of hot gas to be drawn through an alkali collection device. The collector is then analyzed for sodium and potassium content and an alkali gas content derived by calculation. A variety of methods have been developed and tested with varying degrees of success. Lee developed a system known as the Regenerable Activated-Bauxite Sorber Alkali Monitor (RABSAM). The system comprises a fixed bed of activated bauxite through which the hot alkali containing flue gas is passed. The alkali is absorbed on to the bauxite. After exposure the bauxite bed is sectioned and the alkali content determined. The device has been used extensively for measurement of sodium and potassium on small scale PFBC rigs and produced reasonable results.

Other devices use the principle of condensation and filtration or dissolution. Lee et al. use a system in conjunction with the RABSAM, consisting of a stainless cold trap filled with NEXTEL ULTRAFIBRES followed by a series of glass bubblers. The system generally gives results of a similar order to the above sorber. The technique appears unsuitable for measuring ppb wt levels of alkali. This is likely to be due to the use of components such as glass and steel in the sample train. Calibration of the system showed collection efficiencies between 50% and 150 %.

At the Technical Research Centre of Finland (VTT) alkali measurements have been made on an experimental PFBC and Pressurised Fluidised Bed Gasifier (PFBG) using a batch sampling technique (Mojtahedi et al.,1990). The pressurised gas is filtered in a small quartz fibre filter and depressurised before passing through a condenser, with water injection, and a series of bubblers. Measurements have been made at levels as low as 20 ppb wt for each species with apparent variability of around +/- 20 ppb wt.

CRE Development Rationale

Based on experience gained earlier at CRE, and by other workers discussed above, it was decided to develop a batch sampling technique to give flexibility of sampling position and the ability to move rapidly from one test stand to another. Early success was also more likely with a batch system. Before manufacture a list of design and operational criteria was tabulated as follows:-

- The gas must be filtered before sampling to remove particulates which may contain alkali and distort the results.
- The gas must then be maintained at duct conditions of temperature and pressure until it reaches the sampling probe inlet.
- Surfaces of the sampling system in contact with the gas need to be manufactured from alkali "inert" materials ie high purity alumina at high temperatures and high density polymer materials such as PTFE at lower temperatures. Use of material such as steel, low purity ceramics and glass should avoided.
- The probe should draw gas into the sampling system at duct conditions and also be retractable to allow complete recovery of alkali on completion of sampling.
- There should be no valves, connectors or any other devices in the sampling train where alkali could be lost or introduced, until after the last alkali collection device.
- The condenser should cool the gas to below the alkali salt dewpoint in order to condense the alkali vapours and allow collection of the aerosol formed.

- A high efficiency filter is required to collect the aerosol formed and a reliable extraction procedure devised.

- Consideration should also be given to the potential errors that can occur in the sampling procedure (eg measurement of the volume of gas sampled, the accuracy of the technique to measure the alkalis in the extracted solution), so that the minimum required sample volume can be sampled.

- In order to give credibility to the measurement system the device should be calibrated with a known quantity of alkali salts under conditions similar to those proposed for operation.

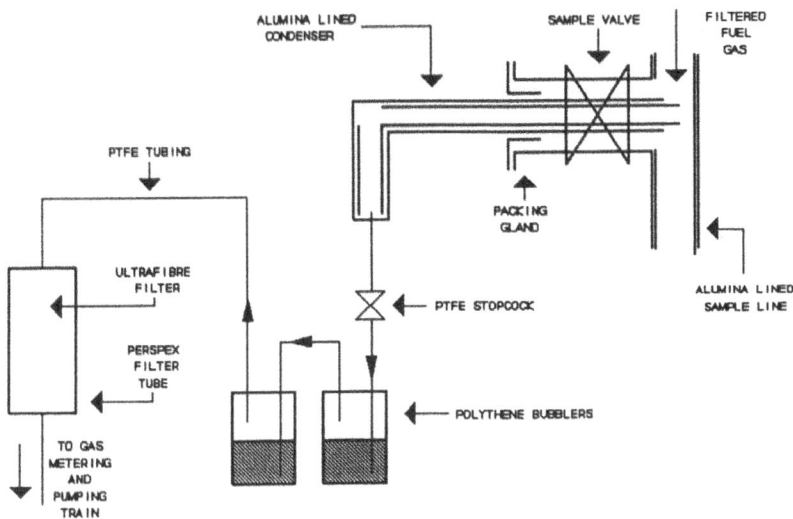

Figure 1   Batch System For Sampling Alkali Metals

During the period 1989-91 a batch sampling probe was designed, tested and refined, using the principles established above, resulting in a system shown in Figure 1. The system comprised a retractable high purity (>99.7% Al$_2$O$_3$) alumina lined air cooled probe followed by two PTFE bubblers and a NEXTEL ULTRAFIBRE filter. The system was designed for use on an atmospheric pressure gasifier. The probe is introduced into the sample gas stream via packing gland, a fixed volume of gas sampled and the probe retracted. Each of the component parts are washed with hot high purity water and the resultant solution analyzed for sodium and potassium using an atomic emission spectrometer (AES). The gas phase concentrations can then be calculated from the solution volume and gas volume sampled. Table 1 shows alkali measurements made during a 24 hour period. The samples were taken immediately after a filter operating at around 950°C. A detailed error analysis of the results was performed and included the errors in measuring the following:-

- volume of gas sampled,
- volume of liquid used in washing,
- gas density,
- accuracy of AES method,
- collection efficiency of system.

The collection efficiency of the system (ie calibration) was measured at Surrey University using a sodium chloride vapour generator. At the ppm wt level collection efficiencies of between 80% and 90% were measured. The system was capable of measuring alkali levels down 50 ppb wt +/- 20 ppb wt for each species. Reproducibility over a number of samples was good with relative standard deviations of between 5 and 17% for each species during a given test period.

TABLE 1
Alkali Vapour Measurements

Sodium		Potassium	
Measured Values ppm wt	Sampling Error %	Measured Values ppm wt	Sampling Error %
1.8	6	0.63	10
1.6	6	0.56	10
1.47	6	0.58	10
1.73	6	0.66	10
1.81	6	0.63	8
1.49	6	0.59	8
1.62	6	0.64	9
1.65	6	0.60	8
Sodium average	1.65	Potassium average	0.61
Standard deviation	0.13	Standard deviation	0.03

Currently, two high pressure sampling systems have been built; one at the Grimethorpe PFBC Establishment to measure alkali vapours after a topping combustor fired on propane and PFBC gas at a pressure of 5 bara (Lee et al.,1990); the other at CRE to measure alkali vapours on a PFBG operating at up to 20 bara. Both systems use a retractable high purity alumina lined condenser probe followed by a deep NEXTEL ULTRAFIBRE filter.

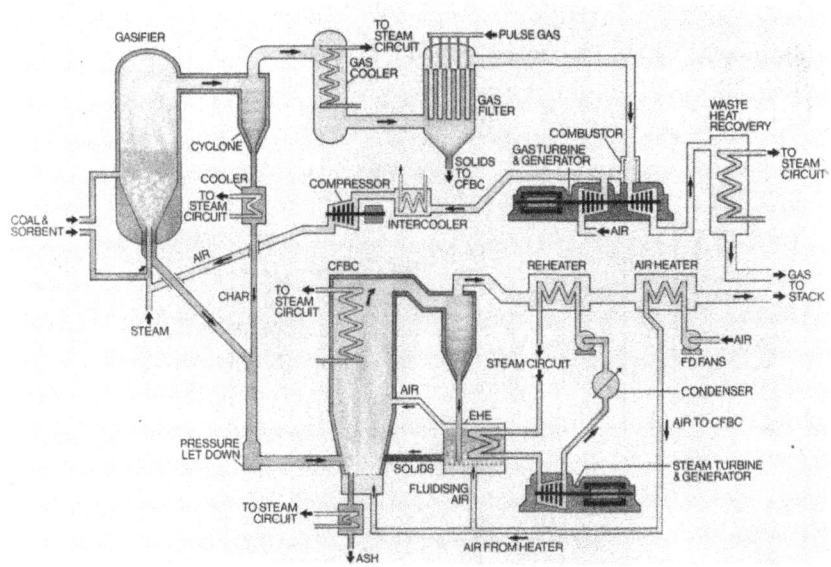

Figure 2   British Coal Topping Cycle

DEVELOPMENT OF ALKALI REMOVAL TECHNOLOGIES

Alkali Control by Condensation and Filtration

The preferred configuration of the British Coal Topping Cycle is shown in Figure 2. The fuel gas is produced in the spouted Pressurised Fluidised Bed Gasifier (PFBG) at around 25 bara and exits the gasifier at between 950-1000°C. Most of the particulates entrained in the gas are removed in series of cyclones. The gas is then cooled to between 400°C and 600°C before filtration to remove the remaining entrained char particulates. Filtration at 600°C offers the optimum cycle efficiency. However, a lower temperature of 400°C is required to reduce the technical risks involved with operating a high performance gas filtration system. The loss of cycle efficiency at this lower temperature is minimal especially if a super critical steam cycle is used. At these temperatures much of the alkali vapour present in the fuel gas is expected to condense out on to the particulates and be removed in the filter. In order to assess the extent of alkali reduction as the gas is cooled alkali vapour measurements were performed before and after a cooler and filter system, operating at 600°C and 950°C, on an atmospheric gasifier. A schematic diagram showing the gasifier, filter system and alkali sampling points is shown in Figure 3.

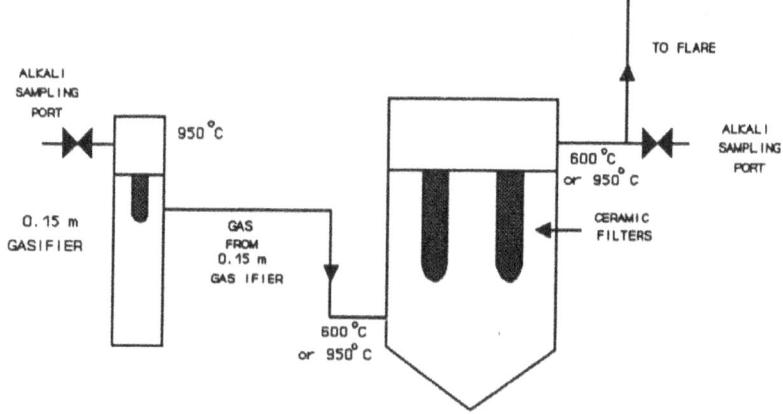

Figure 3 Schematic of Laboratory Filter Unit

TABLE 2
Average Alkali Concentrations in Condensation/Filtration Test

	Na ppm wt	K ppm wt
Gasifier outlet 950°C	1.34	0.52
Filter outlet 950°C	0.52	0.62
Filter outlet 600°C	0.09	0.07

The results of the measurements performed are summarised in table 2. Average values at 950°C, ie the gasifier outlet, were 1.34 pm wt Na and 0.52 ppm wt K. At the filter outlet when operating at 950°C values of 0.52 ppm wt Na and 0.62 ppm wt K were measured. The reduction in sodium could be due to alkali/ash interactions or sodium reacting with the duct walls. However, measurements made when the gas was cooled to 600°C prior to filtration showed a substantial reduction in vapour phase alkali to 0.09 ppm wt Na and 0.07 ppm wt K. The total alkali level of 0.16 ppm wt is close to the target range of 0.07-0.1 ppm wt (Na+K). These measurements were made at atmospheric pressure. In assessing the likely alkali levels at higher pressures it is necessary to consider the possible mechanisms of alkali reduction occurring as the gas cools. Table 3 shows the maximum alkali vapour concentrations at 600°C and 16 bar to be 77 ppb wt Na and 328 ppb wt K (ie the saturation concentrations). Vapour pressures at 400°C are of the order of a few ppb wt or less.

TABLE 3
Alkali Vapour Saturation Levels*

	Sodium ppb wt		Potassium ppb wt	
Pressure Bar a	1	16	1	16
950°C	270000	17000	875000	55000
700°C	30520	1900	111000	7000
600°C	1300	77	4000	328
500°C	25	1	125	8
400°C	0.1	0	1	0

* based on sodium and potassium chloride saturated vapour pressures

However, the measurements made on an atmospheric filter system of 90 ppb wt Na and 70 ppb wt K at 600°C and 1 bar are much lower than the saturated vapour concentrations at these conditions, given in Table 3, of 1300 and 4000 ppb wt for sodium and potassium, respectively (MTDATA). Thus other processes are occurring. These are discussed below.

Homogenous Condensation: Spontaneous homogenous nucleation of salt vapours occurs when the vapour pressure exceeds a critical supersaturation concentration. This value is related to the temperature, pressure and fluid dynamics of the system in question. In theory, in a particle free environment this critical value can be many times the thermodynamic saturation value. However, in the gas cleaning train with a high particulate loading, seeding will occur preventing a supersaturated gas thus leading to heterogenous condensation.

Heterogenous Condensation: Heterogenous condensation occurs normally at the surface of particles or other surfaces where the local concentration just exceeds the saturation concentration. This is normally induced by adsorption of the vapour on to the surface. Heterogenous condensation on to small particulates effectively brings about equilibrium. The vapour phase alkali salts would therefore be in equilibrium with the condensed material on the surface of the particle and equal to the saturated vapour pressure at a specific temperature. In the gas cleaning train the gas is cooled relatively quickly from about 950°C to 600°C and then spends a much longer period in the filter vessel allowing reaction to approach equilibrium. The maximum alkali concentration at the filter exit is therefore expected to be equal to the saturation concentration at the filter pressure and temperature. As indicated above this is equal to 77 ppb wt Na and 328 ppb wt K at the utility conditions proposed of 600°C and 16 bar abs and less than 1 ppb wt for each species at 400°C.

Vapour Deposition: Vapour deposition arises when vapour in a hot gas comes into contact with a cooler surface. Vapour is either adsorbed on to the surface, rapidly exceeding the saturated vapour pressure forming a condensate, or condenses in the boundary layer and deposits as condensate. In a heat exchanger cooling the gas from 950°C to 600°C the wall temperatures are typically between 400°C and 500°C, well below the dewpoint of the alkali vapours. Lower heat exchanger surface temperatures would be expected for the 400°C gas cleaning scenario. Vapour deposition is thus highly likely. The extent of deposition will depend on the rate of mass transfer of alkali vapour across the boundary layer and the residence time of the gas in the heat exchanger.

Reaction with Ash Components: Alkali vapour levels at the gasifier exit are probably controlled by ash/alkali reactions in the gasifier bed and not by alkali chloride vapour/condensate equilibrium. This is apparent since measured values at 950°C are always significantly below saturation concentrations. Equation 1 shows a typical reaction, although the exact forms of the aluminosilicates formed are difficult to determine.

$$2 \, NaCl_{(g)} + Al_2O_3 . 6 \, (SiO_2) + H_2O_{(g)} \rightleftharpoons Na_2O . Al_2O_3 . 6 \, (SiO_2) + 2HCl_{(g)} \quad \textbf{(1)}$$

Similar reactions can be written for potassium. In general these types of reaction are more favourable for potassium than sodium; this partly explains why potassium vapour levels are often lower than sodium levels, even though potassium salts are generally more volatile than analogous sodium salts, and coals often contain more potassium than sodium. Potassium is associated with non-volatile alumino-silicate compounds in the coal whereas sodium exists associated with the organic matrix. As the gas cools in the heat exchanger and filter, reactions between the alkali vapours and the ash component of the char particles becomes thermodynamically more favourable (Scandrett & Clift,1984). Given the long residence time in the filter and high particle density it is reasonable to assume that these reactions come to equilibrium. This may even be the case upstream of the filter, where dust loadings are equally high. Similarly, the char filter cake on the filter surface could act as an effective alkali sorber. It is impossible to predict accurately the theoretical vapour pressure of alkali chlorides above the ash components because the exact form of the aluminosilicate is not known. It is also possible that the alkali vapours react with the carbon matrix of the ash forming interstitial compounds.

Reaction with Filter Media: Several authors suggest that alkali vapours in the fuel gas could react with the ceramic candle filter media (ASME. USDOE,1991). Reaction with the silicon carbide (SiC) granules and the aluminosilicate binder has been postulated. The uncertainty here is similar to that discussed above for reaction with ash.

Alkali Control Using Sorbents

The medium term development of the Topping Cycle is aimed towards the use of intermediate gas cleaning temperatures (400-600°C). At these temperatures it is possible that alkali control will be achieved by simply cooling and filtering the gases. However, it was considered prudent to develop a backup alkali removal system should cooling alone prove insufficient to ensure an economic lifetime of the gas turbine components.

A substantial programme of work was put in place to evaluate alkali removal using sorbents over a range of fuel gas temperatures from 600-950°C. Options for gas/sorbent contacting included sorbent injection prior to filtration or fixed bed sorption before or after filtration.

Alkali measurements have been made at 950°C on an atmospheric pressure fluidised bed gasifier. Vapour levels of between 1 to 3 ppm wt Na and 0.6 to 1.3 ppm wt potassium were measured during many periods of operation with different coals. Alkali removal is required to achieve the target range of 0.07-0.1 ppm wt total alkali. A number of authors have reported developments of alkali control technology focused on the used of alkali sorbents in packed-bed reactors. Bachovchin et al. used Emathlite (a calcium montmorillinite clay native to the USA), Punjak et al. used

Kaolinite, both demonstrated significant retention of alkali vapours from hot alkali laden gases. The alkali was shown to react with the clay producing an alkali aluminosilicate compound. Lee et al.,1990 used activated bauxite as a sorbent; again significant retention of alkali from hot gases was demonstrated. The mechanism of alkali retention was by adsorption; the sorbent could be regenerated by washing out the alkali with water.

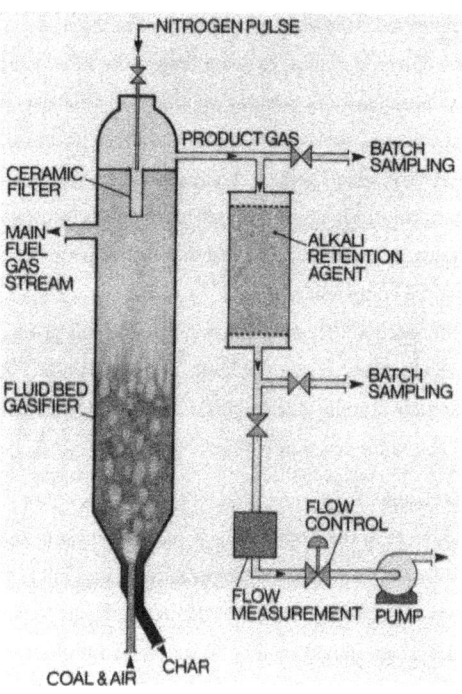

Figure 4 Alkali Retention Test Rig

A number of potential alkali sorbents were identified by Scandrett (Scandrett & Clift,1984 Scandrett,1983) and Singh (Singh at al.,1987) using thermodynamic calculations. These sorbents, including those above, were then screened at Surrey University using a Simultaneous Thermal Analysis technique (STA) (Mclaughlin,1990). The three sorbents discussed above plus Fullers Earth (a calcium montmorillinite native to the UK) were then tested on a 0.8 litre reactor installed on 0.15

552

m atmospheric pressure gasifier at CRE. Figure 4 shows a schematic of the rig. The results are summarised in Table 4. All the sorbents were tested at 950°C, the same fuel gas flow rate and were nominally the same size at around 2-4 mm diameter.

TABLE 4
Alkali retention tests

	Sodium ppm wt*		Potassium ppm wt*	
Sorbent	inlet	outlet	inlet	outlet
Emathlite	1.54	0.25	0.72	0.12
Activated Bauxite	1.26	0.25	0.64	0.17
Kaolinite	1.09	0.19	0.57	0.12
Fullers Earth	1.26	0.17	0.67	0.06

* average of around 8-20 measurements taken at each location during each test.

Each sorbent showed the capability of reducing the alkali vapour concentration significantly. Fullers earth gave the lowest outlet value of 0.17 ppm wt Na and 0.06 ppm wt K. The total value of 0.23 ppm wt is still above the target range of 0.07-1 ppm wt but is predicted to be lower for a high pressure system. Work is continuing to optimise the sorber process. However, the studies are targeted towards providing an alkali sorber system for around 600°C application. This technology will only be implemented if cooling of the gas is shown to be insufficient to control alkali vapour levels.

As part of this process optimisation a chemical engineering model of an alkali removal system, using Fullers Earth, is being developed at Surrey University (Mclaughlin, 1990). The model is being formulated using a series of laboratory sorber tests and post exposure analysis. The successful development of the model will assist in the design of a commercial alkali removal system at an appropriate temperature.

FUTURE WORK

A raw fuel gas cooler and ceramic filter are being built on a high pressure gasifier at CRE. Alkali measurements will be taken after the filter operating at 400°C and 600°C to assess the requirement for additional alkali control in the Topping Cycle.

Work at Surrey University will continue to develop fully an engineering model of the alkali removal process for application at an appropriate location in the fuel gas stream.

CONCLUSIONS

A batch sampling technique has been developed to measure sodium and potassium vapour levels in hot coal-derived gases. Refinement of the technique has reduced the total quantifiable errors in measuring the alkalis to +/- 0.02 ppm wt for each species. The technique has been used to assess alkali control over a range of gas cleaning temperatures. Measurements made after filtration at 600°C suggest alkali vapour levels of 0.16 ppm wt, close to the target range of 0.07-0.10 ppm wt (Na+K). A number of alkali sorbents have been tested on a small scale gasifier. Measurements made at 950°C after a sorbent bed have shown a reduction to 0.23 ppm wt (Na+K) when Fullers Earth was used as a sorbent. Again this is close to the target concentration. Work will continue to develop alkali control technologies for application at pressures of around 25 bar a, typical of the operating pressure in the British Coal "Topping Cycle".

ACKNOWLEDGEMENTS

This work is funded by the British Coal Corporation and the European Coal and Steel Community. The views expressed are those of the authors and do not necessarily represent those of the sponsors.

REFERENCES

ASME. Corrosion and Degradation of Ceramic Particulate Filters in Direct Coal Fired Turbines, ASME paper No. 90-GT-347,.

Bachovchin D. (1986). A Study of High Temperature Removal of Alkali in Pressurised Gasification System, Progress Report, Westinghouse R&D Centre, Chemical Sciences Division, Pittsburgh.

Haas W. et al. (1981). Development of Alkali and Trace Heavy Metal Monitors, Proceedings, High Temperature, High pressure Particulate and Alkali Control in Coal Combustion process Streams, Morgantown, West Virginia DOE/MC/08333-167 pp 599-620.

Helble J., Schrinivasachar S. (1992). Measurement of Vapour Phase Sodium chloride Formed during Pulverised Coal Combustion, Combust.Sci.Tech..

Lee C., Hird W., Schofield P., Fantom I R. (1990), A Novel Probe For Alkali Vapour Sampling on the Grimethorpe PFBC, 11th Int.Conf.Fluid.Bed Comb., p 835, Montreal.

Lee S., Carls, E. (1990). Measurement of Sodium and Potassium in PFBC of Beulah Lignite, J.Inst.Energy, 53(457), pp 203-210.

McLaughlin J. (1990). The Removal of Volatile Alkali Salt Vapours from Hot Coal Derived Gases, Phd. Thesis, University of Surrey.

Mojtahedi W., et al. (1990). J.Inst.Energy, 63, pp95-100.

MTDATA Thermochemical computer package, National Physical Laboratory, Middlesex, England.

Punjak W., Shadman F., (1988). Aluminosilicate Sorbents For the Control of Alkali Vapours During Coal Combustion and Gasification, Energy and Fuels, Vol. 2, p 702.

Scandrett L. (1983). The Removal of Alkali compounds From Hot Gases at High Temperature, PhD. Thesis, University of Cambridge.

Scandrett L., Clift R. (1984). The Thermodynamic Aspects of Alkali Removal From Coal Derived Gases, J.Inst.Energy, pp 391-397.

Singh A. et al. (1987). Thermodynamics Calculations of the effects of Chlorine on Alkali Removal From Hot Coal Derived Gases, Gas Cleaning at High Temperatures, ICHemE, Symp. Ser.No. 99, p167.

USDOE (1986). Hot Gas Cleanup For Electric Generating Systems, US/DOE/METC-86-6038.

USDOE (1991). Degradation of Ceramic Filter Materials in Direct Coal Fired Turbines, Materials and Components in Fossil Energy Applications, DOE/FE-0091/91, No. 91.

Zarchy A.S. (1980). Symp. on Inst. and Control for Fossil Energy Processes, Virginia.

REACTION OF GETTER MINERALS WITH ALKALI SALT VAPOURS

John McLaughlin', Ron Schulz* and Roland Clift

*Department of Chemical and Process Engineering, University of Surrey,
Guildford, Surrey, GU2 5XH, UK*

(*Author to whom correspondence should be addressed)

ABSTRACT

The Integrated Gasifier Combined Cycle appears to offer advantages such as increased efficiency and reduced SO_x emission compared to current coal-fired power generation methods. However gasification of coal with moderate to high chlorine content may result in unacceptably high concentrations of alkali salts (NaCl, KCl) in the gas phase. These lead to catastrophic corrosion of gas turbine components.

Thermodynamic studies have indicated that some alumino-silicate compounds ("getters") may react with these salt vapours at the operating conditions to form non-volatile products. However, limited experimental data are available to determine the reaction mechanism or kinetics. This paper describes an experimental investigation into the effect of HCl upon alkali capture by a high temperature sorbent material; calcium montmorillonite (Fuller's Earth).

A series of experiments was performed using a fixed bed sorption rig at temperature between 1100 and 1200 K. The gas composition included 10-40 ppmv NaCl, 0-160 ppmv HCl and 5% H_2O. The results indicate that alkali uptake by the sorbent is strongly affected by HCl concentration and that the reaction mechanism is complex.

A theoretical model of the experimental system was developed and a "Two Reaction" mechanism proposed. This model was tested against experimental data with encouraging results.

INTRODUCTION

The Integrated Gasifier Combined Cycle potentially offers advantages over current methods of coal-fired power generation, including increased efficiency, lower SO_x emission due to in-bed sulphur retention and compatibility with a wider range of fuel types. However, coal gasification liberates sodium and potassium compounds intrinsic to the coal material (Raask, 1985; Howarth *et al.*, 1987) which quickly form relatively volatile alkali chlorides (NaCl, KCl), at concentrations in the 1 - 10 ppm range (Reed, 1986; Lee & Myles, 1986). These salts, in turn, have been shown to seriously reduce a gas turbine's operational life, due to "catastrophic" corrosion of hot path components (Radcliff, 1987). For a gasifier producing low to medium energy fuel gas (4-20 MJ m^{-3}), high temperature gas clean-up prior to the gas turbine's combustion chamber is therefore essential if high efficiencies are to be attained.

(' Present address: Nederlandse Aardolie Maatschappij b.v., Velsen-Noord, 1950 AA, Netherlands)

In response to this problem several thermodynamic studies have been preformed (Mulik *et al.*, 1983; Scandrett & Clift, 1984; Singh *et al.*, 1986) to investigate the reaction of alumino-silicates with the vapour phase alkali salts to form a non-volatile products. A simplified reaction scheme was formulated as follows:

$$2\ MCl_{(g)} + H_2O_{(g)} + Al_2O_3.xSiO_{2(s)} \rightleftharpoons M_2O.Al_2O_3.xSiO_{2(s)} + 2\ HCl_{(g)} \qquad [1]$$

where M is Na or K.

Singh and Clift predicted that under the anticipated process conditions for gasification of a typical British coal, the equilibrium partial pressure of the volatile alkali salt would be above the target value of 0.1 ppmv. This was due to the effect of the comparatively high HCl levels from these coals, on reaction [1]. They concluded that the target alkali levels could only be met at substantially reduced temperatures.

To study this problem more rigorously, an experimental investigation was initiated to confirm the validity of reaction [1] for alkali salt vapour capture, determine the sorption kinetics and study the effect of HCl on the sorption process. Some results are discussed in this paper.

EXPERIMENTAL METHOD

Various techniques are available for measuring kinetic parameters associated with gas-solid reactions. Unfortunately, although often powerful and elegant, these methods commonly use on-line measurement of gas concentration, a technique not readily applicable to alkali vapour (Luthra & LeBlanc, 1984; Lee & Myles, 1986; Punjak & Shadman, 1988; Fantom, 1989). Consequently, the progress of this gas-solid "gettering" reaction can only be monitored by observing changes in the solid phase, such as changing sample weight or chemical composition.

The technique used in this study was based on a fixed bed system shown schematically in Figure 1. A bed of sorbent pellets (0.425 - 0.5 mm in diameter) was exposed to a reactant gas flow for up to 300 hours. The gas composition and experimental conditions are given in Table 1. After exposure, samples of sorbent were taken from successive layers into the bed, using a vacuum system. These pellets were analysed for total alkali content by digestion in hydrofluoric and nitric acid, and using atomic adsorption spectrophotometery to measure the sodium and potassium concentration in solution. Thus, the average alkali concentration in the solid could be mapped against cumulative weight (distance) to give a concentration profile. This profile provided information on both the sorbent capacity and rate of uptake.

Calcium montmorillonite (Laporte *Fuller's Earth*) was selected as a suitable sorbent for this investigation by a getter screening method based on thermal analysis techniques. It has a composition with a general formula $(Ca_{0.2})(Al_{1.6}Mg_{0.4})(Si_4O_{10})(OH)_2.$, which is not unlike emathlite, a known alkali sorbent (Bachovchin *et al.*, 1986). The powdered material was easily pelletised, forming robust dust free pellets; it proved relatively easy to digest for chemical analysis and was available locally. Mercury porosimetry on calcined pellets measured an average pore radius of 3.84μm and a pellet porosity of 0.2. The surface area was measured to be 3-4 $m^2\ g^{-1}$ using BET nitrogen adsorption. Scanning Electron Microscopy (SEM) showed the pellets to be composed of grains with average diameter 30μm.

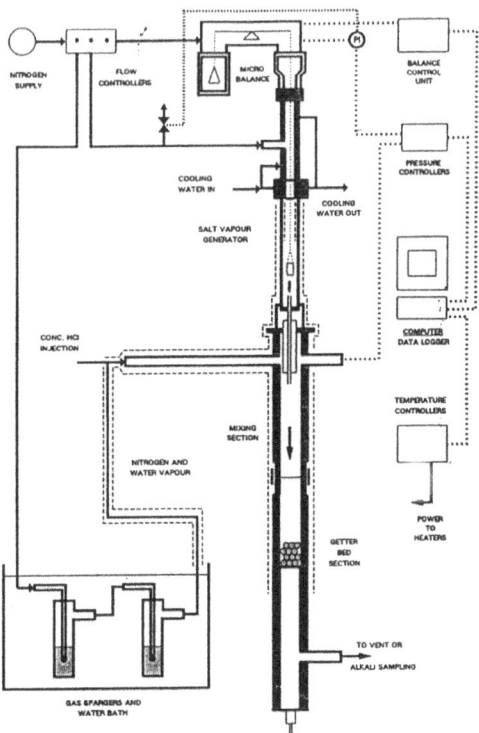

Figure 1: Schematic of
Fixed Bed Rig

TABLE 1
Predicted gasifier operating conditions (Singh *et al.*, 1986) and appropriate
experimental conditions.

Variable	Gasifier	Experimental
Pressure [atm]	16	1
Temperature [K]	1050 - 1250	1100 and 1200
Gas composition:		
Alkali [vpm]	$\approx$ 10 (Na + K)	10, 40 (as NaCl)
H_2O [%]	3.75	5
HCl [vpm]	300	0, 55, 110, 160
CO [%]	18	-
H_2 [%]	10	-
CO_2 [%]	9	-
H_2S [vpm]	200	-
N_2	balance	balance
Superficial Velocity [cm sec^{-1}]		10
Particle Diameter [mm]		0.425 - 0.5

RESULTS OF FIXED BED TESTS

Fixed bed tests have been used as the primary experimental method for the direct study of the effects of HCl addition and temperature on alkali sorption.

Effect of HCl Addition

Adding HCl to the gas stream appears to reduce the rate of sodium uptake by the sorbent dramatically. Experiments with 0, 55, 110 and 160 vpm HCl, 40 vpm NaCl and 5% H_2O at 1100 K were performed for 73.2 hours each. The concentration profiles obtained are shown in Figure 2. The progress of the alkali concentration front down the bed with increasing HCl levels indicates a reduction in the rate of uptake, to the extent that at "high" levels of HCl, alkali breakthrough should occur much sooner than for a similarly sized bed operating in the absence of HCl.

To alter the shape of the concentration profile through the bed, HCl must affect either the kinetics of alkali uptake, or the equilibrium conditions, since changes to the bed hydrodynamics caused by such low levels of additives can be neglected. We can discount the effect of acid addition on the rate of gas film mass transfer, for the same reason. However, the intrapellet diffusivity could be significantly altered if HCl promotes sintering, glass formation, swelling or other gross changes to the sorbent's structure. This may decrease the diffusivity through the pellet by increasing the tortuosity, or by producing an impermeable barrier between the reactant gas and solid, for example through the formation of a glassy phase.

To test this hypothesis, an experiment was performed in which the acid concentration was reduced from 160 vpm to 0 vpm for the final 6 hours of a 200 hour test. The profile obtained is shown in Figure 3, in comparison to a profile obtained with the HCl concentration constant at 160 vpm for the duration of a 100 hour experiment. Both experiments were performed at 1200 K. The shaded area corresponds almost exactly to the amount of sodium added during the final 6 hours, and indicates that the inhibitory effect of HCl is easily reversed once the acid concentration is reduced. Thus it would appear unlikely that gross structural changes are occurring, as these would be difficult, if not impossible, to reverse. This result also excludes loss of reactive surface area through sintering or swelling, for the same reason. Visual and microscopic examination of exposed sorbent pellets showed no glass formation, or any other obvious structural change.

We are therefore drawn to conclude that HCl affects either the reaction itself, the reaction rate or its equilibrium. Unfortunately, given the complexity of the reaction system, and the interaction between kinetic and equilibrium processes it is not possible from these results to identify which particular mechanism is dominant. Later in this paper an attempt will be made to decouple kinetic and equilibrium factors, using a theoretical model of the fixed bed reactor.

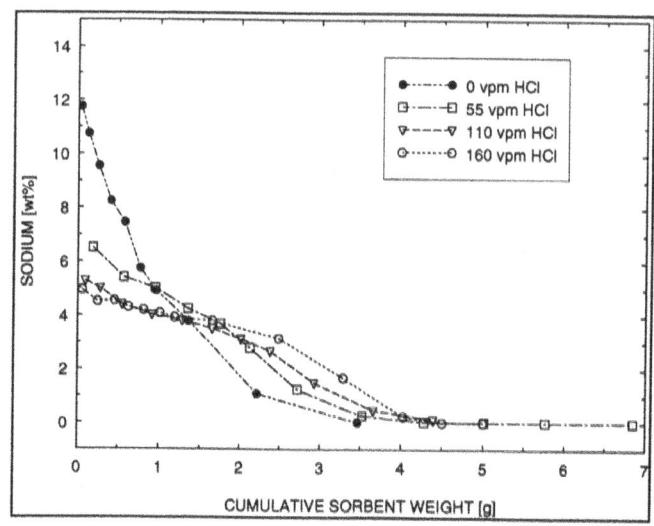

Figure 2: Effect of HCl concentration on alkali uptake by calcium montmorillonite. Experimental conditions: 1100 K, 40 vpm NaCl, 5% H₂O, Pellets: 0.425-0.5 mm diameter

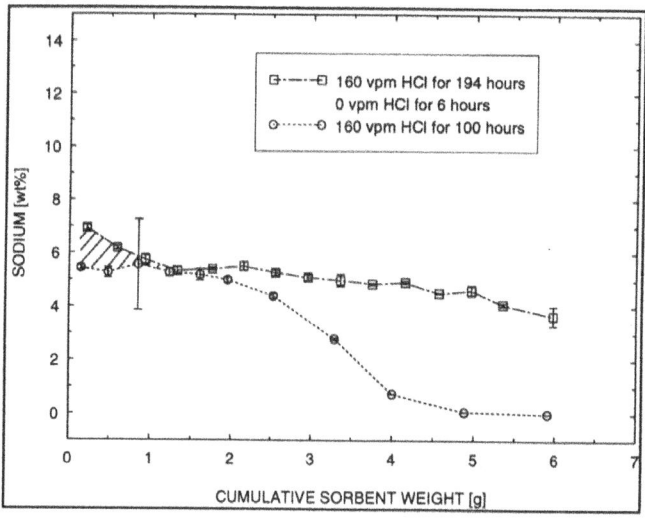

Figure 3: The shaded area coincides with the amount of NaCl added during the final 6 hours of the 200 hour run when HCl concentration was reduced from 160 vpm to 0. Experimental conditions: 1200 K, 40 vpm NaCl, 5% H₂O. Pellets: 0.425-0.5 mm diameter.

Effect of Temperature

Increasing the operating temperature by 100 K, from 1100 K to 1200 K has virtually no effect on the rate for experiments performed without HCl addition, but significantly increases the rate of alkali uptake for experiments with 160 vpm HCl in the inlet gas.

The profile generated from an experiment performed for 100 hours at 1200 K, with 40 vpm NaCl, 5% H_2O and no added HCl is shown in Figure 4, together with a result obtained at 1100 K using the same gas composition and a duration of 98 hours. The similarity of the slope of these two concentration profiles indicates that the rate of uptake is not significantly altered by changing temperature. Saturation capacity however decreases from 12.6%wt sodium at 1100 K to approximately 12%wt at 1200 K. This difference is probably due to the higher temperature enhancing glass formation, thus limiting total utilisation of the solid reactant, or through changes in the stability of the various phases. The saturated pellets had a glassy apperance and had lost their grain structure, becoming a fused mass.

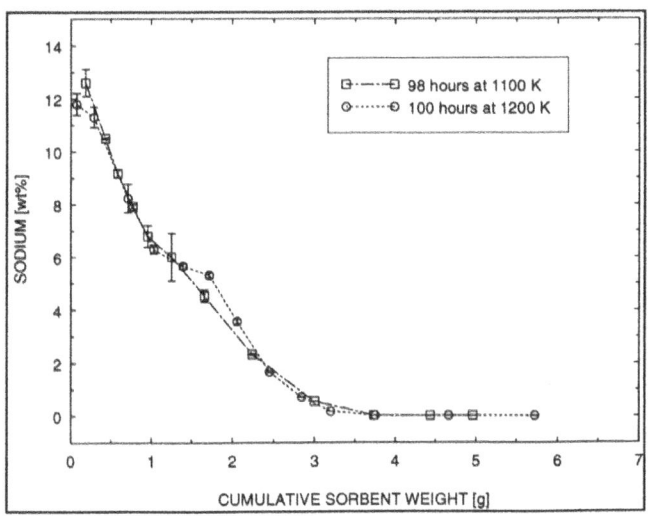

Figure 4: Effect of temperature, in the absence of HCl.
Experimental conditions: 40 vpm NaCl, 5% H_2O,
Pellets: 0.425 - 0.5 mm pellets.

The profiles for experiments performed at these two temperatures with high levels of acid in the inlet gas (160 vpm HCl, 5% H_2O and 40 vpm NaCl) are shown in Figure 5. The experiment at 1100 K lasted 73.2 hours while the test at 1200 K was for 100 hours. The increase in the uptake rate is shown by the increasing slope of the front of the high temperature profile.

561

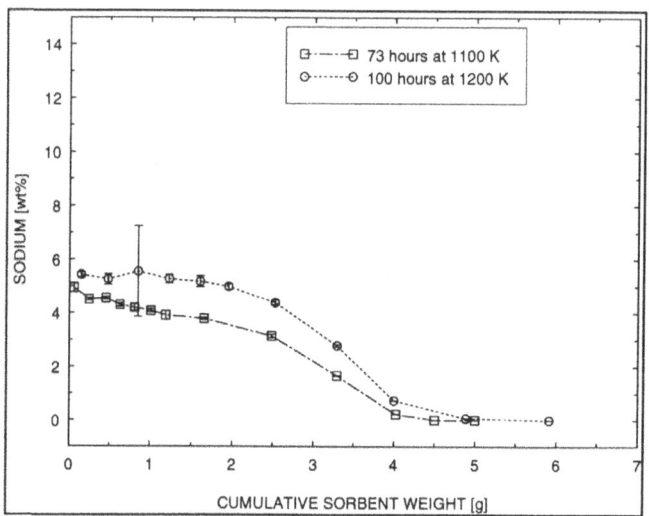

Figure 5: Effect of temperature with 160 vpm HCl addition.
Experimental conditions: 40 vpm NaCl, 5% H$_2$O, 160 vpm HCl,
Pellets: 0.425 - 0.5mm pellets.

The difference between the groups of experiments performed with and without acid addition suggests that the uptake mechanism may be composed of at least two stages. At low HCl concentration a process that is relatively temperature independent dominates the rate of alkali uptake, while at high HCl concentration the dominant process is temperature sensitive. Bachovchin et al. (1976) suggested that the alkali uptake mechanism was consisted of three steps; NaCl dissociation, then an alkali-alumino-silicate reaction, followed by HCl recombination. The overall reaction can still be summarised through an equation similar to reaction [1], but the overall rate would a function of the temperature dependence of each step.

THEORETICAL MODEL OF FIXED BED REACTOR

The alkali concentration profiles from the fixed bed results can be examined qualitatively to indicate the trends in sorbent behaviour with changing reaction conditions. However, gaining further insight into the dominant reaction mechanisms can be assisted by a theoretical description of the behaviour produced in the experimental apparatus. Such a model is comprised of two parts: one describes the physical and chemical processes occurring within a single sorbent pellet, the other deals with the bulk fluid hydrodynamics of the bed.

<u>Single Sorbent Pellet</u>: For the material considered in this study it appears that a particle-pellet or "grain" structure proposed by Shon and Szekely (1972) would be appropriate. The pellets used have a classical particle-pellet nature, with a moderately low surface area and a relatively large pore size. This leads to a visualisation of the pellets as being composed of 30μm, virtually non-porous spheres. These constituent spherical particles are referred to as "grains". A photograph of a sectioned, unexposed pellet imaged by SEM is shown in Figure 6a.

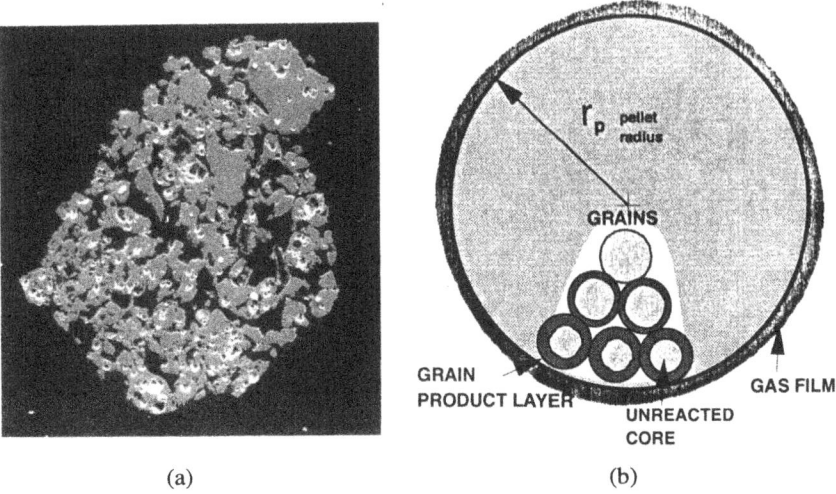

(a)   (b)

Figure 6:   a. SEM image of an untreated calcium montmorillonite pellet.
            b. Idealised representation of 'shrinking core' sorption.

Having selected a physical description it was necessary to select the appropriate model for the reaction behaviour of individual grains. This behaviour falls into two broad classes. A 'homogeneous' model considers the reaction to occur throughout a highly porous grain. Alternatively, low porosity grains at high temperature may display a 'shrinking core' of unreacted material, with a sharp interface at the reaction front. An idealised representation of 'shrinking core' behaviour within a pellet used by Szekely *et al.* (1976) to develop the Grain Model is shown in Figure 6b.

Analysis of sodium levels within exposed pellets and grains were performed using an Electron Probe Micro Analysis (EPMA) system (Link Systems). Typical photographs of a sodium scan superimposed on the image of single grains are shown in Figure 7. These display higher sodium levels at the outer edge of the grain compared to the middle, indicating that a homogeneous model is not applicable. However, due to the poor resolution of these scans it is not possible to determine whether a shrinking core or an alternative approximation (e.g. a zone reaction model) is more appropriate, but given the restrictions of the EPMA system, and for the sake of simplicity, the shrinking core approximation was considered appropriate to describe the reaction's progress within each grain.

Figure 7:   Sodium concentration profile across an individual sorbent grain. The profile was measured by EPMA on a pellet containing approximately 6.5%wt sodium.

The Grain Model: From the above discussion we concluded that the sorbent pellets used in this study could be considered to be composed of an agglomeration of virtually non-porous grains, each about 30μm in diameter. The reaction between the alkali species and the solid was assumed to occur in each grain at the interface between an unreacted core and solid product. Gaseous reactants and products were transported though the pellet by diffusion in the macropores surrounding the grains, and were carried from and to the bulk gas stream by mass transfer across a gas film surrounding the pellet.

The detailed derivation of the grain model for gas-solid reactions given by Szekely *et al* (1976) involved several simplifying assumptions which have been critically assessed by McLaughlin (1990) for the system under consideration. The steps envisaged in this representation are as follows:

a) Diffusion of gaseous reactant from the bulk gas stream to the surface of the pellet.
b) Transport of the reactant in pores in the pellet.
c) Chemical reaction between the alkali vapour and the solid.
d) Transport of the products along the pores in the solid.
e) Diffusion of the products into the bulk stream.

564

For a given gas concentration the model provides a means of calculating the change in conversion within the pellet as a function of time, and hence the change in alkali content in the solid. However, to calculate the solid alkali concentration through the bed, we also need to know how the gas concentration varies with both time and distance through the bed. This is determined by the bulk fluid hydrodynamics.

Bulk Fluid Hydrodynamics: To achieve the required hydrodynamic behaviour of the bulk fluid in the fixed bed, several conditions had to be met:

a) Isothermal operation: The temperature difference from the top to the bottom of the bed was usually less then 2 K, with the average temperature maintained to better than ± 2 K, for up to 300 hours.

b) Dilute gas and negligible pressure drop across the bed, hence no change in gas flow rate: The pressure drop across the bed was monitored continuously using a micromanometer. To ensure constant gas flow rate, the high precision flow controllers and supply pressure regulator were mounted in a temperature controlled enclosure to minimise fluctuations caused by ambient temperature changes.

c) No wall effects and no radial dispersion: To minimise gas channeling near the wall, pellets less then 0.5 mm in diameter were used with a bed diameter of 25 mm. Under these conditions the wall effects should be negligible (Ruthven, 1984).

A 'tanks-in-series' approximation was adopted to account for small to intermediate levels of axial dispersion present in the experimental apparatus, Developed by Deans and Lapidus (1960), this model for the bulk fluid hydrodynamics visualises a packed bed as a series of zones, with uniform gas composition in each zone. Thus each zone is considered to be a well stirred tank. This approximation has been shown to be equivalent to a differential material balance with a term accounting for axial dispersion.

The combination of the grain model describing the behaviour of a single sorbent pellet with the "tanks-in-series" representation of a packed bed produced a description of the non-catalytic gas-solid reactor used in this study. This theoretical description contains two kinetic groups that cannot be measured directly for the system under consideration. These are a reaction rate group, ks, and the effective intragranular diffusivity, De. A third parameter, Ke, arises from the assumption of reversible reactions and represents the equilibrium group for a particular reaction. A method of estimating the value of these groups is to "fit" mathematically generated concentration profiles to those obtained experimentally, involving the systematic variation of the values of these parameter so as to minimise the deviation between experimental and theoretical profiles. A robust, quasi-Newton algorithm for finding the minimum of a function of several variables was used for this purpose.

Experiments at varying levels of HCl and temperature indicate that there may be two dominant reactions occurring; one whose rate is strongly affected by HCl concentration, a second that proceeds virtually independently of HCl. The combination of two simple profiles to form a complex curve is shown in Figure 8. This resultant shape is reminiscent of some of the profiles obtained experimentally, shown in Figure 9. The "knee" in these profiles occurs between 5 and 6%, (coinciding with the capacity of the sorbent at high inlet HCl levels).

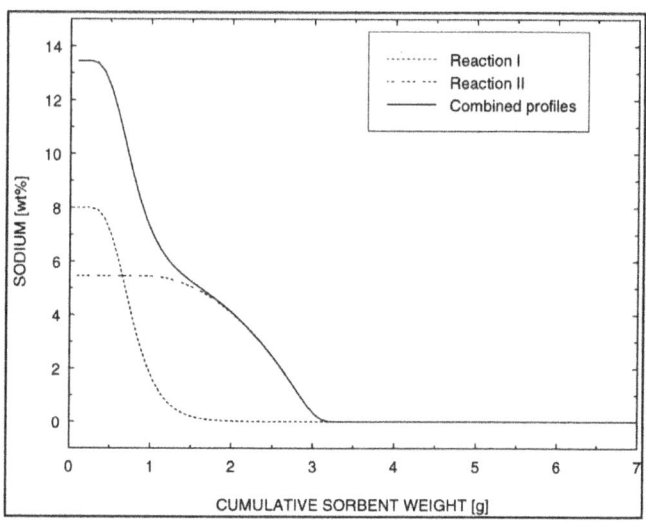

Figure 8:   Combination of two simple proflies to produce a complex curve.

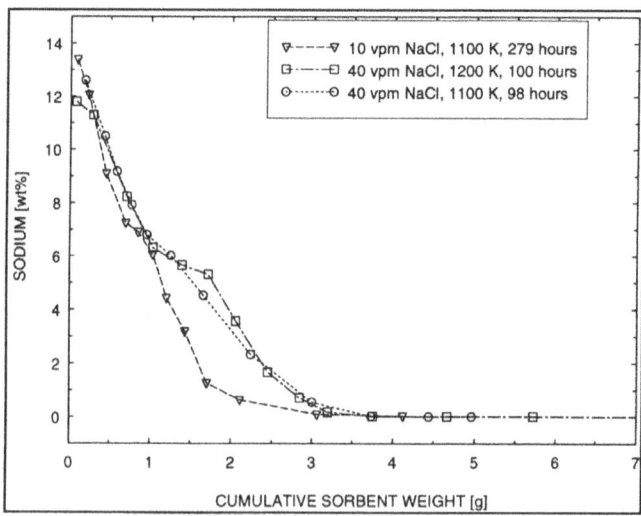

Figure 9:   Experimental data displaying a "knee" at about 6% alkali.

The Two Reaction Mechanism

From the experimental evidence available, the mechanism for alkali uptake could probably be represented by two simultaneous 'reactions': reaction I which is strongly inhibited by HCl and reaction II which proceeds virtually independently of HCl concentration. In the absence of HCl in the inlet gas, both reactions would proceed to completion, while only reaction II would proceed under high HCl concentrations. Using the theoretical description developed previously both reactions are assumed to proceed with "shrinking core" type behaviour and are reversible. The overall alkali uptake is therefore the sum of the uptake due to each reaction. As expected, this model contains many parameters: a reaction rate constant ks, effective diffusivity De and equilibrium group Ke for each reaction (I and II). However, as we develop the postulated mechanism some simplification is possible and the unknown parameters can effectively be reduced to ks_I and Ke_I, the reaction rate and equilibrium group for reaction I, respectively, and De_{II}, the effective diffusivity for reaction II. These can be recovered from experimental data through curve fitting. All other parameters can be assigned.

The experimental data has also provided estimates for the saturation capacity of reactions I and II. The combined uptake is 12.6%wt at 1100 K and about 12%wt at 1200 K without HCl. Under conditions of high HCl, the saturation uptake is about 5.6% at 1200 K. From Figure 5, at 1100 K, it appears that saturation of the first layer is not complete and it is therefore assumed that the saturation uptake is also about 5.6%wt. Also, as the sorbent has not yet reached conditions for the formation of a glassy phase, it is reasonable to assume the temperature effects on ultimate capacity for reaction II are small. In Figure 4 it was shown that increasing temperature had little effect on the rate of uptake for experiments without added HCl. Thus we assume that ks_I (the rate constant for reaction I) is similar for both 1100 and 1200 K experiments.

Parameter Recovery

As described above, the unknown parameters for this model were recovered by "fitting" of mathematically generated profiles to the experimental data, through the systematic variation of the unknown parameters. For the low temperature runs the data from experiments performed with 0 and 160 vpm HCl were used. The recovered values are given in Table 2 and the theoretical and experimental profiles are compared in Figure 10.

TABLE 2
Parameters for the Two Reaction model at 1100 K. The parameters ks_I, Ke_I and De_{II} were recovered from experimental data at 1100 K with 0 and 160 vpm HCl, 40 vpm NaCl and 5% H_2O in nitrogen. Run duration; 98 and 73.2 hours, respectively.

Parameter	Value
ks_I [cm sec^{-1}]	3.0×10^{-2}
*De_I [cm^2 sec^{-1}]	10
Ke_I [mol cm^{-3}]	1.84
*ks_{II} [cm sec^{-1}]	1×10^3
De_{II} [cm^2 sec^{-1}]	1.0×10^{-5}
*Ke_{II} [mol cm^{-3}]	1×10^3

* assigned values

567

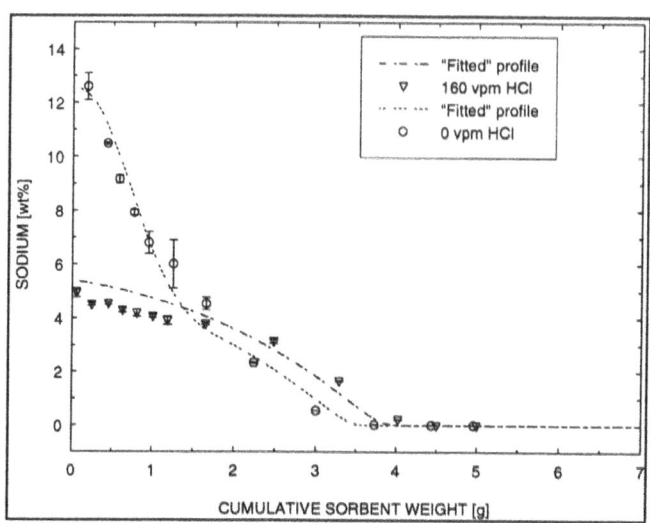

Figure 10: Theoretical profiles generated using the Two Reaction model, fitted to the low temperature (1100 K) data.

This clearly indicates that a Two Reaction mechanism produces theoretical profiles very similar to the experimental data used, and mimics the complex behaviour extremely well.

Using the above parameters the bed performance was predicted for a test at 10 vpm NaCl and zero vpm HCl, at 40 vpm NaCl and 55 vpm HCl. The predicted profiles are shown in Figure 11, together with the experimental data, indicating the versatility of this mechanism to model the sorbent performance over a range of both NaCl and HCl concentrations. Thus the Two Reaction model appears suitable for describing the complex processes occurring in a sorbent pellet, and its applicability over a range of gas concentrations is outstanding. At higher HCl concentrations the convex, diffusion limited profile dominates whereas in the absence of HCl, reaction I (reaction rate limited) is predominant.

To further test this proposed mechanism of alkali uptake, the unknown parameters for the Two Reaction model were recovered from two sets of data obtained at 1200 K (0 and 160 vpm HCl, 40 vpm NaCl). These recovered parameters were then used to predict the behaviour for a 200 hours test at .160 vpm HCl, in which the acid concentration was reduced during the final 6 hours. The recovered parameters are shown in Table 3, and the theoretical profiles versus the experimental data are shown in Figure 12.

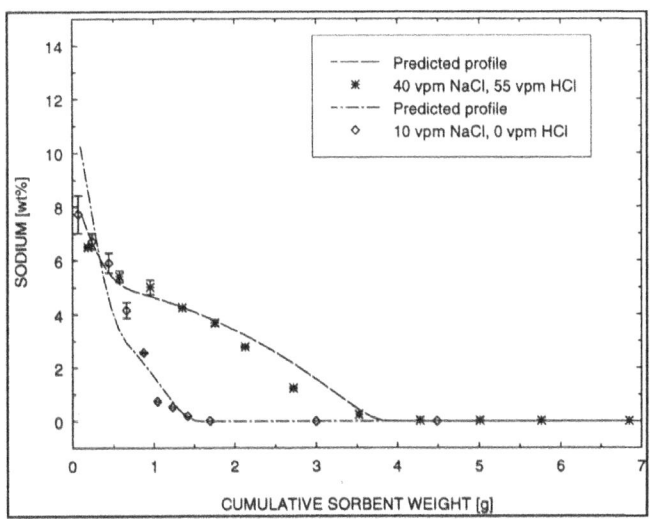

Figure 11: Predicted profiles for a 10 vpm NaCl and 0 vpm HCl experiment, and a 40 vpm NaCl and 55 vpm HCl experiments. Predicted using the Two Reaction model with parameters obtained at 40 vpm NaCl and 0-160 vpm HCl, all at 1100 K.

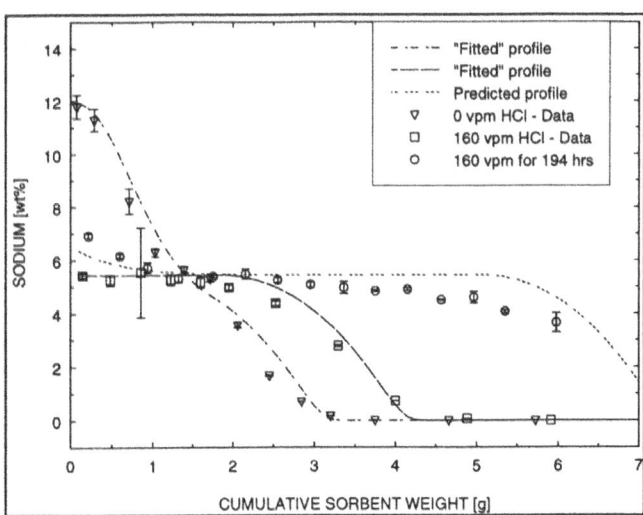

Figure 12: Predicted profiles using the Two Reaction model, for high (1200 K) temperature data. The model parameters were recovered from two 100 hours, of 0 and 160 vpm HCl and used to predict the 200 hours result.

TABLE 3

Rate parameters for Two Reaction model recovered from two data sets for experiments at 1200 K. Experimental details: 40 vpm NaCl, 5% H_2O in nitrogen, 160 vpm HCl for 100 hours, 0 vpm HCl for 100 hours.

Parameter	Value
ks_I [cm sec^{-1}]	3.0×10^{-2}
De_I [cm^2 sec^{-1}]	10
Ke_I [mol cm^{-3}]	1.1
ks_{II} [cm sec^{-1}]	1×10^3
De_{II} [cm^2 sec^{-1}]	2.4×10^{-5}
Ke_{II} [mol cm^{-3}]	1×10^3

Again this model is shown to be remarkably accurate in mimicking the complex trends displayed by the data. The results for the prediction of the 200 hours test are particularly encouraging. The rising concentration in the upper pellet layers of the bed, corresponding to the reduction of HCl, is mirrored by the theoretical profile. Some uncertainty exists at high temperature in that the ultimate sorbent capacity was not as well established as the measurement at low temperature, due to the difficult experimental conditions. This is, however, balanced by the fact that a reasonably accurate estimate is available for the sodium content of the product of reaction II, the HCl-independent reaction.

570

CONCLUSIONS

Investigation into the reaction mechanism and kinetics of alkali "gettering" were performed using pelletised calcium montmorillonite. The results indicate that the uptake mechanism is substantially more complex then assumed in previous thermodynamic studies.

The effect of increased HCl concentration was to drastically alter the overall rate of alkali uptake. It appears that HCl decreases the rate of reaction between the alkali vapour and the solid sorbent, or alters the system's equilibrium. The inhibiting effect of HCl is easily reversed once the acid concentration is reduced.

As the solid sorbent approaches saturation with alkali, a substantial proportion of the reaction products are molten at temperatures as low as 1100 K. This suggests the formation of non-stoichiometric eutectics or glasses. The presence of impurities in the sorbent mineral may alter the stability of possible phases, suggesting that the thermodynamic prediction may be a gross simplification of a complex system.

Increasing the bed operating temperature from 1100 K to 1200 K increased the rate of uptake when high levels of HCl were added to the gas stream, but had no noticeable effect on the rate in the absence of HCl, suggesting a multi-staged reaction mechanism.

Using a theoretical description of the experimental apparatus a model for the reaction mechanism was developed; the Two Reaction model. This model envisages the uptake process to be the combination of two separate, first order reactions, both forming a non-volatile alkali alumino-silicate products via a gas-solid reaction. The rate of one reaction is assumed to be sensitive to the HCl concentration in the gas phase, whereas the rate of the other is virtually independent of it.

This model contains three parameters that must be recovered from the experimental data by systematically varying their values until the sum of squares of error between sets of theoretical and experimental profiles is minimised. These "unknown" parameters were recovered from experimental results at 1100 K and 1200 K. Using the model parameters to predict experimental behaviour produced encouraging results.

ACKNOWLEDGMENTS

The authors would like to acknowledge the funding provided by British Coal Corporation in supporting this work.

REFERENCES

Bachovchin, D.M., Alvin, M.A., DeZubay, E.A. and Mulik, P.A., "A study of high temperature removal of alkali in a pressurised gasification system", Progress Report, Westinghouse Research and Development Center, Chemical Sciences Division, Pittsburgh, PA, (1986)

Deans, H.A. and Lapidus, L., "A computational model for predicting and correlating the behaviour of fixed-bed reactors: 1. Derivation of model for nonreactive systems", AIChE Journal, Vol. 6, No. 4, p 657, (1960)

Evans, J.W. and Ranade, M.G., "The grain model for reaction between a gas and a porous solid - a refined approximate solution to the equations", Chemical Engineering Science, Vol. 35, p 1261, (1980)

Fantom, I., personal communication, Coal Research Establishment, Stoke Orchard, Cheltenham, (1989, 1990)

Froment, G.F. and Bischoff, K.B., Chemical Reactor Analysis and Design, John Wiley and Sons, New York, (1979)

Fuller, E.N., Schettler, P.D. and Giddings, J.C., "A new method for prediction of binary gas-phase diffusion coefficients", Industrial and Engineering Chemistry, Vol. 58, No.5 p19, (1966)

Howarth, O.W., Ratcliffe, G.S. and Burchill, P., "Solid-state nuclear magnetic resonance studies of sodium and aluminium in coal", Fuel , Vol 66, January, p 34-39, (1987)

Lee, S.H.D. and Myles, K.M., "Measurement of alkali vapour in PFBC flue gases and its control by a fixed bed of activated bauxite", Gas Cleaning at High Temperature, IChemE Symposium Series No.99, p 149-166, (1986a)

Mulik, P.A., Alvin, M.A. and Bachovchin, D.M., "Simultaneous High-Temperature Removal of Alkali and Particulates in a Pressurized Gasification System", Progress Report DoE/MC/16372-8, Westinghouse Research and Development Center, Chemical Sciences Division, Pittsburgh, PA, September, (1983)

Punjak, W.A. and Shadman, F., "Aluminosilicate Sorbents for Control of Alkali Vapours during Coal Combustion and Gasification", Energy and Fuels, Vol 2, p702, (1988)

Raask, E., Mineral Impurities in Coal Combustion, Hemisphere, Washington, (1985)

Radcliff, A.S., "Factors influencing gas turbine use and performance", Material Science and Technology, Vol 3, p 554, (1987)

Reed, G.P., "The economics of hot gas cleaning for coal gasification combined cycle power generation", Proceedings of the Filtech Conference, London, p 384, (1983)

Ruthven, D.M., Principles of Adsorption and Adsorption Processes, Wiley and Sons, New York, (1984)

Scandrett, L.A. and Clift, R., "The thermodynamics of alkali removal from coal-derived gases", Journal of the Institute of Energy, Dec., p. 391-397, (1984)

Shon, H.Y. and Szekely, J., "A structural model for gas-solid reactions with a moving boundary - III A general dimensionless representation of the irreversible reaction between a porous solid and a reactant gas", Chemical Engineering Science, Vol 27, p 763 - 778, (1972)

Singh, A., Clift, R., Reed, G.P., Schofield, P. and Bower, C.J., "Thermodynamic calculations of the effect of chlorine on alkali removal from hot coal-derived gases", Gas Cleaning at High Temperature, IChemE Symp. Ser. No 99, p 167, (1986)

Singh, A. and Clift, R., "Thermodynamic aspects of alkali release and removal in combustion and gasification of coal", Report to Coal Research Establishment, Project No. 443 862, (1986)

Szekely, J., Evans, J.W. and Sohn, H.Y., Gas-Solid Reactions, Academic Press, New York, (1976)

Wakao, N. and Funazkri, T., "Effect of fluid dispersion coefficients on particle-to-fluid mass transfer coefficients in packed beds: Correlation of sherwood numbers", Chemical Engineering Science, Vol. 33, p 1375-1384, (1978)

LIST OF SYMBOLS

b   stoichiometric constant for "gettering" reaction
De   effective intragranular diffusivity
Ke   equilibrium group
ks   reaction rate constant

Subscripts
1   Reaction I
2   Reaction II

573

LABORATORY MEASUREMENTS OF METAL ADSORPTION FROM SIMULATED INCINERATOR FLUE GASES: SORBENT SELECTION FOR CESIUM CAPTURE[*]

S. M. Crosley
R. J. Kedl
Oak Ridge National Laboratory
Martin Marietta Energy Systems, Inc.
Oak Ridge, Tennessee 37831-6285

ABSTRACT

The Environmental Protection Agency (EPA) has proposed that the emission of acid gases, particulates, and ten specific metals in the flue gas of incineration be regulated. In addition, the Department of Energy is interested in minimizing the emission of radioactive metals in such flue gas. Dry scrubbing is an attractive alternative to wet scrubbing for the removal of these materials. Laboratory-scale experiments were conducted to measure the adsorption of cesium by silica at 1000-1300°F. Silica was found to be an effective sorbent for cesium. A methodology was developed for pre-selecting suitable sorbents for specific metals for future testing.

INTRODUCTION

It is well known that flue gases from industrial furnaces, boilers, and incinerators can release acids, metals, particulates, and other undesirable materials to the atmosphere. These emissions are toxic and otherwise detrimental to the environment. The Environmental Protection Agency (EPA) intends to regulate the emission of these materials and has issued a proposed rule to that effect (1,2). This rule will set limits on emissions of CO, HCl, particulates, and ten metals. In addition to the ten metals proposed by the EPA for emission standards, the Department of Energy (DOE) is concerned with other metal

---

[*]Research sponsored by the Office of Technology Development, U.S. Department of Energy, under contract DE-AC05-84OR21400 with Martin Marietta Energy Systems, Inc.

emissions from incinerators at their Laboratory sites. Examples of these other metals are Cs, Sr, Co, U, and Th.

Incineration has been identified as an appropriate technology to treat waste materials, such as PCBs. However, the incineration process results in combustion gases that contain noxious components which must be reduced to acceptable levels before release to the atmosphere. Several flue gas cleaning processes are available such as bubbling towers, venturi scrubbers, dry scrubbers, filters, and electrostatic precipitators. Typically a hazardous waste incinerator may require two or more flue gas cleaning processes in order to meet the stringent EPA requirements. Historically, wet scrubbing has been the predominant process used. Wet scrubbing, however, has problems. Often, the corrosive characteristics of the scrubbing solutions lead to material and operational difficulties. In addition, used scrubber solutions become a hazardous waste and must also be disposed of properly. In recent years, dry scrubbing has emerged as an attractive alternative technology that has fewer problems than wet scrubbing.

BACKGROUND

With the advent of more stringent requirements on the release and disposal of toxic and hazardous materials in the environment, interest in incineration is increasing. In parallel with this trend, interest in dry scrubbers for the removal of acids and metals from incinerator flue gas is also increasing. Conceptually, a dry scrubber could take the form of a solid spray dryer, a packed bed, or a fluidized bed. Sorbents that are often considered include limestone for the acids, and silica, alumina, or combinations of these for the metals.

Dry scrubbing is emerging as an appropriate flue gas cleaning technology, but the complete list of regulated hazardous metals has only recently been identified. Consequently, there is not a large body of technical information on appropriate sorbents available in the public literature. However, dry scrubber R&D is currently getting underway on several fronts. Fluidized bed combustors are available commercially from several companies, but much of their technical data and information is proprietary. The following is the summary of the work done in the area of dry scrubbing of incinerator flue gases.

A/S Niro Atomizer of Copenhagen, Denmark, manufactures a spray dry scrubber system that has been installed in several European hazardous waste incinerators (3,4). The spray dryer features a rapidly rotating impeller disk that is fed a slurry of $Ca(OH)_2$, $CaCl_2$, $CaSO_3$, CaF_2, and fly ash. The impeller slings particles into the flue gas that dry almost immediately to a diameter of about 30 μm. The sorbent particles are then effective in removing acid gases, metals, and particulates from the flue gas. Spray dryer scrubber systems have been installed and tested extensively in hazardous waste incinerators located in Denmark, Sweden, and Finland. When these systems were built, the principal concern was the removal of acid gases; nevertheless, they have been found to be also effective for the removal of metals and particulates.

Uberoi and Shadman at the University of Arizona have conducted laboratory experiments comparing the effectiveness of various sorbents for lead (Pb) (5). The experiments consisted of evaporating lead chloride ($PbCl_2$) into a flowing stream of artificial flue gas. The flue gas carried the $PbCl_2$ vapor to a small bed of sorbent, where part of the $PbCl_2$ was adsorbed and the rest went to a vent system. Six different sorbents were used. The results showed that silica had good adsorption efficiency and that virtually none of the adsorbed lead was leachable. Kaolinite showed a better adsorption efficiency but some of the adsorbed lead was leachable. Lime, which is often used to remove acid gases from flue gas, showed low adsorption efficiency, and almost all of the adsorbed lead was leachable. X-ray diffraction analysis of kaolinite particle absorbent indicated the formation of a lead aluminum silicate ($PbAl_2Si_2O_8$) which is water insoluble.

Ho et al., (6) investigated the capture of lead in a laboratory scale fluidized bed incinerator. The three sorbents tested were limestone, bentonite, and alumina. Lead was added by soaking wooden pellets in a solution of lead nitrate for several days and then inserting them into the fluidized bed incinerator. The results showed that all sorbents were effective in the capture of lead and that alumina was the best. The data also showed that, over the range tested, capture of lead by the sorbents decreased with increased temperature,

576

increased with smaller sorption particle sizes, and increased with higher air velocities. The authors concluded that metal capture by sorbent particles in a fluidized bed incinerator is highly promising.

Punjak et al., (7) at the University of Arizona tested three mineral sorbents for their ability to remove alkali metals from hot incinerator flue gases. This research was also directed toward the removal of alkali metals from hot flue gas resulting from the combustion or gasification of coal. In these laboratory-scale experiments, NaCl was vaporized and carried by artificial flue gas (with excess oxygen) to the sorbent samples. The sorbent samples were continuously weighed, and the increase in weight was interpreted in terms of the rate of adsorption. Each run included chemical analysis by atomic emission spectrometry, X-ray diffraction, and scanning Auger microscopy. The conclusions of the study were that kaolinite, bauxite, and emathlite (all minerals composed primarily of SiO_2 and Al_2O_3) are suitable sorbents for removal of alkali metal vapors from hot flue gas, and that the rate of adsorption decreases with alkali metal loading and drops to zero when a saturation limit is reached.

Lee and Johnson (8) at Argonne National Laboratory tested six commercially available sorbents for their ability to remove alkali metals from hot flue gases. This research was directed toward development of the pressurized fluidized-bed combustion (PFBC) concept. In a laboratory-scale experiment, vaporized NaCl and KCl were carried by either air or artificial flue gas (with excess oxygen) to a small packed bed of sorbent sample. Diatomaceous earth (SiO_2 - 92.0% & Al_2O_3 - 5.0%), Burgess No. 10 pigment (SiO_2 - 45.0% & Al_2O_3 - 38.5%), and activated bauxite (SiO_2 - 10.0% & Al_2O_3 - 81.5%) were found to be effective granular sorbents for the gaseous alkali metals. Lee and Johnson concluded that retention of alkali metals by diatomaceous earth occurs by the formation of water-insoluble alkali metal silicates and that activated bauxite captures alkali metals by an adsorption mechanism.

Laboratory testing of a fluidized bed dry scrubbing process for the removal of acidic gases from simulated incinerator flue gases was performed at the Oak Ridge National Laboratory (9). The acidic gases studied were sulfur dioxide, hydrogen chloride, and

577

phosphorus pentoxide. The experiments were conducted with the operating temperatures of the fluidized bed ranging from 540 to 650°C (1000 to 1200°F) using sorbents such as lime, limestone, and pulverized dolomitic quicklime (57% CaO, 40% MgO). Lime was eventually selected to be the primary sorbent because it exhibited a removal efficiency exceeding 98%.

Based on the references discussed above, and others, it is believed that sufficient research has been conducted on dry scrubbers for metal removal from flue gas to show that the concept has potential. It is known, for example, that some metals (e.g., lead) are efficiently adsorbed by silica and alumina sorbents. It is also known that some metals (e.g., mercury) are not efficiently adsorbed by the same sorbents. But, the adsorption characteristics are not known for a large group of metals. There are a number of other technical issues that need to be studied, such as possible effects of the metal chemical species (oxides, chlorides, etc.) and the effects that acid gases, particularly HCl, may have on metal adsorption characteristics. Finally, the chemical nature of the spent sorbent needs to be established because this material has to be classified as hazardous waste. It is important to determine whether the adsorbed metals are chemically or physically bonded to the sorbent and whether or not they are water soluble. Thus, although the concept has considerable merit, there are technical questions that need to be answered before a fluidized bed dry scrubber for metals in flue gas can be confidently designed.

EXPERIMENTAL

The screening experimental effort, reported on here, was directed toward a specific concept for a fluidized bed dry scrubber. The concept was based on assumptions that the fluidized bed sorbent would be a mixture of materials such as lime to adsorb the acid gases and sorbent oxides (SiO_2, Al_2O_3) to adsorb the metals, and that the operating temperature of the fluidized bed would probably be in the range of 538 to 760°C (1000 to 1400°F).

The original intent of this effort was to investigate the adsorption characteristics of multiple metals by a number of different sorbents. However, funding limitations constrained the scope to only cesium adsorption onto silica. The alkali metal cesium is not included in the list of metals proposed by the EPA for emission controls, but it is one of the metals of interest for control to the DOE as a radioactive component of mixed wastes.

A thermogravimetric analyzer (TGA) was selected as the basic apparatus for running the screening tests. A schematic of the system is shown in Fig. 1. The test fixture features up-flow of the carrier gas through two separate furnace chambers. The lower furnace of a Netzsch TGA serves as the metal salt evaporator. This furnace controls the metal salt temperature and, therefore, its evaporation rate. The TGA scales in this chamber continuously weigh the salt. The evaporation rate can be determined from the rate of weight decrease. The measured carrier gas flow rate and the evaporation rate together determine the concentration of metal vapor in the carrier gas. The upper furnace controls the temperature of the sorbent sample which is contained in a cup with a porous bottom. A screen is positioned on top of the sample to hold it in place. The sample temperature control is based on the reading of a thermocouple positioned in the middle (near the sample). The carrier gas was typically manufacturer-prepared artificial flue gas (N_2 − 81%; CO_2 − 11%; O_2 − 8%), although any gas could be used. The carrier gas was humidified by passing it through a water bubbler. The clearance between the cup and the flow tube was about 0.05 mm to keep sample bypass to a low level.

ORNL-DWG 90M-4253R ETD

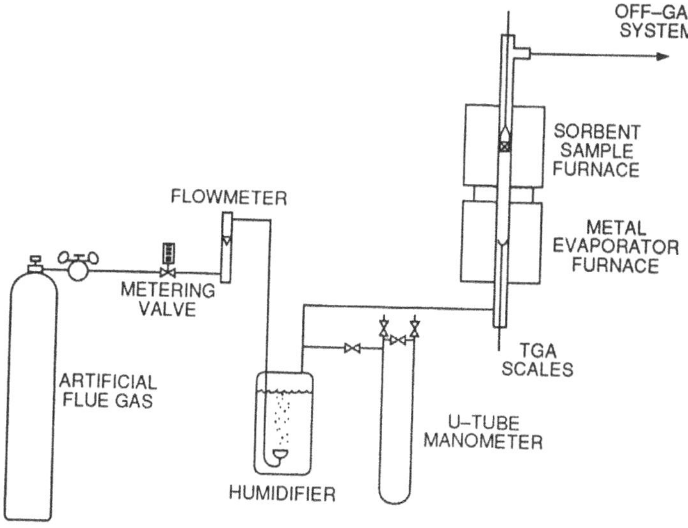

Fig. 1. TGA flow system for metal/sorbent screening test

Typically, an experimental run lasted for one day. In each experiment, the evaporator cup was loaded with metal salt and the sorbent sample holder was loaded with a weighed amount of the sorbent to be tested. The evaporator cup and sorbent sample holder were then installed in the flow tube, and the carrier gas was turned on. The sorbent sample holder furnace was heated first. As the sorbent approached the desired temperature, the evaporator furnace was activated. Typically, the evaporator furnace was maintained at its operating temperature from four to eight hours. The sorbent sample was maintained at its temperature to within 1°C (1.8°F) for the entire run. After terminating the test and allowing the system to cool, the sorbent sample holder was removed for weighing and for chemical analysis by the Inductively Coupled Plasma (ICP) technique.

The cesium source was cesium chloride (CsCl), which has a melting point of 645°C (1193°F). The measured evaporation rate in the experimental system, as a function of evaporator temperature, is shown in Fig. 2. Also shown in this figure is the CsCl vapor pressure (10). Clearly the evaporation rate correlates well with the vapor pressure. Several exploratory runs were conducted by evaporating CsCl and passing it over a sorbent. During the exploratory runs, much was learned about the system and this knowledge resulted in a few modifications. A principal problem was the highly corrosive nature of CsCl. It was necessary to remove the stainless steel screen at the bottom of the sample holder, that supports the silica sample, and replace it with a porous silica disk. The stainless steel screen on top of the sample holder, that prevents blowout of the silica sample, showed virtually no sign of corrosion and did not have to be replaced. This perhaps is an indication that the silica sorbent may have been efficient in removing the vapor.

Following the exploratory runs, three runs were completed with CsCl as the vaporized metal salt and silica as the sorbent. Results are shown in Table 1.

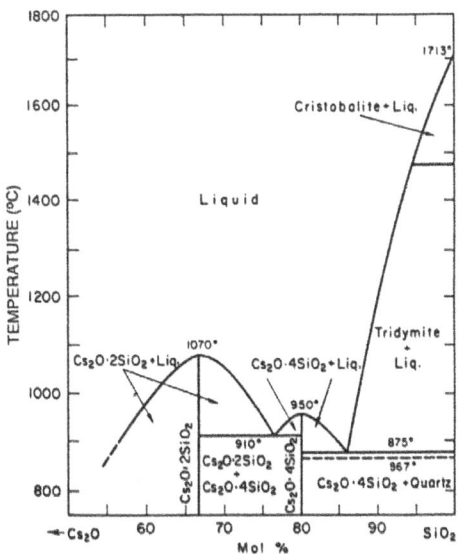

Fig. 2. Vapor pressure and evaporation rate CsCl

TABLE 1

EXPERIMENTAL CONDITIONS FOR CESIUM ADSORPTION

Run No.	CsCl T°C (°F)	Silica T°C (°F)	Evap. Rate (mg/h)	Duration of run (h)
1	850 (1562)	538 (1000)	135	4
2	800 (1472)	649 (1200)	67	8
3	800 (1472)	704 (1300)	67	8

RESULTS AND ANALYSIS

The purpose of these experiments was to show the extent of Cs adsorption onto silica. In Experiment No. 1, the test conditions were such that saturation of CsCl vapors occurred and thus the CsCl condensed onto the sorbent sample. Experiments 2 and 3 were therefore conducted to reduce the CsCl vapor pressure to well under saturation levels at the silica

581

support disk and at the silica sample so that condensation would not occur to interfere with the adsorption.

The condensation of CsCl onto the surface of the silica disk experiment No. 1 was indicated by an observed steadily increasing pressure drop au sample and a considerable decrease in the carrier gas flow rate. On post-test inspection, a layer of solid material was found on the bottom surface of the porous silica sample support disk. This layer appeared to be about 1/32-inch thick and was pale yellow in color. The silica disk had increased in weight by 101 mg, and the silica sample had increased in weight by 13 mg. Since there was also considerable corrosion at the stainless steel flow tube and sample holder, an overall material balance could not be made to determine the percentage of CsCl that had condensed onto the sorbent surface.

In Experiments No. 2 and No. 3, the CsCl vapor pressure was maintained unsaturated and the yellow cake of condensed CsCl was not produced on the bottom of the porous silica support disk. In addition, the artificial flue gas flow rate remained approximately constant for the duration of both runs. Thus, for those two tests, the pickup of CsCl by silica was believed to have resulted from a chemical affinity between the materials and not by simple condensation. A mass balance still could not be calculated because of considerable corrosion of the tube.

Samples of the silica, porous quartz support disk, and the corrosion scale were submitted for analysis by atomic adsorption spectroscopy. Water soluble and water insoluble fractions of each sample were also determined. Most of the cesium that was adsorbed onto the silica disk and the sample was insoluble in water. The greatest amount of insoluble cesium was found on the porous silica disk and the greatest amount of soluble cesium was found in the silica sample. The presence of insoluble cesium (cesium compounds are generally very soluble) may be explainable by the proposed mechanism for metal adsorption described in the next paragraph. The CsCl was likely converted to Cs_2O by the oxygen and water vapor in the carrier gas. The Cs_2O was then adsorbed by the silica and chemically reacted to $Cs_2O \cdot 4SiO_2$. Figure 3 is a phase diagram of the Cs_2O-SiO_2 system. Note that at low Cs_2O concentrations (<20 mole %) and below 875°C a stable solid compound with

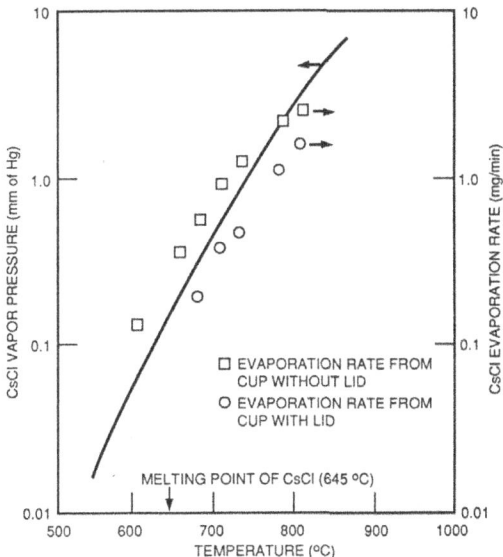

Fig. 3. Phase Diagram for the $Cs_2O - SiO_2$ system

the formula $Cs_2O \cdot 4SiO_2$ is formed. Solubility data for this material are unavailable; however, it is likely to be an insoluble mineral-like material.

PROPOSED MECHANISM FOR METAL ADSORPTION

Based on the research already done on a number of metals and the large amount of data present in the literature concerning metals and metal compounds, a methodology was developed for selecting appropriate sorbents for specific metals as candidates for future testing. The following explains the basic principles of this methodology.

The chemistry of metals in the incineration process is largely unknown. Since incineration operates at high temperature and with excess oxygen, it is generally believed that most metals will be released as oxides. However, if chlorine is also present, some metals may be released as chlorides. Because of their generally high melting points and low vapor pressures, metal oxides are the most desirable form of release. These will more readily condense to form particles, which are amenable to removal from the flue gas by standard filtration methods. In addition, the disposal of oxides is expected to be easier because they are generally insoluble and chemically stable.

The afterburner of a general purpose hazardous waste incinerator operates at about 1204°C (2200°F). This temperature is required for PCB destruction. At this temperature, the vapor pressure of many metal salts and oxides can be high. Flue gas leaving the afterburner could potentially contain metal salt and oxide vapors saturated at this temperature. After leaving the afterburner, the flue gas is generally cooled to the temperature desired for fluidized bed dry scrubbing. By cooling, many metal compounds will condense to form particles. Thus, the metal compounds will pass through the fluidized bed in two forms, as vapor and as particulate. The issue under study here is whether or not the metal vapor compounds have affinity for the sorbent materials. That is, will the sorbent materials remove vapor phase metal compounds from the flue gas? One approach to answering this is to search the literature for stable materials composed of the sorbent oxide and the metal oxide. For example, if PbO is the metal vapor of concern and SiO_2 is the sorbent, one would search to see if there is a stable structure involving both materials at the temperature of the fluidized bed. If there is such a stable structure, then it may be assumed that adsorption of PbO by SiO_2 is likely. Figure 4 is a phase diagram for the PbO-SiO_2 system. It shows that PbO and SiO_2 form solid crystalline structures below 700°C (1292°F) at all concentrations. Thus, if a molecule of PbO from the vapor phase collides with a SiO_2 crystal and forms a PbO-SiO_2 compound, it can be assumed that compound is stable and the PbO molecule would be effectively removed from the flue gas. Although this approach is only qualitative, it does show where the potential exists for an affinity between the metal oxide and the sorbent oxide.

An excellent source of phase diagrams has been compiled by the National Bureau of Standards and published by the American Ceramic Society. This seven-volume series, titled "Phase Diagrams for Ceramists," was searched for phase diagrams involving the oxides of 15 metals and 5 sorbents. Table 2 shows the results of this search. The reference numbers in the table indicate that a phase diagram was found for the indicated metal and sorbent and that a stable crystal structure existed at the temperature of 700°C (1292°F). Our interpretation of Table 2 is that, for a given metal oxide-sorbent oxide combination that is referenced, there is the potential for a chemical affinity between the two oxides and for effective removal of the metal oxide from the flue gas.

From Table 2 it can be seen that some of the metal oxides (e.g., for Ba, Be, Co, Sr, U) have entries for all or most of the sorbent oxides studies. Based on our interpretation of the table, these metals should be removable from flue gas by a dry scrubber containing the sorbent. Other metals (e.g., Sb, As, Hg, Ag, Tl) have no entries for the sorbent oxides studied. These metals may not be removable from the flue gas by the dry scrubber process using the indicated sorbents.

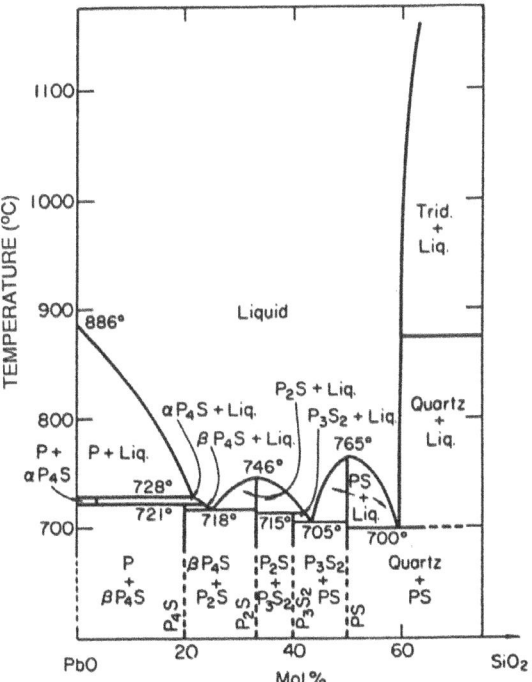

Fig. 4. System PbO - SiO$_2$

P = PbO, S = SiO$_2$, P$_4$S = Pb$_4$SiO$_6$, P$_2$S = Pb$_2$SiO$_4$, P$_3$S$_2$ = Pb$_3$Si$_2$O$_7$, PS = PbSiO$_3$

TABLE 2

Metallic element of the metal oxide compounds	Sorbents				
	SO_2	$Sio_2{\cdot}Al_2O_3$	Al_2O_3	CaO	MgO
EPA Metals of Concern					
Antimony (Sb)					
Arsenic (As)					
Barium (Ba)	A_1	A_1, A_3, A_4	A_1		A_1
Beryllium (Be)	A_1	A_1, A_3	A_1	A_1	A_1, A_3
Cadmium (Cd)			A_3		
Chromium (Cr)	A_1	A_2	A_1	A_1, A_6	A_1
Lead (Pb)	A_1, A_4	A_1	A_1, A_4	A_4	
Mercury (Hg)					
Silver (Ag)					
Thallium (Tl)					
DOE Metals of Concern					
Cesium (Cs)	A_2			A_1	A_1
Cobalt (Co)	A_1, A_2	A_3	A_1, A_6		A_1
Strontium (Sr)	A_1	A_1			
Thorium (Th)	A_1, A_2		A_1	A_1	A_1
Uranium (U)	A_1	A_1			

NOTE: A_1 - E. M. Levin et al., "Phase Diagrams for Ceramists - Vol. I," The American Ceramic Society, Columbus, Ohio, 1964.

A_2 - E. M. Levin et al., "Phase Diagrams for Ceramists; 1969 Supplement - Vol. II, " The American Ceramic Society, Columbus, Ohio, 1969.

A_3 - E. M. Levin et al., "Phase Diagrams for Ceramists; 1975 Supplement - Vol. III, " The American Ceramic Society, Columbus, Ohio, 1975.

A_4 - R. S. Roth et al., "Phase Diagrams for Ceramists - Volume IV," The American Ceramic Society, Columbus, Ohio, 1981.

A_5 - R. S. Roth et al., "Phase Diagrams for Ceramists - Volume V," The American Ceramic Society, Columbus, Ohio, 1983.

A_6 - R. S. Roth et al., "Phase Diagrams for Ceramists - Volume VI," The American Ceramic Society, Columbus, Ohio, 1987.

A_7 - R. S. Roth et al., "Phase Diagrams for Ceramists - Volume VII," The American Ceramic Society, Columbus, Ohio, 1989.

CONCLUSIONS

Some conclusions and observations can be made from the results of these screening experiments and analyses:

1. Silica appears to be a good adsorption media for dry scrubbing of gaseous cesium.
2. Most of the adsorbed cesium was insoluble in water.
3. The adsorption of cesium appears to be based on a chemical reaction between Cs_2O and silica to form $Cs_2O \cdot 4SiO_2$, an insoluble mineral-like material.
4. A method for selecting appropriate sorbents for various metals has been developed, and a mechanism was proposed to explain the chemical adsorption of various metals by sorbent oxides.

REFERENCES

1. 40 CFR Part 260 et al., "Burning of Hazardous Waste in Boilers and Industrial Furnaces; Supplement to Proposed Rule," Published in Federal Register (October 26, 1989).
2. 40 CFR Parts 260, 261, 264 and 270, "Standard for Owners and Operators at Hazardous Waste Incinerators and Burning Hazardous Wastes in Boilers and Industrial Furnaces; Proposed and Supplemental Proposed Rule, Technical Corrections, and Request for Comments," Published in Federal Register (April 27, 1990).
3. J. T. MOLLER and O. B. CHRISTIANSEN, "Dry Scrubbing of Hazardous Waste Incinerator Flue Gas by Spray Dryer Absorption," 77th Annual Meeting of the Air Pollution Control Association, San Francisco, CA (June 24–29, 1984).
4. J. T. MOLLER and O. B. CHRISTIANSEN, "Dry Scrubbing of MSW Incinerator Flue Gas by Spray Dryer Absorption: New Developments in Europe," 78th Annual Meeting of the Air Pollution Control Association, Detroit, MI (June 16–21, 1985).
5. M. UBEROI and F. SHADMAN, "Sorbents for Removal of Lead Compounds from Hot Flue Gases," *American Institute of Chemical Engineers Journal*, Vol. 36, No. 2, pp. 307–309 (February 1990).

6. T. C. HO et al., "Metal Capture During Fluidized Bed Incineration of Solid Wastes," AIChE Symposium Series – Advances in Fluidization Engineering, No. 276, Vol. 86, pp. 51–60 (1990).

7. W. A. PUNJAK, M. UBEROI, and F. SHADMAN, "High Temperature Adsorption of Alkali Vapors on Solid Sorbents," *American Institute of Chemical Engineers Journal*, Vol. 35, No. 7, (July 1989).

8. S. H. D. LEE and I. JOHNSON, "Removal of Gaseous Alkali Metal Compounds from Hot Flue Gases by Particulate Sorbents," *Journal of Engineering for Power*, Vol. 102, (April 1980).

9. W. M. BRADSHAW, R. P. KRISHNAN, and J. M. YOUNG, "Laboratory Testing of a Fluidized-Bed Dry-Scrubbing Process for Removal of Acidic Gases from a Simulated Incineration Flue Gas," *American Institute of Chemical Engineers 1988*, Annual Meeting, Washington, D.C.

10. R. C. WEAST et al., "CRC Handbook of Chemistry and Physics," 70th Edition, 1989-1990, CRC Press, Boca Raton, FL.

11. E. A. BRANDES, "Smithells Metals Reference Book," Sixth Edition, 1983, Butterworths, Boston, MA.

BEHAVIOUR OF VOLATILE MATERIALS IN CEMENT KILN SYSTEMS

C. P. KERTON,
Principal Scientist
Blue Circle Industries PLC
Technical Centre, 305 London Road, Greenhithe, Kent DA9 9JQ, UK

ABSTRACT

Factors governing behaviour of minor amounts of Cl, K, Na and S in cement production kilns are reviewed. Recirculating loads form as a result of partial volatilisation in the burning zone and are often implicated in formation of build-ups, coating and blockages in cooler parts of the process. Differences arise between plants from design features and from characteristics of raw materials and fuels. In a given location, burning zone temperature and atmosphere are dominant driving forces. Key areas for action to gain control lie in selection and preparation of fuels, burner settings and raw materials; also in application and control of secondary firing systems in precalciner plants and process control and diagnosis via appropriate sensors, sampling and chemical analysis. Many Works can "live with" effects of condensation by appropriate modifications and plant maintenance. Prospects for more fundamentally based modelling and detailed understanding of controlling mechanisms have improved in recent years with progress of relevant thermodynamic studies in other domains.

INTRODUCTION

Naturally occurring minerals - principally limestone and shale - are sintered and chemically combined in an inclined rotary kiln to produce clinker, subsequently ground with a minor gypsum addition to produce Portland Cement. Flame temperatures are above 2000°C, with solids reaching over 1400°C. Pulverised solid fuel flames are most frequently encountered (ash being incorporated in the product) and modern plants preheat and partly calcine incoming finely-ground dry feed material in a system of cyclones before it reaches the kiln proper (Kerton & Murray, 1984); some process fuel may be burned in a supplementary furnace below the preheater (Figure 1). The concern is with minor constituents which partly or wholly evaporate in kilns and travel towards cooler zones in the process gases. A number of effects are observed which are similar to those seen in certain other fuel combustion processes where lime-based sorbents are injected to capture vapour and gas phase species.

589

(The usual cement industry convention is followed in expressing the results of many chemical analyses in terms of oxides, e.g., CaO, SO_3, Al_2O_3, etc., usually on a "loss free" basis, i.e., after allowing for the loss in weight eventually experienced due to destruction of carbonates, etc., during heat treatment.)

"VOLATILES" AND MECHANISMS OF VOLATILISATION

The principal volatile elements are K, Na, Cl, S. In the case of raw materials, certain sulfur compounds (sulfides or organics) can readily decompose/volatilise below 600°C, but most volatile compounds in raw materials only evaporate partially and at higher temperatures as the feed passes towards the kiln burning zone. The residue remains in the product, either in solid solution in the principal phases of the clinker or as discrete compounds. Whilst almost all feed chloride will evaporate, lesser amounts of other compounds do so, whilst in contrast, fuel volatiles are almost always entirely evaporated during combustion.

Evaporated volatiles travel back up the kiln with the combustion gases and condense as inorganic compounds (liberating latent heat):

i)   on the feed, forming the basis of a recirculating internal volatile load

ii)   as a fine dust or fume which is finally trapped in the gas cleaning system or raw mill and becomes part of an external volatile cycle, as dust is partly or wholly returned to the system

iii)   on colder surfaces in the system, forming the basis of build-ups.

Pressures to exploit ever more marginal reserves of raw materials and fuels give rise to increasing familiarity with the effects of volatile species on process performance. When condensed volatiles return towards the burning zone, depending on the overall chemical conditions and burning conditions, they form a range of volatile compounds which themselves evaporate partially and the cycle only finds an equilibrium when the total quantity leaving the system (in clinker and non-returned dust) equals that entering the system.

Alkalis and sulfates entering the burning zone in practice largely form a separate molten sulfate phase, immiscible with the principal ferrite flux. The level of volatiles in recirculation is significantly greater than their total rate of introduction and the substances in the vapour phase can be in various states of dissociation and recombination. In general, the preferred chloride compound is potassium chloride and only when there is an excess of chlorine for chemical combination with potassium will sodium chloride be formed as a recirculating volatile species. Alkali sulfates (Na_2SO_4, K_2SO_4) evaporate congruently, disappearing entirely when heated for a long period. $CaSO_4$ decomposes and leaves residual CaO (in oxidising conditions), so that $CaO:CaSO_4$ melts can form. Typical recirculating volatile loads expressed as % of the total quantity introduced are as follows:

	%			%
Chloride(s)	5000		Na_2O	150 - 200
K_2O	200 - 650		SO_3	200 - 800

590

FIGURE 1
FLOW SHEET FOR DRY PROCESS CEMENT KILN
(Various geometrical adaptations are used; some indicative temperatures are listed)

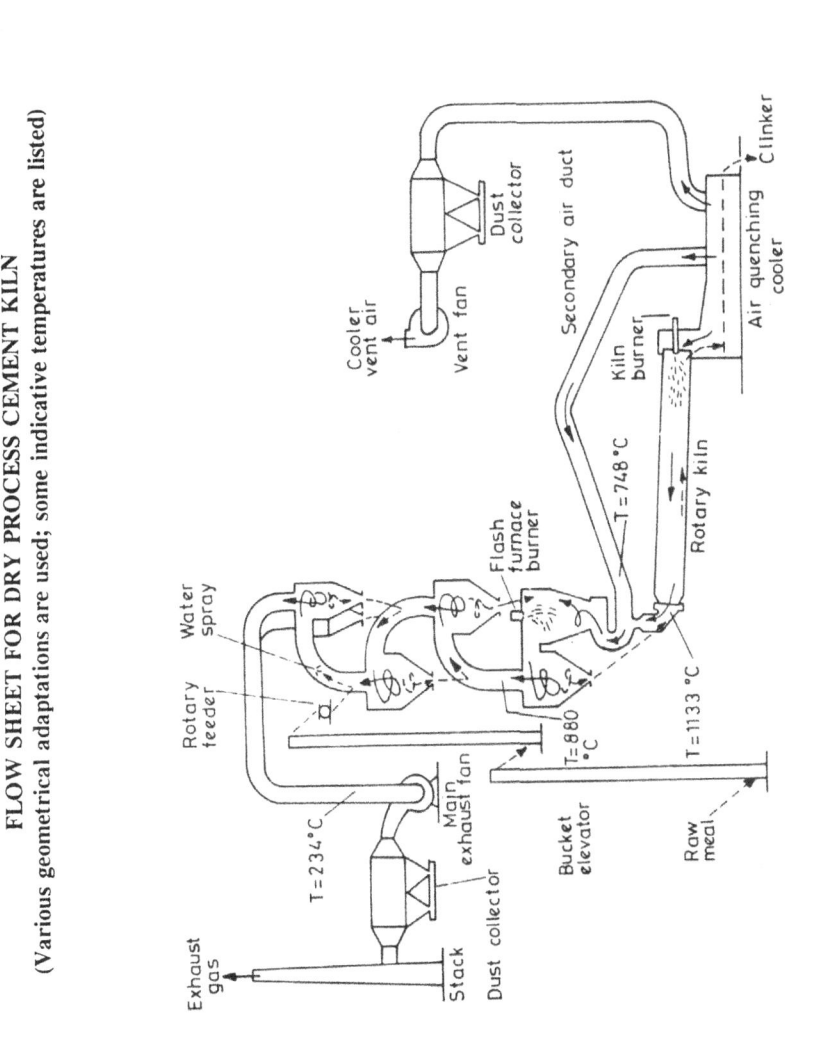

EFFECTS OF PROCESS CONDITIONS ON BEHAVIOUR OF VOLATILES

The principal factors which influence volatile behaviour are summarised in Table 1. Once a plant is in operation, the main parameters available for control are the temperature and the atmosphere in the kiln. The rate of gas flow seems to be of secondary importance. Process design and local chemical conditions also play a part, determining the total quantities introduced to the system, chloride content, relative concentrations, combinability of major oxides (i.e., the intensity of heat treatment needed), fuel and firing system parameters and mixing within the clinker bed (with exposure of nodules to hot gas flow at the surface).

Effect of clinker nodule size on volatilisation seem impossible to separate from those of alkali melt liquid on nodule growth. This is just one example of the difficulty in distinguishing chicken from egg in industrial kiln operation. A number of illustrative cases follow. A full theoretical description should explain all of these: confusion can result from looking at any one case in isolation. Later sections outline physical mechanisms involved.

Case 1: Changing from or to gas as a fuel has effects on cycles. This is to be expected due to the presence of a higher level of water vapour in the combustion products (typically, coals < 10%, gas > 20%) and the higher vapour pressures of alkali hydroxides compared with chlorides and sulfates, but such changes are always accompanied by other alterations to the quantity and composition of volatiles introduced and/or the raw meal chemistry.

Case 2: With additional calcium chloride injection at the kiln flame on a precalciner system (fitted with a bypass, where part of the kiln gases is "bled" or "by-passed" below the preheater for quenching and separate de-dusting to remove part of its volatile load), there was a marked effect on Cl and alkali levels in kiln inlet material but not on sulfate. Full system equilibrium was not reached for three days.

Case 3: Various phenomena were seen when returning to coal firing after the use of gas, including a reduction in sulfate cycles and in the level of decarbonation at the preheater tower exit. This latter effect is attributed to a changed heat release within the preheater, notably from $CaSO_4$ recombination. The thermal contribution was of the order of 75 kcal/kg with the inverse effect in the burning zone. (At another site the magnitude of this heat pump effect has been calculated as about 110 kg/kcal.) The sulfate level in the kiln entry material can be reduced (and that of chloride raised) by increased burner momentum: wear of the burner tip allows the level to rise again in due course. A "non-stick" kiln feed chute lining also gives beneficial results, allowing operation at slightly higher volatile levels. Through frequent and regular sampling and analysis of kiln entry material, a positive statistical correlation has been established between frequency of kiln stoppages due to cyclone blockage and sulfate content (which itself is inversely related to potash content). It is also noted that input of used tyres at the kiln back end was useful for mechanical removal of build-ups.

Case 4: In examining the performance of a new burner with increased flame momentum, it was noted that the CO signal at the upper end of the kiln could be made to disappear (for the same O_2 level) with less decarbonation of material at the preheater exit and a higher level of kiln drive power (indicating an increased and/or more viscous load of solids) and a lower level of gaseous SO_2 detected in gas samples taken at the kiln back end.

Case 5: Where clinker K_2O level serves as an indirect control parameter for burning

TABLE 1

FACTORS INFLUENCING VOLATILE BEHAVIOUR	MAJOR FACTOR	LESSER FACTOR	POSSIBLE CONTROL ACTIONS
Burning zone temp. (vapour pressures) — level	●	○	Soft burn (incl. fluxes & mineralisers).
— variations		○	Controlled burning.
Temp. profile of burning zone		○	Flame/burner settings.
		●	Fuel characteristics.
Alkali composition in burning zone			Selection of raw feed chemistry (sulfate/alkali ratio).
Burning zone atmosphere			
- globally reducing	●	○	Burner/flame settings.
- locally reducing	●	●	Fuel characteristics.
- water vapour	●	●	Coal/coke fineness.
			Choice of fuel (solid, liquid, gas).
Clinker size grading		●	Selection of raw feed chemistry (sulfate/alkali ratio) - incl. flux.
Clinker flux content (density)		●	As above.
Thermal efficiency (gas flow)	●	●	Process design/selection.
Preheater system design			
- vertical (shaft)		●	Process design/selection.
- cold areas/air inleaks		●	Eliminate air inleaks/ Insulate.
- anti-build-up lining		○	Design.
- geometry		○	Type of precalciner.
- solid/gas loading		○	Throughput.
Precalciner: design	○		Selection of precalciner.
operation	○		Control of precalciner.
Composition of preheater alkalis	○	○	Select of raw mix components. Selective quarrying/ Up-grade.
Dust return			Control.
By-pass system	●		Add.

○ = *uncertain*

temperature in a wet process kiln, a considerable data-bank of measurements has been built up. The use of lower ash fuel lowered clinker K_2O level by 0.15%, despite the introduction of a little more potassium to the system (an extra 0.1% on clinker). The dust - returned to the kiln - had become more rich in alkalis, so that the proportion of K_2O brought in by solid fuel fell from 24% to 18% while that brought in by dust return rose from 18% to 29%. This suggests that K_2O incorporation in clinker not only depends on the quantity introduced but also - and above all else - on the type of material which brings it in and perhaps on the position where it is injected. On screening the clinker at 20 mm, chemical analysis showed a K_2O concentration some 10% higher in the coarse fraction. The clinker alkali content has successfully been reduced in trials by calcium chloride addition at the flame.

Case 6: A high chloride coal ($\sim 0.15\%$ Cl) can only be used as a mix with another coal to avoid build-ups with typically 2% Cl at the bottom of cyclone 4 (the lowest in the preheater tower). Analysis of build-ups along the kiln indicated chloride levels up to 30% (at zero loss on ignition) in the coating from the base of cyclone 2 and 20% at 50 m into the kiln (despite its less than 5% level in both hot and cold parts of the kiln).

Case 7: Adding a second preheater cyclone stage to a long dry process kiln (a variant of the Figure 1 process, with a higher kiln length/diameter ratio and a single cyclone stage above it) yielded various build-up problems. To resolve these, the kiln gas exit O_2 level was increased from 0.5 to 1.5%, solid fuel residue at 90 microns was reduced to below 25% and several compressed air "blasters" were installed to dislodge material from the lower regions of the preheater. These actions improved the situation and subsequently additional measures were taken: refractory stirrers were added to the lining near the kiln back end, burner air velocity was increased, and a "non-stick" lining was installed in the kiln exit gas duct and cyclone dip-tubes. These efforts made better output rates possible without build-ups.

Recently, a higher sulfur fuel blend has been used, accompanied by slag (S $\sim 1\%$) among the raw mix components. Preheater blockage problems recurred, but build-ups can be avoided if the SO_3 level in samples taken from the kiln entry material is kept below 2.5% by limiting fuel S content and slag use in the raw mix, provided that in addition the oxygen level at the kiln back end is kept consistently at or above 2%. Computer control of the kiln helps to achieve success, reducing the variability of the O_2 signal.

Case 8: Laboratory data on minor elements confirm effects observed in practice. For example, sulfur volatility in a standard regime (70% N_2, 30% CO) is close to 100% at 0% O_2 but falls in the presence of O_2; nevertheless the effect of O_2 is much less at 1400°C than at 1200°C. The volatility of minor elements in the laboratory is also much greater for powdered material than for granules.

Case 9: A precalciner kiln ran well with a Cl level in kiln entry material of 3 to 4% (about 0.5% less than the K_2O level) provided no trace of CO was indicated. (There was about 1.1% SO_3 in the kiln entry meal in this situation.) If CO was detected, there was about 2% SO_3 and 5% K_2O in the hot kiln inlet feed, accompanied by build-ups based on Spurrite ($2Ca_2(SiO_4).CaCO_3$) and cubic KCl crystals. (It is generally recognised that regular kiln operation helps to minimise the phenomenon of cementation by the freezing/ thawing of chloride-based deposits.)

Case 10: To examine the feasibility of producing a sulfate-rich clinker without installing a by-pass, tests were run for some weeks with the objective of reducing burning zone volatilisation by playing on process parameters and producing a lightly mineralised clinker with higher volatile retention. During changeover there was some tendency to form soft build-ups in the preheater, but with the new regime established these moved towards the kiln feed chute without causing major problems for kiln operation. (Evidently there are phenomena of both short-term and long-term stability: once a stable burning zone volatilisation is established, it takes time for stable conditions to arrive higher up the system and in the large masses of material in the build-ups and coatings already in existence.) The apparent burning zone temperature was reduced by about 120°C, while K_2O volatility dropped from 70% to 60% in the burning zone and that of SO_3 from 80% towards the range 50% to 60%, provided that kiln exit oxygen level was kept above 2%. There were improvements in kiln output rate and fuel consumption and the experimental Works adopted certain of these changes during normal operation for several years.

Case 11: Tests were carried out involving various NO_x levels (to indicate flame temperature) as well as chemically reducing flame conditions.

SO_3 :   The ratio of SO_3 in Stage IV to SO_3 in raw meal varied typically from 1.8 to 2.7 for the higher levels of NO_x and was 3.0 for a low O_2 level. The clinker SO_3 content fell.

K_2O and Na_2O :   In a parallel manner, for K_2O the ratio of the content in Stage IV to that in raw meal varied from 3.8 to 4.4 and for Na_2O from 1.6 to 2.0.

In general, reducing conditions increased SO_3 level at Stage IV by a factor of 2, also giving a lower clinker SO_3.

Case 12: SO_2 has been monitored at the kiln back end to determine local rules for avoiding blockage tendencies. The SO_2 signal is noisy and difficult to interpret without a knowledge of the history of the system, e.g., a recent breakaway of sulfate build-up material arriving in the burning zone can give a high SO_2 signal at the kiln back end despite the presence of a good flame and acceptable levels of volatiles in kiln entry material. Rules have lowered the number of kiln stops per year caused by preheater blockage from over 90 to less than 10, lost time hours having also fallen from around 450 per year to about 100. (There were also major gains in stops caused by rings and breakaways at the kiln entry seal.)

EFFECTS OF CONDENSATION

The effect which recirculating volatiles exert on build-up formation at the kiln (gas) exit depends on composition (governing temperature of liquid formation and thus the position and hardness of build-ups as well as the surface tension and viscosity of the liquid condensate) and on the quantity (which governs rate of formation). The plant geometry and the throughput also play a part in making effects more or less pronounced. In extreme cases, part of the kiln gases are "bled" or "by-passed" for quenching and separate de-dusting to remove volatiles - incurring financial penalties in plant cost, complexity and fuel use.

In the past, several empirical limits have been proposed for concentrations of volatiles admissible in a kiln, e.g., 0.03% Cl on clinker for preheater kilns. Today there is a tendency to prefer to specify concentrations tolerable at the kiln inlet. By way of indication, concentrations of volatiles which can be tolerated in the lower stages of a preheater are typically given in the following ranges (exceptions being of special interest for study):

	%	
Cl	1.2 - 1.8	
SO$_3$	2.5 - 4.5	
Alkalis (eq. Na$_2$O)	2.5 - 3.5	(eq. Na$_2$O = Na$_2$O + 0.658K$_2$O)

It may be possible to operate with a higher level of some individual component, depending on the composition of other materials present and the resources devoted to cleaning the interior of the kiln system.

The F L Smidth encrustation index

$$R = \frac{\text{Total molar S input}}{\text{Total K}_2\text{O input} + 0.5 \text{ (Total Na}_2\text{O input)}}$$

can be used as an indicator of the potential nature of any build-up, as follows:

R > 1	Hard build-ups based on SO$_3$
0.7 < R < 0.9	Relatively soft build-ups (easily removed)
R < 0.5	Carbonate-based build-ups in due course.

(There are older variants of the cement kiln process which involve different manners of kiln feed preparation - wet, semi-wet, semi-dry. These have external cycles of dust return as well as possibilities of internal cycles - although the likelihood of volatile escape from the rotary kiln is greater, given the simpler pre-heating systems employed, with less intimate gas/solid contact than in the multi-stage cyclones of the dry process.)

There is less literature on phenomena governing condensation than evaporation. Potassium chloride alone condenses between 800 and 900°C (sodium chloride at a slightly lower temperature). Build-ups can develop in the kiln feed chute or the riser duct towards the bottom preheater stage. The accumulation of fluorine to around 1% concentration can aggravate build-up problems, aiding the formation of various silicates.

There is an optimum temperature for capture of SO$_2$ by freshly calcined raw meal (e.g., 880°C in one study). Most sulfates condense in the range 900 - 1100°C., most probably initially as liquid alkali sulfates. Melting in the ternary system Na$_2$SO$_4$/ K$_2$SO$_4$/ CaSO$_4$ starts below 800°C (Ritzmann, 1971). It is often interesting to calculate the composition of the sulfate phase in kiln inlet material, with its addition of KCl which increases the range of sulfate compositions liquid at this temperature, allowing formation of liquid melts even below 700°C (Figure 2). Deposits on the feed can provoke chemical reactions; they can equally cause adhesion and - as with deposits on surfaces - initiate build-ups. As the thickness of coatings increases, internal temperature drops and new equilibria may be established and compounds formed, e.g., direct condensation as calcium langbeinite - 2CaSO$_4$.K$_2$SO$_4$ - is not expected on thermodynamic grounds. This and other complex compounds are identified in material removed from kiln systems: examples are spurrite (2Ca$_2$(SiO$_4$).CaCO$_3$); syngenite (CaK$_2$(SO$_4$)$_2$); ellestadite; K$_3$Ca$_2$(SO$_4$)$_3$F; KF.2(Ca$_6$(SO$_4$)$_3$F; Ca$_6$Al$_4$Fe$_2$O$_{15}$.

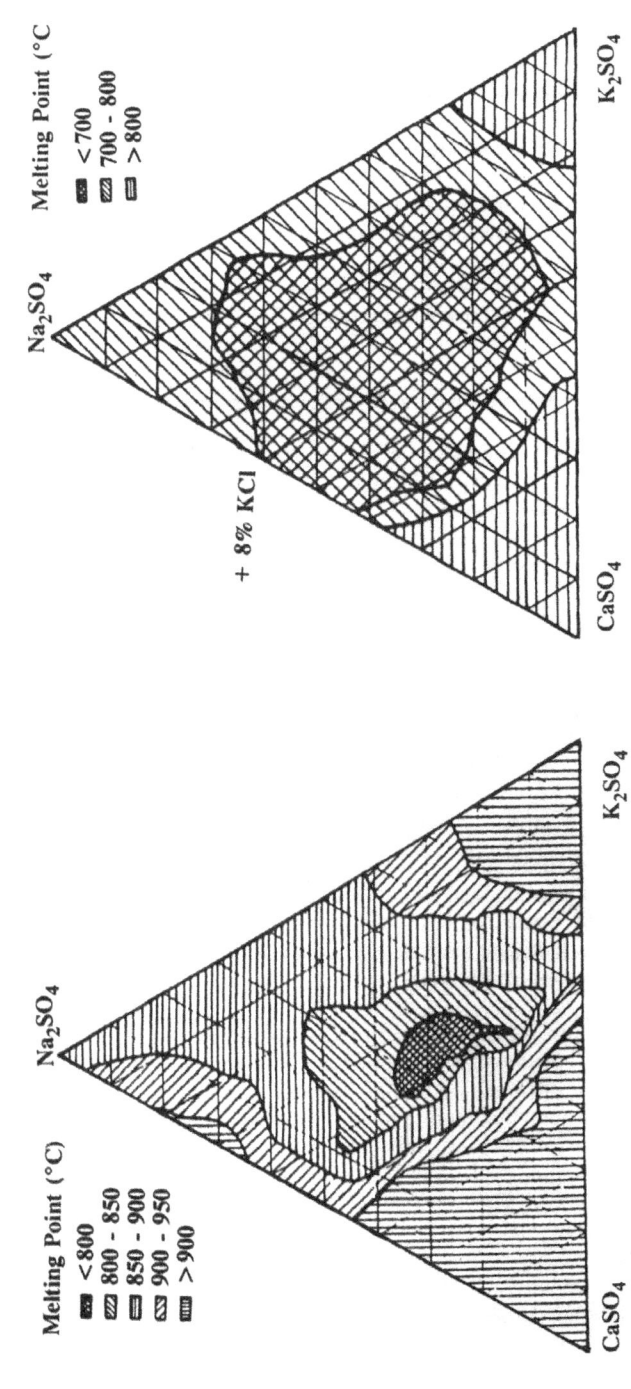

FIGURE 2
MELTING RANGE OF CaSO₄-K₂SO₄-Na₂SO₄ SYSTEM
AND EFFECT OF 8% KCl ADDITION

The interpretation of chemical data is usually further complicated by a lack of precise knowledge of material flow rates other than at the entry and exit of the kiln system. Not only are there the return fluxes of volatiles to consider, but also dust cycles (of various compositions) between the various cyclone stages and especially from the kiln to preheater.

The effects of atmosphere on condensation and the implications of precalciner operation are perhaps worthy of study, especially for sulfate compounds which can be present in various states of oxidation and which can react with water vapour to form bisulfites and bisulfates. The characteristics are as follows (Choi & Glasser, 1988):

Chemical State	Fuel and raw material	Clinker	Vapour
Oxidised S^{4+}, S^{6+}	Sulfates, e.g., gypsum, anhydrite, alum, etc.	Alkali sulfates and alkali/alkali metal sulfites	SO_2 and (at low temp.) SO_3
Neutral, S	-	-	Elemental S
Reduced, S^{2-}	Pyrites, marcasite organic sulfides	Oldhamite, CaS, complex sulfides of calcium, aluminium, etc	Non-volatile

The oxidised sulfur cycle is illustrated in Figure 3. In general, sulfur escaping the preheater is found in the form of dust rather than SO_2 and forms part of the "external cycle". Further, it is expected that SO_2 in the gas phase will be absorbed in the lower stages of the preheater, unless there are very adverse conditions (e.g., chemical reduction due to poor combustion of supplementary fuel).

Literature tends to agree on vapour pressures of pure alkali compounds: potassium salts are more volatile than corresponding sodium salts and their chlorides are very much more volatile than sulfates. This explains some observations of volatile behaviour in kilns, but information for more complex species is rare. Studies at Aberdeen University (Choi & Glasser, 1988) have produced self-consistent results for sulfosilicate, sulfoaluminate and langbeinite. The order of volatility alters with temperature. Liquid alkali sulfate systems have a poor dissolving power for most principal oxides of cement clinker. However, they have low viscosities and low surface tension against silicates and thus cover and englobe particles very effectively. It seems likely that small quantities of dissolved silicate have a high mobility, so that liquids are effective at producing a reaction (for preference towards dicalcium silicate at 700 to 800°C.). Stabilisation of carbonates has been suggested ($CaCO_3$ can dissolve in the liquid phase in the presence of alkali sulfates at 880 - 900°C.). The

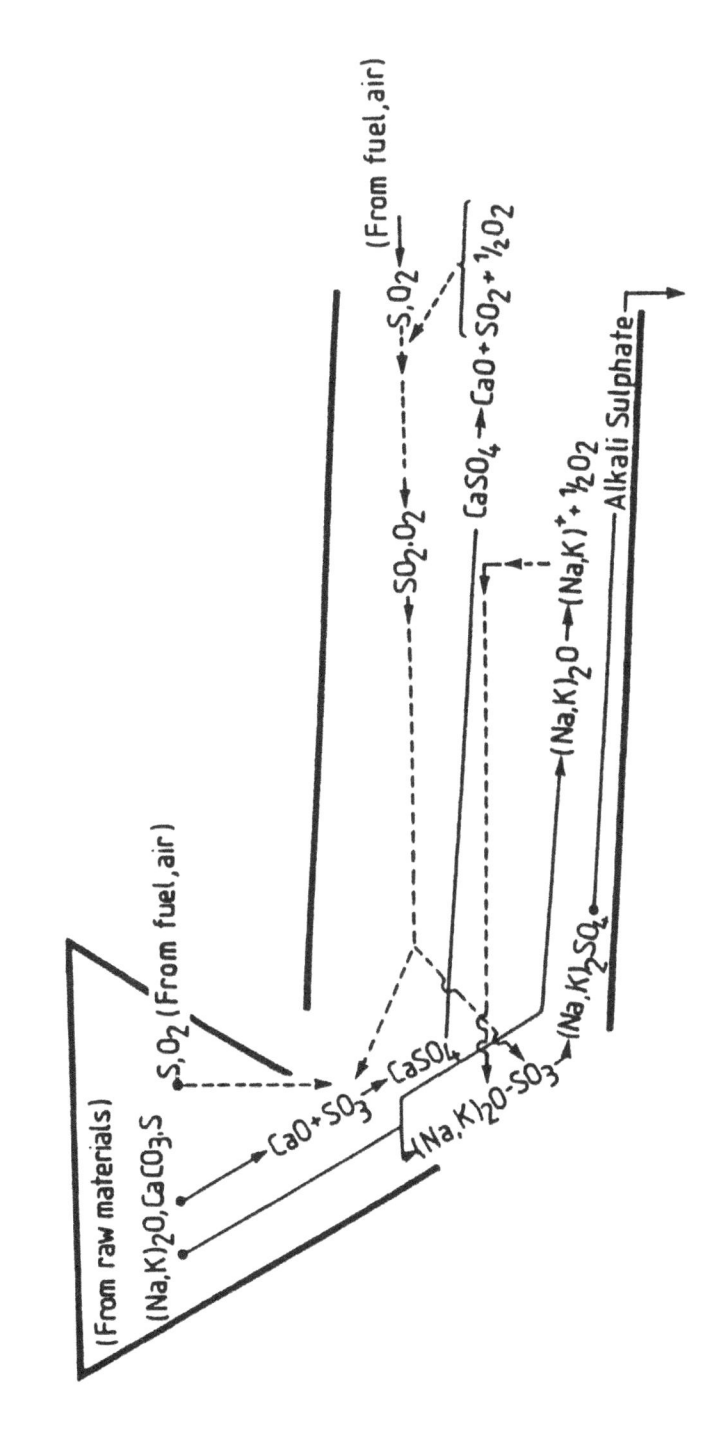

FIGURE 3

OXIDISED KILN SULFUR CYCLE

(........ vapour phase; ——— solid phase)

presence of fluoride can cause further complications for clinker quality - but equally (in combination with certain concentrations of other compounds) certain advantages.

EFFECTS ON PRODUCT CHARACTERISTICS

The effects on clinker may be summarised as follows:

Fluxing action:

- lower temperature of first liquid phase formation
- change of liquid viscosity
- alteration of surface tension of liquid
- modification of crystal morphology.

Phase relations:

- the relative thermodynamic stabilities of the clinker minerals can be altered by solid solution effects.

Hydraulic activity:

- the reactivities of the clinker minerals are altered by solid solution and/or by the effects of crystal symmetry (high temperature stabilisation of polymorphs) and/or effects occurring during hydration (e.g., coating of cement particles by insoluble salts).

(It is difficult to isolate these three classes of effect in practice.)

In general, incorporation of additional clinker sulfate in a situation with excess alkalis yields a more difficult "apparent grindability" with advantages in the market of improved early concrete strength and workability.

The effects of minor components on the viscosity and surface tension of liquid phases can be complex. Lower viscosities encourage alite (calcium trisilicate) formation. Calcium sulfate flux can, however, stabilise belite (calcium disilicate) and/or cause the production of clinker alite with lime inclusions. Further, in clinkers with a low alkali content, belite stabilisation due to excess SO_3 leads to difficult combinability. Strongly chemically reducing conditions in the burning zone can give a cement with poor flow characteristics (due to free K_2O and Na_2O), poor workability (due to the increased content of tricalcium aluminate and its reactivity), poor strength (lower tricalcium silicate content) and variable colour.

Alkalis retained in clinker are present either as stable sulfates or absorbed in the silicate and aluminate structures: these influence the behaviour of fresh concrete and mortar due to their various solubilities. Na_2O has a more marked tendency than K_2O to form solutions in calcium aluminate. For clinkers with (molar) ratios of sulfate:total alkalis below 0.5, almost all the sulfate is combined in water soluble form, K_2SO_4 being predominant. A proportion of the alkalis are in solid solution in the clinker aluminate phase and this has an adverse effect on the initial cement reactivity and thus on concrete and mortar rheology.

For ratios between 0.5 and 1.0, a certain quantity of langbeinite is also formed (and not all the alkalis are soluble). For ratios above 1.0, significant fractions of the sulfates are combined within the silicates and aluminates or as anhydrite ($CaSO_4$), which dissolves more slowly than alkali sulfates, whilst the fractions of K_2O and Na_2O which are soluble in water approach 1.0 and 0.5, respectively, at a ratio of about 1.5. At sulfate:alkali ratios above 1.5 trends are somewhat erratic. For most normal clinkers the principal sulfate phase will be aphthitalite (potassium/sodium sulfate) with a maximum K/Na ratio of 3.0. This phase is accompanied by minor quantities of K_2SO_4 and calcium langbeinite, Na_2SO_4 being found only for unusually low K/Na ratios.

As well as the solid solution effects and the formation of compounds described above, various permutations of volatiles (especially in the presence of fluorine) can influence the structure and behaviour of alite and belite crystals (Moir & Glasser, 1992).

TOWARDS A MODEL OF VOLATILE CYCLES

Various empirical volatility factors have been proposed and used with a certain measure of success. This section considers possible approaches to a more fundamentally based model.

It is generally supposed that (other factors being equal) the extent of volatilisation decreases as the thermal efficiency of the kiln increases. This is probably due to the limiting effect of vapour saturation by alkali compounds, as confirmed by studies of the treatment of kiln dust in a 100 mm diameter fluidised bed (Tettmar et al, 1978) to examine the feasibility of producing clinker from cement kiln flue dust with capture of the alkalis distilled from the bed for possible use in the fertiliser industry.

It is suggested that

$$V^* = \frac{(P^*) \ (M_v)}{(P - P^*) \ (M_g)}$$

V^*	=	saturated vapour concentration in transport gases (kg/kg)
P^*	=	saturated vapour pressure
		of an alkali compound
P	=	gas pressure
M_v	=	molecular weight of vapour
M_g	=	molecular weight of gas.

)
) same units
)

Given mathematical expressions for saturated vapour pressures as functions of temperature and knowledge of kiln system temperature profiles, the saturated vapour concentration can be calculated for each alkali compound and thus the maximum quantities evaporated from the feed per unit mass of gases. Then, considering the amounts of gas passing through the kiln at various temperatures, the true quantity of volatiles transported per unit mass of clinker can be calculated and from this knowledge, "ideal" volatile cycles can be deduced.

For example, saturated vapour pressures at 1200°C are (for the pure substances):

KCl	0.18 atm	
K_2SO_4	0.8×10^{-3} atm	(0.6×10^{-3} atm with decomposition suppressed)
Na_2SO_4	0.13×10^{-3} atm	(0.01×10^{-3} atm with decomposition suppressed)

The transport capacity of air for vapours at 1200°C is thus: KCl - 700 g/g; K_2SO_4 - 4 g/g; Na $_2SO_4$ < 0.5 g/g. The capacity at 1250°C is about two times higher.

It can therefore be foreseen that (unless the equilibrium vapour pressures differ greatly from saturated values) there will be little problem in removing KCl from many kiln flue dusts in a fluidised bed with a gas flow rate of, say, 2 g per gramme of dust, although the capacity for sulfate removal may be limited. The same reasoning applies to kilns, with wet process kilns typically showing a ratio of a little less than 2 g/g gas/solids in the burning zone and perhaps 2.75 g/g at the back end, with corresponding values for the dry process (without precalcination) of 1.4 g/g and 1.94 g/g.

Despite the fact that qualitative differences between two kilns (one dry process and one wet) are reflected in sample calculations, such "ideal" calculated recirculating loads are about 10 times larger than those encountered in practice. The probable reasons include:

a) Incomplete contact between gases and solids in the kiln, where only a small fraction of the solid surface is exposed at a given time. (It is expected that there is better contact in the colder dusty regions of the system.)

b) Volatilisation characteristics of the alkali-containing minerals at a given plant, i.e., combination in more complex silicates and aluminates, not only sulfates.

c) Transport of compounds which are condensed as/on solid dust or fume.

d) Reduction of vapour pressure over solutions of alkali compounds.

e) Inadequate treatment of the transport of heat and of vapour within the bed of clinker nodules in the kiln.

f) Formation of other compounds, e.g., $CaSO_4$, depending on alkali:sulfate ratio.

g) Unstable operation of production kilns: practical conditions are not exactly those expected for very long term stability of temperature and material flow.

The further development of a predictive model will have to take account of such factors, as well as the effects of composition of kiln atmosphere. In recent years there has been much investigation of factors governing the blockage of cyclones and their performance in high temperature coal combustion processes (in the hope of protecting turbine blades in direct cycle electric power generation systems). When lime (or limestone) is injected to absorb SO_2, the compounds and thermodynamic criteria invoked are exactly those encountered in the cement industry - particularly when relatively high chloride coals are used. It is probable that there is now sufficient academic knowledge to better treat our situation and allow improved modelling and understanding. Another aspect to consider is knowledge acquired from study of the regeneration of CaO sorbents used for SO_2 scrubbing: again, data potentially relevant to kiln systems are produced, for example, on pressures of SO_2 in the system $CaSO_4$/ CaS/ CaO in the presence of various concentrations of CO and CO_2. It is hoped that this paper promotes cross-fertilisation between these various fields of work and the body of practical knowledge available in the cement industry.

THANKS

This text, originally based on various internal and external reports, has been adapted on the basis of useful discussion with technical staff from a number of cement companies. Thanks are given to all involved, as well as to the Directors of Blue Circle Industries PLC for their permission to publish this paper.

REFERENCES

Choi, G.-S. & Glasser, F. P. (1988). The Sulphur Cycle in Cement Kilns: Vapour Pressures and Solid-Phase Stability of the Sulphate Phases. **Cement and Concrete Research, 18**, 367-374.

Kerton, C. P. & Murray, R. J. (1984). Portland Cement Production. In **Structure and Performance of Cements**, ed. P. Barnes. Applied Science Publishers Ltd, Barking, pp 205-236.

Moir, G. K. & Glasser, F. P. (1992). Mineralisers, Modifiers and Activators in the Clinkering Process. In **9th International Congress on the Chemistry of Cement, New Delhi, India, 1992: Congress Reports, Volume I**, National Council for Cement and Building Materials, New Delhi, pp. 125-154.

Ritzmann, H. (1971). Cyclic Phenomena in Rotary Kiln Systems. **Zement-Kalk-Gips, 24**, 338-343.

Tettmar, B., Khor, J. H., & Gregory, S. (1979). Processing of Kiln Dust. **Zement-Kalk-Gips, 31**, 288-290.

SORPTION OF SO_2 AND HCl IN GRANULAR BED FILTERS

WOLFGANG PEUKERT
Hosokawa MicroPul
Welserstrasse 9-11, 5000 Köln, Germany
FRIEDRICH LÖFFLER
Universität Karlsruhe
Kaiserstrasse 12, 7500 Karlsruhe, Germany

ABSTRACT

This publication describes experimental and theoretical results for the collection of HCl and SO_2 in granular bed filters in the temperature range between 600 - 1200 K. Sorption of HCl and SO_2 has been characterized by conversion-time curves under differential gas conditions as well as by breakthrough curves in an integral fixed bed reactor. Various sorbents such as raw limestone, limestone- and lime-pellets have been studied. In the gas-solid reaction under consideration a solid product is formed on the reactive surface of the sorbent so that the reaction rate decreases with time. In order to obtain high overall solid conversions and thus maximal exploitation of the sorbent, pellets have been used. In case of sulphation of lime solid conversions of over 80 % could be achieved. Theoretical description of the sorption of HCl can be accomplished by an elementary shrinking core-model. The sulphation reaction, however, requires the application of a model which takes into account the pellet's internal pore size distribution.

Introduction

Emission control of acid gaseous pollutants such as SO_2/SO_3, HCl or HF at high temperatures is based on gas-solid reactions of the specific gaseous component with basic oxides, mainly in form of $CaCO_3$, $Ca(OH)_2$ or CaO. In order to use these reactions for the control of gaseous pollutants the flue-gas must come in contact with sorbent particles. One often used method is the injection of fine dispersed limestone powder into the hot environment, e.g. a coal fired combustor, where the limestone particles undergo calcination. The calcined product CaO reacts with the SO_2 produced during coal combustion by forming a solid product, here mainly calcium sulphate whose molar volume is more than 3 times higher than that of CaO. Development of a voluminous product layer leads to a decrease of reaction rate. After injection sorbent particles will be transported with the flue-gas to an electrostatic precipitator or bag filter where they are collected together with the fly-ash. Because of pore plugging and short residence times of the sorbent particles in the system utilization of the sorbent is far from complete, usually in the range of 30-50%. Application of a high-temperature bag filter opens the additional possibility to use the filter cake as gas-solid reactor [Gäng, 1991].

Another option for contacting flue-gas and sorbent particles is the use of fluidized beds or granular bed filter either fixed or moving. Application of granular bed filters opens the possibility of collecting both gaseous and solid dust particles in one filter unit up to temperatures of 1100 K [Peukert & Löffler 1991]. One advantage of using granular bed filters is the long residence time of the sorbents in the system which allows almost complete exploitation of the sorbents and therefore a minimization of the residues.

This paper describes a theoretical analysis and experimental results of collection of SO_2 and HCl in a fixed bed granular filter in the temperature range between 600 and 1100 K.

PHYSICAL-CHEMICAL FUNDAMENTALS

The chemical reactions between HCl or SO_2 and limestone are

(1) $CaCO_3 + 2\,HCl \rightarrow CaCl_2 + CO_2 + H_2O$ $\Delta_R H_{298} = -\,45.5$ kJ/mole
(2) $CaCO_3 + SO_2 + 1/2\,O_2 \rightarrow CaSO_4 + CO_2$ $\Delta_R H_{298} = -\,501.5$kJ/mole
(3) $CaCO_3 + SO_2 \rightarrow CaSO_3 + CO_2$ $\Delta_R H_{298} = -\,61.5$ kJ/mole
(4) $CaCO_3 \rightarrow CaO + CO_2$ $\Delta_R H_{298} = +178.9$ kJ/mole

These reactions are not a complete set of all possible ones but of only the most important overall reactions. This is true especially for the sorption of SO_2. At temperatures below 900 K, for example, $CaSO_3$ is formed together with $CaSO_4$. With decreasing temperature the amount of $CaSO_3$ is increasing [Balekdjian 1987].

The calcination reaction (4), i.e. the thermal decomposition of limestone to lime which occurs above 900 K, contributes only indirectly to the absorption of gaseous pollutants. Calcination leads to a better reactivity of the sorbent due to an increase of the pore volume and surface area of the sorbent. If lime is used as sorbent at a temperature below 900 K recarbonation of lime may impair sorption of gaseous pollutants.

Reaction rate is governed by transport of the gaseous species to the surface of sorbent particle, pore diffusion into the interior of the porous particle and solid diffusion through product layer and reaction kinetics. The rate determining step depends not only on the physical parameters of the system (e.g. species, temperature, particle diameter or pore structure) but also on time. For short times reaction kinetics or pore diffusion determine the rate of the process. For longer times diffusion through the product layer will be the rate determining step. One common feature of the considered gas-solid reactions is that solid products formed occupy considerably more volume than the sorbent The ratio of molar volumes of $CaCl_2$ (resp. $CaSO_4$) and CaO are 3.07 (3.09), respectively. Formation of a solid product leads to plugging of pores at the surface and the interior of the reacting particles and therefore often to an incomplete utilization of the sorbent. Diffusion coefficients through product layers are in the order of $D_S \sim 10^{-12}$ m^2/s and therefore about 6 orders of magnitudes smaller than the diffusion coefficients in the pores ($De_0 \sim 10^{-6}$ m^2/s).

Most attention has been paid to the sulphation reaction between sulphur dioxide and lime or limestone [Borgwardt & Harvey 1972, Hartman & Coughlin 1976, Simons et al 1987, Zarkanites & Sortichos 1989]. Mostly, these publications deal with the reaction of a small sample at constant gas concentrations. It is stressed that particles inertial pore structure is most important for achieving high solid conversions. Various methods have been proposed to optimize the pore structure and to increase solid conversion, for example by adding various substances [Borgwardt et al 1984, Weisweiler et al 1985]. Few publications consider the sorption of HCl with Ca-based sorbents [Petrini et al 1979, Balekdjian 1987] or with the breakthrough behavior of gaseous pollutants through a fixed bed [Orbey & Dogu 1982].

In this investigation sorption of SO_2 and HCl has been studied under differential gas conditions for single pellets as well as under integral gas conditions, i.e. in a fixed bed reactor. Solid conversion U is defined as the mole ratio of reaction product n_{prod} to inlet sorbent $n_{s,in}$

$$U = \frac{n_{S,in} - n_{S,out}}{n_{S,in}} = \frac{n_{prod}}{n_{S,in}} \tag{1}$$

Breakthrough curves, i.e. the ratio of gas concentrations before and after the filter $C(t)/C_0$ have been measured in dependence of time. Breakthrough curve as a function of time can be transferred to a curve in dependence of solid conversion by using a mass balance for the gaseous species.

$$U = \frac{v_g \dot{V} C_0}{v_s n_{S,in}} \left(t - \int_0^t \frac{C}{C_0} dt \right) \tag{2}$$

This approach gives the mean solid conversion of the bed and allows a better comparison between different reaction conditions.

EXPERIMENTAL SET-UP

Reactivity measurements for both calcination and sulphation experiments under differential gas conditions were conducted in a thermogravimetric analysis system (DTA), which allows continuous recording of the changing weight of the sample during calcination and sulphation. Change of sample weight is directly related to solid conversion. The sample weight used was usually below 10 mg. In order to ensure constant calcination conditions the sample was heated with a constant rate of 10 K/min to the reaction temperature. Sulphation experiments were started after completion of calcination. Flue-gas was simulated by mixing of air, CO_2, water-vapor and SO_2. After each experiment calculated solid-conversion (from weight gain) was checked by wet chemical analysis, agreement was usually better than 5%.

Due to the strong corrosive effect of HCl, sorption of HCl for differential conditions was studied in a quartz reactor, which could be heated up to 1200 K. Here a greater sample of sorbents, usually around 100 mg was used. Flow rate in these experiments was increased until solid conversion became independent from the flow rate of the simulated flue-gas, which was a mixture of air, CO_2, H_2O-vapor, HCl and in some experiments additionally SO_2. Solid conversion was obtained both from increased weight and chemical analysis.

The basic arrangement of the high-temperature apparatus for measurement of breakthrough-curves (integral conditions) and fractional particle collection efficiencies at temperatures up to 1100 K is shown schematically in Fig.1. Collection of particles is, however, not covered in this paper (see Peukert & Löffler 1991). A model flue-gas has been generated by mixing the gaseous components (SO_2, HCl, CO_2) with air. In some experiments inert quartz particles were added. The model flue-gas is heated by electric tube heaters to the desired temperature. In the experiments with SO_2 a quartz tube was additionally inserted into the metal tube of the test apparatus in order to avoid oxidization of SO_2 to SO_3 above 900 K. Test-filter is located at the end of the heating section. The filter-bed is supported by a grid. Temperatures at various locations and pressure drop across the filter have been recorded continuously. After the filter, flue-gas is cooled and cleaned in a wet-scrubber. A vacuum pump - combined with a critical nozzle for adjustment of the flow-rate - is located at the end of the apparatus.

Measurements of SO_2 and HCl concentrations before and after the test-filter have been accomplished by sampling a small gas volume through heated pipes. For continuous analysis photometers have been used (BINOS, Leybold-Heraeus for SO_2, SPECTRAN 677IR,

Bodenseewerk for HCl). A detailed description of both differential reactors as well as of the test-apparatus for the fixed bed reactor is given in [Peukert 1990].

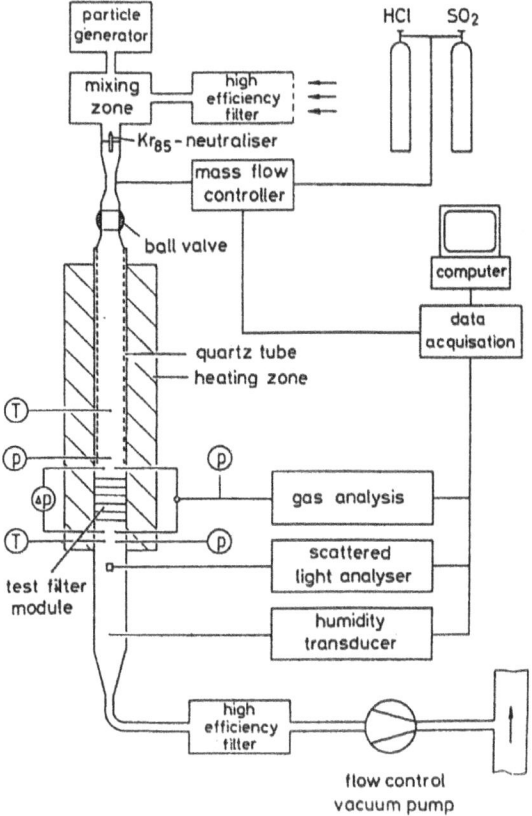

Figure 1   Schematic illustration of the high-temperature apparatus

Sorbents

Experiments have been conducted with raw limestone particles and pellets. Pellets have been agglomerated from fine limestone powders with mean mass-related diameters of 1.8 μm (F1), 9.8 μm (M1000) and 60 μm (G). Agglomeration of the pellets has been accomplished in a rotating disk agglomerator. CaO samples were fabricated by calcination of the respective limestone samples at 1073 K in the fixed bed. Sorbents have been characterized by chemical composition, pellet diameter, porosity ε_s, pore volume distribution and mass-related pore volume V_m (measured by mercury intrusion method) and BET-surface S_m.

EXPERIMENTAL RESULTS

Sorption of HCl

Effect of temperature on the sorption of HCl with lime and limestone was measured between 673 and 973 K for differential and integral conditions. Fig.2 shows for example measured conversions of CaO-pellets at 673 and 773 K for differential gas conditions (HCl concentration C_0 = 1500 ppm) in air as carrier medium. In order to study the effect of water vapor at both temperatures, measurements have been conducted with no vapor (closed symbols) and with 10 vol.-% of water vapor (open symbols).

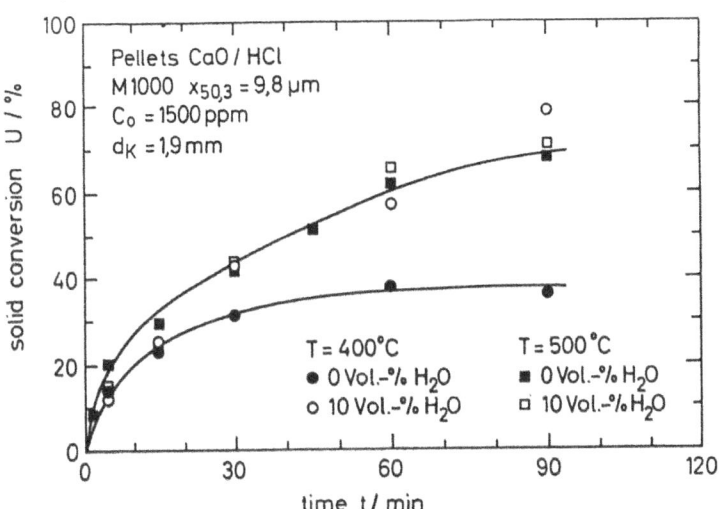

Figure 2   Effect of temperature and water-vapor on the conversion time behavior of CaO-pellets

The results without water-vapor show the expected increase of reaction rate and therefore higher solid conversion with increasing temperature. Reaction rate decreases, however, with time due the product layer of $CaCl_2$. The results with water-vapor indicate that water enhances the reaction at 673 K but is of little influence at 773 K. This effect can be explained by an interaction with $Ca(OH)_2$ that may be formed during the reaction. The chemical equilibrium constant for the system $CaO/Ca(OH)_2/H_2O$ is approximately K_p = 0.05 at 673 K so that formation of $Ca(OH)_2$ as intermediate product is possible. At 773 K, however, chemical equilibrium is at the side of CaO so that the formation of $Ca(OH)_2$ is not probable.

It is known that solid conversions are higher when CaO is formed from dehydratisation of Ca(OH)$_2$ in comparison with CaO obtained from calcination of limestone. Pore structure of CaO samples made from Ca(OH)$_2$ seems to be more favorable for the sorption of HCl [Balekdjian 1987]. One reason for this is the high amount of fine pores with diameters below 40 nm. Adsorption of water-vapor in the fine pores and pore condensation will not occur at these high temperatures. Pore condensation is likely to be effective at temperatures below 373 K.

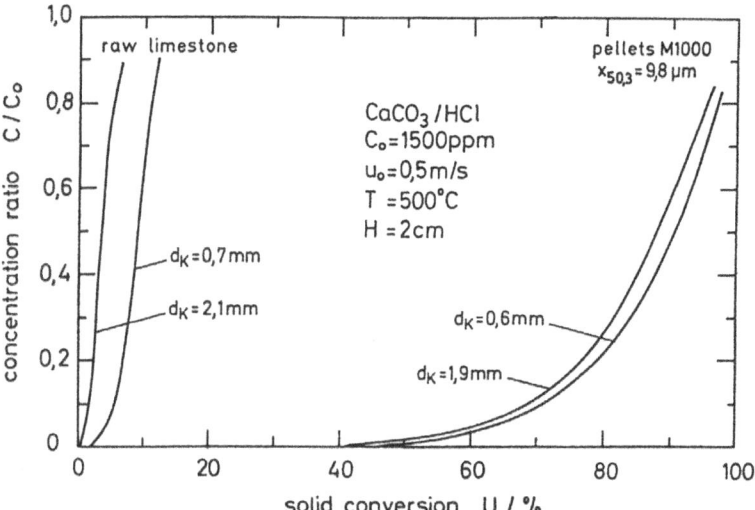

Figure 3   Influence of the grain size on the breakthrough behavior of HCl through granular beds of raw limestone and limestone pellets at 773 K (air as carrier medium)

Fig.3 shows a comparison of measured breakthrough curves of HCl using raw limestone (particle diameter 0.7 and 2.1 mm, respectively) and pellets fabricated from fine limestone powder (M1000, $x_{50,3}$ = 9.8 μm, pellet diameter d_K = 0.6 and 1.9 mm, respectively). The actual size of the pellets themselves is only of subordinate relevance although a certain increase of solid conversion was observed with smaller pellets. In case of raw limestone it is evident that the attainable solid conversions are, by no means, adequate for industrial purposes. This does not stem from a financial assessment (since in comparison with other sorbents such as Ca(OH)$_2$, limestone is a cheap raw material), but more from the problems involved in disposing the residues.

At 773 K a better sorbent exploitation can be attained by utilizing limestone pellets which offer average conversions of over 90%. As a result of the low porosity of raw limestone ($\varepsilon_s = 9.5\%$), the diffusive transport into the grain is not only impaired from the beginning of the reaction, but ceases almost completely at low conversions. The limestone pellets, however, possess a larger, more easily accessible internal surface so that an almost complete conversion occurs. Even at high solid conversions, the large pore volume of the pellets ($\varepsilon_s = 45\%$) still warrants a diffuse transport into the core of each sorbent particle.

Sorption of SO_2

Fig.4 shows breakthrough curves of SO_2 through beds of CaO-pellets (M 1000) in the temperature range between 773 K and 973 K at a superficial gas velocity of 0.5 m/s. CaO-pellets have been produced by heating the bed of limestone pellets to 1173 K and subsequent calcination in air at a gas velocity of 0.5 m/s for 1h. After calcination air flow was stopped and the bed has been cooled until the respective temperature of the sorption experiment was reached. Experiments with air ($C(CO_2) = 0.035$ vol.-%) exhibit the expected improvement of sorption with increasing temperature. Solid conversion increased from 45% at 773 K to 58% at 873 K.

CO_2-concentration influences the calcination of limestone at temperatures above 873 K as well as the recarbonization of CaO to $CaCO_3$ at lower temperatures. Thermogravimetric studies under differential gas conditions showed that SO_2 reacts mostly with CaO. At temperatures of 773 K and 873 K and CO_2-concentrations of 10 vol.-% recarbonization is fast so that only solid conversions of 10 % are possible. The measured carbonate content of the samples was 23.5 and 19.7 weight-% at 773 and 873 K, respectively. At 973 K, however, recarbonization does not occur. Carbonate content in the sorbent particles was less than 1%. Therefore the two curves with low and high CO_2-concentration are almost identical. A similar result was found at 1073 K, although a solid conversion of 85% can be achieved.

The size of the $CaCO_3$ dust particles from which pellets had been agglomerated was a key parameter for the achievement of high solid conversion rates and hence, maximal sorbent exploitation. Fig.5 exemplary illustrates breakthrough curves for SO_2 for various parent grain sizes through beds of limestone pellets as functions of the solid conversion.

Sorption improves with decreasing parent grain size $x_{50,3}$. An important aspect is that the breakthrough curves initially possess flat trends which only increase at higher specific conversions. Therefore, should one wish to reduce the SO_2 gas concentration by 80% (inlet gas concentration $C_0 = 1500$ ppm) then, in the case of the finer fraction F_1, 70% of the solids may be converted before effective filtration ceases after 7 hours of operation.

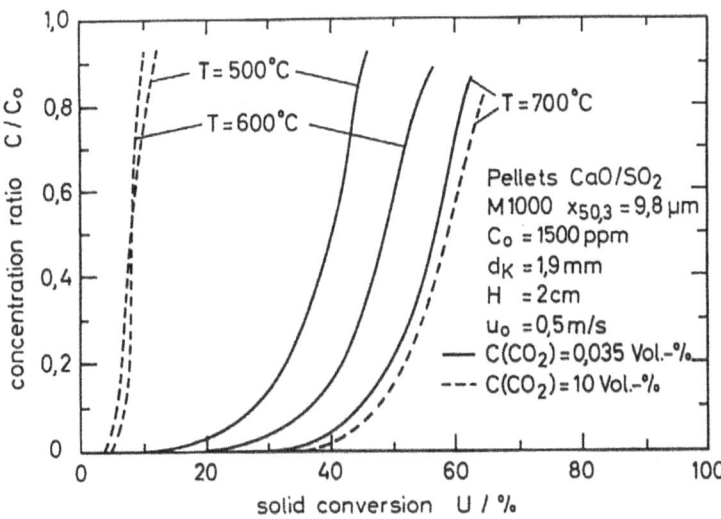

Figure 4   Influence of temperature and CO_2-concentration on breakthrough curves
of SO_2 through beds of CaO-pellets

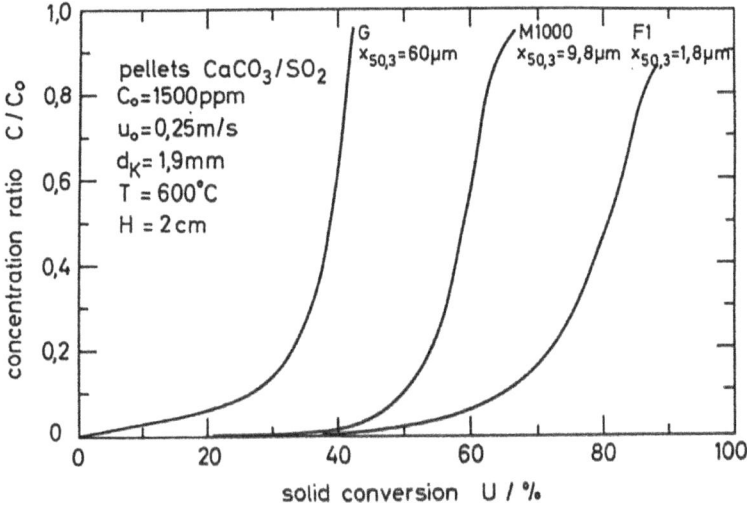

Figure 5   Influence of the parent grain size $x_{50,3}$ on the breakthrough behavior of
SO_2 for three different limestone pellet beds (pellet size $d_K = 1.9$ mm)
at a filter face velocity $u_0 = 0.25$ m/s (air as carrier medium)

Sorption of gaseous components in a granular bed filter can be influenced by dust particles in two possible ways. Inert dust particles collected in the filter bed influence the transport of the gaseous component to the reactive surface. Sorption of the gas can be improved by adding fine disperse reactive sorbent particles to the raw gas with subsequent collection in the filter together with the fly ash. This second case will not be considered in this paper. Influence of inert dust particles (quartz, x < 10 μm) on breakthrough of SO_2 has been investigated at 873 and 1073 K (s. Fig.6).

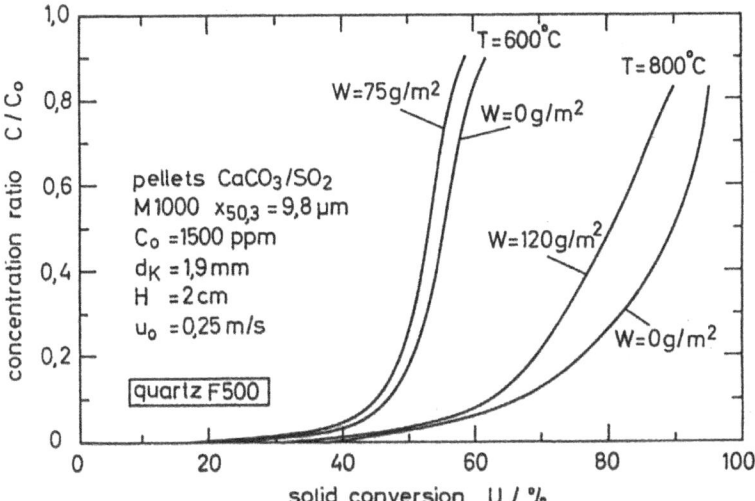

Figure 6   Influence of inert quartz dust on breakthrough of SO_2 through beds of lime pellets (W = mass of dust/filter cross section)

Inert dust particles have only a minor influence on the sorption of gaseous components. The reaction rate is with exception of a short time interval at beginning of the sorption determined by internal pore diffusion and solid state diffusion through the product layer. The rate of the latter is about 6 orders of magnitude smaller than gas phase diffusion. Inert dust in the filter influences only mass transfer from gas phase to the outer surface of sorbent grains. This process, however, is faster than internal pore diffusion or solid diffusion.

Results of simultaneous sorption of HCl and SO_2 at 773 K are shown in Fig.7. For this experiment lime pellets (F1) with a diameter of 1.9 mm have been used. For comparison, breakthrough curves for single-component adsorption experiments (for SO_2 and HCl, respectively) are shown, too. Solid conversion due to HCl is reduced from almost 100% for sorption of HCl alone to less than 45% in case of simultaneous collection of SO_2 and HCl.

For SO₂, however, solid conversion related to sulphation reaction is increased from 15% to about 60% for simultaneous sorption of both components under otherwise similar

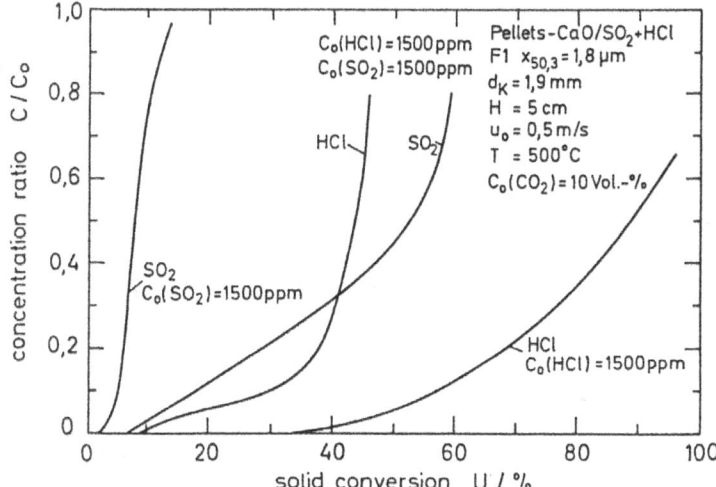

Figure 7   Simultaneous sorption of SO₂ and HCl through beds of CaO-pellets at 773 K (carrier gas: air, C(CO₂) = 10 Vol.-%)

conditions. The calculated overall conversion of over 100% is due to an error in mass balance, which was usually smaller than 15% when compared with chemical analysis of the sorbent particles.

The increased reactivity of SO₂ in the presence of chloride agrees with the experimental observation that small amounts of NaCl or LiCl₂ enhance the sulphation reaction. Similar results have been observed by [Borgwardt et al 1984] and [Peukert & Löffler 1987]. The reason for the observed better sulphation reaction may be seen in a more porous product layer which is a mixture of CaCl₂ and CaSO₄.

Increase of temperature above 900 K for otherwise similar reaction conditions is problematic since CaCl₂ becomes plastic. Melting point of CaCl₂ is 1045 K. Sorption of HCl at temperatures above 900 K led to an increased pressure drop due to changes in the filter structure and in the extreme to the destruction of the filter bed structure.

THEORETICAL DESCRIPTION OF CONVERSION-TIME BEHAVIOR

Various theoretical models for gas-solid reactions are described in the literature. The shrinking core model [Ishida & Wen 1968] has been selected because of its simplicity. The pore model described by [Christman & Edgar 1983] is more sophisticated since it includes the pore size-size distribution of the sorbent. In the following a qualitative description of these models is provided. A detailed derivation of the model equations is given elsewhere [Christman & Edgar 1983, Kocaefe et al 1987, Ishida & Wen 1968, Peukert 1990].

Shrinking-core model

This model allows an analytical calculation of the conversion-time behavior of non-catalytic gas-solid reactions for homogeneous spherical particles. In equ.3 reaction time is given as function of solid conversion U.

$$t = \frac{v_g \rho_S R_0}{v_S M_S k_S C} \left(1 - (1 - U)^{1/3} + \frac{k_S}{3\beta} U + \frac{k_S R_0}{6 D_e} (3 - 3(1 - U)^{2/3}) \right) \qquad (3)$$

(A complete list of symbols is provided at the end of this paper).

The model takes into account the influence of mass transfer from gas phase to the pellets surface (β), diffusion in the pellet (D_e) and reaction kinetics (k_S). Gas phase mass transfer coefficient has been calculated by using the correlation given by [Petrovic and Thodos 1968]. The model assumes D_e to be constant. Actually, for the reactions under consideration D_e is a function of changing grain structure and thus a function of solid conversion U. In order to derive a relationship between internal diffusion coefficient D_e and solid conversion U, we start from the following well-known equation

$$D_e = D_{e0} \frac{\varepsilon_S}{\chi} \qquad (4)$$

where D_{e0} denotes a mean pore diffusion coefficient, ε_s is the mean pellet porosity. Turtuosity χ has been estimated according to [Wakao & Smith 1962] by $\chi = 1/\varepsilon_s$. D_e therefore becomes

$$D_e = D_{eo} \varepsilon_s^2(U) \qquad (5)$$

A volume balance relates sorbent porosity to solid conversion U

$$\varepsilon_S = \varepsilon_{S0} + (1 - \varepsilon_{S0}) \frac{V_P - V_S}{V_P} \qquad (6)$$

V_i denotes the mole volumes of product (index p) and sorbent (index s). ε_{s0} is the porosity at the beginning of reaction. Effective diffusion coefficient D_e decreases therefore with U simulating the growing inner diffusion resistance. The value of D_{e0} for a specific reaction system has to be fitted to the experimental results.

Pore model

This model is more sophisticated than the shrinking core model since evaluation of changing pore size distribution is included in the model formulation. The model starts from the diffusion equation of a porous spherical particle

$$\frac{1}{R^2}\frac{\partial}{\partial R}\left(R^2 D_e(R)\frac{\partial C_K}{\partial R}\right) = \sum_{i=1}^{N} k_S S_{Vk} C_{Sk} \tag{7}$$

C_K is the gas concentration inside the pores, C_{Sk} the local gas concentration at the reaction surface. S_{Vk} denotes the local volume specific surface. $D_e(R)$ is the effective diffusion coefficient in the pores at pellet radius R. The model assumes straight cylindrical pores. Initial pore size distribution measured by mercury intrusion method has been divided in N ranges each with a mean pore radius and a respective pore volume fraction. Since the solid around the pores is converted into a product with higher molar volume, pores will plug with increasing reaction time. Gas concentration C_{Sk} at the reaction surface in the k-th range can be calculated by solving the diffusion equation for a cylindrical pore. The solution is given by:

$$C_{Sk} = \frac{C_K(R)}{1 + r_{2k}\dfrac{k_S}{D_S}\ln\dfrac{r_{2k}}{r_{1k}}} \tag{8}$$

D_S is the diffusion coefficient through the product layer. r_{2k} is the radius of the reaction surface, r_{1k} is the inner radius of the pore in size range k. Solid conversion can be calculated in every section using a volume balance for the plugging pore. Integration over all pore sizes gives finally the overall solid conversion for one pellet. It was found that often 2 pore radii are sufficient for the calculations if the mean radii of the measured bimodal pore size distribution are chosen.

For a given reaction system and temperature solid conversion depends on the characteristics of the sorbent (radius, porosity, pore size distribution) and on kinetic parameters k_S and D_S. Pellet properties as well as reaction rate constant k_S have been measured. Only solid diffusion coefficient D_S must be fitted to the experimental results. It was found, however, that the influence of k_S on solid conversion is small and restricted to short reaction times (t < 30 min). Therefore a mean value of $k_S = 0.2$ cm/s has been assumed for all calculations.

Comparison with experimental results

Reaction of HCl with lime or limestone pellets can be described by the shrinking core model and the pore model. Description of sulphation reaction of SO_2 with CaO-pellets, however, is only possible with the pore model. The shrinking core model does not allow to describe the sharp decrease in the measured conversion-time curves after about 20 or 30 min which is due to blocking of small pores of diameters between 10 and 30 nm developed during calcination of limestone. Fig.8 shows a comparison of measured and calculated conversion-time curves in the temperature range between 773 K and 1273 K for sulphation of CaO-pellets with 1.9 mm diameter. SO_2 gas concentration was 1500 vol.-ppm.

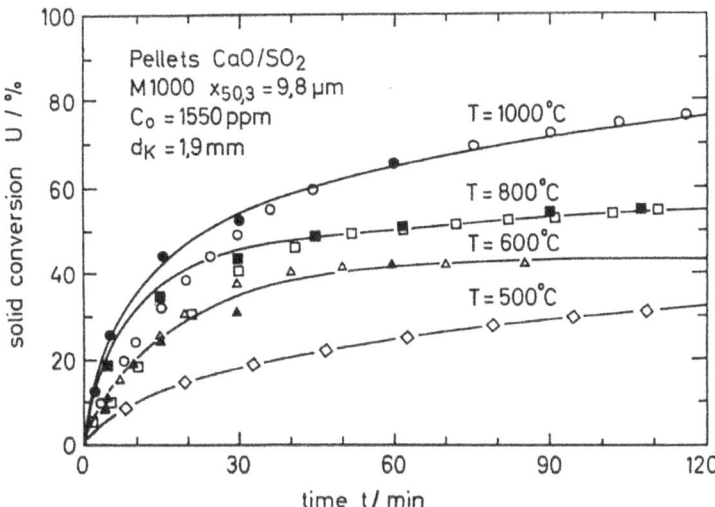

Figure 8   Comparison of measured and calculated conversion-time curves for the sulphation of CaO-pellets between 773 and 1273 K (o DTA-system, • quartz reactor, — calculated)

Measurements have been performed in a quartz reactor (closed symbols) and a thermogravimetric reactor (open symbol). Solid-conversions measured with the DTA system are, for short reaction times (t < 30 min), lower than the results from the quartz reactor. This effect is due to the influence of mass transfer from gas phase to pellet surface which could not be fully excluded in the DTA measurements. Data from both reactors coincide for temperatures below 873 K and for higher temperatures for long reaction times. Experimental results from the quartz reactor could be well described by the pore-model in the whole temperature range. The diffusion coefficient D_S of SO_2 through product layer of $CaSO_4$ has been fitted to the experimental results. All other physical parameters have been measured. D_S in dependence of temperature is given by the following Arrhenius equation

$$D_S = D_0 \exp (-E_A/RT) \qquad (9)$$

The calculated activation energy of $E_A = 114.1$ kJ/mole is in good agreement with values of 116 kJ/mole and 120 kJ/mole published by [Bhatia & Perlmutter 1983] and [Christman & Edgar 1983], respectively. Frequency factor D_0 is $5.35 \cdot 10^{-7}$ m^2/s. The model allows not only to describe the influence of temperature but also the effect of grain size and SO_2-concentration.

For comparison, the solid diffusion coefficient for the reaction of HCl with $CaCO_3$ is given by the values $E_A = 122$ kJ/mol and $D_0 = 0.138$ m^2/s. The frequency factor is much larger for diffusion of HCl in comparison to SO_2 resulting in a faster diffusion of HCl.

Theoretical description of fixed bed reactor

Theoretical modeling of the breakthrough behavior through a fixed bed reactor starts with the following mass balance:

$$\varepsilon \frac{\partial C}{\partial t} + u_0 \frac{\partial C}{\partial z} - \varepsilon D_{ax} \frac{\partial^2 C}{\partial z^2} = -k_e C = -r_V \qquad (10)$$

The initial and boundary conditions are

$$
\begin{array}{ll}
t = 0 & C = C_0 \ (z = 0), \ C = 0 \ (z > 0) \\
z = 0 & u_0 \ (C-C_0) = -\varepsilon \ D_{ax} \ \partial C/\partial z \\
z = H & \partial C / \partial z = 0
\end{array}
$$

The reaction term r_V denotes the converted mass in the volume element per time unit. r_V is defined by the product of the number of grains per volume and the reaction rate of a single pellet. In order to calculate r_V for the shrinking core model or the pore model, r_V has been defined as the product of an effective reaction rate constant k_e and the bulk gas concentration C. In the case of the shrinking core model, k_e is given by

$$k_e = \frac{3(1 - \varepsilon)R_S^2 k_S}{R_0^3 \left[1 + \dfrac{k_S}{\beta} \left(\dfrac{R_S}{R_0} \right)^2 + \dfrac{k_S R_S}{D_e} \left(1 - \dfrac{R_S}{R_0} \right) \right]} \qquad (11)$$

Using the pore model, k_e can be calculated from

$$k_e = \frac{v_g(1-\varepsilon)(1-\varepsilon_S)}{v_S V_S C}\frac{dU}{dt} \tag{12}$$

For the calculation of the change of solid conversion with time the concentration C_{Ph} at the surface of the pellets is needed. These can be evaluated from

$$\beta(C - C_{ph}) = -D_e\left(\frac{\partial C_K}{\partial R}\right)_{R=R_0} \tag{13}$$

i.e. the flux given by the mass transfer from the gas phase to the sorbent surface equals the diffusive flux into the pellet. C_K is the gas concentration in the pellet.

Solution of the mass balance has been accomplished with two different methods. At first a differential procedure was applied incorporating the effect of axial dispersion. The algorithm is implemented at the computer centre of the University of Karlsruhe. The axial dispersion coefficient was calculated according to [Gunn 1987]. The mass transfer coefficient β has been estimated according to [Petrovic & Thodos 1968]. A second solution was obtained by neglecting axial diffusion. The quasi-linear partial differential equation can then be solved by applying the method of the characteristics [Franck & v. Mises 1961]. The solution is given by

$$C = C_0 \exp\left(-\frac{k_e}{u_0}z\right) \tag{14}$$

The change of the concentration with time is given implicitly by reaction rate k_e which is a function of the solid conversion and thus of time. The concentration profile was calculated by evaluating solid conversion for a constant filter height for a timestep dt i.e. from t to t+dt by using the equations for the single pellet. Cyclic input of the result in equ. (14) gives the concentration profile for a constant time. Results obtained with both methods were found to be nearly identical. The influence of axial dispersion is therefore of minor importance and has been neglected in the following calculations.

A comparison between experimental and theoretical results is given in Fig.9 for the reaction of HCl with limestone at 773 K. In these calculations the shrinking core model has been used. Kinetic parameters are identical to those obtained for differential gas conditions. (The diffusion coefficient and the reaction rate constant have been fitted to the experimental

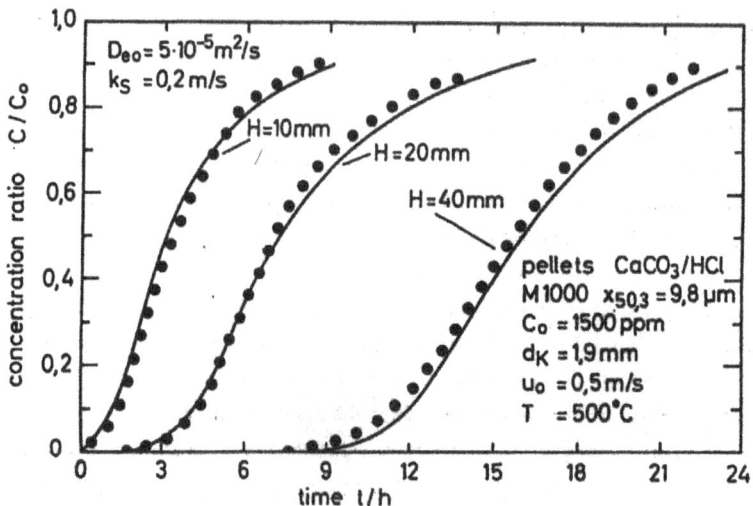

Figure 9   Comparison of calculated and measured breakthrough curves of HCl
            through beds of limestone pellets at 773 K

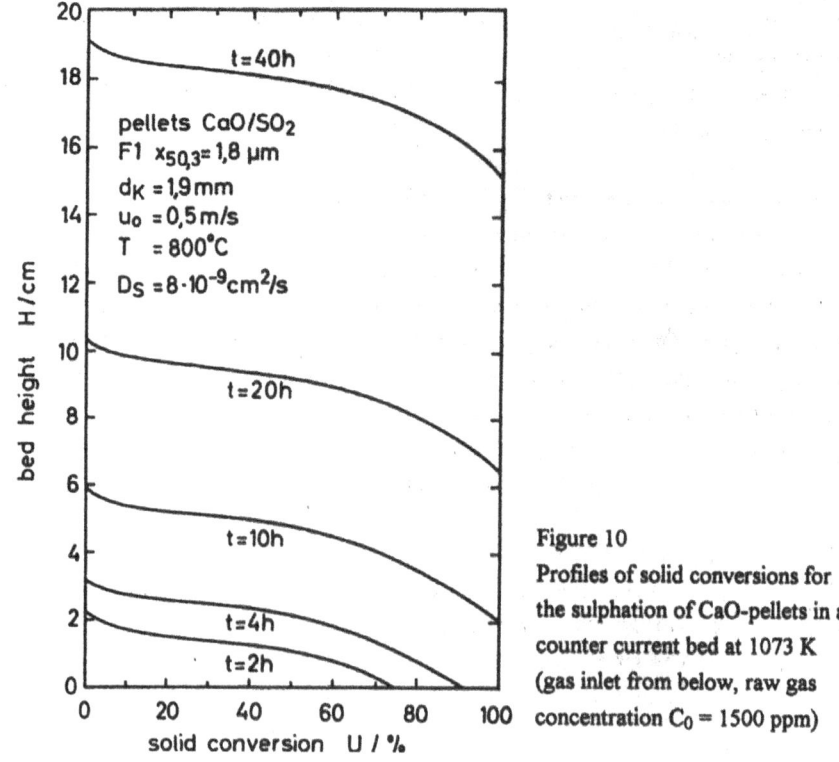

Figure 10
Profiles of solid conversions for
the sulphation of CaO-pellets in a
counter current bed at 1073 K
(gas inlet from below, raw gas
concentration C_0 = 1500 ppm)

results obtained for differential gas conditions.) Agreement between measured and calculated data is good. Parameter is the height of the fixed bed reactor. Even for very thin layers (0.01 m) of limestone time for breakthrough of HCl is in the order of hours.

Calculated solid conversions profiles for sulphation of CaO-pellets within a 0.2 m deep bed (inlet concentration C_0 = 1500 ppm) are demonstrated in Fig.10. The kinetic parameters for the pore model (i.e. reaction rate constant and diffusion coefficient through the product layer) were implemented with those values with which both the temporal single grain solid conversion behavior and the breakthrough characteristics could be described.

The operational life-time is in the order of many hours. Due to the long particle residence time in the filter high solids conversions become possible despite the fact that the $CaSO_4$ product layer progressively retards the reaction. Counter-current operation allows an almost complete sorbent exploitation in combination with low emission values. This is specially important with respect to the increasing necessity of minimizing industrial wastes.

CONCLUSIONS

Sorption of SO_2 and HCl in granular bed filters has been studied in the temperature range between 673 and 1073 K. In order to achieve maximal sorbent utilization pellets have been used which have been agglomerated from fine limestone powders. Particle diameter of these powders proved to be a key parameter for the achievement of high solid conversions. Solids conversions of over 80 % for the sulphation and over 90 % for the sorption of HCl could be achieved. Results of two-component sorption of both SO_2 and HCl showed that sorption of SO_2 is enhanced by the presence of chloride.

Theoretical description of sorption of HCl can be accomplished with a relatively simple shrinking core model. For the sulphation reaction, however, a more sophisticated pore model which takes into account the sorbent's internal pore has to be used. Calculated profiles of solid conversions through reactive beds showed that an almost complete sorbent exploitation is possible in counter current beds. This is of special importance when the deposition of the residues is considered.

SYMBOLS

C	gas concentration
C_P	gas concentration in the product layer
C_K	gas concentration in the pellet
C_S	gas concentration at the reaction surface
C_{Ph}	gas concentration at pellet surface

D_s	solid diffusion coefficient
d_K	sorbent diameter
D_{ax}	axial dispersion coefficient
D_S	solid diffusion coefficient
E_A	activation energy
H	bed height
k_S	reaction rate constant
$n_{s,in}$	moles of sorbent before reaction
$n_{s,out}$	moles of sorbent after reaction
M_s	mole weight of sorbent
r	pore radius
R	sorbent radius (subscript 0: at time t = 0)
R_S	radius of reaction surface in the pellet
S_V	volume specific surface
t	time
T	temperature
u_0	superficial air velocity
U	solid conversion
V_i	molar volume of component i
$\dot{V}$	flow rate
z	length coordinate
β	mass transfer coefficient
χ	tortuosity
ε	bed porosity
ε_S	sorbent porosity (index 0: at the beginning)
v_i	stochiometric coefficient (s: sorbent, g: gas)
ρ_s	density of sorbent

REFERENCES

Balekdjian, O. (1987). Abscheidung von Chlorwasserstoff und Schwefeldioxid an calciumhaltigen Absorbentien. Ph.-thesis, University Karlsruhe, Germany.

Bhatia, S.K., Perlmutter D.D. (1981). The Effect of Pore Structure on Fluid-Solid Reactions: Application to the SO_2-Lime Reaction. AIChE Journal 27, 226-234.

Borgwardt, R.H.& Harvey, R.D. (1972). Properties of Carbonate Rocks related to SO_2 Reactivity. Env.Sci.&Techn. 6, 350-360.

Borgwardt R.H., Bruce K.R., Blake J. (1984). Experimental Studies of the Mechanisms of Sulphur Capture by Limestone. 1st Joint Symposium on Dry SO_2 and Simultaneous SO_2/NO_x Control Technology. San Diego, USA

Christman, P. & Edgar G. (1983). Distributed Pore Size Model for Sulfation of Limestone. AIChE Journal 29, 388-395

Frank P., v.Mises F. (1961). Die Differential- und Integralgleichungen der Mechanik und Physik. Vieweg Verlag, Braunschweig, Germany

Gäng P. (1991). Die kombinierte Abscheidung von Stäuben und Gasen mit Abreinigungsfiltern bei hohen Temperaturen. PhD-thesis, University of Karlsruhe, Germany

Gunn D.J. (1987). Axial and Radial Dispersion in Fixed Beds. Chem.Eng.Sci. 42, 363-373

Hartman, M.&Coughlin,R. (1976). Reaction of Sulphur Dioxide with Limestone and the Grain Model. AIChE Journal 22, 490-498

Ishida M., Wen C.Y. (1968). Comparison of Kinetic Models for Gas-Solid Reactions. AIChE Journal 14, 311-317

Kocaefe D., Karman D., Stewart F.R. (1987). Interpretation of the Sulfation of CaO, MgO, and ZnO with SO_2 and SO_3 AIChE Journal 33, 1835-1843

Orbey N., Dogu T. (1982). Breakthrough Analysis of Noncatalytic Solid-Gas Reactions of SO_2 with Calcined Limestone. Can.J.Chem.Eng. 60, 314-318

Petrini, S., Eklund, H., Bjerle, J. (1979). HCl Adsorption durch Kalkstein. Aufbereitungstechnik, 309-315

Petrovic L.J., Thodos G. (1968). Mass Transfer in the Flow of Gases through Packed Beds. I&EC Fundamentals 7, 274-280

Peukert W., Löffler F. (1987). Zur Abscheidung von Staub und gasförmigen Stoffen in Schüttschichtfiltern. Staub-Reinhaltung der Luft 48, 379-386

Peukert W. (1990). Die kombinierte Abscheidung von Stäuben und Gasen in Schüttschichtfiltern. PhD-thesis, University Karlsruhe, Germany

Peukert W., Löffler F. (1991). Influence of Temperature on Particle Separation in Granular Bed Filters. Powder Technology 68, 263-270

Simons G.A., Garman A.G., Boni A.A. (1987). The Kinetic Rate of SO_2 Sorption by CaO. AIChE Journal 33, 211-217

Wakao, N., Smith, J.M. (1962). Diffusion in Catalyst Pellets. Chem. Eng. Sci. 17, 825-834

Weisweiler W., Hoffman R., Stein R. (1985). Trockene Rauchgasentschwefelung mit Kalkstein oder Dolomit: Verbesserung der Entschwefelungswirkung durch Optimierung der Feststoff-Porenstruktur, Pelletierung und chemische Aktivierung. PEF Kernforschungszentrum Karlsruhe, Germany

Zarkanites S. & Sortichos St. (1989). Pore structure and Particle Size Effects on Limestone Capacity for SO_2 Removal. AIChE Journal 35, 821-830

ACID GAS TREATMENT AT A CERAFIL PILOT PLANT

A. J. STARTIN,* P. H. DYKE,** C. J. WITHERS^

*Foseco International Limited, 285 Long Acre, Nechells, Birmingham, B7 5JR, UK
**Energy Technology Support Unit, B154, Harwell, Oxon., OX11 0RA, UK
^Cerel Limited, Catherine House, Coventry Road, Hinckley, Leics., LE10 0JT, (present address Glosfume Environmental Controls, 1 Nup End, Ashleworth, Glos., GL19 4JJ, UK)

ABSTRACT

A number of technologies are available for the incineration of clinical waste. Similarly several methods can be employed for the abatement of gaseous pollutants generated by the incineration of such material. One such method, targeted at the removal of acid gases and particulate matter, is the use of low density ceramic filtration elements in conjunction with a basic powder such as hydrated lime.

This paper presents results from a pilot plant fitted with 36 low density ceramic elements treating off-gases from a waste burning boiler incinerating clinical waste. Gas analysis, upstream and downstream, of the pilot plant was carried out using an on-line infrared gas analyser. Following characterisation of the gas entering the pilot plant, the outlet gases were characterised while feeding varying quantities of lime into the inlet duct and adjusting the reverse pulse cleaning interval and unit temperature.

HCl removal efficiencies in excess of 90% were achieved. HCl removal appeared to be highest at moderate temperatures, lower than 170°C. Increasing the lime feedrate and pulse interval also proved beneficial but an upper limit, beyond which improvements were small, was evident. The results obtained from the on-line analyser were supplemented by an established wet chemical technique which showed a comparable HCl removal efficiency for the chosen conditions.

1. INTRODUCTION

High temperature incineration is the preferred disposal option for clinical wastes (HSAC, 1992). Following the introduction of the Environmental Protection Act (EPA) in 1990 incineration is required to comply with stringent environmental standards. However, the majority of clinical waste incinerators in the UK are small scale and very old. These units will not meet the new requirements and will have to be upgraded or replaced.

Under the EPA, plant operators are required to utilise the *Best Available Techniques Not Entailing Excessive Cost* (BATNEEC) to eliminate or minimise environmental emissions. Larger, potentially more polluting, plant is regulated by Her Majesty's Inspectorate of Pollution (HMIP) and environmental discharges to all media are controlled. Smaller plants are regulated by Local Authorities for atmospheric emissions only. There is a series of guidance notes available from HMSO giving the limits considered achievable by application of BATNEEC to the prescribed processes. For clinical waste incineration the guidance notes are:-

Chief Inspectors Guidance to Inspector's IPR5/2: Clinical Waste Incineration (HMIP, 1992).

Secretary of State's Guidance - Clinical waste incineration processes under 1 tonne an hour PG5/1 (92) (DoE, 1992).

A variety of incineration systems can be applied to clinical waste, all of these will require the addition of comprehensive pollution abatement equipment in order to comply with the new legislation. There is limited experience in the UK with small scale pollution abatement equipment for waste incineration plant. One of the designs that shows potential for cost effective application is "dry scrubbing". Dry scrubbing removes gaseous and solid pollutants by injection of a suitable dry reagent into the flue gas and collection of the reaction product and particulate matter in a high efficiency filter or electrostatic precipitator.

The ability of low density ceramic elements to control particulate matter emissions has been demonstrated both in laboratory tests and in the field (Seville et al., 1989; Elliott, 1992; Koch et al., 1992).

1.1 Objective

The objectives of this work were:

- to investigate the performance of low density ceramic filter elements (Cerafil S) as the collection device for a dry scrubbing system;

- to investigate some of the controlling parameters for dry scrubbing in a system not employing any reaction vessel i.e. the reactions are on the filters;

- to investigate the effect of having a rigid support for a filter cake as opposed to flexible fabric bags.

1.2 The Clinical Waste Incinerator

The incinerator at which the trial programme was conducted is a waste burning boiler with a design capacity of 750 kg/hr. A flow diagram of the incineration process is shown in figure 1. Waste from hospitals, doctors' surgeries, veterinary practices and other generators of waste classified as "clinical" is incinerated. The waste is highly variable in composition and can contain anything from empty biscuit tins through used disposable nappies to human tissue. Due to the nature of the waste being incinerated and process variability the off-gases from it are also highly variable both in composition and concentration. However the off-gases can be characterised as containing significant quantities of H_2O, CO_2, particulates, CO, NO_x, HCl and SO_2 with the HCl concentration being somewhat higher than SO_2 and NO_x. In addition there are trace quantities of heavy metals and "dioxins" which require abatement.

The plant used for these tests was fitted with a dry scrubbing system using conventional fabric filtration. A variety of problems had been encountered in the operation of the plant. As a result of process variability the fabric filters had not performed adequately and their lifetime was severely reduced.

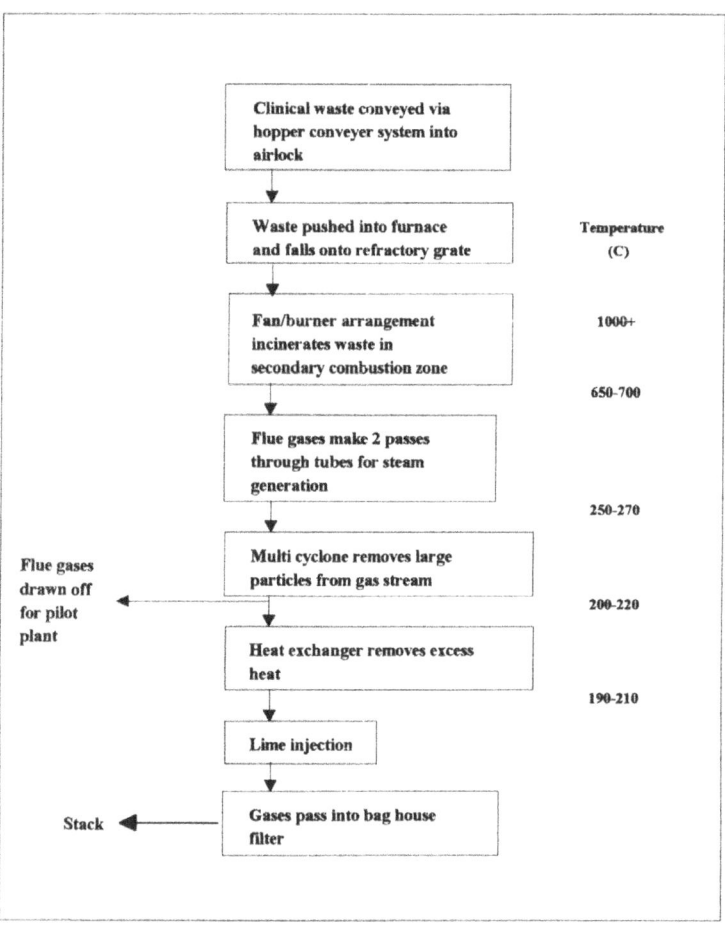

Figure 1. Schematic of the Incineration Process

2. EQUIPMENT EMPLOYED

Figure 2 shows a schematic diagram of the pilot plant.

The induced draught (ID) fan draws incineration off-gases into the pilot plant from the post cyclone stage of the incineration process. The pilot plant was installed by Environmental Air Filtration (EAF) and housed 36 elements arranged in 6 vertical rows of 6. Gases are drawn through the elements from outside to inside with each row of elements being cleaned by a reverse pulse of air in turn to remove dust build-up.

Lime was introduced into the gas in the inlet duct directly adjacent to the pilot plant using a Rospen screw feed unit. This was calibrated such that the addition rate could be related directly to the rotational speed of the feed screw. The cleaning cycle pulse interval can be selected.

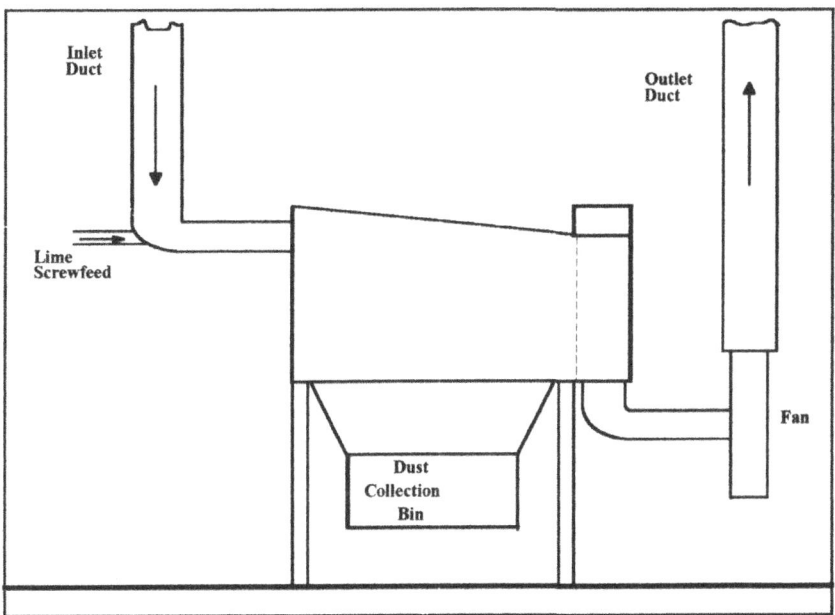

Figure 2. Schematic of the Pilot Plant

3. MEASUREMENTS TAKEN

Figure 3 shows the measurement scheme employed throughout the trials.

3.1 Gas Analysis

A key requirement to understanding the nature of the process under investigation is a real time, continuous gas analysis. Traditional measurement methods for acid gases rely on an extractive "wet chemical" test in which a sample of flue gas is bubbled into an absorbing solution. This gives an average acid gas concentration over a 2-3 hour period. The very erratic nature of the emissions from clinical waste incineration (Fells, et al, 1992; Loader & Scott, 1992) makes this testing procedure inappropriate for plant optimisation or for giving a true picture of what is occurring.

For these trials a continuous HCl meter was utilised. The meter was manufactured by Perkin Elmer and uses gas filter correlation techniques utilising a hot cell system to minimise losses in the sampling system. The meter has been extensively tested by TuV in Germany and, as far as the authors are aware, is the only continuous HCl meter approved by TuV.

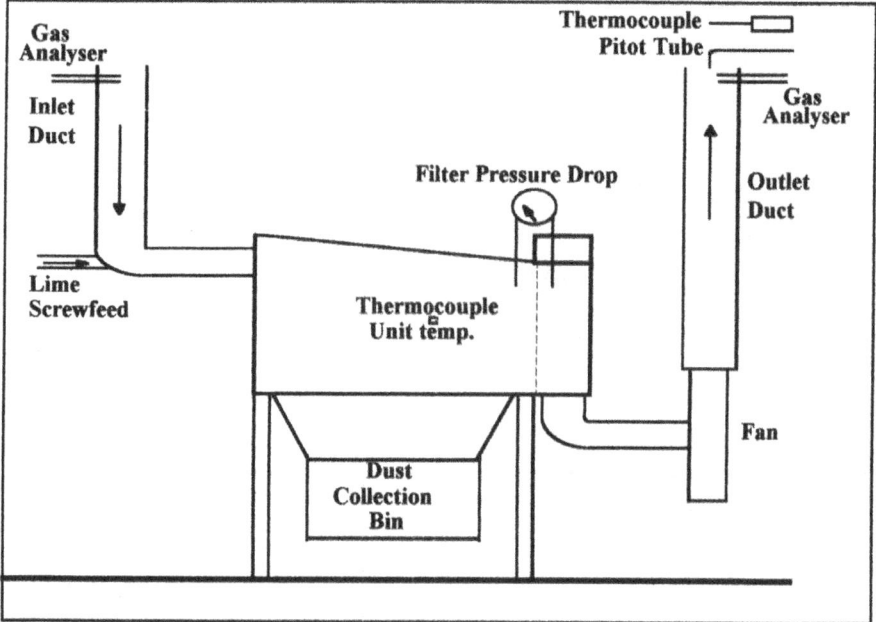

Figure 3. Measurement Positions

The analyser has 6 measurement channels as follows:

1) HCl; 0 - 1500 mg/m³

2) NO; 0 - 1000 mg/m³

3) CO; 0 - 1000 mg/m³

4) SO_2; 0 - 400 mg/m³

5) CO_2; 0 - 20 vol. %

6) H_2O; 0 - 40 vol. %

The concentration readings were automatically corrected by the analyser to 273 K (0°C), 101.3 kPa (1 atmosphere) and dry gas conditions. The reference conditions stipulated by HMIP for "concentrations of substances in emissions to air from contained sources" also require correction to 11 % v/v oxygen; however this feature was not available from the analyser. Readings were generated every 30 seconds and then averaged over 2 minutes.

The output from the analyser was in 2 forms:

•   A digital readout giving an actual concentration which could be manually scrolled through all 6 channels.

•   A data output to a data logger which comprised the date, time and percentage of full scale for each channel.

3.2 Temperature Measurement

Unit temperature was measured using a K type thermocouple positioned in the element array. Outlet duct temperature was also measured using a K type thermocouple positioned some 2 metres above the analyser probe and adjacent to the pitot connection.

3.3 Gas Flow

Gas flow through the pilot plant unit was measured using a fixed pitot tube connected to a micro-manometer. Flow was measured as velocity pressure (mm H_2O).

3.4 Pressure Drop

Pressure drop across the filter elements was measured using an analogue Magnehelic gauge and recorded manually. An electronic pressure transducer which was fitted to allow continuous recording proved to be unsuccessful.

4. DATA ANALYSIS

Figure 4 shows the data analysis scheme used throughout the trial programme.

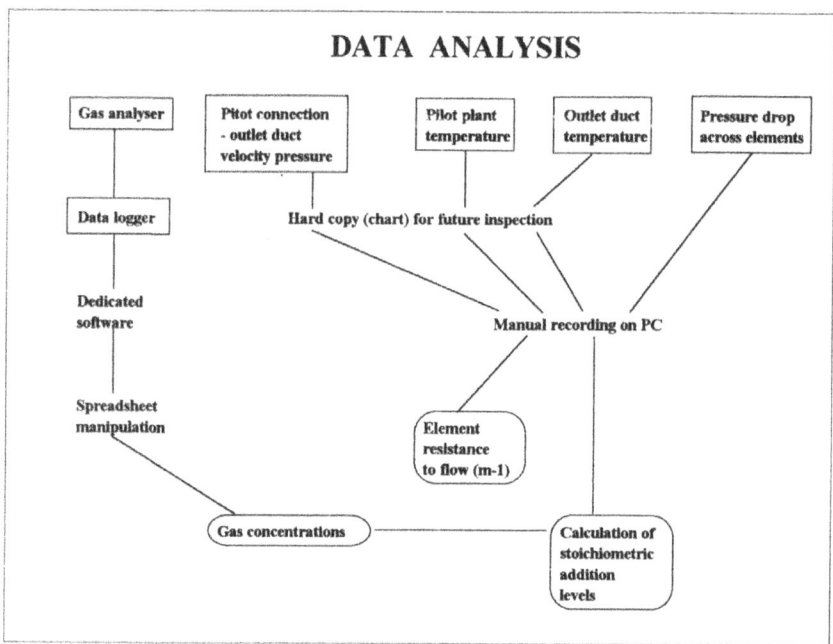

Figure 4. Data Analysis Flowchart

Gas analysis data was recorded on a "Squirrel" data logger which was down-loaded using dedicated (Squirrelwise) software. A further piece of dedicated software, Filewise, allowed for the preparation of Lotus readable files which were subsequently manipulated in Excel. Spreadsheet analysis allowed the calculation of average, maximum and minimum concentrations and standard deviation.

Monitoring of unit and outlet duct temperature and gas flow allowed the theoretical stoichiometric lime feed-rates to be calculated i.e. to neutralise the acid gases (HCl and SO_2) entering the unit with no excess addition. Pressure drop readings, in combination with the flow rate and temperature readings, were used to calculate the resistance to flow (K value) of the elements - the principal indicator of their performance.

5. EXPERIMENTAL PROGRAMME

The trial programme was divided into 4 distinct stages during which the output gases from the incinerator were characterised and the ability of Cerafil S elements, in conjunction with lime, to remove acid gases was explored. At the start of the trial programme a new set of 36 elements was installed in the pilot plant unit.

5.1 Stage I

Stage I of the trial was concerned with the characterisation of the off-gases from the incineration process and therefore the demands on the pilot plant.

In order to analyse the inlet gases to the pilot plant the gas analyser was positioned on the inlet duct and calibrated with standard gases. Monitoring and data collection was allowed to continue for a period of 10 days. This prolonged monitoring period was chosen due to the requirement to characterise the extremely variable off-gases as accurately as possible.

5.2 Stage II

Stage II of the trial was designed as an initial assessment of the performance of the pilot plant unit with respect to acid gas removal. The gas analyser probe was re-positioned onto the outlet duct, downstream of the pilot plant, and the analyser re-calibrated.

Each trial run comprised:

1. Running the pilot plant without lime for a period of approximately 20 minutes to effect a "standard" starting condition.

2. Setting the desired pulse interval.

3. Setting the desired lime feedrate.

4. Commencing lime feed and data logging.

5. Running the trial for a minimum of 24 hours.

Each trial ran for a minimum period of 24 hours in order to minimise the effect of the frequent incineration stoppages.

5.3 Stage III

Following Stage II of the trial the incineration plant was shut down for approximately 1 month while a new bag house filtration unit was installed. During this period the Stage I trial results were analysed and a chart recorder was installed adjacent to the pilot plant to provide continuous recording of velocity pressure and temperatures.

The trial scheme employed for Stage III was broadly similar to that used for Stage II but with the added variable of unit temperature. Different unit temperatures were achieved by varying the amount of insulation on the pilot plant unit and the inlet duct, or by bleeding in ambient air. Due to the frequent stoppages of the incineration plant which occurred during this stage of the experimental programme each individual trial was reduced to a minimum of 1 hour with standardisation of starting conditions not always possible.

5.4 Stage IV

Acid gases are typically measured using extractive wet chemical techniques. Previous experience has shown that there are often discrepancies between wet and on-line tests. Due to the reliance throughout the trial on an on-line gas analyser it was felt that a one-off trial should be carried out during which wet chemical testing should be carried out in order to provide further information on the performance of the system. Based on the results from Stage III suitable conditions were chosen with respect to unit temperature, lime feedrate and pulse interval. Wet chemical tests were carried out simultaneously on the inlet and outlet ducts.

6. RESULTS

The incineration plant operated only intermittently due to the many stoppages. In analysing the results only data collected when the incinerator was "on waste" (burning waste) have been used to calculate average, maximum and minimum concentrations.

The output channels from the gas analyser of particular interest were HCl, SO_2 and CO_2. HCl and SO_2 are the principal acid gases to be controlled. The CO_2 data was used to standardise the data for comparison. A correction to 9% CO_2 is regarded as being broadly equivalent to a correction to 11% O_2 for most waste fired applications.

In general the HCl and CO_2 data down-loaded from the data logger were intact and therefore suitable for spreadsheet manipulation. The SO_2 data was however largely corrupted, possibly due to the interference effects of unburnt organic compounds, and little was usable. The bulk of the results that follow are therefore centred on HCl.

6.1 Stage I

Figure 5, shows a typical, as measured, HCl inlet concentration profile. The HCl concentration depended on many factors but principally the composition of the waste being incinerated and the prevailing combustion conditions.

HCl Concentration vs. Time

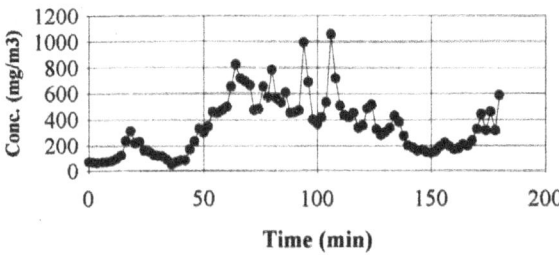

Figure 5. HCl concentration for the raw gas

For the particular time period shown in Figure 5 analysis of the HCl concentration indicates a maximum concentration of 1050 mg/m³; minimum concentration of 50 mg/m³ and a mean concentration of 348 mg/m³. These values are typical of the variability in input HCl concentration experienced throughout the trials.

Table 1 shows time weighted average concentrations for the gases measured by the gas analyser. Time weighted average concentration readings obtained from an investigation on similar plant at Ryhope hospital (Fells, et al., 1992) are shown for comparison. For this trial the values shown are averages for the complete 10 day period during which analysis took place and cover only those periods during which the incinerator was on waste.

TABLE 1
Input Gas Composition

	HCl	NO	CO	CO_2	SO_2
	mg/m³	mg/m³	mg/m³	vol. %	mg/m³
Time weighted average	306	57.7	91.6	3.02	62.8
Time weighted average*	912	172	272		187
Ryhope Test 2 Readings*	1020	145	314	-	89.0

* Standardised to 9% CO_2.

The large volumes of excess air employed during the incineration process have resulted in considerable corrections being necessary in order to correct to standard conditions. The corrected values are consistent with the previous investigation cited. For comparison purposes the as measured figures from the Ryhope investigation have been corrected to 9% CO_2 although O_2 concentration figures were available.

6.2 Stage II

During Stage II of the trial, acid gas concentrations were measured in the outlet duct from the pilot plant. The gas temperature inside the pilot plant ranged from 130°C to 160°C. The extent of the temperature range reflects the variable operating conditions of the incineration plant and the extended time period over which the trials took place. Table 2 shows the summary of the Stage II results.

To allow a comparison between inlet and outlet loads a surrogate inlet HCl concentration has been calculated based on the average concentration found during Stage I adjusted to the CO_2 value measured during Stage II. In view of the variable operating conditions there may have been higher or lower inlet loads at the time.

1) The time weighted average outlet CO_2 concentration recorded during each test is shown in the table.

2) An adjusted inlet concentration has been calculated for each run. This is the average inlet load found in stage I restated to the CO_2 concentrations measured in stage II. This is by the following process:

$$HCl_{adj} = HCl_I \times CO_2 / CO_{2I} \text{ where-}$$

i) HCl_{adj} is the adjusted inlet HCl concentration.

ii) HCl_I is the time weighted average HCl concentration recorded during Stage I of the trial.

iii) CO_2 is the time weighted average CO_2 concentration measured during the test.

iv) CO_{2I} is the time weighted average CO_2 concentration measured during Stage I of the trial (see table 1).

3) Removal efficiency has been calculated using the expression:

$$\text{Removal efficiency (\%)} = 100 \times (HCl_{adj} - HCl) / HCl_{adj}$$

			Pulse	Interval	(seconds)				
			24			60			
Lime Feedrate (gm/min)	Outlet HCl conc. (mg/m3)	Outlet CO₂ conc. (vol%)	Adjusted inlet HCl conc. (mg/m3)	Inferred Efficiency (%)	Outlet HCl conc. (mg/m3)	Outlet CO₂ conc. (vol%)	Adjusted inlet HCl conc. (mg/m3)	Inferred Efficiency (%)	
0.00	306					306			
12.73						15.2	1.31	132	88
13.08						52.9			
15.50						75.7	2.22	225	66
17.15						27.0	1.89	191	86
29.38	113	2.59	263	57					
29.60						47.4	2.02	205	77
30.00						45.3	1.64	166	73
43.10	71.2	2.60	264	73					
43.44						23.4	1.86	189	88
60.20	46.6	3.15	320	85		15.7	2.33	236	93

Figure 6 shows a plot of as measured HCl outlet concentration vs. Lime feedrate for the 24 and 60 second pulse intervals employed.

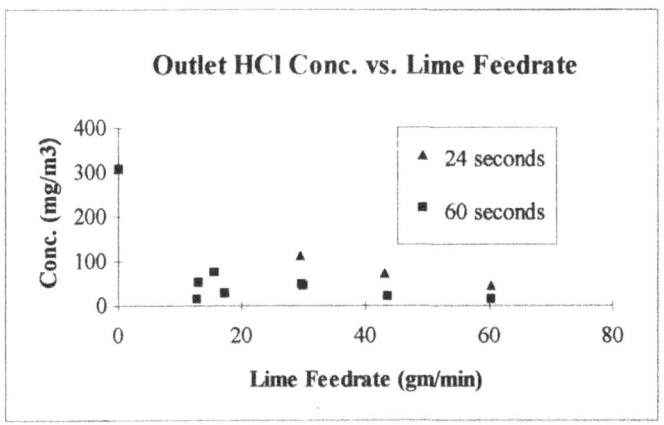

Figure 6. Outlet HCl concentration against lime feedrate

6.3 Pilot Plant Operating Conditions

Table 3 gives a guide to the pilot plant operating conditions based on accumulated manual monitoring data collected during Stage III of the trial. The high standard deviations reflect the variability in the incineration process and changes to the operating conditions of the pilot plant.

TABLE 3
Stage II Results- Pilot Plant Performance

		Maximum	Minimum	Mean	Stand. Dev.
Flowrate (actual)	**(m3/min)**	25.7	15.1	19.2	2.1
Face Velocity (actual)	**(m/sec)**	0.063	0.037	0.047	0.005
Pressure Drop	**(N/m2)**	3300	1990	2880	269
Stoichiometric Lime Feedrate	**(gm/min)**	6.09	3.62	4.79	0.58
Resistance (K)	**(m-1)**	3.82E+09	1.29E+09	2.67E+09	4.72E+08

The flowrate is approximately 10% of the total off-gases generated by the incinerator. The face velocity shown at 4.7 cm/sec compares favourably with typical bag house installations.
The stoichiometric lime feedrate has been calculated based on the generalised reactions:

$$2\ HCl + Ca(OH)_2 \rightarrow CaCl_2 + 2H_2O$$

$$SO_2 + Ca(OH)_2 \rightarrow CaSO_3 + H_2O$$
and the expression:

$$1\ /\ purity \times gas\ flow \times (HCl\ conc. \times 74/73 + SO_2\ conc. \times 74/64)$$

The resistance to flow (K) of an element is a useful expression for characterising an element and its associated filter cake and is calculated by using the expression:

$$K = Pressure\ drop\ across\ the\ elements\ /\ gas\ viscosity \times face\ velocity.$$

6.4 Stage III

Stage III of the trial was an extension of Stage II and designed to investigate further the principal operating variables i.e. pulse interval, lime feedrate and unit temperature. Tests were carried out with the unit running at 3 different temperature levels i.e. 130-140°C, 150-170°C and 180-210°C. The first temperature level was achieved by bleeding a small amount of air at ambient temperature into the inlet duct to the filter unit. The higher temperature levels were achieved by improving the degree of lagging to the required amount.

Due to the advantages shown in Stage II of increasing pulse interval and therefore residence time of the lime on the filter elements it was decided to investigate an extended pulse interval of 110 seconds during the 150-170°C tests. Unfortunately time constraints did not allow such a pulse interval to be tested in conjunction with the other temperatures.

TABLE 4
Stage III results for trials carried out at 150-170°C. The data has been manipulated as for the Stage II analysis.

					Pulse	Interval	(seconds)					
			30				60				110	
Lime	Outlet	Outlet	Adjusted	Inferred	Outlet	Outlet	Adjusted	Inferred	Outlet	Outlet	Adjusted	Inferred
Feedrate	HCl conc.	CO$_2$ conc.	inlet HCl	Eff.	HCl conc.	CO$_2$ conc.	inlet HCl	Eff.	HCl conc.	CO$_2$ conc.	inlet HCl	Eff.
(gm/min)	(mg/m3)	(vol%)	conc.(mg/m3)	(%)	(mg/m3)	(vol%)	conc.(mg/m3)	(%)	(mg/m3)	(vol%)	conc.(mg/m3)	(%)
0.00	306				306				306			
22.60	229	2.41	244	6	67.5	1.88	190	65				
23.80									40.0	2.12	215	81
30.00					55.9	2.07	210	73				
33.50									30.6	2.63	267	88
34.50	108	2.68	271	60								
42.50	66.9	2.35	238	72	28.0	2.13	216	87				
43.50									52.0	2.29	232	78
50.50									33.1	2.69	273	88
52.00					31.6	2.01	204	85				
52.25	36.8	2.33	237	84								
56.50									37.7	2.94	298	87
57.50	43.4	3.30	335	87								
58.00					35.5	2.43	247	86				

636

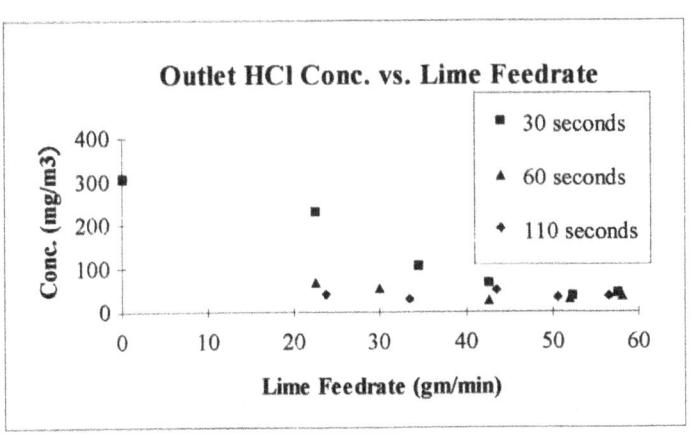

Figure 7 . Plot of HCl outlet concentration vs. lime feedrate for the 3 pulse intervals employed.
(Values as measured by the gas analyser)

TABLE 5

Results for tests carried out with the pilot plant operating at 130-140°C.

| Lime Feedrate (gm/min) | Pulse | Interval | (Seconds) | | | | | |
| | 30 | | | | 60 | | | |
	Outlet HCl conc. (mg/m3)	Outlet CO_2 conc. (vol %)	Adjusted inlet HCl conc. (mg/m3)	Inferred Efficiency (%)	Outlet HCl conc. (mg/m3)	Outlet CO_2 conc. (vol %)	Adjusted inlet HCl conc. (mg/m3)	Inferred Efficiency (%)
0.00	306							
16.50	87.3	2.15	218	60				
17.00					26.6	1.74	177	85
22.00	45.0	2.09	212	79				
24.00					14.7	1.45	147	90
31.00	50.5	2.33	236	79				
34.00					26.9	1.57	159	83
40.00					38.2	1.91	194	80
41.00	102	2.54	258	61				
50.00					15.3	1.88	191	92
51.00	37.6	2.06	209	82				
58.00	56.7	2.14	217	74	19.4	1.40	142	86

The final phase of Stage III involved running the pilot plant at a temperature of 180-210°C. This was achieved by sealing leaks into the unit and providing complete lagging in order to prevent heat loss. The results from this phase of stage III are shown in table 6.

637

TABLE 6
Results for Tests at 180-210°C

Lime Feedrate (gm/min)	HCl Outlet Concentrations (mg/m3) Pulse Interval (Seconds)	
	30	60
16.50	186	
17.00		626
23.50	121	
24.00		267
33.50	202	
34.00		432
39.50	87.7	
41.50		279
49.00		522
58.00	244	

The results obtained for the 180-210°C temperature range are high and variable and therefore merit limited further investigation.

The three sets of results can be combined into a single chart showing outlet HCl concentration (corrected to inlet CO_2) against the function lime feedrate x pulse interval, which has a unit of grams. This function can be considered a measure of the amount of lime allowed to accumulate on the filter elements.

It can be seen that outlet concentrations are lowest in the range 130-140°C, slightly higher at 150-170°C and very high and variable at 180-210°C.

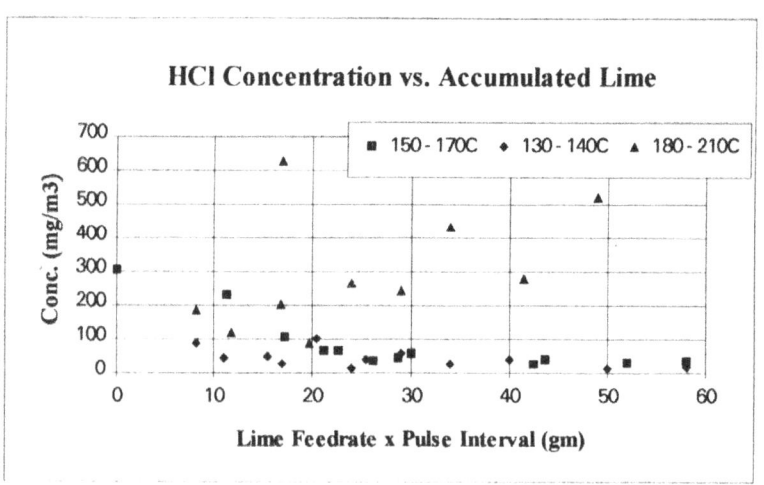

Figure 9: HCl outlet concentration vs. accumulated lime at 3 temperature ranges.

Figure 10 shows an uncorrected plot of HCl concentration data recorded by the data logger during the 130-140°C tests. The trace to the left of the graph shows typical HCl concentration behaviour for the incineration plant. The point at which the lime feed was switched on can be clearly seen. The HCl concentration was brought under control and remained under control while the lime feedrate was reduced from 40 gm/min to 34 gm/min and finally to17gm/min.

HCl Concentration vs. Time

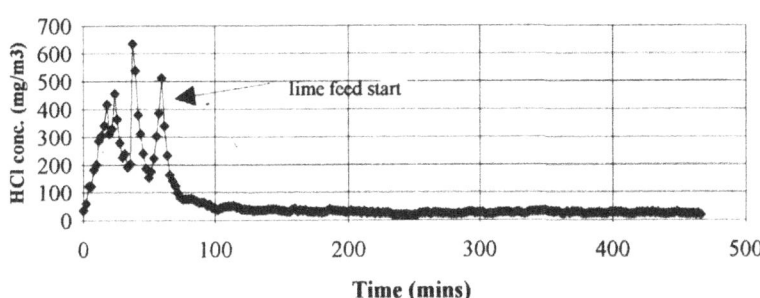

Figure 10. HCl output with start of lime feeding

6.5 Stage IV

Stage IV of the trial involved simultaneous wet chemical analysis on the inlet and outlet ducts to the pilot plant. Prior to the trial the pilot plant was thoroughly cleaned out to remove as much lime build-up as possible. The following list gives the experimental parameters employed:

- Wet chemical analysis on inlet and outlet ducts

- Analysis time- 1.5 hours

- Pulse interval- 60 seconds

- Lime feedrate- 41 gm/min

- Pilot plant operating temperature- 160-170°C

The results from the trial are shown in table 7:

TABLE 7
Wet Chemical Trial Results

	Inlet HCl conc. (mg/m3)	Outlet HCl conc. (mg/m3)	Efficiency %
As measured^	225	6	
Corrected*	237	32.7	86

^ Dry Gas Basis, 273K, 101.3 kPa
* Correction to 273K, 101.3 kPa, 11% O_2, dry gas

7. DISCUSSION AND CONCLUSIONS

The first point to stress is the variability of the clinical waste incinerator at which the trials took place. The incinerator ran intermittently and variably, making the collection of meaningful data difficult. However this situation is typical of clinical waste incinerators and thus the conditions were difficult but realistic for assessing the duty of gas cleaning equipment for such applications. The trial programme was largely completed despite the many delays. A vast amount of data was collected during the trial programme and much further analysis is possible.

Stage I of the trial involved an extended characterisation of the gas being treated. The low CO_2 concentration was the result of the large quantities of excess air employed in the incineration process. The time weighted average CO_2 figure of 3.02% results in a 3 fold correction factor to allow for a correction to the equivalent of 11% O_2. This level of correction can introduce errors due to the amplification of inaccuracies which are inherent in the measurement systems. The corrected HCl and SO_2 figures are consistent with those quoted from similar studies. The great variation in the levels of all the pollutant species is typical of this type of process. It leads to significant difficulties in the design of suitable gas cleaning equipment and of making a detailed assessment of the performance of the plant.

Stage II of the trial involved the collection of data over a prolonged period in order to compensate for the variability in the incineration process and thereby collect reliable time weighted concentration figures. The variable operation of the incinerator and the limited insulation on the pilot plant resulted in a low and variable operating temperature between 130 and 160 °C. The relatively low outlet CO_2 figures shown in table 2, and particularly those associated with the 60 second pulse interval, are thought to be a result of leaks into the pilot plant and its associated duct work. These have been compensated for in the calculation of the HCl removal efficiency figures.

The HCl removal efficiency shows an upward trend as the lime feed rate is increased for both the 24 and 60 second pulse intervals. The neutralisation of the HCl depends on the available lime. It is apparent that there is a benefit from increasing the pulse interval between cleaning cycles. This may be due to the possibility of more complete lime coverage on the elements but may also be due to the formation of a deeper cake and hence increased reaction time. A removal efficiency of 85% was achieved with a 24 second pulse interval and a lime feedrate of nominally 12.6 times stoichiometric while at the same lime feedrate a 93% removal efficiency was achieved with a 60 second pulse interval. It can be deduced from the effect that the reaction in this case is occurring on the elements themselves and not in the gas stream. The improvement in HCl removal efficiency at the 60 second pulse interval (6 minutes between elements) can be utilised to drive down the absolute outlet concentration or reduce the lime consumption. Since simultaneous monitoring was not available, and the inlet HCl concentration could be different from the assumed figure, the stoichiometric ratios can only be regarded as a guide.

Stage III of the trial was an expansion of Stage II. As a result of observations made during Stage II the pilot plant was modified to seal leaks where possible. The effect of temperature and pulse interval on HCl removal efficiency was then further investigated. At the lowest temperature range investigated, 130-140°C, consistently good removal efficiencies were achieved even at low lime feedrates and particularly for the 60 second pulse interval. Results for the 150-170°C operating temperature were also encouraging showing similar trends to those already indicated. It can also be seen that little benefit was accrued from increasing the pulse interval to 110 seconds or increasing the lime feedrate above 50 gm/min (10.4 times stoichiometric). It is likely that a useful maximum lime addition rate is reached beyond which excess lime does not contribute to HCl removal.

The results obtained for an operating temperature of 180-210 °C are disappointingly erratic and further investigation is required to understand the process. Other work has shown significant reductions in acid gas concentrations in dry scrubbers at these temperatures. These systems would involve a significant amount of reaction occurring in a reaction vessel, hence in the gas phase. In this system it has been suggested that the majority of the acid removal is occurring on the filter elements and the poor removal efficiency at these temperatures may be due to a different reaction mechanism being dominant for the gas phase and solid phase reactions.

Stage IV of the trial was carried out, by an independent laboratory, in order to provide a confirmatory test at fixed, pre-specified conditions using an established wet chemical technique. The corrected inlet and outlet concentrations, as determined by the wet chemical route, are extremely encouraging and indicate good HCl removal (86%) for the chosen conditions.

The principal conclusions that can be drawn from the trial are:

- Low density ceramic media, such as Cerafil, can provide a support for a renewable bed of basic powder. Neutralisation of acid gases can be improved by increasing the time between cleaning cycles.

- HCl gas removal with efficiencies in excess of 90% can be achieved without the addition of a reaction vessel.

- Wet chemical tests showed that outlet HCl levels can be maintained at a level well below that specified in PG5/1(92).

- Increasing the pulse interval increases the removal efficiency but an upper limit was evident.

- Increasing the lime feedrate improved HCl removal efficiency but high efficiencies required high lime feedrates suggesting that recirculation might be desirable.

- A moderate i.e. <170°C operating temperature was found to be most effective.

- Only one set of elements was used throughout the trials. The elements were only in use intermittently and then under highly variable conditions. The elements remained completely intact throughout the trials with no visible deterioration.

- Further work to optimise the process would be valuable in order to increase understanding.

Acknowledgement. The authors wish to express their thanks for the support of the parties involved in the test work and the DTI for providing funds for the monitoring equipment and the independent laboratory tests.

References

The Health Services Advisory Committee (HSAC), Safe Disposal of Clinical Waste, HMSO, ISBN 0 11 88 6355X.

Her Majesty's Inspectorate of Pollution, (HMIP) (1992). Environmental Protection Act 1990, Process Guidance Note IPR5/2, Waste Disposal and Recycling, Clinical Waste Incineration, HMSO, ISBN 0 11 752652 5.

Department of the Environment, (DoE) (1992). The Scottish Office, Welsh Office, Environmental Protection Act 1990, Part 1, Secretary of State's Guidance- Clinical waste incineration processes under 1 tonne an hour, PG5/1, HMSO, ISBN 0 11 752688 6.

Seville, J., Clift, R., Withers, C., Keidel, W., (1989). Rigid Ceramic Media For Filtering Hot Gases, Filtration and Separation, July/August 1989.

Elliott, G. K., (1992). Ceramic Filtration Accepted as Particulate Removal Technique, Pollution Prevention, June 1992.

Koch, D., Cheung, W., Seville, J., Clift, R.,(1992). Effects of Dust Properties on Gas Cleaning Using Rigid Ceramic Filters, <u>Filtration and Separation</u>, July/August 1992.

Fells, A. R., Jackson, P. M., King, P. G., (1992). An Emissions Audit From A Small Clinical Waste Incinerator, <u>ETSU B 1313 - P1</u>.

Loader, A., Scott, D. W.,(1992). Emission Investigation at an Advanced Starved Air Clinical Waste Incinerator, <u>Warren Spring Laboratory Report LR890</u>.

The views expressed in this paper are those of the authors and not necessarily those of ETSU or the Department of Trade and Industry.

PREDICTION OF DRY SCRUBBING PROCESS PERFORMANCE

W. Duo, J.P.K. Seville, N.F. Kirkby, and R. Clift
Department of Chemical and Process Engineering
University of Surrey, Guildford, GU2 5XH, UK

ABSTRACT

A mathematical model is presented to simulate dry scrubbing processes for sorption of acid gases in filter cakes. The model can deal with both fixed and moving cake boundary problems. The data for major kinetic parameters available in the literature for reactions of Ca-compounds with HCl are reviewed and analysed. Numerical solutions for the $Ca(OH)_2$/HCl reaction at 373-523 K demonstrate that solid state diffusion through the product layer is the rate controlling step in this temperature range. Chemical reaction rate is shown to be controlling at higher temperatures. The model is used to compare the effectiveness of different strategies for sorbent injection. Compared with continuous addition, the injection of at least part of the sorbent at the beginning of the operation cycle immediately following filter cleaning can significantly improve the overall efficiency.

Notation

A_{cb} reaction interfacial area per unit volume of packed bed, m^2/m^3
A_{cp} reaction interfacial area per unit volume of sorbent particles, m^2/m^3
b stoichiometric coefficient in reaction (6)
C concentration of reactant gas, mol/m^3
C_1 gas concentration within pores of a particle, mol/m^3
C_2 gas concentration in product layer, mol/m^3
C_c gas concentration at the unreacted core of a grain, mol/m^3
C_e equilibrium concentration of the reactant, mol/m^3
C_i inlet concentration of reactant gas, mol/m^3
C_m time-average concentration at $z = 0$, defined by Eq. (30), mol/m^3
C_{Pg} concentration of gas product P_g, mol/m^3
C_s gas concentration at the exterior surface of a particle, mol/m^3
C_w water content, volume fraction
D_{ef} effective diffusion coefficient within a sorbent particle, m^2/s
D_{efi} initial effective diffusion coefficient, m^2/s
D_k Knudsen diffusion coefficient, m^2/s

D_m molecular diffusion coefficient, m^2/s
D_s coefficient of solid state diffusion through product layer, m^2/s
D_z axial dispersion coefficient, m^2/s
F_1 dust fraction in filter cake containing sorbent
H current thickness of the filter cake with moving boundary, m
H_0 initial thickness of the filter cake due to batch injection, m
k_g exterior mass transfer coefficient, m/s
K_p equilibrium constant of reaction (1), m^3/mol
k_s reaction rate constant per unit area of reaction interface, m/s
k_{-s} reverse reaction rate constant per unit area of surface, m/s
M_A molar weight of reaction gas A, kg/mol
M_s molar weight of sorbent, kg/mol
P pressure, Pa
Q increasing rate of cake thickness, m/s
r radius coordinate in solid product layer, m
R radius coordinate in a porous particle, m
r_A net reaction rate of reaction gas A, $mol/(m^2.s)$
r_c current radius of unreacted core of a grain, m
r_g grain radius, m
R_{Gas} gas constant, J/(mol.K)
R_p particle radius of sorbent, m
r_t total radius of a grain, defined by Eq. (25), m
r_{vb} reaction rate for a unit volume of packed bed, $mol/(s.m^3)$
r_{vp} reaction rate based on unit volume of sorbent particles, $mol/(s.m^3)$
S total flux of sorbent injected in a full cycle, kg/m^2
S_i mass fraction of sorbent initially injected to that added in full cycle
S_g specific surface of particles, m^2/kg
t time, s
T temperature, K
t_f duration of the operation cycle, s
u_0 face velocity of flue gas to filter, m/s
v volume, m^3
V_{mb} molar volume of solid sorbent, m^3/mol
V_{mp} molar volume of solid product, m^3/mol
V_p volume of a particle, m^3
W_d mass concentration of dust particles, $kg/(m^3$ gas)
W_s mass concentration of sorbent particles, $kg/(m^3$ gas)
X local conversion of sorbent
X_p mean conversion of sorbent particles
X_{pm} average conversion of sorbent in filter cake, defined by Eq. (32)
z length coordinate along fixed bed thickness, m
Z_1 thickness of sorbent-containing cake for the case of batch injection, m

β stoichiometric ratio of sorbent to reactant gas
ε interparticle voidage of filter cake
ε_p initial porosity of sorbent particle
ε_s porosity of the reaction product
ε_x porosity of reacting particles
ρ_s true density of sorbent, kg/m^3

ρ_{ps} particle density of sorbent, kg/m^3
ρ_{pd} particle density of dust, kg/m^3
τ_t tortuosity

INTRODUCTION

Fossil fuels such as coal and heavy oil and wastes, including municipal and hospital wastes and refuse-derived fuels, contain significant concentrations of sulphur, nitrogen, halogens (particularly chlorine) and heavy metals. A considerable portion of these components will be converted into harmful emissions, such as sulphur dioxide (SO_2), hydrogen chloride (HCl), and metal fume or vapour, during combustion of the fuels at high temperatures. The gases from stationary combustion installations and incinerators can make an important contribution to local and global environmental pollution and therefore need to be cleaned before emission.

Dry scrubbing has been proposed as an effective method for the removal of acid gases from flue gases. In this method, a solid sorbent is injected into the flue gas at appropriate temperatures up-stream of a barrier filter (e.g. a rigid ceramic filter). The spent sorbent and dust are collected together as a filter cake. Much of the gas-solid reaction is expected to take place in the deposited cake because, in normal operation, the residence time of the sorbent particles in the cake is much longer than that in the entrainment duct. Under appropriate conditions, heavy metals can also be collected by adsorption (Heap, 1992).

The kinetics of the reaction between SO_2 and CaO or $CaCO_3$ has been extensively investigated in fixed bed reactors (Borgwardt et al., 1986; Dam-Johansen & Østergaard, 1991; Peukert & Löffler, 1993) or by means of thermogravimetric analysis (Iisa & Hupa, 1990; Tullin & Ljungstrom, 1989). Sorption of hydrogen chloride with lime or limestone has also been studied (Weinell et al., 1992; Daoudi & Walters, 1991) but not so extensively. Kinetic models have been proposed based on sorbent particle structure models known as the "grain model" (including the "grain-micrograin model") (Wen & Ishida, 1973; Hartman & Coughlin, 1976; Dam-Johansen et al., 1991; Peukert & Löffler, 1993) and the "pore model" (Christman & Edgar, 1983; Hartman & Coughlin, 1974; Peukert & Löffler, 1993). Most of the existing models deal with the gas solid reactions occurring in a single particle or in a fixed bed of particles with a constant bed thickness. Keener and Biswas (1989) proposed a dry scrubbing model for SO_2 removal with Na_2CO_3 in filter cakes, which considered the movement of the cake boundary. However, this model did not incorporate a sorbent particle structure model but used an assumed expression to describe the particle reaction rate.

A mathematical model is presented in this paper for simulation of the dry scrubbing process occurring in a filter cake containing sorbents for removal of acid gases. The model is formulated to be general, including both cases of current practical interest: fixed boundary, where the sorbent containing cake is formed instantaneously, and then exposed to the gas;

moving boundary, where the cake builds up continuously from particles of dust and sorbent entrained in the gas. The data for major kinetic parameters, such as the reaction rate constant and the solid state diffusion coefficient, available in the literature for reactions of Ca-compounds with HCl are reviewed, analysed and used in the dry scrubbing model. The predictions are reported for different strategies for sorbent feeding, including batch, continuous, and semi-continuous feeding.

DESCRIPTION OF THE MODEL

Take HCl as an example of the acid gas of interest. HCl may be removed by calcium compounds through the following gas solid reactions:

$$Ca(OH)_2 + 2HCl = CaCl_2 + 2H_2O \qquad (1)$$
$$CaO + 2HCl = CaCl_2 + H_2O \qquad (2)$$
$$CaCO_3 + 2HCl = CaCl_2 + H_2O + CO_2 \qquad (3)$$
$$CaCO_3 = CaO + CO_2 \qquad (4)$$
$$Ca(OH)_2 = CaO + H_2O \qquad (5)$$

The above reactions must eventually reach global equilibrium and should in theory be considered simultaneously. However, under certain conditions, one or more of them are dominating. For instance, if $Ca(OH)_2$ is added as the sorbent into a flue gas containing water vapour at temperatures below 400 ^{0}C, reactions (2) -(5) will be insignificant. For simplicity, a single reaction is considered in the development of the mathematical model, which is represented in a general form as follows:

$$A(g) + bB(s) = cPg(g) + dPs(s) \qquad (6)$$

Dry scrubbing with filtration of dust involves a multicycle process of deposition of particles and intermittent removal of filter cake from the filter medium. In some circumstances, the cake may be detached from the whole surface of the filter (complete cleaning). In other circumstances, the cake detaches from patches of the surface only; this is known as "patchy cleaning" (e.g. Cheung et al., 1988; Seville et al., 1989). This work is focused on a dry scrubbing process with complete cleaning for which we need to consider only one cycle.

For a constant concentration of the acid gas, C_i, the overall stoichiometric ratio of the sorbent to the gas, β, is determined by

$$\beta = \frac{S/M_s}{bC_i t_f u_0}$$

where S (kg/m^2) is the total amount of sorbent to be injected in a full cycle with the cycle duration of t_f. It may be so arranged (Duo et al., 1993) that part of this amount of sorbent is injected in batch at the beginning of the cycle and the rest of it added continuously at a constant rate thereafter in the cycle. S_i is defined in this paper as

$$S_i = \frac{amount\ of\ sorbent\ injected\ at\ the\ beginning\ of\ the\ cycle}{total\ amount\ of\ sorbent\ added\ during\ the\ full\ cycle}$$

This arrangement is illustrated in Fig. 1. If $S_i = 1$, the process is referred to as batch injection with a fixed boundary; if $S_i = 0$, as continuous addition with a moving boundary; and if $0 < S_i < 1$, the process is referred to as semi-continuous addition with a moving boundary.

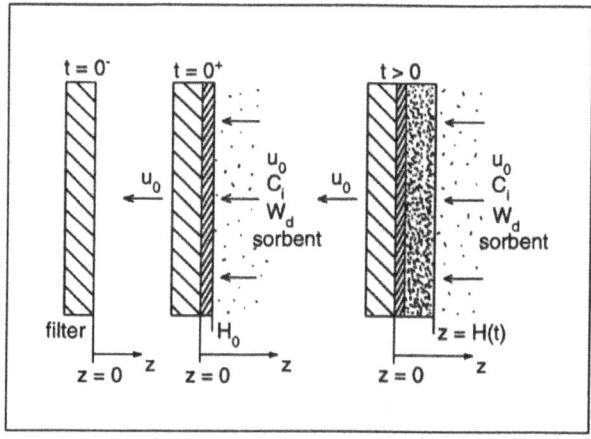

Figure 1. Schematic diagram of feeding approaches.

In the development of the model, it is also assumed that

* process dust is inert to the acid gas;
* filtration velocity, u_0, and mass concentration of the dust W_d, are constant;
* particles of the sorbent and dust are mixed completely and distributed uniformly in the filter cake without particle agglomeration
* voidage of the cake, ε, is constant;
* radial dispersion of gas in the filter cake is neglected.

The governing equation in each of the two sections of the filter cake is expressed as

$$\varepsilon \frac{\partial C}{\partial t} - u_0 \frac{\partial C}{\partial z} - \varepsilon D_z \frac{\partial^2 C}{\partial z^2} = -r_{vb} \tag{7}$$

where r_{vb} represents the reaction rate for a unit volume of the packed bed. The boundary conditions for Eq. (7) are:

$C(z,0) = 0$, for $0 < z < H$

$C(z,0) = C(H,0)$, for $z = H$

$u_0[C_i - C(H,t)] = \varepsilon D_z(\partial C/\partial z)_{z=H}$

$(\partial C/\partial z)_{z=0} = 0$

$(\partial H/\partial t) = Q$ and $H(0) = H_0$

where Q, the rate at which the cake is being thickened, is determined by the rate of arrival of sorbent and dust particles:

$$Q = \frac{W_s/\rho_{ps} + W_d/\rho_{pd}}{1 - \varepsilon} u_0 \tag{8}$$

Symbols used are listed at the beginning of the paper.

It is assumed that both the forward and reverse reactions are first order with respect to the concentration of the gas at the reaction surface. For reaction (6), the specific reaction rate (that is based on a unit area of surface) is expressed by

$$r_A = k_s C_A - k_{-s} C_{Pg} = k_s[C_A - (k_{-s}/k_s)C_{Pg}] \tag{9}$$

At equilibrium, $r_A = 0$, thus,

$$k_{-s}/k_s = (C_A/C_{Pg})_e \tag{10}$$

where subscript e is used to indicate the equilibrium state. For HCl sorption as represented by reactions (1-3), $C_{Pg} = (C_{Pg})_e$, since H_2O and CO_2 are usually present in flue gases at a content far in excess of the content of HCl. Inserting Eq. (10) into Eq. (9) gives

$$r_A = k_s(C - C_e)_A \tag{11}$$

where C_A and C_e represent the actual gas concentration and the equilibrium concentration at the reaction surface, respectively. The value of C_e is a function of temperature and can be calculated from the equilibrium constant.

The reaction rate is proportional to the area of the reaction surface. To determine r_{vb}, the reaction interfacial area per unit volume of the bed, A_{cb}, must be known. Different particle structure models may be incorporated to determine the values of A_{cb}. The grain model or pellet model, which is relatively simple and well defined (Hartman and Coughlin, 1976; Szekely et al., 1976), is adopted in the present paper. The model treats each sorbent particle

649

as a spherical pellet, made up of spherical grains of uniform size. Each grain is taken to be impermeable, and to react by a "shrinking core" process as the reaction front advances into it. This model has also been shown to apply to sorption of alkali salt vapours from hot gases (McLaughlin et al., 1993).

The following equation describes the exterior and interior mass balances for the particles:

$$\varepsilon_x \frac{\partial C_1}{\partial t} - \frac{1}{R^2} \frac{\partial}{\partial R}(R^2 D_{ef} \frac{\partial C_1}{\partial R}) = -r_{vp} \tag{12}$$

with the boundary conditions:

$C_1(R,z,0) = 0$
$C_1(R_p,z,t) = C_s(z,t)$
$(\partial C_1/\partial R)_{R=0} = 0$
$D_{ef}(\partial C_1/\partial R)_{R=R_p} = k_g[C(z,t) - C_s(z,t)]$

where C_1 is the gas concentration in intraparticle pores and D_{ef} the effective diffusion coefficient within the particles. The value of D_{ef} is dependent on the values of tortuosity, the molecular and Knudsen diffusion coefficients, and the particle porosity (Hartman and Coughlin, 1976; Dam-Johansen et al., 1991). The following correlation is adopted:

$$D_{ef} = \frac{\varepsilon_x}{\tau_t}(\frac{1}{D_m} + \frac{1}{D_k})^{-1} \tag{13}$$

A method provided by Reid et al. (1977) was used to calculate the molecular diffusion coefficient, D_m, and the Knudsen diffusion coefficient, D_k, was estimated based on the dusty gas model (Mason and Malinauskas, 1983):

$$D_k = \frac{4}{3}(\frac{8R_{gas}T}{\pi M_A})^{1/2} K_0 \tag{14}$$

where

$$K_0 = [\frac{128}{9}n_d r_t^2(1 + \frac{\pi}{8})]^{-1} \tag{15}$$

and

$$n_d = \frac{3(1 - \varepsilon_p)}{4\pi r_g^3} \tag{16}$$

The variations of ε_x, D_k and hence D_{ef} with sorbent conversion due to the difference in molar volumes between the sorbent and the product are considered in the model:

$$\epsilon_x = \epsilon_p + (1 - \epsilon_p)(1 - V_{mp}/V_{mb})X \tag{17}$$

where V_{mb} and V_{mp} are the molar volumes of solid sorbent and solid product, respectively. According to the grain model, the degree of local conversion of sorbent in a sorbent particle is defined by the extent of conversion of the grains at that position:

$$X = 1 - r_c^3/r_g^3 \tag{18}$$

The grain model assumes that reaction occurs at the surface of the unreacted cores of the grains. According to Eq. (11), the reaction rate per unit volume of particles is expressed as

$$r_{vp} = A_{cp}k_s(C_c - C_e) \tag{19}$$

where C_c represents the reactant gas concentration at the surface of the unreacted core in a grain. The reaction interfacial area per unit volume of sorbent particles, A_{cp}, is given by

$$A_{cp} = 3(1 - \epsilon_p)r_c^2/r_g^3 \tag{20}$$

The reaction rate per unit volume of bed, r_{vb}, is calculated from the averaged values of r_{vp}, the reaction rate per unit volume of sorbent particles, according to the following relation:

$$r_{vb} = n \cdot \int_{V_p} r_{vp} dv \tag{21}$$

where n is the number of sorbent particles per unit volume of the filter cake, i.e.

$$n = \frac{3(1 - \epsilon)(1 - F_1)}{4\pi R_p^3} \tag{22}$$

Combining Eqs (19), (20), (21) and (22), the expression for r_{vb} is obtained:

$$r_{vb} = \frac{9(1 - \epsilon)(1 - F_1)(1 - \epsilon_p)k_s}{(r_g R_p)^3} \int_0^{R_r} (C_c - C_e)r_c^2 R^2 dR \tag{23}$$

The gas concentration at the surface of the unreacted core of a grain, C_c, can be determined by carrying out a mass balance within the product shell outside the unreacted core:

$$\epsilon_s \frac{\partial C_2}{\partial t} - \frac{1}{r^2}\frac{\partial}{\partial r}(r^2 D_s \frac{\partial C_2}{\partial r}) = 0 \tag{24}$$

with the following boundary conditions:

651

$C_2(r,R,z,0) = 0$

$C_2(r_t,R,z,t) = C_1(R,z,t)$

$D_s(\partial C_2/\partial r)_{r=r_c} = k_s[C_c(R,z,t) - C_e)]$

where r_t, the total radius of a grain allowing for volume change on reaction, is defined by

$$\frac{4}{3}\pi r_t^3 = \frac{4}{3}\pi r_c^3 + \frac{4}{3}\pi(r_g^3 - r_c^3)\frac{V_{mp}}{V_{mb}} \tag{25}$$

Inserting Eqs (17) and (18) into Eq. (25), the expression for r_t is obtained:

$$r_t = (\frac{1 - \varepsilon_x}{1 - \varepsilon_p})^{1/3} r_g \tag{26}$$

The rate at which the reaction interface moves towards the centre of a grain is given by

$$\frac{\partial r_c}{\partial t} = -\frac{bM_s}{\rho_s}k_s(C_c - C_e) \tag{27}$$

with the following boundary conditions:

$r_c(R,z,0) = r_g$

PARAMETERS IN THE MODEL

As in previous studies (Hartman & Coughlin, 1976; Weinell et al., 1992), a pseudo steady state approximation is applied to the present partial differential equations with respect to the gas concentration. That is, the terms of $\partial C/\partial t$ in Eq. (7), $\partial C_1/\partial t$ in Eq. (12), and $\partial C_2/\partial t$ in Eq. (24) are set to zero. By integrating Eq. (24) from r_t to r_c, $(C_c - C_e)$ is expressed explicitly as

$$C_c - C_e = \frac{D_s(C_1 - C_e)}{D_s + k_s r_c(1 - r_c/r_t)} \tag{28}$$

The preceding system of equations has been solved numerically. The value of the derivative $\partial D_{ef}/\partial R$ was approximated to zero in the calculations, as Hartman and Coughlin (1976) suggested that the radial gradient of D_{ef} is small and may be negligible. This is also confirmed by our study for dry scrubbing using small sorbent particles (the results not presented here).

The solutions for the reaction system $Ca(OH)_2 + 2HCl \Rightarrow CaCl_2 + 2H_2O$ are presented as an example in the following. The interference of other reactions is neglected. The values of the model parameters listed in TABLE 1 were used in the calculations, except where otherwise stated.

The value of grain radius, r_g, was calculated from the specific surface area, S_g:

$$r_g = 3/\rho_s S_g \qquad (29)$$

C_e was estimated from the equilibrium constant given by Balekdjian (1987) for HCl. The value of k_g was estimated using the correlations of Wakao and Funazkri (1978) and Szekely et al. (1976). The value of D_z was estimated according to the correlation given by Ruthven (1984).

It has been shown that k_s and D_s are kinetically the most important parameters for the gas-solid reactions at dry scrubbing conditions (Duo et al., 1993). Depending on the values of k_s and D_s, the overall rate may be controlled by chemical reaction or by diffusion through the product layer. A few experimental and estimated data for D_s and k_s are available for reactions of Ca-compounds with HCl. A review of these data is presented in the following.

Weinell et al. (1992) suggested solid state diffusion through the product layer as the controlling step for the $Ca(OH)_2/HCl$ reaction at 373-523 K. Uchida et al. (1979) showed for the $CaCO_3/HCl$ reaction that the rate was controlled by chemical reaction at temperatures below 673 K and by diffusion through the product layer at temperatures above 773 K. Peukert and Löffler (1993) reported that for the $CaCO_3/HCl$ reaction at about 773 K the influence of k_s on sorbent conversion was small and restricted to short reaction times. By simulation, Duo et al. (1993) predicted that the rate controlling step for the CaO/HCl reaction at 800 K was chemical reaction. The available data for k_s and D_s are represented in Figs 2 and 3, respectively.

It is still unknown whether the values of k_s and particularly D_s for the reactions $Ca(OH)_2/HCl$, $CaCO_3/HCl$, and CaO/HCl are the same or different at the same conditions. Given, as widely accepted, that the solid product of all these reactions is the same ($CaCl_2$), the values of D_s should be the same unless the structure of the product layers varies with the reaction system. A conclusion in this regard may not yet be obtained from the limited and scattered data in the two figures.

The figures show the temperature dependencies of k_s and D_s. The activation energies vary for k_s from 0 to 22 kJ/mol (Daoudi and Walters, 1991) and for D_s from 19 (Weinell et al.,

TABLE 1. Parameter values used in calculations

C_e	3.42×10^{-7}	mol/m^3
C_i	1.22×10^{-2}	mol/m^3
C_w	12%	
D_{efi}	2.51×10^{-6}	m^2/s
D_s	2.05×10^{-13}	m^2/s
D_z	3.48×10^{-5}	m^2/s
k_g	4.62	m/s
K_p	7.31×10^{13}	kg/m^3
k_s	4.75×10^{-4}	m/s
P	1	atm
r_g	6.8×10^{-8}	m
R_p	1.23×10^{-6}	m
S_g	1.97×10^{4}	m^2/kg
T	500	K
t_f	40	min
u_0	5.0×10^{-2}	m/s
W_d	1.66×10^{-2}	kg/m^3
W_s	9.03×10^{-4}	kg/m^3
β	2.00	
ε	0.600	
ε_p	0.519	
ρ_s	2.24×10^{3}	kg/m^3
ρ_{ps}	1.08×10^{3}	kg/m^3
ρ_{pd}	2.50×10^{3}	kg/m^3
τ_t	3.0	

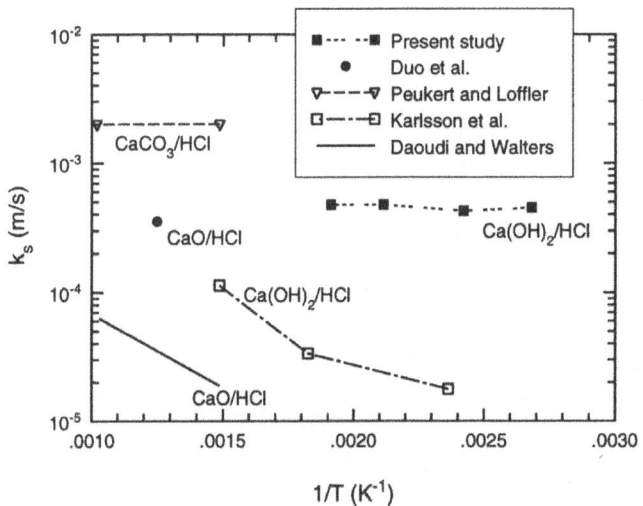

Figure 2. Reaction rate constant from different sources

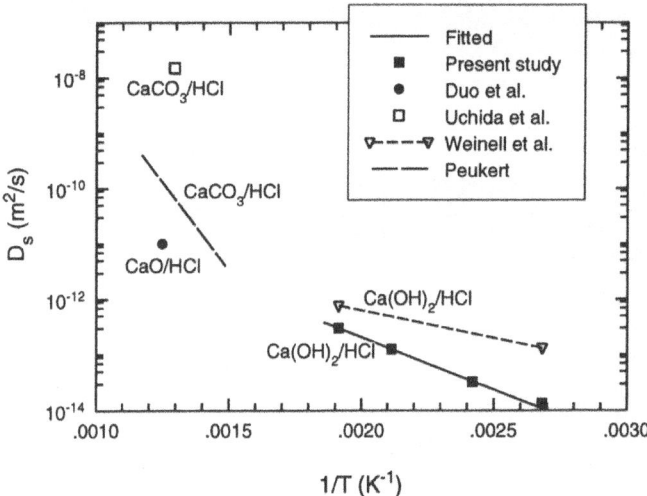

Figure 3. Solid state diffusion coefficient from different sources

1992) to 123 kJ/mol (Peukert, 1990). Taken broadly, in the temperature range between 373 and 1000 K, the values of k_s vary by less than 3 orders of magnitude, whereas the values of D_s vary by more than 6 orders of magnitude. It is thus concluded that the activation energy for solid state diffusion is significantly higher than that for chemical reaction. Consequently, the HCl sorption rate is controlled by solid state diffusion at lower temperatures but by chemical reaction at higher temperatures. This is in agreement with the results of Weinell et al. (1992) and Duo et al. (1993), but not with Uchida et al. (1979).

Further evidence to support this conclusion is available. Bandt et al. (1992) observed that a product layer in the reaction between HCl and calcined dolomite was formed within 0.75 s at 300 K and within 0.1 s at 580 K. However, the solid conversion was low, suggesting a faster chemical reaction and slower solid diffusion through the product layer at lower temperatures. Similar results were reported by Bandt et al. (1991). Iisa and Hupa (1990) showed that the activation energy of the reaction of SO_2 with $CaCO_3$ was considerably lower than the activation energy of solid state diffusion of SO_2 through the $CaSO_4$ layer.

In Figs 2 and 3, the data labelled "Present study" were obtained by fitting the present model to the experimental results of Weinell et al. (1992), with input values of the model parameters taken from their experimental conditions. As shown in Fig. 4, the experimental data are fitted well by the model. The solid state diffusion coefficient thus obtained may be described by the Arrhenius correlation, as shown in Fig. 3 (labelled "Fitted").

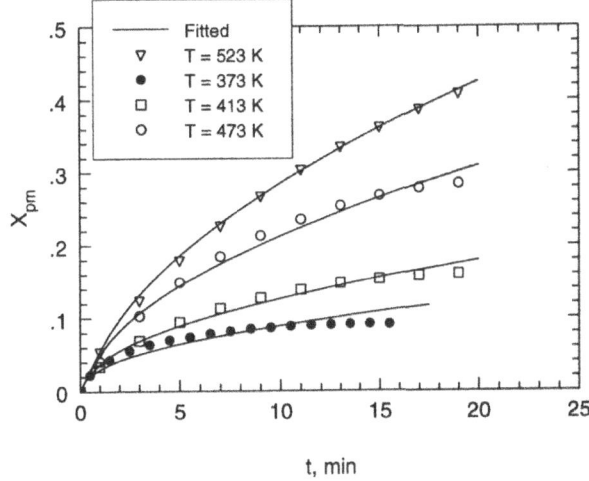

Figure 4. Sorbent conversion curves. points: experimental data of
Weinell et al. (1992). lines: fitted by the present model

Qualitatively, we agree with Weinell et al. (1992) that solid state diffusion through the product layer is the rate controlling step for the $Ca(OH)_2$/HCl reaction between 373 and 523 K. However, the present fitted data for D_s are different from those obtained by Weinell et al. (1992). This is mainly because their model contained a parameter called "final degree of conversion", X_{max}, which is absent in our model. The values of X_{max} vary with temperature, as shown in Fig. 4. Therefore, the presence of this parameter would result in an alteration of the values of the other parameters. In the following calculations, the present fitted values of D_s are used, as given in TABLE 1.

Since, as shown in Figs 2 and 3, the available data for k_s and D_s scatter over a large range, it is necessary to know whether large or small errors will be caused if incorrect values of the parameters are taken. For this purpose, calculations were performed in this study with the values of k_s and D_s varied over a large range, under otherwise identical conditions given in TABLE 1, to examine the sensitivity of the predicted results to variation in these parameters.

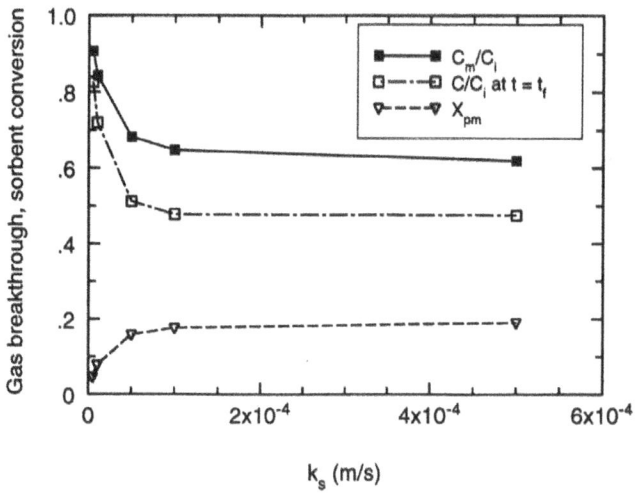

Figure 5. Effect of the reaction rate constant on predicted results

Some of the results are shown in Fig. 5 for the effect of k_s and in Fig. 6 for the effect of D_s, including HCl breakthrough at $z = 0$ and $t = t_f$, C/C_i, the time-average HCl breakthrough from $t = 0$ to t_f, C_m/C_i, where C_m is defined as

$$C_m = \frac{1}{t_f} \int_0^{t_f} C(0, t) \, dt \tag{30}$$

and the average sorbent conversion from $z = 0$ to $H(t_f)$, X_{pm}, defined as

$$X_p = \frac{3}{R_p^3} \int_0^{R_p} R^2 [1 - (r_c/r_g)^3] \, dR \tag{31}$$

$$X_{pm} = \frac{1}{H(t_f)} \int_0^{H(t_f)} X_p(z, t_f) \, dz \tag{32}$$

It is demonstrated that the HCl adsorption results represented by these quantities are sensitive to the variation of k_s only if k_s is smaller than 1.0×10^{-4} m/s, above which the influence of k_s becomes negligible. D_s has a similar but more pronounced influence. The value of D_s above

656

which the influence becomes insignificant is about 1.0×10^{-11} m^2/s. These results confirm that the overall rate of the HCl sorption reaction is controlled by the solid state diffusion through the product layer at lower temperatures (see TABLE 1 for k_s & D_s at 500 K).

In Figs 5 and 6, the values of C/C_i at $t = t_f$ are smaller than the values of C_m/C_i. That is because the simulations were performed with the assumption of continuous feeding of sorbent, i.e. $S_i = 0$. As will be seen later, for continuous feeding the gas breakthrough decreases with operation time and reaches its lowest value at $t = t_f$.

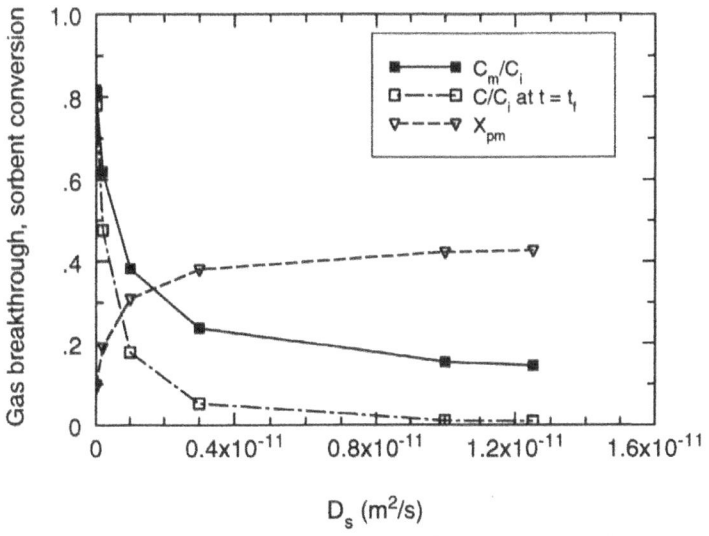

Figure 6. Effect of solid state diffusion coefficient on predicted results

ALTERNATIVE STRATEGIES FOR SORBENT FEEDING

As mentioned earlier, there may be alternative arrangements for the injection of sorbent particles. It is of practical interest to know how the sorbent feeding strategy influences the overall efficiency of gas sorption.

Gas concentration profiles and sorbent conversions have been simulated for different values of S_i. Some results predicted under alternative strategies were reported in an earlier paper (Duo et al., 1993) for the CaO/HCl reaction system at 800 K. The results for the Ca(OH)$_2$/HCl reaction system at 500 K are presented in Figs 7 to 10 of this paper.

Figure 7 shows the concentration variations with time at $z = 0$ (gas exit). The gas breakthrough for $S_i = 0$ peaks at the beginning of the cycle and continuously decreases with operation time because insufficient sorbent has been injected to adsorb the gas in the earlier period of the cycle. A similar tendency is shown for $S_i = 0.25$ and 0.50 but the concentration peaks at later times and the peak concentrations are much lower than for $S_i = 0$. This may be expected because a significant portion of the gas has been adsorbed by the sorbent initially present. For $S_i = 1$, the breakthrough concentration continuously increases with operation time due to sorbent conversion.

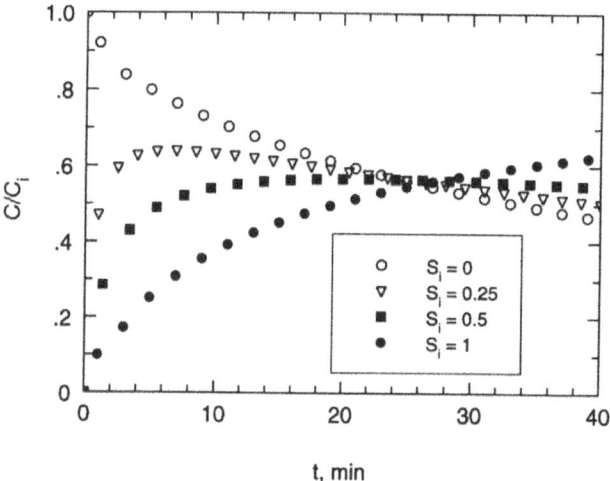

Figure 7. Concentration variations with time at $z = 0$ for alternative sorbent feeding strategies. $S_i = 0$: continuous addition; $S_i = 1$: batch injection.

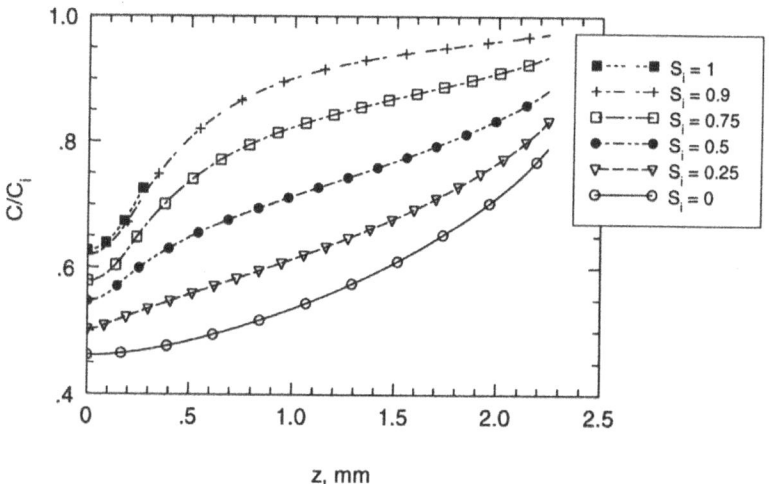

Figure 8. HCl concentration distribution in cake at $t = t_f$.

Figure 8 shows the concentration distributions within the filter cake at $t = t_f$. The thickness of the sorbent-containing cake for $S_i = 1$, Z_1, is much smaller than that for other strategies, $H(t_f)$, since it contains a much lower fraction of inert dust. There is a concentration gradient in each of the cakes with the highest concentration at the boundary which faces the incoming gas flow: $C(t_f,Z_1)$ for $S_i = 1$ or $C(t_f,H)$ for other strategies. However, the highest concentrations are lower than C_i in the bulk flow. This is due to axial dispersion. Since axial dispersion is more effective for a thinner bed under otherwise identical conditions, $C(t_f,Z_1)$ is lower than $C(t_f,H)$.

It is interesting to note the gradual change of the shape of the curves in Fig. 8 with different values of S_i. For $S_i < 0.5$, the shape of the curves resembles that for $S_i = 0$, particularly at the z positions approaching $H(t_f)$ (the maximum cake thickness reached at the end of the cycle). For $S_i > 0.5$, the shape of the curves resembles that for $S_i = 0$, particularly at the z positions close to 0 (surface of filter medium). For $S_i = 0.5$, the shape near the left end of the curve resembles that for $S_i = 1$, whereas the shape near the right end of the curve resembles that for $S_i = 0$.

Sorbent conversion, X_p, is defined by Eq. (31). Figure 9 shows the distribution of sorbent conversion in the filter cake at $t = t_f$. The zero conversion degree is expected at $z = H(t_f)$ where the sorbent particles have just been deposited and hence their residence time is zero. There is a maximum conversion for each of the curves situated at $z = 0$ for $S_i = 0$, $z = Z_1$ for $S_i = 1$, and between $z = 0$ and some position greater than Z_1 for other feeding strategies. This is due both to the differences in residence time and to the differences in gas concentration to which the sorbent particles are exposed.

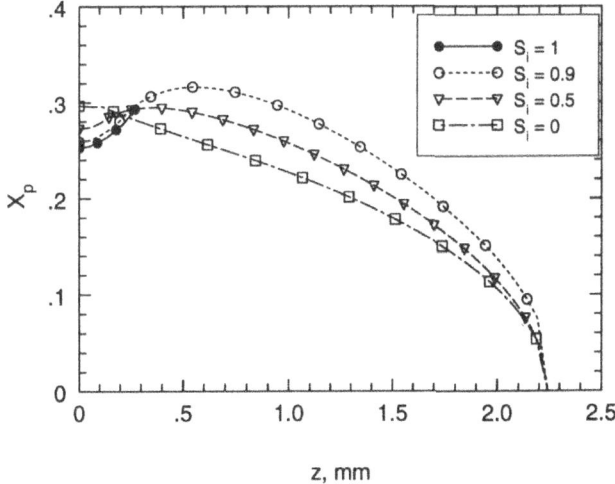

Figure 9. Distribution of sorbent conversion degree in cake at $t = t_f$.

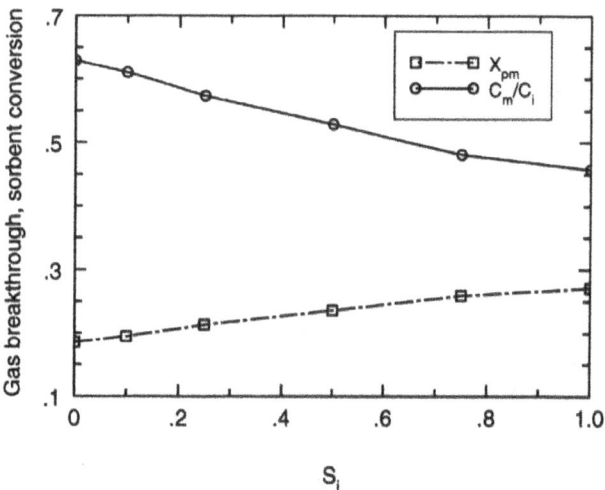

Figure 10. Effect of sorbent feeding strategies on overall efficiency
of HCl removal and sorbent conversion.

Figure 10 shows the effect of sorbent feeding strategies on the overall efficiency of gas removal and sorbent conversion. C_m and X_{pm} are defined by Eqs (30) and (32). It is seen that the efficiency of HCl removal and sorbent conversion improve almost linearly with increasing S_i. The linear relationship may only be obtained for reaction systems in which solid state diffusion is the rate controlling step. A non-linear relationship was shown for the CaO/HCl reaction system at 800 K (Duo et al., 1993). This study confirms the conclusion drawn by Duo et al. (1993) that batch injection of all the sorbent at the beginning of a cycle ($S_i = 0$) or semi-continuous addition ($0 < S_i < 1$) is better than continuous feeding ($S_i = 0$).

Currently, the common practice in dry scrubbing is to feed sorbents continuously. This and our previous studies, however, suggest that alternative feeding strategies should be considered in order to optimise the design and operation of a practical system for dry scrubbing.

CONCLUSIONS

A mathematical model has been developed for simulation of dry scrubbing of acid gases in filter cakes. The model can deal with both fixed and moving cake boundary problems. Numerical solutions for the $Ca(OH)_2$/HCl reaction at 373-523 K demonstrate that the overall sorption is sensitive to the coefficient of solid state diffusion through the product layer but less sensitive to the rate constant of the chemical reaction, leading to the conclusion that solid state diffusion is the rate controlling step in this temperature range. However, review and analysis of literature data for major kinetic parameters for reactions of Ca-compounds with HCl has suggested that the HCl sorption reactions are controlled by solid state diffusion at lower temperatures but by chemical reaction rate at higher temperatures. The results for acid gas sorption and sorbent conversion predicted under alternative strategies for sorbent feeding show that compared with continuous addition, injection of at least part of the sorbent at the beginning of the operation cycle immediately following filter cleaning can significantly improve the overall sorption efficiency.

Acknowledgements

The authors gratefully acknowledge financial support from the Process Engineering Committee of the Science and Engineering Research Council. The authors would also like to thank C. E. Weinell, K. Dam-Johansen and their colleagues in the Department of Chemical Engineering, the Technical University of Denmark for providing their experimental data.

REFERENCES

Balekdjian, O. (1987). "Abscheidung von Chlorwasserstoff und Schwefeldioxid an Calciumhaltigen Absorbentien", Dissertation, University of Karlsruhe.

Bandt, G., M. Lodder, & D. Hesse (1991). "Investigations on the Kinetics of the Reactions HCl/Dolomite and HCl/Limestone". 1991 UK ICHEME Research Event, 9-12.

Bandt, G., K. Schaper, & D. Hesse (1992). "Adsorption of HCl with Calcined Dolomite: Investigations on the Kinetics of the Formation of the Product Layer". 1992 UK ICHEME Research Event, 305-307.

Borgwardt, R.H., N.F. Roache & K.R. Bruce (1986). "Method of Variation of Grain Size in Studies of Gas-Solid Reactions Involving CaO", Ind. Eng. Chem. Fundam., 25, 165-169.

Cheung, W., J.P.K. Seville, R. Clift, C.J. Bower, & A.N. Twigg (1988). "Filtration and Cleaning Characteristics of Ceramic Media", 4th International Conference on Fluidised Bed Combustion, Inst. Energy, London, pp. II/9/1-14.

Christman, P.G. & T.F. Edgar (1983). "Distributed Pore-Size Model for Sulphation of Limestone", AIChE Journal, 29(3), 388-395.

Dam-Johansen, K., P.F.B. Hansen, & K. Østergaard (1991). "High-Temperature Reaction between Sulphur Dioxide & Limestone - III", Chem. Eng. Sci., 46(3), 847-853.

Dam-Johansen, K. & K. Østergaard (1991). "High-Temperature Reaction between Sulphur Dioxide and Limestone - I", Chem. Eng. Sci., 46(3), 827-837.

Daoudi, M. & J.K. Walters (1991). "A Thermogravimetric Study of the Reaction of Hydrogen Chloride Gas with Calcined Limestone: Determination of Kinetic Parameters", Chem. Eng. Journal, 47, 1-9.

Duo, W., N.F. Kirkby, J.P.K. Seville & R. Clift (1993). "Modelling of Dry Scrubbing of Acid Gases in Filter Cakes", Proceedings of the 1993 International Incineration Conference, Knoxville, Tennessee, U.S., May 3-7, pp. 207-212.

Hartman, H. & R.W. Coughlin (1974). "Reaction of Sulphur Dioxide with Limestone and the Influence of Pore Structure", Ind. Eng. Chem. Process Des. Develop. 13(3), 248-253.

Hartman, M. & R.W. Coughlin (1976). "Reaction of Sulphur Dioxide with Limestone and the Grain Model", AIChE Journal, 22(3), 490-498.

Heap, B.M. (1992). "The Scrubbing of Combustion Off-Gases to Remove Acids and Other Pollutants", paper presented at Meeting on "Recent Advances in Gas Filtration", Filtration Society, University College London, UK, 9th September.

Iisa, K. & M. Hupa (1990). "Sulphur Absorption by Limestone at Pressurized Fluidized Bed Conditions", 23rd Symposium (International) on Combustion, The Combustion Institute, pp. 943-948.

Karlsson, H.T., J. Klingspor, & I. Bjerle (1981). "Adsorption of Hydrochloric Acid on Solid Slaked Lime for Flue Gas Clean Up". J. Air Pollution Control Assoc., 31(11), 1177-1180.

Keener, T.C. & P. Biswas (1989). "A Dry Scrubbing Model for SO_2 Removal", Chem. Eng. Communication, 81, 97-108.

Mason, E.A. & A.P. Malinauskas (1983). "Gas Transport in Porous Media: the Dusty-gas Model", Elsevier, Amsterdam.

McLaughlin, J., R.A. Schulz, & R. Clift (1993). "Reaction of Getter Minerals with Alkali Salt Vapours", Proceedings of the Second International Symposium on "Gas Cleaning at High Temperatures", University of Surrey, Guildford, UK, 27th - 29th September.

Peukert, M. (1990). "Die Dombinierte Abscheidung von Partikeln und Gasen in Schüttschichtfiltern", Dissertation, University of Karlsruhe.

Peukert, W. & F. Löffler (1993). "Sorption of SO_2 and HCl in Granular Bed Filters", Proceedings of the Second International Symposium on "Gas Cleaning at High Temperatures", University of Surrey, Guildford, UK, 27th - 29th September.

Reid, R.C., J.M. Prausnitz, & T.K. Sherwood (1977). "The Properties of Gases and Liquids", 3rd ed., McGraw-Hill, New York.

Ruthven, D.M. (1984). "Principles of Adsorption and Adsorption Processes", Wiley and Sons, New York.

Seville, J.P.K., W. Cheung, & R. Clift (1989). "A Patchy Cleaning Interpretation of Dust Cake Release from Non-woven Fabrics", Filtration and Separation, 26(3), 187-190.

Szekely, J., J.W. Evans, & H.Y. Sohn (1976). "Gas-Solid Reactions", Academic Press, New York.

Tullin, C. & E. Ljungstrom (1989). "Reaction between Calcium Carbonate and Sulphur Dioxide", Energy and Fuels, 3, 284-287.

Uchida, S., S. Kageyama, M. Nogi, H. Karakida, T. Kakizaki, & K. Tsukagoshi (1979). "Reaction Kinetics of HCl and Limestone". J. Chinese Inst. Chem. Eng., 10, 45-49.

Wakao, N. & T. Funazkri (1978). "Effect of Fluid Dispersion Coefficients on Particle-to-Fluid Mass Transfer Coefficients in Packed Beds: Correlation of Sherwood Numbers", Chem. Eng. Sci., 33, 1375-1384.

Weinell, C.E., P.I. Jensen, K. Dam-Johansen, & K. Livbjerg (1992). "Hydrogen Chloride Reaction with Lime and Limestone: Kinetics and Sorption Capacity", Ind. Eng. Chem. Res., 31, 164-171.

Wen, C.Y. & M. Ishida (1973). "Reaction Rate of Sulphur Dioxide with Particles Containing Calcium Oxide", Environmental Science and Technology, 7(8), 703-708.

MEMBRANE DEVELOPMENT FOR THE SEPARATION OF H₂ AND CO₂: Pd/Ag ALLOY MEMBRANES

I R Summerfield, J Duxbury, and G Dennison
British Coal Corporation, Coal Research Establishment,
A P Davidson, Davidson and Associates

ABSTRACT

As a fall-back response to the possibility of Global Warming due to increasing concentrations of greenhouse gases in the atmosphere, British Coal is investigating Low-CO_2 Power Generation options. One of the most promising options is an Integrated Gasification Combined Cycle (IGCC) system using a membrane separation unit to separate H_2 and CO_2. However, the performance requirements cannot be met by currently available commercial membranes. A small-scale membrane permeation test rig has been designed and built at CRE to aid development of novel membranes.

This paper describes tests carried out on Pd/Ag alloy membranes, which are the most promising for the separation of H_2 and CO_2. The tests examined the variation of permeation rate with pressure, temperature, and pressure drop. Results of the tests are reported, with recommendations for further work to improve performance.

INTRODUCTION

Global Warming is a threat causing widespread concern. The complex natural mechanisms involved are difficult to quantify adequately. They may combine to accelerate the effect or reduce it to negligible proportions. Thus the extent and significance of any warming will not be clear for at least a decade. Meanwhile the world population will continue to grow exponentially, accompanied by demands for improved living standards. This will place increased demands on energy production and agriculture, both of which are major sources of greenhouse gases.

A number of responses to this threat should be implemented now. These include increased efficiency of power generation and end use as well as technology transfer to developing nations. As a fall-back option, development of 'Low CO_2 Power Generation' systems featuring integral carbon dioxide removal and disposal is required so that the technology is ready for deployment if appropriate. In 1990 British Coal initiated a research and development programme at the Coal Research Establishment (CRE) aimed at contributing to international efforts to combat the threat of Global Warming.

The thermal efficiency of a number of coal-fired power systems has been estimated using British Coal's ARACHNE process simulation system to determine the reduction in efficiency which would occur if a CO_2 separation stage was introduced to remove 90% of the CO_2 (Goldthorpe et al 1992).

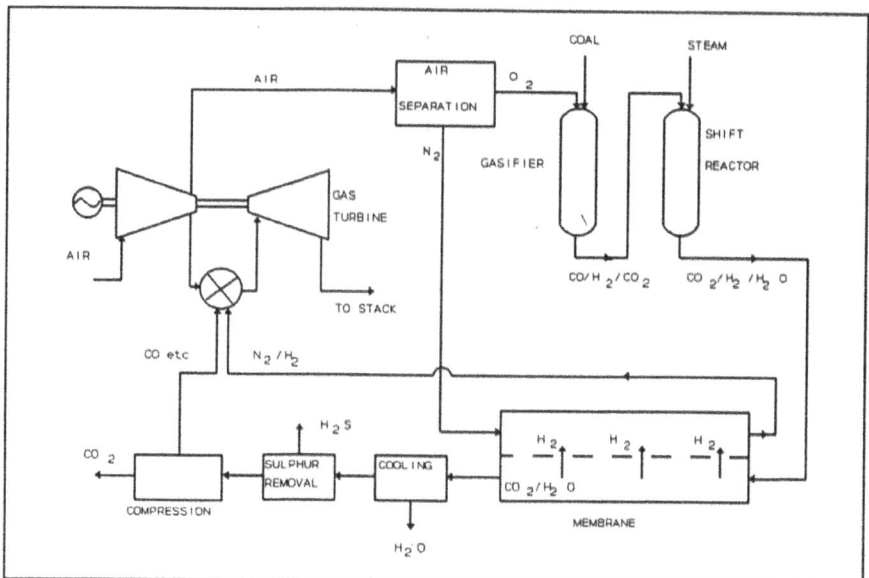

Figure 1. Flowsheet of IGCC with membrane gas separation.

This allowed the selection of the most promising options for follow-on costing studies. The highest efficiency was predicted for an Integrated Gasification Combined Cycle (IGCC) scheme in which the carbon monoxide component of synthesis gas is converted by shift reaction with steam to produce a mixture of CO_2 and H_2, leaving a small remaining proportion of CO. The most efficient gas separation process was membrane gas separation. The flowsheet is shown in Figure 1.

Several options were costed. In the cheapest scheme, the CO_2 recovered from a 500 MWe IGCC power plant in the East Midlands is piped 140 km to a disused gas production platform in the southern North Sea. The liquefaction of the CO_2 and its compression to 136 bar is included in the power plant design so that no other energy input is required for disposal of the CO_2.

The addition of CO_2 separation and disposal plant increases the capital and operating costs of the power station. For membrane gas separation the increase in electricity cost would be 34%. This provided the economic case for the membrane development programme.

MEMBRANE TEST RIG

A high temperature, high pressure experimental unit for testing gas permeable membranes was built. The equipment is shown schematically in Figure 2. The membranes are in the form of flat discs, nominally 50 mm in diameter. Feed gases are metered from cylinders, mixed, preheated, and piped to the high pressure side of the membrane holder. The membrane holder is installed in a fluidised sandbath furnace to ensure good temperature control. Steam can be added to the feed gas via a water metering and steam generation system. Gas separation takes place at the membrane. Permeable species pass through the membrane to the low pressure permeate side while non-permeable species remain on the high pressure retentate side. After pressure let-down, cooling and metering, the two exit flows are analysed and vented to atmosphere. Temperatures, pressures, and flows are recorded using a data logging computer. The equipment is designed to operate at temperatures up to 500 °C and pressures up to 60 bar(g), with gas flowrates of up to 17 l/min. The flow regime across the membrane is very important. Different permeation rates will be obtained depending on whether the flow is counter-, co-, or cross-current, or perfectly mixed. Therefore the membrane holder was designed to establish perfectly mixed flow, a flow regime for which the permeation coefficient can be easily calculated. The resulting design is shown in Figure 3. The long inlet chamber ensures complete mixing of the gas arriving at the membrane, and the annular retentate offtake

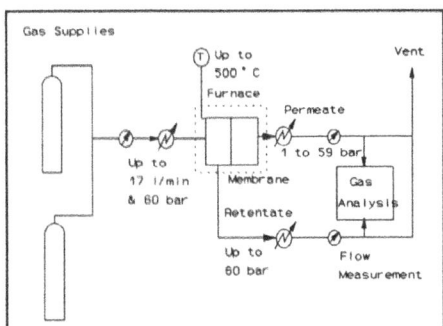

Figure 2. Schematic drawing of experimental test unit.

ensures that the gas concentration is similar at all points on the membrane surface.

The experimental procedure was based on carrying out mass balances around the permeator. Following leak testing, gas flows are started and the required pressures are set on the permeate and retentate line pressure controllers. The membrane separator is then brought up to the required temperature using the fluidised sandbath furnace. Once the required temperature is reached, the rig is left for flows and pressures to stabilise, this normally takes between 30 to 60 minutes. When the

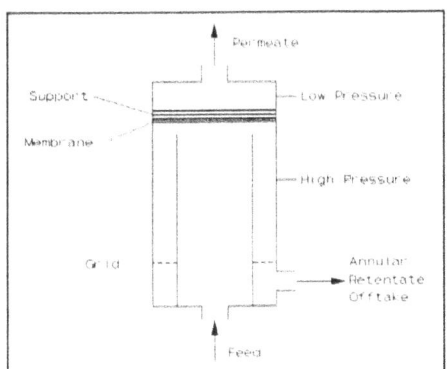

Figure 3. Schematic drawing of membrane holder.

flows, temperatures and pressures are stable, the mass balance period is started. During the (typically 1 hour) mass balance periods the plant operating parameters are logged every two minutes. These readings are then averaged. Screening tests were carried out to measure the permeation and separation performance of a range of non-porous polymeric and metal membranes (Summerfield et al 1993). These tests showed that palladium/silver alloys performed best in terms of H_2 permeation and H_2/CO_2 separation. Therefore, four different thicknesses of Pd/Ag alloy foil (200, 100, 50,25μm) were tested at a range of temperatures, pressure differences and feed gas compositions. Before each experiment, the palladium/silver membranes were cleaned with dichloromethane in an ultrasonic bath for 15 minutes. This procedure was then repeated using cyclohexane and then methanol as the solvent.

RESULTS AND DISCUSSION

The basic equation predicting the permeation rate of a diatomic gas through a membrane in which it is soluble atomically is given by:-

$$J = P_c A \left(\sqrt{p_f} - \sqrt{p_p} \right) / \delta \qquad (1)$$

J is the permeation rate, P_c the Permeability coefficient, A the membrane area, p_f and p_p the feed and permeate partial pressures, and δ the membrane thickness. The equation assumes that the process is diffusion controlled and driven by the concentration difference between the feed and permeate side of the membrane, also that the concentration in solution at the surfaces is proportional to the square root of the partial pressure (this results from Sievert's Law (Smithells, 1976)). It would be expected that the expression would break down when the membrane is thin enough for surface dissociation or recombination kinetics to become rate-determining.

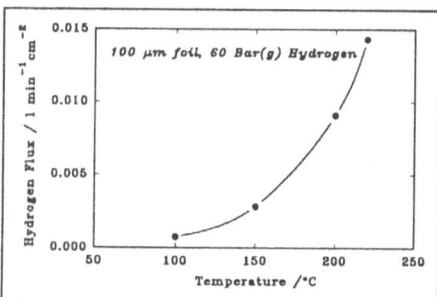

Figure 4. Graph showing variation of permeation rate with temperature.

An experiment was done to assess the temperature-dependence of permeation rate for hydrogen from a pure hydrogen feed gas through a 200μm thick foil. The temperature was varied in the range 100 to 220°C, the feed gas pressure was held at 60 Bar(g) and the permeate pressure was 1.05 Bar(a)he results are shown in Figure 4. The permeation rate increased monotonically by an order of magnitude over this temperature range. Figure 5 shows the data plotted in Arrhenius format, ln(rate) vs. 1/T, and they show a reasonable straight-line relationship. Therefore, in this temperature range, the rate of hydrogen permeation followed an Arrhenius-type relationship of the form:

$$J = J_0 \exp(-\Delta H/RT) \qquad (2)$$

J is the experimental hydrogen permeation rate, J_0 is the rate constant, ΔH is the apparent activation enthalpy, R is the Gas Constant and T is the absolute temperature. The experimental estimate of ΔH was 37 ± 5 kJ/mol. It should be noted that, in all experiments reported here, the CO_2 permeation rate was zero, ie the Pd/Ag membranes behaved as perfect separators.

The Effect of Pressure on Permeation Rate.

Experiments were carried out to investigate the effect of change of hydrogen pressure in the feed gas on permeation rate. The single

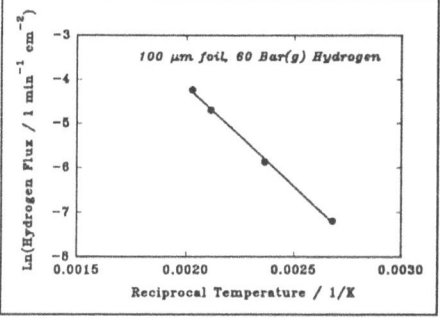

Figure 5. Arrhenius plot of data shown in figure 4.

component or partial pressure of hydrogen was controlled in three different ways:

(i)   by variation of total pressure of a pure hydrogen feed.

(ii)   by variation of total pressure of a hydrogen/carbon dioxide mixture of constant composition.

(iii)   by variation of the composition of a hydrogen/carbon dioxide mixture at constant total pressure.

Figure 6 shows the experimental permeation rates achieved for foils with thicknesses ranging from 25μm to 200μm as hydrogen partial pressure was varied at constant temperature. In all cases the permeate pressure was 1 atmosphere + 0.04 Bar.

Curves I and II show data for, respectively, 25μm and 50μm thick foils at the same temperature, 240°C, and with 60 Bar(g) pure hydrogen feed gas. As would be expected, the permeation rate was higher for the thinner foils. The 25μm foil was less permeable than would have been expected on the basis of results obtained for thicker foils. From (1) it would have been expected that it would have been twice as permeable as the 50μm foil at a given hydrogen feed pressure; instead, it was only ~20% higher. This effect has been observed previously (Ackerman & Koskinas, 1972); when the metal is very thin, the

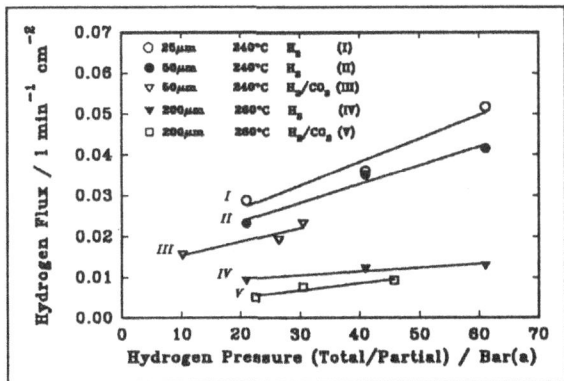

Figure 6. Graph showing the variation of permeation rate with hydrogen feed pressure (total/partial) for membranes with different thicknesses.

permeation rate ceases to be diffusion-controlled, instead, it becomes surface-controlled. Curve IV shows data for a 200μm foil with pure hydrogen at 260°C.

Curves III and V show the effect of change of partial pressure of hydrogen in a hydrogen - carbon dioxide mixture feed on the permeation rate through, respectively, a 50μm foil at 240°C and a 200μm foil at 260°C. Curve III should be compared with curve II and curve V should be compared with curve IV. In both cases the permeation rate with the gas mixture appeared to be lower than would have been expected for a pure hydrogen feed gas at the same pressure as the partial pressure in the gas mixture. These results indicate that the carbon dioxide inhibited the permeation of hydrogen through the foil. It is likely that this was due to mass transfer effects: the Reynolds No for the system was very low (~5) resulting in poor mass transfer of H_2 to the membrane on the feed side. This Reynolds No was calculated using the average radial velocity of the gas stream across the membrane surface.

The Effect of Nitrogen Purge on the Permeation Rate.

Experiments were done to investigate the effect of purging the permeate side of the membrane on the hydrogen permeation rate. Nitrogen was fed into the permeate side of the cell at a N_2:permeate H_2 ratio of ~1. This should reduce the H_2 partial pressure on the permeate side of the membrane, thereby increasing the driving force and thus the H_2 flux. A range of permeate pressures was used (5 to 40 Bar(g)), and the pure hydrogen feed was maintained at 60 Bar(g). Figure 7 shows the experimental permeation rate as a function of the permeate pressure; it decreased by a factor of ~4 as the permeate pressure was increased. Also shown on this figure are predictions of the H_2 flux for two cases:

(i)   the N_2 and H_2 on the permeate side are well mixed, and

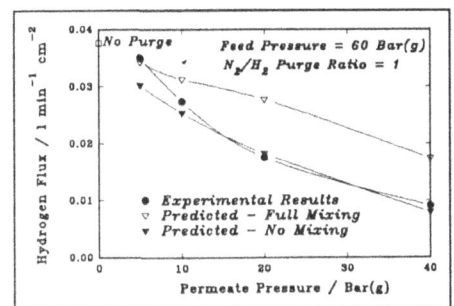

Figure 7. The effect of permeate purging on the permeation rate.

(ii)   the N_2 and H_2 are not mixed (i.e. the N_2 does not reduce the H_2 partial pressure at the membrane surface).

The best prediction was for the no mixing case. It is likely that the sintered support plate inhibited diffusional mixing of the nitrogen purge and the permeated hydrogen at the permeate side of the membrane. This effect is likely to be reduced in larger scale equipment where the mixing would be much better.

Comparison of the Present Results with Previous Data.

The best available experimental data for the same type of palladium alloy as that used in this work are in the papers of Ackermañ and Koskinas (1972) and Wileman et al (1989). These data were obtained using, respectively, tube and unsupported sheet, and were established using pure hydrogen feed and permeate pressures and temperatures that were significantly different from those used in the present experiments. Therefore, it is necessary to normalise them before comparison can be made with the present data. This can be done by calculating the Permeation coefficient, P_c.

Rearranging (1):-

$$P_c = J\delta \left(\sqrt{p_f} - \sqrt{p_p}\right) / A \qquad (3)$$

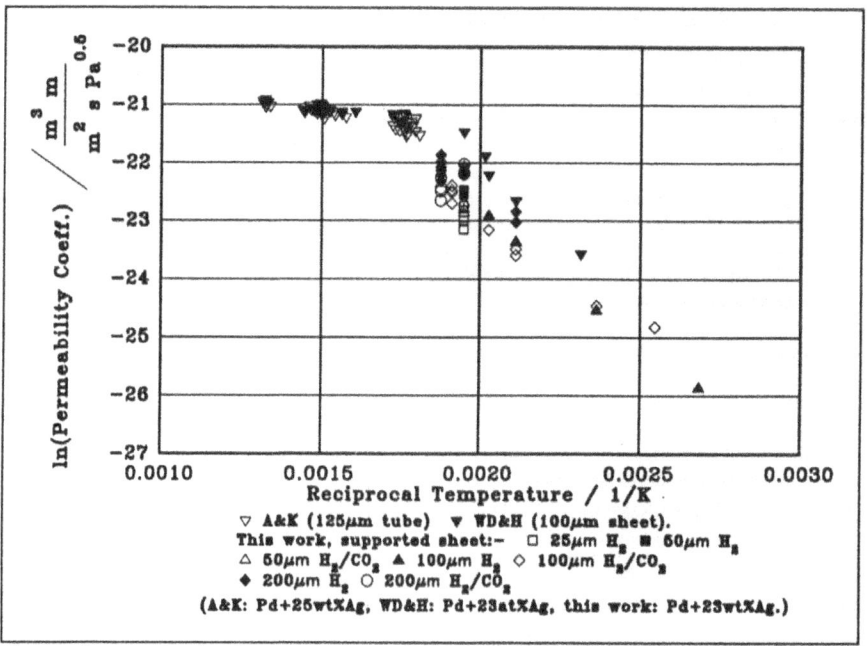

Figure 8. Graph showing experimental Permeability Coefficients plotted in Arrhenius form.

Values of P_c are plotted against temperature in Arrhenius form ($\ln(P_c)$ vs. $1/T$), in Figure 8. Data shown in this figure include A &K, W, D & H and the present data. The following points are apparent:

(i) there are low temperature ($<$ c. 310°C) and high temperature regions. The apparent activation enthalpy for the low temperature region is c.35-40 kJ/mol. The apparent activation enthalpy for the high temperature region is c. 4-5 kJ/mol.

(ii) all of the present data were in the low temperature region.

(iii) the present data were lower than the low temperature data of Wileman et al by factor ranging from ~ 1.5 at very low temperature, to ~ 2 at higher temperature.

(iv) for the present experiments the 25μm and hydrogen-carbon dioxide mixture feed gas results were systematically lower than the pure hydrogen feed results.

The general form of the results was very similar to that of previously-determined data; the temperature-dependence was very similar, but the magnitude of the permeability coefficient was lower than other workers' data. This may have been due to surface poisoning of the palladium. It might also have been a result of the reduction of the effective surface area of the permeate side of the membranes by the sintered stainless steel plate used to support them. For all of the temperatures used in this study it would be expected, and it was confirmed experimentally, that the permeation of hydrogen was in the 'low-temperature' region determined previously. The data in Figure 8 are sufficient to predict the area of palladium-silver alloy of a given thickness and for a given operating temperature needed to provide a selectively-permeable membrane for hydrogen separation.

CONCLUSIONS AND RECOMMENDATIONS

1. Permeation and separation tests have been carried out on palladium/silver alloy membranes.

2. The work has demonstrated that a Pd/Ag alloy membrane system is a practical means of separating H_2 and CO_2.

3. The design of the permeation cell may have limited the H_2 permeation rate and alternative designs should be considered.

ACKNOWLEDGEMENTS

The work was funded by British Coal Corporation and the European Coal and Steel Community. The testwork was carried out by Mr. S Hale. The CO_2 control studies are part of an international collaborative project in which Deutsche Montan Technologie GmbH (Germany), CERCHAR (France) and ENDESA and CIEMAT (Spain) are also participating. The views expressed are those of the authors and not necessarily those of the British Coal Corporation, other sponsors or collaborators.

REFERENCES

Ackerman, F.J., and Koskinas G.J. (1972). J. Chem and Eng Data, Vol 17, No. 1.

Goldthorpe, S.H., Cross P.J.I. and Davison J.E. (1992). 'System Studies on CO_2 Abatement from Power Plants'. First International Conference on Carbon Dioxide Removal, Amsterdam.

Smithells, C.J. (Editor) (1976). 'Metals Reference Book', 5th Ed., Butterworths London.

Summerfield, I.R., Duxbury J. and Dennison G. (1993), 'Membrane Development for the Separation of H_2 and CO_2: Screening Tests'. IChemE Research Event, Birmingham, 6-7 January.

Wileman, R.C.J., M. Doyle, and I.R. Harris, Z. fur Physikalische Chemie Neue Folge, Vol 164, pp 797-802, 1989.

SIMULATION OF HYDROGEN SEPARATION FROM HYDROGEN SULFIDE DECOMPOSITION GASES USING INORGANIC MEMBRANES

J. Zaman[+] and A. Chakma[*]
Dept. of Chemical and Petroleum Engineering
University of Calgary, Calgary, Alberta, T2N 1N4

ABSTRACT

One dimensional isothermal models were used to simulate the thermal decomposition of hydrogen sulfide in palladium and ceramic membrane reactors. The operational characteristics of the reactors were investigated in terms of dimensionless numbers relating the physical parameters of the membranes, the dimensions of the reactor, the reaction parameters and the operating variables. The results obtained suggest that conversions as high as 50% can be achieved with the palladium membrane reactor at high sweep gas flow rates, compared to an equilibrium conversion of 8.84%. Ceramic membranes on the other hand were found to lead to limiting values of conversion because of poor selectivity of permeation.

+ Dept. of Chemical Engineering
  Bangladesh University of Engineering and Technology
  Dhaka-1000, Bangladesh

* Author to whom correspondence should be sent.

INTRODUCTION

Hydrogen sulfide is produced in large quantities in the sweetening of sour natural gas and in the removal of organically bound sulfur from petroleum and coal (Fletcher et al., 1984). The Claus process is the most extensively used method to treat this hydrogen sulfide, recovering sulfur and generating steam of low energy value (Kappauf et al., 1985). The recovery of both sulfur and hydrogen sulfide is increasingly coming to focus because of the extensive requirement for hydrogen in the synthetic fuel industry and the importance of the reaction in the hydrogen energy cycle (Raymont, 1975; Fukuda et al., 1978; Bowman and Du Plessis, 1986).

The thermal decomposition of hydrogen sulfide to produce sulfur and hydrogen suffers from the twin disadvantages of high endothermicity and low equilibrium conversion even at fairly high temperatures. The problem of low equilibrium yield can be partially resolved by operating at low pressure (since the reaction leads to an increase in the number of mols) and by removal of the products of the reactions as they are formed. Bandermann and Halder (1982) observed that the decomposition reaction proceeds at a fast rate and to a significant conversion level of 20 to 30% at low pressure (0.13 to 0.25 bar) and high temperature (ca 1230 K) and suggested commercial exploitation at these conditions. The product gases will be cooled to remove sulfur, and hydrogen will be separated in pressure swing adsorption units before the unreacted H_2S is recycled back. Fletcher et al. (1984) discounts problems of material of construction at high temperatures suggesting the use of ceramic materials and proposes of a solar system at 1600 K and 0.1 to 0.5 atm. pressure. Kiuchi et al. (1982), Chivers and Lau (1987) and Al-Shamma and Naman (1990) suggest a two-step process for the production of hydrogen and sulfur from hydrogen sulfide at relatively moderate temperatures. They used a lower sulfide which takes up the sulfur and liberates hydrogen and subsequently gives up the sulfur on being heated. Fukuda et al. (1978) worked on the catalytic decomposition of hydrogen sulfide at 800°C and achieved a conversion of 95% by continuous removal of sulfur and intermittent removal of hydrogen. The continuous removal of the products keep the reaction tilted in the forward direction leading to the high conversion. However, this requires sequential cooling and heating of gases which would be impractical in a commercial plant. Raymont (1975) suggested the use of membranes to effect the removal of the products at the reaction temperature. The removal of hydrogen by a selective membrane will also cause an equilibrium shift resulting in higher conversions of hydrogen sulfide.

The proposition to use membrane reactors to achieve separation and gain equilibrium shift generated considerable interests in recent times for a number of reactions (Gryaznov et al., 1976; Nagamoto and Inoue, 1981; Nagamoto and Inoue, 1985; Gryaznov, 1986; Itoh, 1987; Schmitz et al., 1988; Okubu et al., 1991; Zaspalis et al., 1991; Itoh et al., 1992; Lin and Burggraaf, 1992). Dehydrogenation of cyclohexane was carried out in a palladium membrane reactor and nearly 100% conversion was achieved at conditions where the equilibrium yield was only about 18% (Gryaznov, 1986). However, the same reaction carried out in a microporous glass reactor could reach a conversion level twice the equilibrium value (Itoh et al., 1988). Kameyama et al. (1981; 1983) carried out the decomposition of hydrogen sulfide at 600-800°C in a porous Vycor glass membrane reactor. The yield of hydrogen was found to increase to about twice the equilibrium value in a process without permeation. Schmitz et al. (1988) used a palladium silver alloy membrane in the lower part of a reformer tube at 973 K and 10-35 bar pressure. A 25% increase of conversion was achieved as compared to the thermodynamic value. Champagnie et al. (1992) recently used a multilayered alumina membrane reactor impregnated with 0.5 wt% Pt for the dehydrogenation of ethane at 773-973 K and obtained significant equilibrium shift.

Extensive simulation work was done on palladium membrane reactors (Itoh et al., 1990) and on microporous reactors (Mohan and Govind, 1988a; 1988b; Ito and Govind, 1989) in order to characterize the operating features of the respective reactors. The dense membranes like Pd and its alloy membranes have high selectivity for permeation but suffer from low permeation rates while the microporous membranes like glass and ceramics have high permeability but poor selectivity. Simulation in such cases can be used as an effective tool in defining the system boundaries, fixing the design parameters and carrying out an economic evaluation.

The decomposition of hydrogen sulfide has long been considered as a candidate reaction for membranes (Raymont, 1975), but surprisingly little is available in the literature on this reaction. In this work, the reaction is investigated in two reactor systems: a palladium membrane reactor and a ceramic membrane reactor. Using mathematical models, the operating features of these reactors are investigated in order to identify the operating limits as well as to evaluate the membrane properties needed for the successful use of membrane reactors for this reaction.

MODEL DEVELOPMENT

Two types of membrane reactor have been considered in this investigation: palladium and ceramic. The palladium membrane acts by dissolving hydrogen and is very selective. On the other hand, the microporous membranes such as ceramic membranes are controlled by Knudsen diffusion and hence they allow all molecules to permeate at different rates.

Figure 1 shows the schematic of a membrane reactor. It essentially consist of an outer shell which houses an inner tube. The walls of the inner tube act as the membrane barriers. Catalysts are placed inside the inner tube. Feed gases are passed through the inner tube and H_2S is converted into sulphur and hydrogen. The gases then permeate through the membrane walls of the inner tube into the outer shell. Inert gases passing through the annular space between the inner tube and the outer shell carries the permeated gas away from the reactor.

The model equations for the palladium membrane reactor are given in Table 1 and those for the ceramic membrane reactor in Table 2. The models are based on isothermal co-current, plug flow assumptions with negligible pressure drop and constant permeability coefficients. In the case of palladium, only hydrogen permeates to the separation side and it

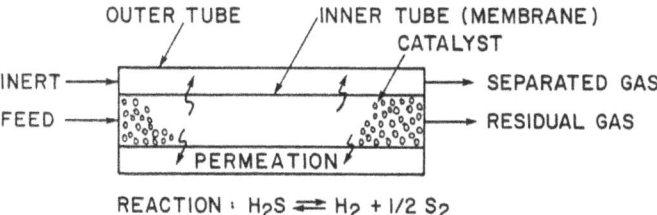

Figure 1:   Schematic diagram of the membrane reactor for H_2S decomposition.

was assumed that the permeation rate was proportional to the difference between the square roots of the hydrogen partial pressures in the reaction and separation sides (Nagamoto and Inoue, 1979). In the case of the ceramic membrane, all components permeate and the permeation rate was proportional to the partial pressure difference of the respective components on the permeation and reaction sides. The permeation rate coefficients were estimated by equations developed in the literature (Shindo et al., 1983a; 1983b). The model equations are presented in dimensionless form. Three dimensionless parameters Da^o, Tu^o and h were obtained as a result of the normalization procedure and they are defined in the nomenclature. The Damkohler Da^o is a measure of the forward rate constant at the inlet conditions and is a function of reactor volume, operating pressure and the feed flow rate. It appears in the case of both the reactors. The dimensionless number Tu^o is a measure of the ratio of the permeation rate of hydrogen to the feed rate of reactant. It is also a function of the membrane area and thickness. The dimensionless number h in the case of the ceramic membrane reactor similarly deals with the permeation to feed rate ratio for hydrogen while a_i values give the ratios of the permeability of a component gas with respect to hydrogen. The terms U_I and V_I represent the dimensionelss flow rates of inert gas on the reaction and separation sides, respectively.

The parameter values used in the simulation study are given in Table 3. The expression for the equilibrium constant for the decomposition reaction was obtained by fitting the equilibrium conversion values from the literature (Kaloidas and Papayannakos, 1987). The rate equation has been used from the work of the same authors (Kaloidas and Papayannakos, 1989), but the activation energy value obtained by them leads to a very low value of Da^o at 1073 K. The simulation runs presented in the results were carried out at 1073 K when the equilibrium conversion with $U_i = 1.0$ was 8.84%. The value of k_1^o, the forward rate constant at 1073 K is embedded in the value of Da^o and its value would be within the range given by

Table 1: Model equations for palladium membrane reactor

Reaction side:	Permeation side:
$\dfrac{dU_{H2S}}{dZ} = - Da^o{}_r\, f_a$ $\dfrac{dU_{H2}}{dZ} = - Da^o{}_r\, f_a - Tu^o\ [\ \sqrt{\dfrac{P_{H2}}{P_o}} - \sqrt{\dfrac{P_{H2}}{P_s}}\ ]$ $U = U_{H2S} + U_{H2} + U_{S2} + U_I$ $U_{S2} = \dfrac{1 - U_{H2S}}{2}$	$V_{H2} = 1 - U_{H2S} - U_{H2}$ $V = V_{H2} + V_I$ ------------------------------- Initial Conditions (indicated by superscript ^{o}): $U^o{}_{H2S} = 1$; $U^o{}_{H2} = 0$; $V^o{}_{H2} = 0$ $U_I = U^o{}_I$, $V_I = V^o{}_I$

activation energy of 195.8 kJ/mol obtained for noncatalytic reaction (Kaloidas and Papayannakos, 1989) and 112 kJ/mol for catalytic reaction (Fukuda et al., 1978). The model equations were solved by a modified Runge-Kutta method using an IMSL routine.

Table 2: Model equations for Ceramic Membrane Reactors

Reaction side:	Permeation side:
$\dfrac{dU_i}{dZ} = \pm a_i\, Da^o\, f_a - \alpha_i\, h\, (x_i - y_i\, P_r)$ + for reactants - for products $a_i = 0$ for inert $\alpha_i = 1$ for the fastest permeating gas (H_2)	$\dfrac{dV_i}{dZ} = \alpha_i\, h\, (x_i - y_i\, P_r)$ ------------------------------- Initial Conditions (indicated by superscript ^{o}): $U_i{}^o = 1$ for reactant $U_i{}^o = 0$ for product $U_i{}^o = U_I{}^o$ for inerts $V_i{}^o = 0$ for reactants and products $V_i{}^o = V_I{}^o$ for inerts

Table 3: Parameter Values Used in the Simulation

Parameter	Value	Units
Reaction temperature, $T = T_O$	1073	K
Reaction side pressure, P_O	1.01325×10^5	Pa
Permeation side pressure, $P_S = P_O$	1.01325×10^5	Pa
Equilibrium constant, K	$2.77 \times 10^4 \exp \dfrac{-9350}{T}$	$Pa^{1/2}$
Rate constant, $k_1{}^O$	$7.738 \times 10^3 \exp \dfrac{-E}{RT}$	$\dfrac{mol}{m^3.s.Pa}$
Activation energy, E	$112 \times 10^3 - 195.8 \times 10^3$	$\dfrac{J}{mol}$
Gas constant, R	8.34	$\dfrac{J}{mol.K}$
α_{H2S}	0.246	-
α_{S2}	0.174	-
α_I	0.224	-

RESULTS AND DISCUSSIONS

The computational results are presented in Figures 2 to 7. The discussions for the two types of membrane reactors are made separately followed by general assessments and conclusions.

Palladium Membrane Reactor

Figures 2, 3 and 4 show the conversion profiles for the palladium membrane reactor at different settings of the parameters Da^O, Tu^O and V_I. Figure 2 shows that large flow rates of sweep gas on the permeation side, denoted by V_I, can drive the reaction towards completion. However, with lower values of V_I such as 10, the reaction can quickly reach an asymptotic value. For a reactor system with given parameters only a sweep gas flow rate of 100 could push the conversion beyond 50% completion, compared to the equilibrium value of 8.84%. This is a flow, by definition, 100 times the flow rate of the reactant, H_2S.

676

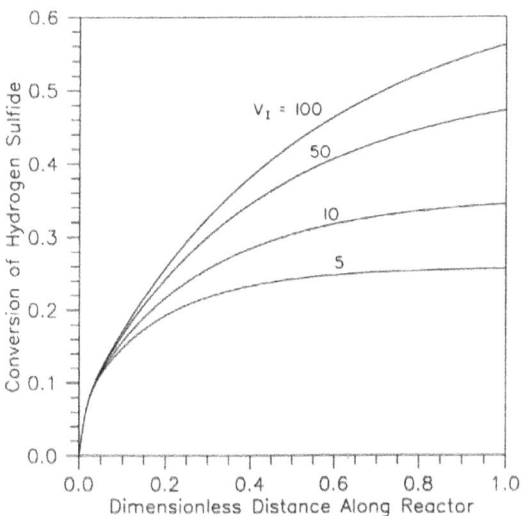

Figure 2:   Conversion Profiles of H$_2$S in a Pd Membrane Reactor:
Effect of V$_I$ (U$_I$ = 1, Da^O = 10, Tu^O = 10)

Figure 3 shows conversions at a range of values of **Da^O**. For a reactor of a given size and flow rate operating at a specified temperature and pressure the magnitude of **Da^O** is governed by the activation barrier for forward rate constant k$_1^O$. The same figure also shows that very low values of **Da^O** result in poor conversion. Even lower values of **Da^O** will be obtained in the case of noncatalytic decomposition of hydrogen sulfide using the activation energy values obtained by Kaloidas and Papayannakos (1989). It is therefore imperative to have a catalytic process for the decomposition, so that significant conversion with equilibrium shift can take place in a membrane reactor of reasonable size operating around a temperature of 1073 K. Only operation at substantially higher temperatures (beyond 1250 K) will render the influence of the catalytic effect on the equilibrium shift insignificant.

The conversion level increases with increases in the rate ratio, **Tu^O**. At low values of **Tu^O**, the equilibrium shift is small, as shown in figure 4. For a membrane of reasonable surface area, the rate ratio can be increased by decreasing the flow rate of the reactant or decreasing the thickness of the film. Since decreasing the flow rate also means low reactor throughputs, the desirable way to achieve higher **Tu^O** is by making thin membranes. The preparation of thin membranes is limited by the techniques of preparation as well as by the mechanical strength and durability considerations.

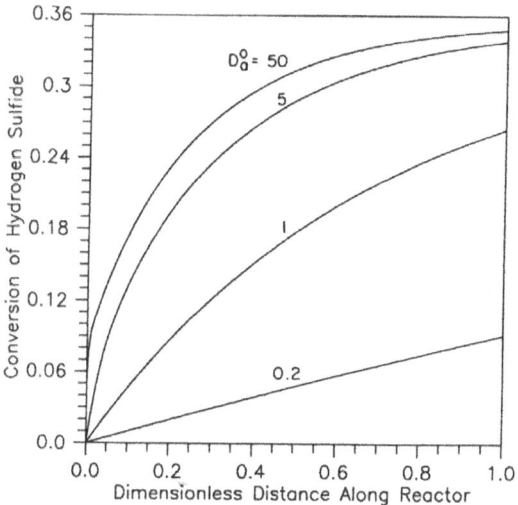

Figure 3:   Conversion Profiles of H_2S in a Pd Membrane Reactor:
              Effect of Da^0 ($U_I = 1$, $V_I = 20$, $Tu^0 = 10$)

On the whole, palladium membranes are very effective in enhancing conversion by equilibrium shift. For a given system, the physical parameters are fixed and a high level of conversion can be obtained by using very high sweep gas flow rates, V_I. This can cause difficulty in a practical system in terms of separation of the desired gases from a very dilute mixture as well as significant cost in circulating large volumes of inert gases. The large sweep gas volumes will also create a large heat load on the system. At a specific value of V_I, the conversion in the reactor reaches a limiting value because the driving force for hydrogen permeation can become equal to zero. However, by increasing the value of V_I, the reaction can be driven towards completion as could be obtained in the case of dehydrogenation of cyclohexane (Itoh, 1987; Itoh et al., 1992).

Ceramic Membrane Reactor

The computational results for a ceramic membrane reactor are shown in figures 5 to 7. The major difference between this and the palladium membrane reactor arises because of the permeability of this class of membranes to all the gases in the reacting mixture. Not only hydrogen gas permeates, large amounts of hydrogen sulfide and sulfur also permeate. The depletion of the reactant due to reaction and permeation causes the reaction rate to fall and

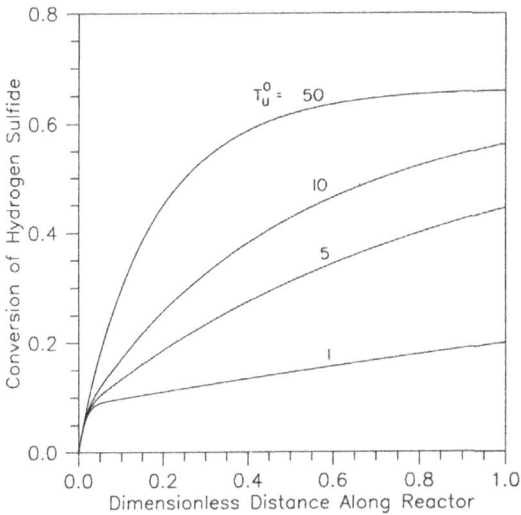

Figure 4:   Conversion Profiles of H_2S in a Pd Membrane Reactor:
Effect of Tu^0 ($U_I = 1$, $V_I = 100$, $Da^0 = 10$)

hence the conversion in a given system can rise to a limiting maximum value only. Figure 5 shows the concentration profiles of hydrogen sulfide on the permeation and reaction sides as well as the conversion. It is clear that as permeation increases, the conversion tends to level off and would reach a value lower than in the case without permeation of reactant. In fact, it was difficult to drive the reaction beyond about 2 to 2.5 times the equilibrium conversion values. This is in agreement with what has been obtained experimentally for the decomposition of hydrogen sulfide (Kameyama et al, 1981; 1983), and computationally for the decomposition of hydrogen iodide (Itoh et al., 1984). Similar results were also obtained for the dehydrogenation of cyclohexane (Itoh et al., 1988).

The effect of **Da^0** on conversion is shown in Figure 6. Higher values of **Da^0** increase the conversion level. The effect of rate ratio **h** is also similar. But in both cases, as in the case of **Da^0** and **Tu^0** for the palladium membrane reactor, the conversion approaches a limiting value, other parameters remaining constant.. The sharp contrast to the palladium reactor comes in the case of the effect of sweep gas flow rate, V_I. While in the case of the palladium reactor, increased V_I could always push the reaction towards completion (see Figure 2), in the case of ceramic membrane reactor, the conversion reaches a limiting value because of loss of reactant to the permeation side.

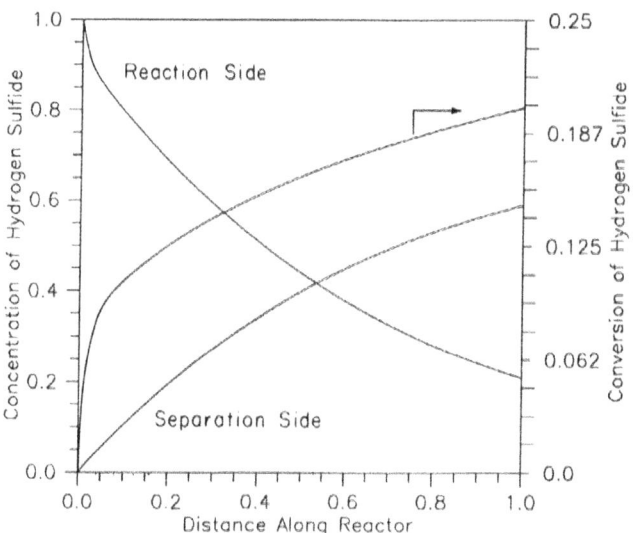

Figure 5:   Concentration and Conversion Profiles of H_2S in a
              Membrane Reactor ($U_I = 1$, $V_I = 100$, h =100, $Da^o = 10$)

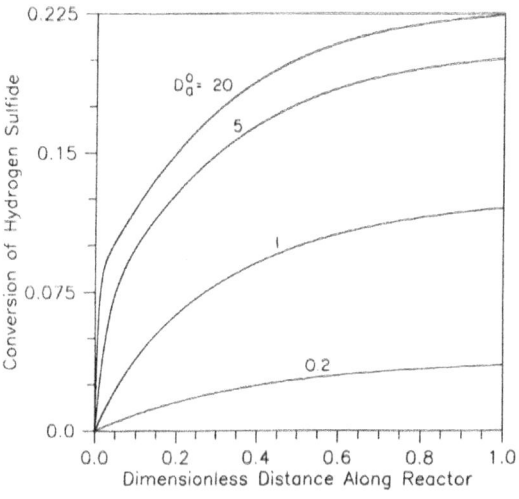

Figure 6:   Conversion Profiles of H_2S in a Ceramic Membrane Reactor:
              Effect of Da^o ($U_I = 1$, $V_I = 100$, h =20)

In Figure 7, a low sweep gas flow rate shows little equilibrium shift, the conversion increasing as the sweep gas flow rate is increased from $V_I = 1$ to $V_I = 50$. It has been seen that the concentration profile and the final conversion remain nearly unaffected beyond $V_I = 50$.

It is thus clear that limiting values of conversion are attained with specific choices of V_I, **h** and **Da^0** in the case of ceramic membrane reactors, while such limiting values could be overcome by choice of V_I in the case of palladium membrane reactors. This finding is consistent with results obtained both experimentally and computationally (Itoh et al., 1988; Mohan and Govind, 1988a; 1988b). The performance of ceramic membrane reactors is thus intrinsically poorer than palladium membrane reactors. In addition, the permeation of all gases produces two dilute streams as shown for illustrative purposes in Table 4. The separation of the desired gases from the two dilute streams can be a formidable task.

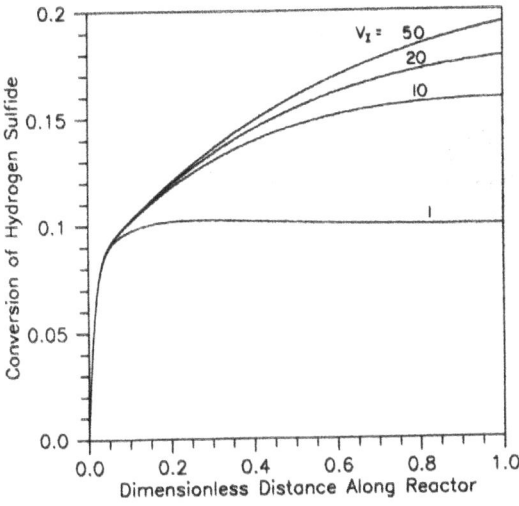

Figure 7:   Conversion Profiles of H$_2$S in a Ceramic Membrane Reactor:
            Effect of V_I ($U_I = 1$, h =10, Da^0 = 10)

681

Table 4: Mixture Composition on the Reaction and Permeation sides in a Ceramic Membrane Reactor

Parameter Values: $U_I = 1$, $Da^O = 10$, $h = 10$

Values of V_I	Mixture Composition Reaction Side	Mixture Composition Permeation Side	Conversion
10	14.81% H_2S, 1.46% H_2, 2.09% S_2, 81.6% I	5.46% H_2S, 1.29% H_2, 0.38% S_2, 92.8% I	15.89 %
100	10.9% H_2S, 0.9% H_2, 2.74% S_2, 85.4% I	0.59% H_2S, 0.18% H_2, 0.05% S_2, 99.18% I	20.05 %

The analysis of the results above show that high performance demands high values of Da^O, Tu^O and V_I in the case of the palladium membrane reactor while the ceramic membrane reactors require high values of Da^O, h and V_I and low values of α_j. High values of Da^O can be obtained by operating at higher temperatures or by developing catalysts to lower the activation barrier of the forward reaction. High temperatures naturally put a severe strain on the materials of construction and the development of a suitable catalysts for the decomposition of hydrogen sulfide should therefore be an active area of research. High values of Tu^O or h are dependent on the thickness of the membrane, given a required throughput and membrane surface area. In the case of palladium membrane reactors, developments are still occurring with new techniques and new materials whereby very thin palladium and its alloy membranes can be supported on porous membranes (Kikuchi et al., 1989; Govind and Atnoor, 1991). Various types of ceramic membranes are in developmental stages which may allow thinner membranes to be manufactured (Hsieh, 1991), while developments are taking place to produce nanosized pores in ceramic membranes which will decrease α_j values by molecular sieving effects (Tsapatsis et al., 1991; Lin and Burggraaf, 1992; Lin et al., 1992).

The problem of handling the large volumes of gases because of high values of V_I in terms of recycle and separation have not been addressed in the literature. The accompanying issue of heat load on the system has also not been investigated. These factors will significantly affect the overall economics of a membrane reactor system and demands immediate attention.

CONCLUSIONS

The following conclusions can be drawn as a result of this study:

1. High conversions in the palladium membrane reactor can only be achieved at high sweep gas flow rates. For a given conversion, lowering the sweep gas flow rate requires that an effective catalyst be found for the decomposition reaction to raise **Da^O** and thinner membranes be prepared to have high **Tu^O**.

2. Ceramic membranes lead to limiting values of conversion even with high sweep gas flow rates because of poor selectivity in permeation. To improve selectivity in permeation and reduce the values of α_i, ceramic membranes having nanosized pores have to be produced. Higher values of **h** will enhance conversion and this requires thinner membranes.

3. The high flow rates of sweep gases can have significance in terms of circulation, separation and heat load. An economic evaluation of membrane systems should be made, taking these factors into consideration.

NOMENCLATURE

A	Surface Area of membrane, m^2.
A_R	Cross sectional area of reactor, m^2.
a_i	Stoichiometric coefficients, 1 for H$_2$S and H$_2$ and $\frac{1}{2}$ for S$_2$.
Co	Concentration of H$_2$ dissolved in Pd at temperature T$_0$ and pressure P$_0$, mol/m^3.
D	Diffusion coefficient of hydrogen in palladium, m^2/s.
Da^O	Damkohler number, $\dfrac{k_1{}^O V_R P_0}{u^O{}_{H2S}}$, (-)
f_a	Dimensionless rate expression for reaction, $\dfrac{1}{P_0}$ [P$_{H2S}$ - $\dfrac{1}{K}$ P$_{H2}$ P$_{S2}{}^{0.5}$], (-)
h	$\dfrac{K_{H2} A P_0}{u^O{}_{H2S} t_m}$ (-)
$k_1{}^O$	Forward rate constant at inlet temperature, $\dfrac{mol}{m^3.s.Pa}$
K	Equilibrium constant, Pa$^{1/2}$
K_{H2}	Permeability of hydrogen, $\dfrac{mol}{m.Pa.s}$
K_i	Permeability of component i, Pa$^{1/2}$
L	Total length of the membrane reactor, m

P_O	Reaction side pressure, Pa
P_S	Permeation side pressure, Pa
P_r	Pressure ratio, P_S/P_O
p_i	Partial pressure of component i, Pa
t_m	Thickness of the membrane, m
T	Reaction temperature, K
T_O	Reactor temperature at inlet, K
Tu^O	$\dfrac{DC_O A}{u^O_{H2S} t_m}$ (-)
u	Total flow rate on reaction side, mol/s
u^O_{H2S}	Inlet flow rate of H_2S to the reaction side, mol/s
U	$\dfrac{u}{u^O_{H2S}}$, dimensionless flow rate of component i on the permeation side
V	$\dfrac{v}{u^O_{H2S}}$, dimensionless total flow rate on the permeation side
V_R	Volume of the reactor, m^3
x_i	Mol fraction of component i on the reaction side
y_i	Mol fraction of component i on the permeation side
z	Length along reactor, m
Z	Dimensionless reactor length, $\dfrac{z}{L}$, (-)
α_i	Permselectivity of component i, $\dfrac{K_i}{K_{H2}}$

REFERENCES

Al-Shamma, L.M. and Naman, S.A. (1990). The Production and Separation of Hydrogen and Sulfur from Thermal Decomposition of Hydrogen Sulfide over Vanadium Oxide/Sulfide Catalysts. Int. J. Hydrogen. 15, 1, 1-5 .

Bandermann, F. and Harder, K.B. (1982). Production of H_2 via Thermal Decomposition of H_2S and Separation of H_2 and H_2S by Pressure Swing Adsorption. Int. J. Hydrogen. 7, 6, 471-475

Bowman, C.W. and Du Plessis, M.P. (1986). The Canadian Synthetic Fuel Industry - A Major User of Hydrogen. Int. J. Hydrogen. 11, 1, 43-59 .

Champagnie, A.M., Tsotsis, T.T., Minet, R.G. and Wagner, E. (1992). The Study of Ethane Dehydrogenation in a Catalytic Membrane Reactor. J. Catal. 134, 713.

Chivers, T. and Lau, C. (1987). The Thermal Decomposition of Hydrogen sulfide Over Vanadium and Molybdenum Sulfides and Mixed Sulfide Catalysts in Quartz and Thermal Diffusion Column Reactors. Int. J. Hydrogen. 12, 4, 235-243.

Fletcher, E.A., Noring, J.E. and Murray, J.P. (1984). Hydrogen Sulfide as a Source of Hydrogen. Int. J. of Hydrogen Energy . 9, 7, 587-593.

Fukuda, K., Dokiya, M., Kameyama, T. and Kotera, Y. (1978). Catalytic Decomposition of Hydrogen Sulfide. Ind. Eng. Chem. Fundam. 17, 4, 243-248 .

Govind, R. and Atnoor, D. (1991). Development of a Composite Palladium Membrane for Selective Hydrogen Separation at High Temperature. Ind. Eng. Chem. Res. 30, 591.

Gryaznov, V.M., Smirnov, V.S. and Slinko, M. (1976). Binary Palladium /Alloys as Selective Membrane Catalysts. In Proc. Sixth Int. Congr. on Catalysis. 2, 894-902.

Gryaznov, V.M. (1986). Hydrogen Permeable Palladium Membrane Catalysts. Platinum Metals Rev. 30, 2, 68-72.

Hsieh, H.P. (1991). Inorganic Membrane Reactors. Catal. Rev. Sci. Eng. 33, 1&2, 1-70.

Itoh, N., Shindo, Y., Hakuta, T. and Yoshitome, H. (1984). Enhanced Catalytic Decomposition of HI by using a Microporous Membrane. Int. J. Hydrogen Energy. 9, 10, 835-839.

Itoh, N.(1987). A Membrane Reactor Using Palladium. AIChE J, 33, 9, 1576-1578.

Itoh, N., Shindo, Y., Haraya, K. and Hakuta, T. (1988). A Membrane Reactor Using Microporous Glass for Shifting Equilibrium of Cyclohexane Dehydrogenation. J. Chem. Eng. Japan. 21, 4, 399-404.

Itoh, N. and Govind, R. (1989). Development of Novel Oxidative Palladium Membrane Reactor. In Membrane Reactor Technology. eds. Govind, R. and Itoh, N. AIChE Symposium Series 268, 85, 10-17.

Itoh, N., Shindo, Y. and Haraya, K. (1990). Ideal Flow Models for Palladium Membrane Reactors. J. Chem. Eng. Japan. 23, 4, 420-426.

Itoh, N., Xu, W. and Haraya, K. (1992). Basic Experimental Study on Palladium Membrane Reactors. J. Membrane Sci. 66, 149-155.

Kaloidas, V.E. and Papayannakos, N.G. (1987). Hydrogen Production from the Decomposition of Hydrogen Sulfide. Equilibrium Studies. Int. J. Hydrogen Energy. 12, 6, 403-409.

Kaloidas, V.E. and Papayannakos, N.G. (1989). Kinetics of Thermal Non Catalytic Decomposition of Hydrogen Sulfide. Chem. Eng. Sci. 44, 11, 2493-2500.

Kameyama, T., Dokiya, M., Fujishige, M., Yokokawa, H. and Fukuda, K. (1981). Possibility for Effective Production of Hydrogen from Hydrogen Sulfide by Means of a Porous Vycor Glass Membrane. Ind. Eng. Chem. Fund. 20, 1, 97-99.

Kameyama, T., Dokiya, M., Fujishige, M., Yokokawa, H. and Fukuda, K. (1983). Production of Hydrogen from Hydrogen Sulfide by Means of Selective Diffusion Membrane. Int. J. Hydrogen Energy. 8, 1, 5-13.

Kappauf, T., Murray, J.P., Palumbo, R., Diver, R.B. and Fletcher, E.A. (1985). Hydrogen and Sulfur from Hydrogen Sulfide. Energy . 10, 10, 1119-1137.

Kiuchi, H., Nakamura, T., Funaki, K. and Tanaka, T. (1982). Recovery of Hydrogen from Hydrogen Sulfide with Metal or Metal Sulfides. Int. J. Hydrogen Energy . 7, 6, 477-482.

Kikuchi, E., Uemiya, S., Sato, N., Inoue, H., Ando, H. and Matsuda, T. (1989). Membrane Reactor Using Microporous Glass Supported Thin Film of Palladium. Chemistry Letters . 489-492.

Lin, Y.S. and Burggraaf, A.J. (1992). CVD of Solid Oxides in Porous Substrates for Ceramic Membrane Modification. AIChE J . 38, 3, 445-454.

Lin, Y.S., de Vries, K.J., Brinkman, H.W. and Burggraaf, A.J. (1992). Oxygen Semipermeable Solid Oxide Membrane Composites Prepared by Electrochemical Vapor Deposition. J. Membrane Sci. 66, 211-226.

Mohan, K. and Govind, R. (1988a). Studies on Membrane Reactor. Separation Sci. Technol. 23, 1715-1733.

Mohan, K. and Govind, R. (1988b). Analysis of Equilibrium Shift in Isothermal Reactors with a Perselective Wall, AIChE J, 34, 9, 1493-1503.

Nagamoto, H. and Inoue, H. (1979). Permeation Rate of Hydrogen through Palladium Membrane and Hydrogenation Rate of Ethylene by Permeate Hydrogen. J. Chem. Soc. Japan . 12, 327-332.

Nagamoto, H. and Inoue, H. (1981). Analysis of Mechanism of Ethylene Hydrogenation by Hydrogen Permeating Palladium Membrane. J. Chem. Eng. Japan 14, 5, 377-382.

Nagamoto, H. and Inoue, H. (1985). A Reactor with Catalytic Membrane Permeated by Hydrogen. Chem. Eng. Commun. 34, 315-323.

Okubu, T., Haruta, K., Kusakabe, K., Morooka, S., Anzai, H. and Akiyama, S. (1991). Equilibrium Shift of Dehydrogenation at Short Space-Time with Hollow Fiber Ceramic Membrane. Ind. Eng. Chem. Res. 30, 614-616.

Raymont, M.E.D. (1975). Make Hydrogen from Hydrogen Sulfide. Hydrocarbon Proc. 54, 7, 139-142.

Schmitz, J., Lucke, L., Herzog, F. and Glaubitz, D. (1988). Permeation Membranes for the Production of Hydrogen at High Temperatures. In Hydrogen Energy Progress vii, Int. Assoc. Hydrogen Energy, vol. 2, 819-830.

Shindo, Y., Hakuta, T., Yoshitome, H. and Inoue, H. (1983a). Gas Diffusion in Microporous Media in Knudsen Regime. J. Chem. Eng. Japan . 16, 2, 120-126.

Shindo, Y., Hakuta, T., Yoshitome, H. and Inoue, H. (1983b). A Dimensionless Equation for Gas Diffusion in Microporous Media in Knudsen Regime, J. Chem. Eng. Japan. 16, 6, 521-523.

Tsapatsis, M., Kim, S., Nam, S.W. and Gavalas, G. (1991). Synthesis of Hydrogen Permselective S_iO_2, T_iO_2, Al_2O_3 and B_2O_3 Membranes from the Chloride Precursor. Ind. Eng. Chem. Res. 30, 2152-2159.

Zaspalis, V.T., Van Praag, W., Keizer, K., Von Ommen, J.G. and Burggraaf, A.J. (1991). Reactions of Methanol over Catalytically Active Alumina Membrane. J. App. Catal. 74, 205-222.

The manufacturer's authorised representative in the EU is Springer
Nature Customer Service Centre GmbH, Europaplatz 3, 69115 Heidelberg,
Germany. If you have any concerns regarding our products, please
contact ProductSafety@springernature.com

Printed and bound by CPI Group (UK) Ltd, Croydon, CR0 4YY

23/04/2026

02095593-0010